Molecular Biology
Biochemistry and Biophysics

32

Chemical Recognition in Biology

Edited by F. Chapeville and A.-L. Haenni

With Contributions by
M.J. Anderson · A. Boiteux · H. Boman · M.S. Bretscher · E.T. Chance
H. Chantrenne · F. Chapeville · L. Chayet · P.B. Chock · O. Cori
J.N. Davis · S. Ebashi · F. Egami · H. Ernst · L.A. Fernandez · J.L. Fox
M. de la Fuente · I.H. Goldberg · H. Grosjean · G.R. Hartmann
U. Hashagen · T. Hatayama · C. Hélène · B. Hess · R.J. Hohman
W.P. Jencks · M.E. Jones · F. Jurnak · M. Kamen · L.S. Kappen
Y. Kaziro · T. Kitazawa · H. Kleinkauf · D.G. Knorre · K. Kohama
H. Koischwitz · D.E. Koshland · Ch. Leib · A. McPherson · T. Mikawa
I. Molineux · R. Monro · M.A. Napier · D. Nathans · J. Ninio · Y. Nishizuka
Y. Nonomura · G.D. Novelli · T. Oshima · L.M. Perez · M.F. Perutz
G. Portilla · S.G. Rhee · A. Rich · J.M. Robertson · C. Rojas · G. Sanchez
E.R. Stadtman · J.Sy · T. Uchida · M.V. Vial · V.V. Vlassov · Ch. Walsh
A. Wang · M.D. Williams · W. Wintermeyer · M. Wittenberger
H.G. Wittmann · R. Wolfenden · H.G. Zachau

With 210 Figures

Springer-Verlag
Berlin Heidelberg New York 1980

Prof. Dr. F. CHAPEVILLE
Dr. A.-L. HAENNI
CNRS, Inst. de recherche en biologie moléculaire
Université Paris VII
Tour 43
2 Place Jussieu
F-75221 Paris Cedex 05

QP
601
,C43

ISBN 3-540-10205-1 Springer-Verlag Berlin Heidelberg New York
ISBN 0-387-10205-1 Springer-Verlag New York Heidelberg Berlin

Library of Congress Cataloging in Publication Data. Main entry under title: Chemical
recognition in biology. (Molecular biology, biochemistry, and biophysics; v. 32).
Bibliography: p. Includes index. 1. Enzymes-Congresses. 2. Molecular association-
Congresses. 3. Binding sites (Biochemistry)-Congresses. 4. Chemical affinity-Congresses.
I. Anderson, M.J. II. Chapeville, François. III. Haenni, Anne Lise. IV. Series.
QP601.C43   574.19'25   80-23968.

Offsetprinting and bookbinding: Brühlsche Universitätsdruckerei, Giessen.
2131/3130-543210

Preface

Studies of chemical recognition in biology were initiated about half a century ago with the first kinetic data obtained on enzyme catalysis and inhibition. They led to a rather static representation of the recognition process illustrated by the lock and key model that still continues to influence our overall image of recognition and its specificity. In several cases, crystallographic studies of enzyme-substrate complexes have supported this model. Indeed, in a crystallized ligand-enzyme complex, a close fitting is observed between the active center of the enzyme and the functional groups of the ligand. However, this does not necessarily result from a direct recognition process between rigid structures, but may result from a progressive adaptation during which the initial structures of the enzyme and the ligand are modified (induced-fit mechanism).

Recently, a great deal of work has been devoted to the study of recognition in more complex systems such as the replication or the translation machineries; clearly, the extraordinary precision of such systems cannot be explained solely in terms of physical matching between enzymes and their substrates. This has led to a noticeable change of perspective in these areas. As a result of the new kinetic viewpoint, one rather focuses on the time-course of the processes, on the kinetic balance between steps of the reaction, on the energy-accuracy relationships and on the strategies which permit the achievement of high precision using relatively error-prone components in an appropriate dynamic interplay.

New developments in the study of various systems and their specificity justified the organization of a meeting on chemical recognition in biology. The contribution of F. Lipmann in the discovery and study of many very important biological systems, and the celebration this year of his 80th anniversary promted us to bring together, with other experts, many of Lipmann's collaborators interested in the field. He was the central figure of this Symposium organized in Grignon (France) in July 1979. In this book, the lectures presented at this meeting have been assembled.

Several Foundations and Pharmaceutical Companies contributed generously to the financial support of this meeting. We wish to thank the European Molecular Biology Organization, Carlsberg Foundation, Institut National de la Recherche Agronomique, Foundation for Microbiology, Philippe Foundation, Rhône-Poulenc-Santé, Unilever Research, The Upjohn Company, Merck, Sharp & Dohme, Miles Laboratories, G.D. Searle, Boehringer Mannheim-France, Mitsubishi-Kasei Institute, Takeda Pharmaceutical Company, Pharmacia Fine Chemicals, Smith Kline & French Laboratories, E.R. Squibb, Syntex Research and Lipmann's Alumni in Japan for their generosity. We thank M.J. Bourdin of the Institut National Agronomique, Centre de Grignon, for his help and hospitality. We are grateful to Mrs. S. Srodogora and to Mrs. M. Garafoli who helped us all through the preparation of the meeting.

Paris, December 1979   Anne-Lise Haenni
                                                            Francois Chapeville

Fritz Lipmann

Contents

A. Recognition of Ligands – Enzymic Catalysis   1

  1. What Everyone Wanted to Know About Tight Binding and Enzyme
     Catalysis, but Never Thought of Asking.
     W.P. JENCKS. With 8 Figures   3
  2. The Cytochromes c: Paradigms for Chemical Recognition.
     M.D. KAMEN. With 3 Figures 26
  3. Recognition of Ligands by Haem Proteins.
     M.F. PERUTZ 38
  4. Influences of Solvent Water on the Transition State Affinity of Enzymes,
     Protein Folding, and the Composition of the Genetic Code.
     R. WOLFENDEN. With 21 Figures 43
  5. Suicide Substrates: Mechanism-Based Inactivators of Specific Target
     Enzymes.
     C. WALSH 62
  6. Recognition: the Kinetic Concepts.
     J. NINIO and F. CHAPEVILLE 78
  7. Coupled Oscillator Theory of Enzyme Action.
     M.D. WILLIAMS and J.L. FOX. With 3 Figures. 86
  8. Stereochemical Aspects of Chain Lengthening and Cyclization Processes
     in Terpenoid Biosynthesis.
     O. CORI, L. CHAYET, M. DE LA FUENTE, L.A. FERNANDEZ,
     U. HASHAGEN, L.M. PEREZ, G. PORTILLA, C. ROJAS, G. SANCHEZ,
     and M.V. VIAL. With 6 Figures 97

B. Enzyme Regulation 111

  1. Three Multifunctional Protein Kinase Systems in Transmembrane Control.
     Y. NISHIZUKA. With 25 Figures 113
  2. Effect of Catabolite Repression on Chemotaxis in *Salmonella typhimu-*
     rium.
     D.E. KOSHLAND, Jr. and M.J. ANDERSON. With 2 Figures. 136
  3. Subunit Interaction of Adenylylated Glutamine Synthetase.
     E.R. STADTMAN, R.J. HOHMAN, J.N. DAVIS, M. WITTENBERGER,
     P.B. CHOCK, and S.G. RHEE. With 9 Figures. 144

4. Dynamic Compartmentation.
B. HESS, A. BOITEUX, and E.M. CHANCE. With 3 Figures 157
5. The Genes for and Regulation of the Enzyme Activities of two Multifunctional Proteins Required for the De Novo Pathway for UMP Biosynthesis in Mammals.
M.E. JONES. With 4 Figures 165
6. Regulation of Muscle Contraction by Ca Ion.
S. EBASHI, Y. NONOMURA, K. KOHAMA, T. KITAZAWA, and
T. MIKAWA. With 5 Figures 183
7. Why is Phosphate so Useful?
M.S. BRETSCHER 195
8. ppGpp, a Signal Molecule.
J. SY. 197
9. Gramicidin S-Synthetase: On the Structure of a Polyenzyme Template
in Polypeptide Synthesis.
H. KLEINKAUF and H. KOISCHWITZ. With 9 Figures 205
10. A Molecular Approach to Immunity and Pathogenicity in an Insect-
Bacterial System.
H.G. BOMAN. With 6 Figures 217

C. Nucleic Acid – Protein Interactions; Mutagenesis. 229

1. Structure of the Gene 5 DNA Binding Protein from Bacteriophage fd and
its DNA Binding Cleft.
A. McPHERSON, A. WANG, F. JURNAK, I. MOLINEUX, and A. RICH.
With 7 Figures 231
2. Recognition of Nucleic Acids and Chemically-Damaged DNA by Peptides
and Proteins.
C. HELENE. With 1 Figure 241
3. Specific Interaction of Base-Specific Nucleases with Nucleosides and
Nucleotides.
F. EGAMI, T. OSHIMA, and T. UCHIDA. With 8 Figures 250
4. Structural and Dynamic Aspects of Recognition Between tRNAs and
Aminoacyl-tRNA Synthetases.
D.G. KNORRE and V.V. VLASSOV. With 6 Figures 278
5. Recognition of Promoter Sequences by RNA Polymerases from Different
Sources.
C. LEIB, H. ERNST, and G.R. HARTMANN. With 3 Figures. 301
6. DNA as a Target for a Protein Antibiotic: Molecular Basis of Action.
I.H. GOLDBERG, T. HATAYAMA, L.S. KAPPEN, and M.A. NAPIER.
With 11 Figures. 308
7. Site-Specific Mutagenesis in the Analysis of a Viral Replicon.
D. NATHANS. With 5 Figures 323

D. Protein Biosynthesis. 331

  1. Molecular Mechanism of Protein Biosynthesis and an Approach to the
     Mechanism of Energy Transduction.
     Y. KAZIRO. With 2 Figures 333
  2. On Codon — Anticodon Interactions.
     H. GROSJEAN and H. CHANTRENNE. With 7 Figures 347
  3. Fluorescent tRNA Derivatives and Ribosome Recognition.
     W. WINTERMEYER, J.M. ROBERTSON, and H.G. ZACHAU. With
     5 Figures 368
  4. Structure and Evolution of Ribosomes.
     H.G. WITTMANN. With 11 Figures 376

E. Philosophical Reflexions 399

  1. Molecular Biology, Culture, and Society.
     R. MONRO. With 10 Figures 401
  2. Personal Recollections of Fritz Lipmann During the Early Years of
     Coenzyme A Research.
     G.D. NOVELLI. 415

List of Contributors

You will find the addresses at the beginning of the respective contribution

Anderson, M.J. 136
Boiteux, A. 157
Boman, H.G. 217
Bretscher, M.S. 195
Chance, E.T. 157
Chantrenne, H. 347
Chapeville, F. 78
Chayet, L. 97
Chock, P.B. 144
Cori, O. 97
Davis, J.N. 144
Ebashi, S. 183
Egami, F. 250
Ernst, H. 301
Fernandez, L.A. 97
Fox, J.L. 86
Fuente, M. de la 97
Goldberg, I.H. 308
Grosjean, H. 347
Hartmann, G.R. 301
Hashagen, U. 97
Hatayama, T. 308
Helene, C. 241
Hess, B. 157
Hohman, R.J. 144
Jencks, W.P. 3
Jones, M.E. 165
Jurnak, F. 231
Kamen, M.D. 26
Kappen, L.S. 308
Kaziro, Y. 333
Kitazawa, T. 183
Kleinkauf, H. 205
Knorre, D.G. 278
Kohama, K. 183

Koischwitz, H. 205
Koshland, D.E. Jr. 136
Leib, C. 301
McPherson, A. 231
Mikawa, T. 183
Molineux, I. 231
Monro, R. 401
Napier, M.A. 308
Nathans, D. 323
Ninio, J. 78
Nishizuka, Y. 113
Nonomura, Y. 183
Novelli, G.D. 415 *
Oshima, T. 250
Perez, L. 97
Perutz, M.F. 38
Portilla, G. 97
Rhee, S.G. 144
Rich, A. 231
Robertson, J.M. 368
Rojas, C. 97
Sanchez, G. 97
Stadtman, E.R. 144
Sy, J. 197
Uchida, T. 250
Vial, M.V. 97
Vlassov, V.V. 278
Walsh, C. 62
Wang, A. 231
Williams, M.D. 86
Wintermeyer, W. 368
Wittenberger, M. 144
Wittmann, H.G. 376
Wolfenden, R. 43
Zachau, H.G. 368

A. Recognition of Ligands – Enzymic Catalysis

1. What Everyone Wanted to Know About Tight Binding and Enzyme Catalysis, but Never Thought of Asking

W. P. JENCKS

A. Introduction

The notions that an enzyme increases the reaction rate of substrates by bringing them together at the active site and by destabilizing them are among the earliest hypotheses for explaining the extraordinary catalytic power of enzymes toward their specific substrates (Haldane 1930; Pauling 1946). These ideas were eclipsed in recent years by advances in our understanding of chemical mechanisms of catalysis and by the elucidation of the three-dimensional structures of enzymes and enzyme-substrate complexes by X-ray diffraction. However, they have received renewed attention in the last few years as it has become apparent that ordinary chemical mechanisms do not provide an adequate explanation for the observed magnitude of enzymic catalysiś. An important role for physical interactions that are mediated through utilization of the binding energy of specific substrates is suggested, in particular, by the large increases in reaction rate of up to $10^{10}-10^{13}$ that are brought about by *nonreacting* portions of substrates, such as the sugar ring and phosphate of glucose-1-phosphate with phosphoglucomutase and the coenzyme A moiety of succinyl-CoA with coenzyme A transferase (Ray and Long 1976; Ray et al. 1976; Moore 1978; Jencks 1980a). These and other findings have led to the hypothesis that the principal difference between enzymes and chemical catalysts is, in fact, that chemical catalysts can use particular catalytic groups to accelerate a reaction, such as imidazole and serine hydroxyl groups in hydrolytic reactions, whereas an enzyme can use the same catalytic groups but can also utilize the binding energy from the interaction of nonreacting groups of specific substrates with the active site to bring about rate increases (Jencks 1975; Fersht 1977).

This hypothesis has been examined in detail from a number of aspects (see, for example, Page and Jencks 1971; Jencks 1975; Ray and Long 1976; Ray et al. 1976; Fersht 1977; Jencks 1980a) and has also been applied to the operation of coupled vectorial systems, in which the binding energy is used to make the hydrolysis of "energy-rich" phosphates reversible (Jencks 1980b); it will not be reviewed again here. Instead, some questions will be examined that frequently come up regarding the utilization of binding energy for catalysis and that are relevant to an understanding of the manifestations and limitations of the hypothesis. It appeared worthwhile to consider these in one place, even though some of them have been discussed previously from various points of view (Jencks 1975; Fersht 1977; Jencks 1980a).

Publication No. 1327 from the Graduate Department of Biochemistry, Brandeis University, Waltham, Massachusetts 02254, USA

B. The Total Advantage from Approximation and Binding

What is the simplest and most direct way of estimating the contribution to catalysis that can be obtained just from bringing the reactants together at the active site in the right position to react, without invoking chemical mechanisms of catalysis or destabilization mechanisms?

The situation is described by Eq. (1):

$$
\begin{array}{c}
A + B \xrightleftharpoons{K_N} \quad \Big| \quad A \cdot B \xrightleftharpoons{K^{\ddagger}} A \cdot B^{\ddagger} \\[2mm]
\pm E \searrow K_E K_E{}' \\[2mm]
\underbrace{A \cdot B}_{E} \xrightleftharpoons{K^{\ddagger}} \underbrace{A \cdot B^{\ddagger}}_{E}
\end{array} \tag{1}
$$

Suppose that, in a "thought experiment," the nonenzymic reaction can be separated into two parts: first, the bringing together of the reactants into exactly the right position to react, with the equilibrium constant K_N, and second, the activation process to reach the transition state with a barrier that is defined by the pseudo-equilibrium constant $K^{\ddagger}$. Then suppose that binding of the two substrates to the enzyme, with the equilibrium constants K_E and $K_E{}'$, also brings them into the right position to react and that the subsequent activation process is the same as in the nonenzymic reaction; i.e., it is described by the same $K^{\ddagger}$ and does not involve any chemical catalysis by the enzyme. Since $K^{\ddagger}$ is the same for the enzymic and nonenzymic processes, the advantage for the enzymic reaction is simply $K_E K_E{}'/K_N$. Taking typical values of $K_E = K_E{}' = 10^4$ M^{-1} (these binding constants are the reciprocal of dissociation constants, K_s) and a value of $K_N = 10^{-8}$ M^{-1}, the advantage for the enzymic reaction is $10^4 \times 10^4/10^{-8} = 10^{16}$ M^{-1}. This comparison refers to reactions in dilute solution in which the enzymic rate is described by k_{cat}/K_m; the units of M^{-1} reflect the dependence of the rate on the concentration of added enzyme. An analogous comparison may be made for the reaction of a substrate with some group on the enzyme, compared to the same group in solution; this gives an advantage of 10^{12} that is dimensionless (Jencks 1980a).

These advantages are, in a sense, maximal values and certainly will not apply to every reaction. However, they show that the maximal advantage that can be obtained simply by bringing reactants together to increase their probability of reaction is considerably larger than had generally been thought a few years ago. The important number is the limiting value of $K_N = 10^{-8}$ M^{-1} for bringing the reactants together in the nonenzymic reaction. This small number means that the probability of two molecules, A and B at a concentration of 1 M, finding themselves in the right position to react with each other can be as small as 10^{-8}. This quantity is based on the observed rate increases for intramolecular compared with intermolecular reactions, in the absence of strain, and on observed and calculated entropy losses that are required for bimolecular reactions in the gas phase and in solution (Page and Jencks 1971; Page 1973).

These advantages are brought about by the binding of both the reacting and the nonreacting parts of a specific substrate to the enzyme. This binding must be very tight and

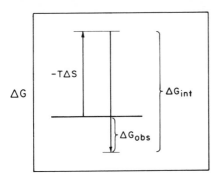

Fig. 1. A large entropy loss upon binding substrates must be paid for by utilizing binding energy

exact to give a complex that is in exactly the right position to react, and the loss of entropy that describes this tight binding must be *paid for* by the binding energy of the substrate. This matter of economics, an unpleasant subject, has been a major cause of confusion in considerations of this subject. One must pay for what one gets as illustrated in Fig. 1. The entropy loss upon bringing molecules together, $-T\Delta S$, corresponds to an increase in Gibbs free energy and must be paid for by the utilization of binding energy; the observed binding energy, ΔG_{obs}, is what is left over after this entropy loss has been paid for. The total binding energy has been called the intrinsic binding energy, ΔG_{int}; some of this is used to pay for the loss of entropy upon "freezing" the molecules, and some appears as observed binding. The intrinsic binding energy cannot be measured directly for whole molecules, but limiting values can be obtained for groups that are added to a molecule from observed increases in binding energy (Jencks 1975).

C. Is Transition State Stabilization Sufficient for Catalysis?

Can efficient catalysis by an enzyme be regarded as simply a stabilization or tight binding of a transition state to the enzyme?

No. Suppose (Fig. 2) that a nonenzymic reaction of S has a certain barrier that must be overcome to reach the transition state, $S^{\ddagger}$. Now suppose that in the course of several billion years of evolution a molecule appeared that bound this transition state very tightly to give $ES^{\ddagger}$, with the binding energy ΔG_{bind}, and bound the substrate and product equally tightly to give ES and EP, respectively. It is evident that this molecule would not be an enzyme because the barrier for reaction of the substrate bound to the enzyme, ES, is just as large as for free S. In order for this molecule to become an enzyme it would have to evolve in such a way as to decrease the binding of substrate and product, as shown on the right side of Fig. 2. It can do this by destabilizing or straining the bound substrate by an amount ΔG_D and by freezing it with a loss of entropy by an amount $-T\Delta S$. This raises the Gibbs free energy of the ES complex and, with similar changes for the EP complex, gives an active enzyme (Jencks 1980a).

   Figure 2 provides a simple way of showing why destabilization mechanisms and entropy loss are essential in order to obtain much catalysis from this kind of utilization

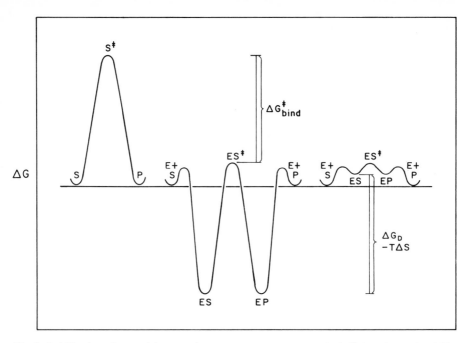

Fig. 2. Stabilization of a transition state by an enzyme causes no catalysis if there is equal stabilization of the substrate (*center*). Catalysis requires a higher, less favorable Gibbs free energy of the ES complex (*right-hand diagram*); the intrinsic binding energy is not expressed because of destabilization, ΔG_D, and entropy loss, $-T\Delta S$

of binding energy. A quantitative statement of the problem is complicated by the problem of choosing standard states for the equilibria and Gibbs energies (Jencks 1975). Figure 2 represents standard Gibbs energy changes that are based on a standard state concentration below the K_m value of the substrate for the active enzyme. This is a useful way of describing the catalytic activity of enzymes, corresponding to k_{cat}/K_m, and is in accord with the fact that the physiological concentrations of most substrates are below their K_m values (Fersht 1974; Jencks 1975).

D. The Role of Strain and Destabilization

If the introduction of strain or other destabilization mechanisms has no effect on k_{cat}/K_m, why is it useful for catalysis?

It has been pointed out that the introduction of strain into an enzyme-substrate complex, holding everything else constant, neither increases nor decreases the rate constant for catalysis in dilute solution, k_{cat}/K_m (Fersht 1974; Jencks 1975; Ballardie et al. 1977). This is shown for the destabilization of ES to ES*, by the amount ΔG_D, in Fig. 3. The value of k_{cat}/K_m is determined by the difference in Gibbs free energy between free E + S and the transition state, ES‡, and this difference is the same whether or not the intermediate ES complex is destabilized.

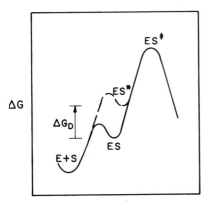

Fig. 3. Destabilization of the ES complex has no effect on k_{cat}/K_m, the second-order rate constant for the reaction of E and S. The diagram is drawn for a standard state concentration of substrate that is below K_m

The question may be dealt with in two parts and may conveniently be illustrated by considering the effect of adding a substituent to an incomplete substrate to make it a good, specific substrate.

(a) The driving force for catalysis is the intrinsic binding energy that can be derived from the interaction of the substituent with the active site. To get catalysis from this interaction, it must be used to stabilize the transition state; i.e., the binding energy must be expressed in the transition state. For small changes in energy it does not matter whether or not it is also expressed in the ES complex if one is interested in catalysis at low substrate concentrations, below K_m.

(b) For large amounts of binding energy it is essential that this binding energy *not* be expressed in the ES complex, in order to avoid a situation like that illustrated in the center of Fig. 2. The argument at this point necessarily becomes dependent on the choice of standard state or substrate concentration, but it is clear from Fig. 2 that the binding energy must be utilized for destabilization or entropy loss in the ES complex, rather than expressed as binding, for catalysis to occur. If large amounts of binding energy were expressed in the ES complex, the substrate concentration would have to be so low in order to stay below K_m and measure k_{cat}/K_m that the absolute rate of the reaction would increase little or not at all in the presence of enzyme (Jencks 1975).

This point may be made more explicitly from the rate equations for the reactions, as follows. If the rates of the nonenzymic and enzymic reactions are given by

$$v = k_N [S]$$
$$\text{and} \quad v = k_{cat} [ES]$$
$$= k_{cat} [E]_{tot} \text{ at saturation}$$

respectively, the advantage for the enzymic reaction is given simply by k_{cat}/k_N at a given concentration of enzyme. Under these conditions any increase in the Gibbs energy of the ES complex by destabilization or entropy loss will be expressed directly as an increase in k_{cat} and an advantage for the enzymic reaction (Fig. 2). Although the second-order rate constant k_{cat}/K_m may be large, it is apparent from the equation for the observed velocity under conditions in which k_{cat}/K_m describes the rate,

$$v = k_{cat} [E]_{tot} \times [S]/K_m$$

that no advantage is obtained from this large rate constant because $[S]/K_m$ must be $\ll 1$ and the observed rate of the enzyme-catalyzed reaction cannot be larger than k_{cat} $[E]_{tot}$. Thus, it is necessary to increase both K_m and k_{cat} by destabilization and entropy loss in order to obtain increases in reaction rate. Starting with an active enzyme, the addition of a substituent that provides additional stabilization (binding) of the transition state must be accompanied by some destabilization or entropy loss mechanism for the ES complex that is sufficient to prevent K_m dropping below $[S]$, in order to give an optimal increase in catalysis.

E. Differences in the Roles of Destabilization and Entropy Loss

Is there an essential difference between strain-destabilization and probability-entropy loss mechanisms, if they both serve to prevent too much expression of the intrinsic binding energy in the enzyme-substrate complex; i.e., to increase the Gibbs free energy of the ES complex?

Strain-destabilization mechanisms introduce an unfavorable energy in the ES complex that must be relieved in the transition state in order that the binding energy can be expressed fully to stabilize the transition state and increase the rate. In contrast, entropy must be lost to bring the substrates and catalytic groups together in the transition state, and if this can be done in the enzyme-substrate complex it need not be done later, so that a large part of the overall free energy barrier for reaction may already be overcome in the ES complex. A large amount of entropy must be lost to form a covalent bond between two molecules; ideally it is already lost in the ES complex but is not gained back in the transition state. The term entropy sink (Westheimer 1962) is apt for this effect, because enzymes utilize binding energy to cause a large but temporary decrease in entropy; they are true catalysts because they suffer no permanent change in entropy or energy during the turnover of substrates to products.

F. Separation of Destabilization and Entropy Contributions

Can entropy and strain-destabilization mechanisms be clearly and quantitatively separated in practice?

In some cases, yes, if the effects are small, but not if large rate increases are brought about by these mechanisms. Some separation is necessary for analysis, but it is not possible to bring about a really large loss of entropy without holding the substrate very firmly indeed in the correct position for reaction and this will necessarily involve some degree of strain or distortion.

   The entropy of a molecule in a crystalline solid is very large, on the order of 20–30 cal mol^{-1} K^{-1}, and most of this entropy must be lost so as to form a new covalent bond between two such molecules. In order to bring about a really large rate increase from loss of entropy, an enzyme must hold the substrate more tightly than it is held in a solid. This is not as unlikely as it might seem at first sight. Enzymes have had several billion years to evolve in such a way as to fit substrates snugly, rather than through the more or less random van der Waals contacts that exist in a crystalline solid. Furthermore, enzymes are held together by a chain of covalent peptide bonds and by disulfide cross links, which do not have the large entropy of a crystalline solid of small molecules.

It is a surprising fact that rate accelerations of at least 10^5 are found for succinate half esters and thiol esters compared to the analogous bimolecular reactions at a concentration of 1 M, although the two reacting groups in succinates are not held tightly next to each other (Bruice and Pandit 1960; Gaetjens and Morawetz 1960; Page and Jencks 1971; Moore 1978). There are three bond rotations that allow relative movement of these groups (I).

I

This means that, in spite of these three rotations, the succinate reaction is more probable (requires less loss of entropy) than a reaction between two molecules in a solid. The covalent backbone of succinate must provide a severe restriction on the number of positions in space that the reacting groups can take up, so that a relatively small further loss of entropy is required for the reaction to occur. This provides a rationale for the large size of coenzymes — the reacting group is attached by covalent bonds to the rest of the coenzyme, which is in turn bound to the protein, so that there should be a severe restriction to the number of positions in space that the reacting groups can take up. The covalent bonds of a protein cannot be this restrictive with respect to the substrate, unless the reaction involves groups that are covalently bonded to the protein, but they do reduce the number of degrees of freedom in the active site of an ES complex.

X-ray diffraction studies of the structures of enzymes have shown that nearly all active sites are in clefts, pockets, chasms, holes, or boundaries between domains, so that there is an opportunity for tight fit of substrates. In several enzymes the substrate is completely surrounded by protein after binding. Such an arrangement appears optimally designed to bring about a loss of most of the overall rotational entropy, which corresponds to some 20 eu or a factor of 10^4 for a typical nonlinear organic molecule, and much of the entropy of low frequency motions that arose from the translational entropy (Page and Jencks 1971). The optimum fit and entropy loss will occur if the substrate has to force open the binding site in order to enter it; this requires utilization of binding energy to force open the "jaws" of the active site and gives an ES complex in which the substrate is held tightly in the correct position to react by compression.

G. The Meanings of "Tight Binding"

If binding of a good substrate with a large loss of entropy means that the substrate is held very firmly in position, why is it that many good substrates do not bind particularly tightly, as measured by observed dissociation constants?

The question describes a paradox: how can tight binding not be tight binding? The paradox arises from the two different meanings that are assigned to "tight binding" — the first is a state of low entropy with restricted movement, and the second is an *observed* tight binding, as measured by a small dissociation constant. The first describes a state that would be reached spontaneously with a small probability and a correspondingly small equilibrium constant and unfavorable ΔG; the second describes a favorable binding in which binding energy is expressed to give a strongly favorable ΔG. The difference between the two in the binding of a good substrate to an enzyme is again an expression of the economics of the situation — it is necessary to *pay for* tight fixation and low entropy by utilizing the intrinsic binding energy (Fig. 1). Consequently, the observed binding energy may be low and good substrates often bind no more tightly, and may even bind less tightly, than poor substrates.

One obvious reason for relatively weak binding is strain in the ES complex, some of which may be used to hold the substrate firmly in position. However, an important reason arises from the nature of entropy itself, as illustrated schematically in Fig. 4. Suppose that a substrate can bind to a flat binding region in the active site, such as might be provided by the indole group of a tryptophan residue, in n_1 different positions with a certain binding energy and with no energy of interaction with the edges of the binding site (Fig. 4A). The quantity n_1 can be a very large number for a small substrate and a large site because it includes every distinguishable translational and rotational position in the plane of the binding site. Now suppose that the edges of the binding site are moved in so that n_2 becomes very small (Fig. 4B). The intrinsic binding energy is the same, but the observed binding constant and free energy are much less favorable because the probability of binding is smaller by the factor n_2/n_1, and the entropy loss upon binding is correspondingly larger.

An analogous situation occurs with a real enzyme when a substituent is added to a nonspecific, incomplete substrate that converts it to a specific substrate that can bind in only a small number of positions. The additional binding energy that is provided by

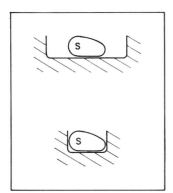

Fig. 4. *Top:* Binding of a substrate to a surface with a certain binding energy and a large number of possible equivalent states, n_1. *Bottom:* The less probable binding of the same substrate with the same binding energy, but a much smaller number of states, n_2

the substituent will be used to pay for the smaller probabolity and increased entropy loss that results from the increased fixation of the specific substrate. Consequently, the observed binding constant of the specific substrate may be no tighter, and may be weaker than the binding of the nonspecific substrate. The same situation holds when the binding of a second substrate restricts the number of positions in which the first substrate can bind — the observed binding constant of the first substrate will often be decreased in the presence of the second substrate when this happens.

The utilization of binding energy to cause binding with a large entropy loss and a small change in Gibbs free energy represents still another example of *compensation* of thermodynamic parameters, one of the most basic and widespread phenomena in chemistry. Almost any situation in chemistry that brings about a more favorable energy, such as the formation of a chemical bond, involves a degree of fixation and, hence, an unfavorable change in entropy and a correspondingly less favorable Gibbs free energy. A well-known example is the fact that the greater acidity of *p*-nitrophenol than of phenol is manifested in a more favorable entropy of ionization of *p*-nitrophenol; the enthalpies of ionization are identical. The anion of phenol is less stable, because it lacks the electron-withdrawing nitro group, but the more localized charge of the phenolate anion requires more solvation by water molecules. This solvation causes a negative change in the entropy and enthalpy of the water molecules, so that the relative instability of the phenolate anion is manifested mainly in the entropy of ionization (Hepler 1963). Similarly, a maximal loss of entropy in an enzyme-substrate complex can be paid for by a strong interaction energy, to give a small net change in free energy.

H. Must Destabilization be Observable in ES?

If strain-destabilization and entropy loss are important in catalysis, should one expect to see them in the enzyme-substrate complex?

Not necessarily. In fact, for the most efficient utilization of binding energy in catalysis, there can be an advantage if strain-destabilization and entropy loss do not occur in the ES complex (Fersht 1974; Jencks 1975).

The important point is that the binding energy should not be utilized to give tight binding in the ES complex. If the strain-destabilization and entropy loss upon binding are so large that one or more groups on the substrate are unable to bind at all, but simply hang out in space, this requirement is met. It is only necessary that the destabilization be removed and that the entropy be lost when the transition state is reached.

The point can be illustrated by the Gibbs free energy diagram of Fig. 5, choosing a standard state concentration of substrate that is equal to its physiological concentration. Figure 5A shows the useless enzyme of Fig. 2 in which the substrate and transition state are equally stabilized by tight binding. The enzyme-substrate complex can be destabilized so as to make a useful enzyme, perhaps by introducing a negatively charged carboxylate group and "freezing" the substrate in the active site, as shown in Fig. 5B. Since the Gibbs energy of the ES complex is still lower than that of free E + S, the enzyme will exist mainly as the ES complex and the observed turnover will be given by k_{cat}. Now suppose that a small additional destabilization is introduced which is

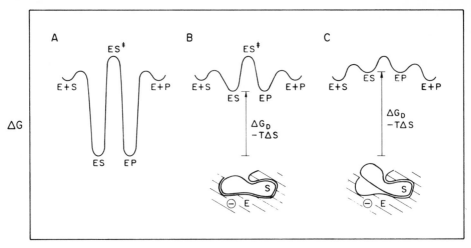

Fig. 5. Gibbs free energy diagrams for an enzyme with (*A*) no destabilization of the ES complex, (*B*) destabilization of the ES complex sufficient to give an active enzyme, and (*C*) more destabilization of the ES complex, so that some of the substrate is forced out of the active site

enough to overcome the intrinsic binding energy of a part of the substrate, so that it prefers to extend into solution rather than bind in the site, and the Gibbs energy of the ES complex becomes higher than that of E + S, as shown in Fig. 5C. Under these conditions the Gibbs energy of activation is smaller, so that catalysis will be faster than in the case of Fig. 5B. The enzyme will be predominantly in the free form and the rate of turnover will be determined by k_{cat}/K_m. From the limiting equations for the rate at high and low substrate concentrations, respectively,

$$v = k_{cat} [E]_{tot}$$
$$v = (k_{cat}/K_m) [E]_{tot} [S]$$

it is apparent that the numerical disadvantage from a K_m value that is smaller than the substrate concentration is simply the ratio $K_m/[S]$ in this situation.

An important function of destabilization mechanisms is to increase the chemical reactivity of groups at the active site. For example, the pK of a carboxylate group might easily be increased by 5, or even 10, pK units when it is covered by substrate and excluded from water so that it would become a strong base to remove a proton from a carbon acid or a strong nucleophile to attack an acyl group. This is the kind of destabilization that would be illustrated by Figs. 5B and C. However, in the case of Fig. 5C the pK of the exposed carboxylate group would be close to normal upon titration, so that the destabilization mechanism would not be detected by titration of the ES complex. *Thus, it is likely that some of the most important destabilization mechanisms are not manifested directly in enzyme-substrate complexes.* This raises a difficult challenge for their experimental demonstration, but should not be allowed to lead to neglecting their potential significance for catalysis.

It may be useful in thinking about or visualizing this situation to consider a second, metastable ES complex, which can be denoted ES*, in which the substrate is completely

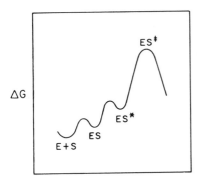

Fig. 6. Entropy loss and destabilization mechanisms of the bound substrate may be manifested only in a metastable enzyme-substrate complex, ES*, that does not represent a significant fraction of the ground state complex ES. This may represent the most efficient way of utilizing binding energy (Fig. 5)

bound and the destabilization effects are fully manifested. Such a complex may be described by Eq. (2)

$$E + S \underset{K_S}{\rightleftharpoons} ES \xrightarrow{K^*} ES^* \longrightarrow [ES]^{\ddagger} \tag{2}$$

and Fig. 6. Since ES* is of higher Gibbs energy than ES, it will not represent a significant fraction of the observed ES complex. It will be formed along the path to the transition state (Fig. 6) and may be useful in examining the balance of binding and destabilization energies in the system.

I. Indirect Stabilization of the Transition State

Must a group that destabilizes the ES complex stabilize the transition state in order to bring about catalysis?

No. Consider, for example, the negative charge that destabilizes the ES complex in Fig. 5. This charge need not have a net favorable interaction with a positive charge in the transition state so long as the destabilization in the ground state is relieved in the transition state.

This situation may well hold for lysozyme, for example. Although calculations have suggested that there is a net stabilization of a carbonium ion-like transition state by an aspartate carboxylate group in lysozyme catalysis (Warshel 1978), there could be a large catalytic effect of this carboxylate group even if its interaction with the positive charge of the transition state were no larger than in water (which is not large). The *destabilization* of the uncharged carbohydrate substrate by this carboxylate group is likely to be large if its solvation by water is prohibited when it is covered by substrate. If this destabilization is relieved in the transition state, the noncovalent interactions of nonreacting groups of the polysaccharide substrate will then be realized and will give rise to a net stabilization, or "binding," of the transition state.

This property of destabilization mechanisms has the important consequence that it makes possible the utilization for catalysis of binding energy from parts of the substrate that have nothing to do directly with the chemical process that is going on in the transition state. This is the aspect of transition state binding and stabilization that makes enzymes so different from most chemical catalysts. It means that the binding energy from nonpolar, hydrogen bonding, and other nonreacting groups of a specific substrate can be used to provide stabilization in a transition state simply by counteracting their favorable interaction with the enzyme through some destabilization mechanism in the ES complex. Calculations of the rate-accelerating effects of interactions with groups on the enzyme, such as the carboxylate group of Asp-52 in lysozyme, are incomplete unless account is taken of their destabilizing effect in the ES complex as well as the stabilizing effect in the transition state, and these energies are added to the binding energy from other parts of the substrate.

J. Observed Thermodynamic Activation Parameters

If entropy loss and strain-destabilization are important in enzymic catalysis, why can their importance not be evaluated by measuring the entropies and enthalpies of binding and activation for enzymic reactions?

Because it is almost impossible to reach unambiguous conclusions from measurements of activation parameters in enzyme-catalyzed reactions (Jencks 1975). For example, hydrophobic and electrostatic contributions to substrate binding are largely entropy-driven and are likely to mask entropy losses upon binding. Destabilization by desolvation of charged groups in the substrate or enzyme will contribute a positive entropy change from the release of bound water molecules. Furthermore, both binding and the activation process are likely to be accompanied by changes of enzyme conformation that have large intrinsic entropy changes and additional entropy changes resulting from changes in exposure of hydrophobic and charged groups to the solvent.

An observed entropy change upon substrate binding or activation may be thought of as the sum of an intrinsic entropy change, ΔS_{int}, which measures the contribution to the enzymic rate increase from the fixation of the substrate molecule in a particular position in space and rotational state, and the solvation contribution, ΔS_{solv}, which represents contributions from changes in the interaction of substrate and active site with solvent molecules. These terms are difficult or impossible to separate experimentally. The solvent effects can themselves provide a significant change in the Gibbs free energy, but this is likely to be small compared with large, compensating changes in entropy and enthalpy that will tend to obscure the intrinsic entropy changes. The same problem exists, but has not been widely recognized, in the analysis of thermodynamic parameters for the binding of ligands to proteins in order to determine the driving force for this binding.

The point may be made clear by a simple numerical example. Suppose that a substrate, S, binds to an enzyme by an enthalpy-driven mechanism with $\Delta G = -4$ kcal mol^{-1}, $\Delta H = -4$ kcal mol^{-1} and $T\Delta S = 0$ kcal mol^{-1} (Fig. 7). Now add a substituent, A, to this substrate that contributes only a small amount of additional binding, with $\Delta G = -1$ kcal mol^{-1}, but exhibits compensating changes in enthalpy and entropy of $\Delta H = +4$ kcal mol^{-1} and $T\Delta S = +5$ kcal mol^{-1}. These numbers are based on the observed thermodynamic parameters for the transfer of ethyl acetate from water to 0.3 mol frac-

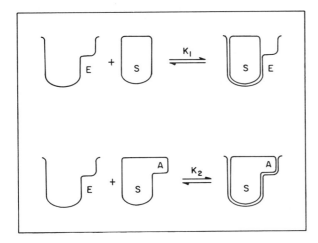

Fig. 7. Binding to an enzyme of a substrate, S, and the same substrate with the substituent A. Compensating thermodynamic changes from binding of the substituent can lead to incorrect conclusions about the driving force for the binding of SA

tion aqueous dimethylsulfoxide of $\Delta G = -0.4$ kcal mol^{-1}, $\Delta H = +4.2$ kcal mol^{-1} and $-298 \, \Delta S = -4.6$ kcal mol^{-1} (Cox 1973). The observed thermodynamic parameters are then $\Delta G = -5$ kcal mol^{-1}, $\Delta H = 0$ kcal mol^{-1}, and $T\Delta S = +5$ kcal mol^{-1}, from which it would be easy to draw the incorrect conclusion that substrate binding was principally entropy-driven.

Similarly, Larsen (1973) has pointed out how the large entropy loss of -31 cal mol^{-1} K^{-1} for the dimerization of cyclopentadiene in a Diels-Alder reaction may appear in aqueous solution largely as an unfavorable enthalpy term with a $\Delta S^{\ddagger}$ of only -4 cal mol^{-1} K^{-1}, because of large, compensating changes in thermodynamic parameters from destruction of the "solvent cage" around the separated molecules upon reaction.

Intrinsic entropies of activation and of an overall reaction can be determined directly only from the observed thermodynamic parameters of nonionic reactions that show little dependence on solvent. Diels-Alder reactions, for example, show entropies of activation and of reaction in the range -30 to -40 cal mol^{-1} K^{-1}, both in the gas phase and in (nonhydroxylic) solvents (Benford and Wasserman 1939; Wassermann 1965). They may be estimated indirectly from the observed rate increases of $10^5 - 10^8$ M for a number of unstrained intramolecular reactions, compared to corresponding bimolecular reactions. Such rate increases provide the most straightforward experimental evidence for the entropic advantage of bringing reacting groups together in ionic reactions (Page and Jencks 1971; Page 1973). If there is no enthalpic advantage from strain or destabilization mechanisms, the observed rate increase must reflect a more favorable intrinsic entropy change for the intramolecular reaction.

If there are no perturbations from compensating solvation effects, the thermodynamic parameters for these reactions ought to show the entropic advantage of intramolecularity. The best example of this that is available at the present time is an examination of the thermodynamic parameters for cyclization of o-ω-bromoalkylphenoxides in 75% ethanol [Eq. (3); Illuminati et al. 1975]. The entropy of

$$(3)$$

activation for this reaction shows a regular decrease of approximately 4 eu for each additional methylene group that is added to the chain, which agrees well with the expected loss of approximately 4.5 eu upon freezing each rotation of a hydrocarbon chain. The observed rate increase of 6.5×10^3 for the intramolecular formation of a 6-membered ring, compared to the comparable bimolecular reaction at 1 M, is accounted for entirely by the more favorable entropy of activation of +5.3 eu compared with -12.2 eu for the bimolecular reaction. The intramolecular formation of a 5-membered ring shows a larger rate increase of 1.3×10^5 M compared with the intermolecular reaction that must reflect the more favorable entropy from the smaller loss of internal rotations in this reaction, but the observed $\Delta S^{\ddagger}$ is +4.1 eu, essentially the same as for the 6-membered ring. It is possible that this represents a compensating solvation effect: the less basic phenoxide anion of this compound is presumably solvated less strongly by ethanol, so that there is less release of water into the bulk solvent and a smaller compensating entropy gain in the transition state than with the longer chain compounds.

K. Pathways for the Formation of ES

What is the pathway by which an ES complex with destabilization and a low entropy is formed? Do the necessary structural changes take place before or after binding?

The limiting paths for the formation of an enzyme-substrate complex can be described by Eq. (4),

$$(4)$$

in which ES* is the final complex in which the necessary conformational changes have taken place after binding of substrate in the upper pathway and before binding in the lower pathway. In any real situation it is probable that both happen — some changes take place before and some after binding. Rapid first-order relaxations in enzyme-substrate complexes are often attributed to the k_2 step of the upper pathway but can sometimes arise from the kinetically equivalent situation in which ES is a nonproductive complex and ES* is formed by a small fraction of free enzyme that reacts through the k_4 path.

The binding of substrates to enzymes often occurs with a rate constant that is one to two orders of magnitude smaller than the diffusion-controlled limit, or even less, although there are a number of instances in which small molecules react with proteins at a diffusion-controlled rate (Fersht 1977, p. 130). These relatively slow rates may be explained by a strict entropy requirement for binding, as well as by strain (Burgen et al. 1975). If the collision that gives ES* must have the substrate in just the right conformation and position to fit into the active site, there will be many collisions that are nonproductive, with a correspondingly slow rate. Thus, if k_4 is diffusion-controlled, the observed rate constant will be smaller than diffusion controlled by the equilibrium constant K_3; i.e., $k_{obsd} = K_3 k_4$. This situation describes a reaction that is diffusion-controlled in one sense, but has a significant unfavorable entropy of activation.

An extreme example of this situation in chemistry is the transfer of hydrogen between tri-*tert*-butylphenol and its radical [Eq.(5)], which proceeds with k = 200 M^{-1} s^{-1},

$$\text{(5)}$$

$\Delta H = 1-2$ kcal mol^{-1} and $\Delta S^{\ddagger} = -40$ e.u. (Kreilich and Weissman 1966). The slow rate of this reaction is caused by the severe steric and orientational requirements of the highly hindered reactants, which are manifested in the unfavorable entropy of activation. Acetylcholinesterase provides a possible example of this situation in enzymology. The specific substrate acetylcholine binds to acetylcholinesterase with a rate constant of 2×10^8 M^{-1} s^{-1}, whereas cationic inhibitors, which bind to the same site but without the restrictions to binding of acetylcholine, bind with rate constants of $1-2 \times 10^9$ M^{-1} s^{-1}, essentially at the diffusion-controlled limit (Rosenberry 1975; Rosenberry and Neumann 1977).

Most substrates and coenzymes can be presumed to bind sequentially, with one group fitting into a binding site for that group, followed by binding of the remaining groups (Burgen et al. 1975). Any steps that take place after the initial binding correspond to the upper path of Eq. (4). If a nonspecific substrate lacks a binding group, it may bind more slowly because this initial step is not possible and the reaction must proceed by the lower path, with a larger orientational and conformational requirement before combination with the enzyme and, consequently, a slower rate. This might account for the relatively slow k_1 step for the binding of nonspecific substrates to acetylcholinesterase and other enzymes. This step shows some of the characteristics of a diffusion-controlled reaction for the acetylcholinesterase-catalyzed hydrolysis of phenyl acetate and amyl acetate, with values of $k_{cat}/K_m = 8 \times 10^6$ M^{-1} s^{-1} and 6×10^5 M^{-1} s^{-1}, respectively. These reactions show no solvent deuterium isotope effect for this rate constant, although normal isotope effects are observed for other substrates and for k_{cat} with these substrates, and exhibit a pH dependence that is not consistent with equilibrium binding of the substrates before reaction (Rosenberry 1975).

L. Rigid versus Flexible Enzymes

Should an enzyme be mobile and flexible, like a liquid or a disordered polymer, or rigid, like a solid? Are conformational changes necessary for catalysis?

Ideally, the enzyme should be rigid with an active site that is complementary to the transition state, so that the maximum interaction energy with the substrate can be expressed in the transition state and a minimum of entropy need be lost to go from the enzyme-substrate complex to the transition state. But a real enzyme is not ideal in this sense and would not work if it were. It is necessary that the enzyme be flexible enough so that the substrate can enter and the product can dissociate from the active site easily and the force constants of many noncovalent interactions in proteins give a molecule that, in some respects, is extremely flexible (McCammon et al. 1977). Furthermore, conformational changes that are brought about by specific substrates play an important role in controlling enzyme activity, as in the folding up of two domains of hexokinase over glucose, so that the enzyme is activated to phosphorylate the hydroxyl group of glucose much faster than the hydroxyl group of water (Koshland 1959; DelaFuente and Sols 1970; DelaFuente et al. 1970; Bennett and Steitz 1978).

But an enzyme is nothing like a liquid. The fact that crystalline enzymes diffract X-rays with sufficient regularity so that their structure can be solved means that the atoms of each molecule in the crystal are held in the same relative position with extraordinary precision, in some cases to within one angstrom or even less. Schrodinger (1967) has emphasized order and negative entropy of biological systems as a prime characteristic of life; the disorder of a liquid or a flexible, denatured protein is characteristic of death.

The flexibility of a native protein involves the loss of only a small fraction of its order and negative entropy. Many of the motions involve movement of large domains as a unit and the gain in entropy from such movements, corresponding to the number of states that can be taken up, is negligible compared to the increase in entropy and number of states upon unfolding a protein. Furthermore, many motions of a group within a protein are possible only when one group moves out of the way to permit the movement — the motions are highly correlated so that the number of possible states and entropy of the system is relatively small (Careri 1974; McCammon et al. 1977). The important thing is that there should be a maximal degree of fixation of the substrate and surrounding groups when the substrate is bound in the active site. The presence of the substrate will itself tend to decrease random motions, especially if it is completely covered by groups on the enzyme.

M. Is Strain Reasonable?

Can significant strain be applied to a substrate by an enzyme in view of the small force constants for some motions within proteins?

It is far more difficult to develop large destabilization energies from mechanical distortion than from other mechanisms, such as desolvation of charged groups, and it is easy to think of situations in which the development of a large strain energy is unlikely. For example, it is difficult to develop a strong squeeze on a substrate that is bound between two domains that are normally in an open state, and it is unlikely that a protein can generate a motion that will pull apart two parts of the substrate.

On the other hand, it is not unreasonable that the maximum binding interactions of two parts of a substrate should be obtained only when the parts are pulled down over a central obstruction, which bends the intervening regions in a kind of "rack" mechanism, or that bond angles must be distorted in other ways in order to obtain optimal fit. It is likely that dispersion forces are important in substrate binding and the r^6 dependence of the dispersion energy means that a precise fit is needed to obtain optimal binding. The binding of a substrate into a site that is complementary to the transition state should generate large, unfavorable energy changes if it disrupts the structure of the protein by breaking dispersion interactions and is hindered simply by the space-filling properties of close-packed regions of the protein. Relatively large forces may be developed if the substrate has to force open the tightly closed "jaws" of an active site in order to bind or if a small substituent on the substrate binds to an empty or partly filled cavity on the enzyme that interacts very weakly with solvent.

The important point is that enzymes almost certainly make use of binding energies for catalysis through the combined effects of a number of different interactions that are individually of small or moderate strength but are collectively strong. There is experimental precedent for a moderately large destabilization energy in the binding of reduced flavin mononucleotide to *Clostridium MP* flavodoxin in the almost planar structure of the oxidized flavin, rather than the normal bent structure, which is associated with a low E_1 potential and an observed binding energy that is 5.3 kcal less favorable than that of the oxidized flavin (Ludwig et al. 1976).

An attractive, though still circumstantial case has been made for the hypothesis that essentially all of the catalysis that is brought about by chorismate mutase, by a factor of 2×10^6, can be accounted for by forcing the substrate into a ring conformation in which the reacting substituents are axial (Andrews et al. 1973; Gorisch 1978). This conformation is required for the Claisen rearrangement that is catalyzed by the enzyme. The nonenzymatic reaction has $\Delta H^{\ddagger} = 20.7$ kcal mol^{-1} and $-T\Delta S^{\ddagger} = 3.9$ kcal mol^{-1}, in which the entropy term represents the loss of internal rotations and other motions required for reaction, and the activation parameters are reduced to $\Delta H^{\ddagger} = 14.5$ kcal mol^{-1} and $-T\Delta S^{\ddagger} < 0.4$ kcal mol in the enzymic reaction. The reaction is nonionic and has a small solvent dependence, so that these activation parameters may reflect the intrinsic properties of the transition states. The reduction in $\Delta H^{\ddagger}$ in the enzymic reaction is close to the value of 7 kcal mol^{-1} that has been estimated for conversion of the diequatorial to the diaxial substrate and the smaller entropy term may be attributed to freezing the motions of the bound substrate into the correct position for reaction.

N. Do Entropy Increases Cause Catalysis?

Is it likely that binding energies are utilized to decrease $\Delta G^{\ddagger}$ by causing an increase in entropy in the transition state compared with the bound substrate?

No. A certain amount of entropy must be lost to bring reactants together in the correct position to react and, to the extent that this can be done in the ES or ES* complexes, a minimum further loss is required to reach the transition state. Unusual (though not impossible) mechanisms must be invoked to develop a net gain in entropy in the substrate or enzyme in the transition state. There are reactions that have large positive entropies of activation, such as the racemization of benzylsulfoxides through a loose, diradical transition state that resembles an encounter complex, with $\Delta S^{\ddagger} = +25$ cal mol^{-1} K^{-1} (Mislow 1967). However, it is difficult to imagine how an enzyme would convert a transition state that normally requires a precise fixation of the reacting groups to one of lower Gibbs energy that has a much smaller requirement for fixation. The simplest thing that an enzyme can do is to bring substrates together in position to react, so that it is likely that it would select transition states for catalysis that are at least as closely positioned as those for the corresponding nonenzymic reactions.

O. Experimental Approaches

How can experimental evidence be obtained to support or refute these notions?

It is necessary first to define more completely the facts that need explanation. There are a few examples of large rate accelerations that are brought about by nonreacting parts of substrates, such as the factor of 10^{10} for hexose phosphates with phosphoglucomutase and 10^{12} for the chain of coenzyme A with coenzyme A transferase, but much more information of this kind is needed, with a detailed analysis of the particular structural features that bring about the rate increase. It is important, therefore, to report upper limits for the rate constants of unreactive, incomplete substrates, rather than to describe their reaction rates as "zero." Comparisons of this kind of information with structural information from X-ray diffraction should help to elucidate the mechanisms for utilization of binding energy.

It will be useful to collect enough information about the effects of substituents to compile a catalog of the maximal intrinsic binding energies that can be obtained from the interaction of different groups with proteins. This is obtained from the change in k_{cat}/K_m in enzyme-catalyzed reactions and the increase in observed binding of ligands to other proteins that is brought about by the addition of a particular substituent group.

Some surprisingly large specificities and intrinsic binding energies have already been observed. The addition of a methyl group to incomplete substrates for aminoacyl-tRNA synthetases, to form isoleucine or valine, gives an increased specificity corresponding to a more favorable intrinsic binding energy of -3.3 kcal mol^{-1} (Loftfield and Eigner 1966; Holler et al. 1973). This is several times larger than can be obtained from the usual "hydrophobic" interactions and suggests that dispersion forces probably contribute to binding. Comparison of cysteine and alanine as substrates for cysteinyl-tRNA synthetase gives the extraordinary value of -9 kcal mol^{-1} for substitution of the thiol group of cysteine for hydrogen on alanine (Fersht and Dingwall 1979a, b). This value may reflect dispersion interactions with the polarizable sulfur atom and weak hydrogen bonding of the thiol to a group on the enzyme that is destabilized by the methyl group of alanine. However, serine, which would be expected to form a stronger hydrogen bond, interacts less strongly than cysteine by 11 kcal mol^{-1}. This suggests that binding of the amino acid at the active site is sufficiently exact so that hydrogen bonding to serine is prevented by the shorter distance between carbon and the proton in COH compared with CSH, of 0.9 Å. The energy of hydrogen bonds has a large dependence on distance (Fersht 1974). A similar requirement for exact positioning is suggested by the inactivity of a modified papain in which the cysteine residue at the active site has been converted to serine; evidently, the shorter COH distance of the serine prevents hydrogen bonding to imidazole and nucleophilic attack on a substrate (Clark and Lowe 1978).

It may be possible to observe a "freezing" of reacting groups of a substrate at the active site by physical techniques, such as nmr and X-ray diffraction; some progress has already been made in this direction. However, there always remains the possibility, and even the likelihood, that the maximum loss of entropy does not occur until the transition state or ES complex is reached, as indicated in Section H.

Transition state analogs have the potential for identifying particular modes of stabilization and destabilization, although the compounds examined so far have not come close to accounting for the entire rate acceleration brought about by any enzyme. This is not surprising for several reasons, including the fact that most such analogs have been designed to probe mechanical distortion, which is probably one of the less important mechanisms for rate acceleration. A large effect has been observed with the uncharged analog of thiamine pyrophosphate, II

II

which was designed to probe destabilization of the positive charge of thiamine pyrophosphate in the active site of pyruvate dehydrogenase and binds with $K_i < 5 \times 10^{-10}$ M, compared with $K_{diss} = 10^{-5}$ M for thiamine pyrophosphate itself (Gutowsky and Lienhard 1976). An ideal bisubstrate analog would provide a measure of the entropic advantage of binding two substrates together at the active site (Wolfenden 1972; Lienhard 1973) but, again, very large increases in binding have not yet been observed. One reason for this is presumably that if there is a very small probability that two substrates will come together in exactly the right position to react and the active site will only accommodate substrates in just this position, there is a comparably small probability of synthesizing a bisubstrate analog with exactly the right structure to bind very tightly to such a discriminating active site.

The most important need is to keep in mind the question of how binding energy can be utilized for catalysis. Many, perhaps most, scientific questions are answered from unexpected directions when a new technique or information provides a hint that is identified and followed up.

P. Other Mechanisms for Rate Increases

Are there other mechanisms by which substituent groups on specific substrates could cause a rate increase?

Yes. The binding energy of a substituent group can be used to induce a conformational change that converts the enzyme from an inactive to an active state. This can serve as a control mechanism to bring about high enzyme specificity without directly contributing to the catalytic process if the active site in the initial state is not complementary to either the substrate or the transition state of the reaction. At least part of the binding energy of sugars to hexokinase is used in this way to bring the enzyme from an open to a closed, active conformation (Bennett and Steitz 1978). It does not seem plausible that the development of enzyme activity toward specific substrates through evolution would not have given enzymes that obtained catalytic activity as well as specificity through the development of destabilization and entropy-loss mechanisms, but the existence of such mechanisms must be demonstrated experimentally in individual enzymes, especially when the rate increases that are observed in a particular case are not large. Even with hexokinase it would be surprising if some of the binding energy of ATP and glucose were not used directly to facilitate the catalytic process.

Nonproductive binding can give changes in k_{cat}, but not k_{cat}/K_m (Haldane 1930; Bernhard and Gutfreund 1958), when substituents are added or deleted. However, nonproductive binding of a specific substrate means that there is some free energy barrier for productive binding and suggests that productive binding involves a destabilization mechanism, a large entropy loss, or both. As noted in Section H, it is not unlikely that such a barrier will be large enough to prevent productive binding of more than a fraction of the substrate.

McCammon et al. (1979) have recently suggested that the dissipation of binding energy from the active site to the remainder of the enzyme may be slow enough after binding a specific substrate so that the binding energy could be used directly to facilitate the activation process by making a "hot spot" in the protein. This interesting suggestion deserves further examination.

Q. The Enzyme as a Two-State Machine

Is there a simple way to summarize the utilization of binding energy for catalysis?

Hypotheses always take on added life if they are given a name (or, better still, several competing names). Accordingly, the hypothesis that *enzymes utilize strong attractive forces to lure specific substrates into an active site in which they undergo an extraordinary transformation of form and structure* may be called the Circe effect (Jencks 1975).

A more informative answer is that, from the point of view considered here, an enzyme may be regarded as a simple machine that exists in two states. In one state, the transition state, it expresses the intrinsic binding energy of a specific substrate directly as binding, whereas in the other state, the enzyme-substrate complex, it does not. In the ES complex the binding energy is not expressed because it is *utilized* to overcome loss of entropy and through destabilization mechanisms.

		Intrinsic Binding Energy
State I	Transition state	Expressed
State II	ES Complex	Not expressed

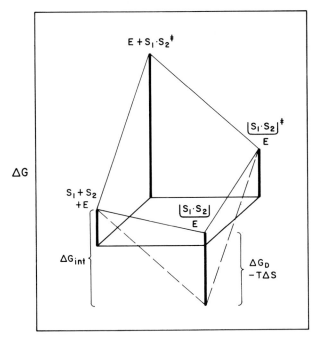

Fig. 8. Gibbs free energy diagram for the uncatalyzed and enzyme-catalyzed reactions of S_1 with S_2. The destabilization of ES that makes the enzyme work can be regarded as an interaction Gibbs energy that results from destabilization, ΔG_D, and entropy loss, $-T\Delta S$

Any machine that has these properties will be an enzyme. The same mechanisms must account for the ready reversibility of the synthesis of high energy compounds in enzymes that catalyze coupled vectorial processes, such as the reversible synthesis of ATP in myosin ATPase and the synthesis of an acyl phosphate in the sodium and the calcium ATPases (Jencks 1980b).

This situation is illustrated by the Gibbs energy diagram of Fig. 8. This diagram shows how the lower energy reaction path through the transition state of the enzymic reaction, $ES_1S_2{}^{\ddagger}$, compared with that of the nonezymic reaction, $S_1S_2{}^{\ddagger}$, necessarily corresponds to a tighter (more downhill) binding to the enzyme of the transition state, $S_1S_2{}^{\ddagger}$, than of the substrates, S_1 and S_2. If the binding of the substrates and transition state were equally tight (equally downhill), the solid and dashed lines describing this binding would be parallel (a plane would pass through the energies of the four species) and the Gibbs energies of activation for the enzymic and nonenzymic reactions would be equal.[1] In this situation the intrinsic binding energy of the substrates, ΔG_{int}, would be expressed in both the transition state and the enzyme-substrate complex. The enzyme works only to the extent that the substrates do *not* bind tightly to it because of a de-

---

1 It is unlikely that the binding energies would be exactly equal even in this hypothetical situation, because some entropy must always be lost to bring two substrates up to the active site of the enzyme from dilute solution at some standard state. However, this entropy loss is small compared with the maximum entropy loss of some -35 e.u. for the formation of a covalent bond between two substrates, based on a standard state of 1 M (Page and Jencks 1971). This diagram is more realistic than that of Fig. 2 because a large entropy loss will contribute to catalysis and give a large increase in the Gibbs energy of the ES complex only if two substrates react with each other or a substrate reacts with a group on the enzyme

stabilization [or interaction, or linked (Wyman 1948)] Gibbs energy, which is composed of the destabilization energy *per se*, ΔG_D, and the entropy loss that occurs upon formation of the ES complex, $-T\Delta S$.

References

Andrews PR, Smith GD, Young IG (1973) Transition state stabilization and enzymic catalysis. Kinetic and molecular orbital studies of the rearrangement of chorismate to prephenate. Biochemistry 12: 3492–3498

Ballardie FW, Capon B, Cuthbert MW, Dearie WM (1977) Some studies on catalysis by lysozyme. Bioorg Chem 6: 483–509, and references therein

Benford GA, Wassermann A (1939) The mechanism of additions to double bonds. Part VII. Chemical equilibrium in solution and in the gaseous state. J Chem Soc 367–371

Bennett WS, Steitz TA (1978) Glucose-induced conformational change in yeast hexokinase. Proc Nat Acad Sci USA 75: 4848–4852

Bernhard SA, Gutfreund H (1958) Some considerations bearing on the mechanism of action of proteolytic enzymes and transferases. Proceedings of the International Symposium on Enzyme Chemistry, Tokyo and Kyoto, 1957, p 124. Maruzen, Tokyo

Bruice TC, Pandit UK (1960) Intramolecular models depicting the kinetic importance of "fit" in enzymatic catalysis. Proc Nat Acad Sci USA 46: 402–404

Burgen ASV, Roberts GCK, Feeney J (1975) Binding of flexible ligands to macromolecules. Nature 253: 753–755

Careri G (1974) The fluctuating enzyme. In: Kursunoglu B, Mintz SL, Widmayer SM (eds) Quantum statistical mechanics in the natural sciences, pp 15–35. Plenum, New York

Clark PI, Lowe G (1978) Conversion of the active-site cysteine residue of papain into a dehydroserine, a serine and a glycine residue. Eur J Biochem 84: 293–299

Cox BG (1973) Free energies, enthalpies, and entropies of transfer of non-electrolytes from water to mixtures of water and dimethyl sulphoxide, water and acetonitrile, and water and dioxan. J Chem Soc Perkin Trans II, 607–610

Fersht A (1977) Enzyme structure and mechanism. Freeman, San Francisco

Fersht AR (1974) Catalysis, binding and enzyme-substrate complementarity. Proc R Soc London Ser B 187: 397–407

Fersht AR, Dingwall C (1979a) Cysteinyl-tRNA synthetase from *Escherichia coli* does not need an editing mechanism to reject serines and alanine. High binding energy of small groups in specific molecular interactions. Biochemistry 18: 1245–1249

Fersht AR, Dingwall C (1979b) An editing mechanism for the methionyl-tRNA synthetase in the selection of amino acids in protein synthesis. Biochemistry 18: 1250–1256

Fuente de la G, Sols A (1970) The kinetics of yeast hexokinase in the light of the induced fit involved in the binding of its sugar substrate. Eur J Biochem 16: 234–239

Fuente de la G, Lagunas R, Sols A (1970) Induced fit in yeast hexokinase. Eur J Biochem 16: 226–233

Gaetjens E, Morawetz H (1960) Intramolecular carboxylate attack on ester groups. The hydrolysis of substituted phenyl acid succinates and phenyl acid glutarates. J Am Chem Soc 82: 5328–5335

Görisch H (1978) On the mechanism of the chorismate mutase reaction. Biochemistry 17: 3700–3705

Gutowsky JA, Lienhard GE (1976) Transition state analogs for thiamin pyrophosphate-dependent enzymes. J Biol Chem 251: 2863–2866

Haldane JBS (1930) Enzymes. Longmans Green, London

Hepler LG (1963) Effects of substitutents on acidities of organic acids in water: Thermodynamic theory of the Hammett equation. J Am Chem Soc 85: 3089–3092

Holler E, Rainey P, Orme A, Bennett EL, Calvin M (1973) On the active site topography of isoleucyl transfer ribonucleic acid synthetase of *Escherichia coli B*. Biochemistry 12: 1150–1159

Illuminati G, Mandolini L, Masci B (1975) Ring-closure reactions. V. Kinetics of five- to ten-membered ring formation from o-ω-bromoalkylphenoxides. The influence of the O-heteroatom. J Am Chem Soc 97: 4960–4966

Jencks WP (1975) Binding energy, specificity, and enzymic catalysis: The circe effect. Adv Anzymol 43: 219–410

Jencks WP (1980a) Intrinsic binding energy, enzymic catalysis, and coupled vectorial processes. In: Kaplan NO (ed) From cyclotrons to cytochromes: A Symposium in Honor of Martin Kamen's 65th Birthday Aug 27–31, 1978, La Jolla, CA. (in press)

Jencks WP (1980b) The utilization of binding energy in coupled vectorial processes. Adv Enzymol 51: 75–106

Koshland DE Jr (1959) Mechanisms of transfer enzymes. In: Boyer PD, Lardy H, Myrbäck K (eds) The enzymes, 2nd ed, vol I, pp 305–346. Academic Press, New York

Kreilich RW, Weissman SI (1966) Hydrogen atom transfer between free radicals and their diamagnetic precursors. J Am Chem Soc 88: 2645–2652

Larsen JW (1973) Entropy contributions to rate accelerations of intramolecular reactions in water vs. non-structured solvents. Biochem Biophys Res Commun 50: 839–845

Lienhard GE (1973) Enzymatic catalysis and transition-state theory. Science 180: 149–154

Loftfield RB, Eigner EA (1966) The specificity of enzymic reactions. Aminoacyl-soluble RNA ligases. Biochim Biophys Acta 130: 426–457

Ludwig ML, Burnett RM, Darling GD, Jordan SR, Kendall DS, Smith WW (1976) The structure of *Clostridium MP* flavodoxin as a function of oxidation state: some comparisons of the FMN-binding sites in oxidized, semiquinone and reduced forms. In: Singer TP (ed) Flavins and flavoproteins, pp 393–404. Elsevier, Amsterdam New York

McCammon JA, Gelin BR, Karplus M (1977) Dynamics of folded proteins. Nature 267: 585–593

McCammon JA, Wolynes PG, Karplus M (1979) Picosecond dynamics of tyrosine side chains in proteins. Biochemistry 18: 927–942

Mislow K (1967) On the stereomutation of sulfoxides. Rec Chem Prog 28: 217–240

Moore SA (1978) Model and enzymic studies on the mechanism and modes of rate acceleration of succinyl-coenzyme-A:3-oxo-acid CoA transferase. Ph.D. Thesis, Brandeis University, June

Page MI (1973) The energetics of neighbouring group participation. Chem Soc Rev 2: 295–323

Page MI, Jencks WP (1971) Entropic contributions to rate accelerations in enzymic and intramolecular reactions and the chelate effect. Proc Nat Acad Sci USA 68: 1678–1683

Pauling L (1946) Molecular architecture and biological reactions. Chem Eng News 24: 1375–1380

Ray WJ Jr, Long JW (1976) Thermodynamics and mechanism of the PO_3 transfer process in the phosphoglucomutase reaction. Biochemistry 15: 3993–4006

Ray WJ Jr, Long JW, Owens JD (1976) An analysis of the substrate-induced rate effect in the phosphoglucomutase system. Biochemistry 15: 4006–4017

Rosenberry TL (1975) Catalysis by acetylcholinesterase: Evidence that the rate-limiting step for acylation with certain substrates precedes general acid-base catalysis. Proc Nat Acad Sci USA 72: 3834–3838

Rosenberry TL, Neumann E (1977) Interaction of ligands with acetylcholinesterase. Use of temperature-jump relaxation kinetics in the binding of specific fluorescent ligands. Biochemistry 16: 3870–3878

Schrödinger E (1967) What is life? University Press, Cambridge

Warshel A (1978) Energetics of enzyme catalysis. Proc Nat Acad Sci USA 75: 5250–5254

Wassermann A (1965) Diels-Alder reactions. Elsevier, Amsterdam New York

Westheimer FH (1962) Mechanisms related to enzyme catalysis. Adv Enzymol 24: 441–482

Wolfenden R (1972) Analog approaches to the structure of the transition state in enzyme reactions. Acc Chem Res 5: 10–18

Wyman J Jr (1948) Heme proteins. Adv Protein Chem 4: 407–531

2. The Cytochromes C: Paradigms for Chemical Recognition

M. D. KAMEN

A. Abstract

The cytochromes c include subgroups which present a variety of redox functions based on well-defined changes in the basic three-dimensional structure exemplified by the mitochondrial and certain bacterial forms, in particular cytochromes c_2. These proteins exhibit overlapping functionality and a graded sequence of structures which provide paradigms well suited for clarification of recognition mechanisms. The character and distribution of cytochromes c will be discussed and approaches to relatedness of structure and function will be described, based on kinetic analyses of cross reactivities of cytochromes c_2 with mitochondrial cytochrome c oxidase.

B. Introduction

Cytochromes c are proteins associated with a redox function based on reversible one-electron oxidation of the ferrous-ferric couple of their prosthetic heme moiety, which is covalently bound usually by thioether linkages, and characterized by distinctive absorption spectra (Florkin and Stotz 1965). In eukaryotes they are localized in the energy conserving organelles — mitochondria and chloroplasts — in which they function in electron transport linked to ATP production. In prokaryotes complex arrays of such proteins occur, the precise functions of which are ill-defined or unknown. Data on their structure and function are still fragmentary (Bartsch 1968; Kamen and Horio 1970; Horio and Kamen 1970; Kamen et al. 1972; Lemberg and Barrett 1973; Kamen 1973; Yamanaka and Okunuki 1974; Salemme 1977; Bartsch 1978). Hence, classification schemes remain tentative, nevertheless providing working schemes (Ambler 1978).

Among the cytochromes c now available for investigation there are certain relatively well-defined subgroups (Ambler 1978) in which the primary polypeptide chain is folded in a distinctive fashion (Dickerson and Timkovich 1975) involving as extraplanar ligands for the central heme iron the sulfur atom of methionine, and the imidazole nitrogen of histidine. These two ligands occur usually at positions 18 (histidine) and 80 (methionine) — as numbered in mitochondrial cytochromes c — brought into appropriate juxtaposition to the heme by the protein folding pattern. Thus, there arises a surface topology to support functional interaction (electron insertion and withdrawal) with the character-

Department of Chemistry, University of California at San Diego, La Jolla, CA 92093, USA

istic enzyme complexes in the redox reaction chain. In mitochondria, these are the so-called complexes II, III and IV, the first two comprising the cytochrome c reductase system linked either to NADH or succinate oxidation, the last being the cytochrome c-oxidase which effect terminal reduction of oxygen to water. In chloroplasts, the function of the corresponding c-type cytochrome — called "cytochrome f" — is less well-defined, but appears analogous to that of mitochondria in that there is also mediation of coupled electron transport, differing in that the redox process is wholly anaerobic with reduction of a photoactive "reaction center P-700" (a specialized chlorophyll a), rather than being aerobic with molecular oxygen as H-acceptor. Both processes support coupled phosphorylation, which in mitochondria is termed "oxidative phosphorylation", and in chloroplasts "photophosphorylation".

Analogous to these eukaryotic systems are those of the photosynthetic heterotrophes together with those of certain facultative nonphotosynthetic prokaryotes. Examples of the former are the nonsulfur purple photosynthetic bacteria, or *Rhodospirillaceae* (Pfennig and Trüper 1973) in which photophosphorylation is confined to membrane complexes, isolated as vesicles or "chromatophores". In these a c-type cytochrome subclass, known as the cytochromes c_2, functions in the coupled redox processes linked to energy transduction and ATP synthesis. Examples of the latter types of prokaryotes are certain pseudomonads which live as facultative denitrifiers, reducing either oxygen or nitrate in coupled oxidative phosphorylation. In at least one instance — that of *Paracoccus denitrificans* — a c-type protein, structurally similar to some of the cytochromes c_2, is an active component of the energy-transduction process.

In all these prokaryotic systems, the reaction partners of the c-type cytochromes, i.e., the associated reductases and oxidases, remain to be adequately characterized. In general, the oxidases bear little resemblance to the mitochondrial cytochrome c oxidase, although some bacterial oxidases appear to be of the type "cytochrome a" as encountered in eukaryotes (Bartsch 1968; Horio and Kamen 1970; Kamen and Horio 1970; Lemberg and Barrett 1973).

The particular aspect of cytochromology relevant to this Symposium is the means whereby the various cytochromes c — whether of mitochondria, chloroplasts or chromato-phores, designated as "c", "f"[1], and "c_2", respectively — may be recognized by their redox partners or, conversely, excluded from reaction in varying degrees when confronted by nonhomologous reductases and oxidases. It will be seen that the cytochromes c of mitochondria, and c_2 of prokaryotic chromatophores, exhibit overlapping functionality and a graded sequence of structures which permit construction of paradigms well suited for clarification of chemical recognition mechanisms at the simplest molecular level. In the discussion which follows I will refer for convenience to the subclass of mitochondrial cytochromes c simply as "cytochromes c".

C. Cytochrome c; Character and Distribution

The official definition of cytochromes c as a class (Florkin and Stotz 1965) is so broad as to include proteins which are in no sense related, functionally or structurally. How-

---

1 Recent reports show that the soluble monomeric cytochrome c-552 (formerely called "f") is not the functional electron receptor of the chloroplast reductase but rather a membrane-bound cytochrome "c_1", which corresponds to "f", *vide* Wood (1979), Wood and Willey (1980)

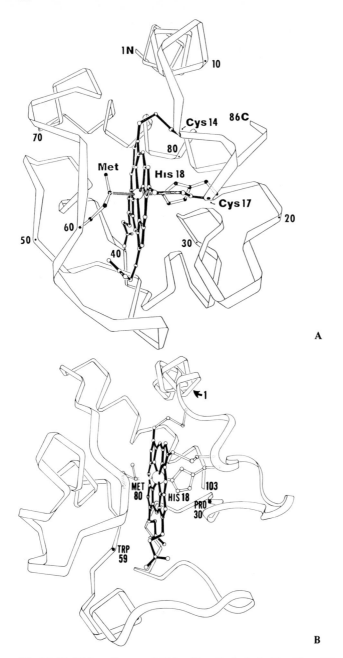

A

B

Fig. 1A, B. Ribbon diagram of *Chlorobium limicola f. thiosulfatophilum* cytochrome *c*-555 showing typical "cytochrome c fold" (after Z.R. Korszun and F.R. Salemme 1977). **A** Beginning at
the amino terminus (*1N*) the first ten residues are folded in a helical section perpendicular to the
plane of the paper. Then, the polypeptide chain makes an excursion to form the *right side* of the
molecule which includes the heme-binding cysteines (*Cys-14* and *Cys-17*) as well as the fifth extraplanar ligand (*His-18*). The chain continues onward to produce the *bottom* portion (from residues
30 to *40*), then extends to eventually construct the *left side* containing the sixth extraplanar ligand

ever, if consideration is restricted to proteins with the same basic tertiary structure, or "cytochrome *c* fold" (Dickerson and Timkovich 1975), vide Fig. 1, the variety of proteins included still remains sufficiently rich so that many test systems can be chosen to serve as paradigms of chemical recognition. A brief summary is offered here.

The cytochromes *c* of mitochondria share the unique function of mediation of terminal electron transport from cytochrome c_1 in "complex III" to the oxidase. At one time it was thought that any cytochrome *c* could be interchanged from species to species and even from any eukaryotic mitochondrion to another, as for instance yeast cytochrome *c* replacing beef heart cytochrome *c* in beef heart mitochondria. Now, as kinetic experimentation has become more sophisticated along with establishment of more demanding criteria for purity of preparations, it is clear that functionality in such cross-reactivity can vary markedly (Margalit and Scheiter 1970; Davis et al. 1972; Banga and Borza 1975; Smith et al. 1975; Errede et al. 1976; Ferguson-Miller et al. 1976; Errede and Kamen 1978).

In the cytochrome system of the protozoan *Tetrahymena pyriformis* (Kilpatrick and Erecinska 1977), as an extreme case, the oxidase is not of the classical "a" type, but appears to be related to the prokaryotic "d" type. The corresponding cytochrome *c* is totally unreactive with beef heart oxidase (Errede and Kamen 1978). Nevertheless, as a subclass the mitochondrial cytochromes *c* exhibit remarkable homology in primary structure (Dickerson and Timkovich 1975). Some 80 such proteins have been sequenced and, while substitutions may occur involving up to 60% difference in sequence most of these are conservative in character and practically no deletions or additions are found in the main body of the sequences. Such changes as do occur are confined to the N-termini. A high content of basic residues (mainly lysine) is characteristic.

Coming to prokaryotes, one notes immediately a structurally closely related subclass – the "cytochromes c_2", distributed, as remarked previously, mainly among *Rhodospirillaceae* and denitrifying pseudomonads – in which a strong homology with the mitochondrial forms is evident. Thus, as in Table 1, the extreme difference among eukaryotic forms (horse *vs Euglena*) closely approaches that between cytochrome c_2 and tuna cytochrome *c*. In the cases of the cytochromes c_2 from *Rhodomicrobium vannielii* and *Rhodopseudomonas viridis,* only a single deletion at position 11 in horse heart cytochromes is needed to achieve complete alignment of the sequences, resulting in some 84 or 86 identities (including conservative substitutions) out of the total 104 residues. In two recent cases – those of the proteins from *Rhodopseudomonas globiformus* and *Rhodospirillum photometricum* – we have found even this single deletion to be unnecessary (R.P. Ambler, T.E. Meyer, and M.D. Kamen, in preparation). In others, massive insertions and some deletions are needed to achieve homology, but these modifications all occur in neutral portions of the structure (as in sections in the left and/or right sides, and at the bottom of the front face) which leave the basic folding pattern unaltered. All these cytochromes c_2 show little or greatly diminished ability to bind and react with the mitochondrial oxidase, while showing good reactivity with the reductase (Errede and Kamen 1978).

(*Met 60*) and ending in another helical portion at the carboxyl terminus. **B** Tuna cytochrome *c* structure. Deletions at **right side** and **bottom** with rotation inward of **left side** would produce structure similar to that of **A** (Courtesy: Prof. F.R. Salemme)

Table 1. Homologies between cytochromes c and c_2

Source c		c_2	I	S	Σ
Tuna	vs.	R. Rubrum	36	42	78
Tuna	vs.	Rps. capsulata	36	34	70
Tuna	vs.	P. denitrificans	43	34	77
Tuna	vs.	Rps. palustris	34	34	68
Tuna	vs.	Rps. sphaeroides	31	43	74
Tuna	vs.	Rm. vannielii	51	35	86
Tuna	vs.	R. molischianum	39	40	79
Horse	vs.	Rm. vannielii	51	35	86
Horse	vs.	R. viridis	55	29	84
Horse	vs.	R. rubrum	42	39	81
Horse	vs.	E. gracilis	56	34	90

Numbers shown are identities (I), identities obtainable by one-base substitutions in codons (S), and summations of both (Σ). Abbreviations: *R, Rhodospirillum; Rps, Rhodopseudomonas; Rm, Rhodomicrobium; E, Euglena; C, Chlorobium; P, Paracoccus*

It has been suggested (Ambler 1978) that the subclass of cytochromes c_2 be subdivided further on the basis of sequence into types. Type "1A", exemplified by the *R. rubrum* or *P. denitrificans* proteins, shows the aforementioned structural affinities to cytochromes c at both primary and tertiary levels but contains certain insertions in similar loops of the peptide chain. Type "1B" is exemplified by the cases of *R. vannielii* and *Rps. viridis* and includes at least half a dozen other examples in which the insertions of type 1A do not occur.

Among other classes of soluble prokaryotic cytochromes c so far identified no such degree of similarity appears. One may note cytochromes c which include: (1) the "pseudomonad cytochromes c-551", found mostly in *Pseudomonas sp.* but also observed in the strictly aerobic *Azotobacter vinelandii* with some related proteins reported to occur in certain nonsulfur purple photosynthetic bacteria; (2) cytochromes "f" (chloroplast and algal cytochromes c associated with photosynthetic function) together with cytochromes "c_5" (distribution as in pseudomonad cytochrome c-551). All of these retain the characteristic "cytochrome c fold", but diverge in primary structure and function.

Further subclasses exist but need not be considered in this discussion. I should mention, however, the particularly interesting cases of the c-type cytochromes obtained from two strains of strictly anaerobic green sulfur photosynthetic bacteria (*Chlorobacteriaceae*), genera *Chlorobium* and *Prosthecocloris*. One of these (*C. limicola* f. *thiosulfatophilium* cytochrome c-555), while exhibiting the cytochrome fold (Fig. 1) and some sequential homology with the above subclasses reacts rather well with mitochondrial oxidase, but not with reductase, while the homologous cytochrome c from *P. aestuarii* reacts not at all with either preparation (Errede and Kamen 1978). The former is less reactive than would be expected from its ability to bind oxidase (Errede and Kamen 1978). Another fascinating case urgently in need of investigation is the cytochrome c from the strictly aerobic sulfur oxidizer, *T. novellus*, which reacts well with mitochondrial oxidase (Yamanaka 1972).

D. Structure and Function in Cytochromes c and c_2

It is well established that in cytochromes c interaction with the mitochondrial oxidase or reductase is controlled by the distribution of positively charged lysine or arginine residues exposed over the "front" face of the protein, through which an edge of the heme prosthetic group projects (Smith and Conrad 1956; Takemori et al. 1962; Wada and Okunuki 1968; Wada and Okunuki 1969; Margoliash et al. 1973; Nicholls 1974; Staudenmayer et al. 1977; Koppenol et al. 1978; Maurel et al. 1978). Quantitative measures of the extent of interaction as a function of this surface topography are required. These must be based on adequate kinetic analyses, involving determinations of general rate laws over a range of concentrations of cytochromes c sufficient to define precisely the kinetic parameters. Hopefully, unambiguous deductions as to reaction mechanisms may also be possible, but they are not necessary merely for the purposes of comparisons of reactivity.

Such kinetic studies are in relatively short supply, but a few have been performed using a wide selection of both eukaryotic and prokaryotic cytochromes c and c_2, with standardized mitochondrial oxidase preparations as the test systems (Errede and Kamen 1978; Nicholls 1974; Brautigan et al. 1978). High resolution X-ray structure determinations of several cytochromes c (Takano et al. 1973; Swanson et al. 1977), as well as a cytochrome c_2 from one strain of the *Rhodospirillaceae* (Salemme 1977; Salemme et al. 1973a, b) and from *P. denitrificans* (Almassy and Dickerson 1978) enable extrapolations using available primary structures to others as yet to be determined. Nuclear magnetic resonance spectroscopy (Kowalsky 1965; Wüthrich 1970; Redfield and Gupta 1971; Keller et al. 1973; Smith and Kamen 1974) combined with electron spin resonance (Aasa et al. 1970; Dogson et al. 1974; Brautigan et al. 1979) and Mössbauer studies (Huynh et al. 1978) give further basis for such extrapolations. [Indeed one study leading to a proposed structure for a prokaryotic "cytochrome c_3" has been published (Dobson et al. 1974) for which the direct structure determination by X-ray diffraction analysis has yet to be performed[2].]

Thus, there are available sufficiently precise definitions of surface charge topography of these selected cytochromes c and c_2 to enable correlations to be made with kinetic behavior when interacting with the test oxidase. The proteins chosen exhibit function in this system ranging from almost nil to over several orders of magnitude greater, up to the full functionality of the homologous mitochondrial beef heart cytochrome c. A full account of these experiments is given elsewhere (Errede and Kamen 1978).

Briefly, a single rate law, as in Eq. (1)

$$\text{Velocity} = \frac{\alpha_1 + \alpha_2 \,[c]}{1 + \beta_1 \,[c] + \beta_2 \,[c]^2} \; (\text{oxidase}) \, (\text{ferrocytochrome } c) \qquad (1)$$

is demonstrated to hold over all relevant concentrations and for cytochromes of greatly differing reactivities. The kinetic constants can be related to reaction mechanisms, using a reaction scheme which is a modification of that proposed earlier by Minnaert. This scheme is shown in Table 2 (Errede and Kamen 1978). In this scheme, known as the dependent site mechanism, active complexes occur in which first one cytochrome c

---

2 Note added in proof: R. Haser et al. (1979) have published recently the molecular structure of a cytochrome c_3

Table 2. "Modified Minnaert" mechanism (Errede and Kamen 1978)

Mechanism:

$$\text{ferro } c \quad k_1 \qquad\qquad k_3 \qquad\qquad\qquad k_5 \qquad \text{ferri } c$$

ferro c	k_1		k_3		k_5	ferri c
+	$\underset{k_2}{\overset{}{\rightleftarrows}}$	[ferro c-ox]	$\longrightarrow$	[ferri c-ox]	$\underset{k_6}{\overset{}{\rightleftarrows}}$	+
ox						ox

ferro c	k_1		k_9		k_{11}	ferri c
+	$\overset{}{\rightleftarrows}$	[ferro c-ox-ferro c]	$\longrightarrow$	[ferri c-ox-ferro c]	$\overset{}{\underset{k_{12}}{\rightleftarrows}}$	+
[ferro c-ox]	k_8					[ferro c-ox]

ferro c	k_{13}		k_{15}		k_{16}	ferri c
+	$\overset{}{\rightleftarrows}$	[ferro c-ox-ferri c]	$\longrightarrow$	[ferri c-ox-ferri c]	$\overset{}{\underset{k_{17}}{\rightleftarrows}}$	+
[ferri c-ox]	k_{14}					[ferri c-ox]

Rate equation:

$$\text{Velocity} = \frac{k_1{}^0 + k_1 k_2{}^0 \,[c]}{1 + K_1[c] + K_1 K_2[c]^2}\,[\text{oxidase}]\,[\text{ferro } c] = \frac{\alpha_1 + \alpha_2[c]}{1 + \beta_1[c] + \beta_2[c]_2}\,[\text{oxidase}]\,[\text{ferro } c],$$

and [c] is *total* concentration

Definition of constants:

Electron transfer:
$$k_1{}^0 = \frac{k_1 k_3}{(k_2 + k_3)}$$

$$k_2{}^0 = \frac{k_1 k_9}{(k_8 + k_9)}$$

Binding:
$$K_1 = \frac{K_1}{K_2}$$

$$K_2 = \frac{K_7}{K_8}$$

"Back" reaction constants (K_4, K_{10} and K_{16}) assumed negligible

reacts with oxidase to form a productive reaction complex, then, as concentrations increase, a second cytochrome c reacts with this complex to form a second complex. Thus, binding and reaction of the second cytochrome c depends on precomplexing of the protein. At steady state, which occurs under the experimental conditions employed, the kinetic parameters α_i and β_i are related to specific rate constants k_i, as shown in Table 2. The approximations used and justifications for this scheme on an experimental basis have been fully discussed elsewhere (Errede and Kamen 1978).

It is seen that binding leading to the formation of productive reaction complexes can be set forth in quantitative terms by the binding constants K_1 and K_2, while the limiting reaction rates at low concentration $k_1{}^0$, and at high concentration $k_2{}^0$, are correlated with expressions involving both the reaction rate constants for complex formation and dissociation together with the forward reaction rate constants.

Dr. Errede in our laboratory has determined values for these constants; a few are displayed in Table 3. Thus, we find that for horse heart cytochrome c and the *P. denitri-*

Table 3. Some values for the kinetic constants of the in vitro oxidase reaction (Errede and Kamen 1978)

Cytochrome	I (mM)	k_1^0 (M^{-1})	K_1 (M^{-1} s^{-1})
Horse heart c	44	2.50×10^8	6.21×10^6
$Cr.$ $fasciculata$ c-555	44	7.94×10^7	1.40×10^6
$C.$ $thiosulfatophilum$ c-555	44	3.65×10^6	9.03×10^4
$P.$ $denitrificans$ c-550 ("c_2")	44	6.05×10^5	1.29×10^5
$R.$ $rubrum$ c_2	44	9.70×10^4	2.56×10^4
$Rm.$ $vannielii$ c_2	44	2.55×10^4	6.98×10^3

Abbreviations: Cr, $Crithidia$; P, $Paracoccus$
I = ionic strength

ficans c_2 the high affinity complex (formed at low concentrations of cytochrome c_2) dominates the reaction process almost completely, so that the ordered binding postulated actually appears to occur, while for *R. rubrum* cytochrome c_2 the second site binding reaction accounts for a considerable fraction of the total complex formed. Hence, another form of this mechanism, in which two independent binding sites occur, may equally well rationalize these kinetic data for the *R. rubrum* cytochrome c_2.

There is a close resemblance in tertiary structures for these three proteins, as seen in Fig. 2, wherein topographies of positive charge distribution can be visualized as well as some other topographies also shown (Fig. 3). These structural features appear to pro-

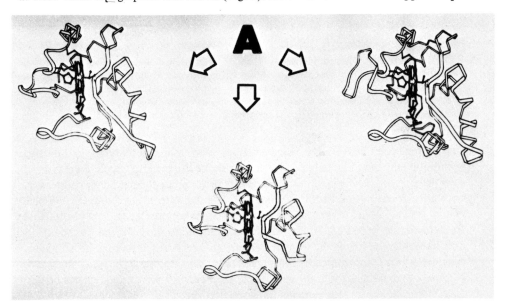

Fig. 2. Ribbon diagrams of three c-type cytochromes, possessing the characteristic "cytochrome c fold". On the *left* is the 134-residue *P. denitrificans* cytochrome c-550, in the *center* the 103-residue cytochrome c of tuna, and on the *right* the 112-residue cytochrome c of *R. rubrum* cytochrome c_2. (*A*) denotes a possible prototype from which these proteins could have originated by divergent evolution. (Courtesy: Prof. F.R. Salemme)

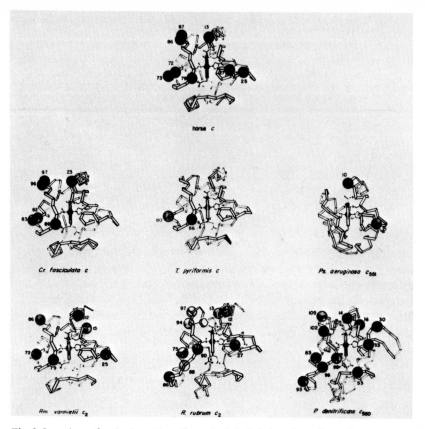

Fig. 3. Locations of cationic residues (*large shaded circles*) on front faces of selected cytochromes *c*. The front view for the folded polypeptide chain, with the heme seen on edge as well as the heme-iron extraplanar ligands (histidine nitrogen and methionine sulfur), is shown schematically (vide ref (9), adapted from figures for tuna *c*, *R. rubrum* c_2, *P. denitrificans* *c*-550, and *P. aeruginosa* *c*-551; Courtesy: Prof. R.E. Dickerson). Note: "Lys 23" of *C. fasciculata* cytochrome *c* corresponds to lysines 13 (tuna) 10 (*P. aeruginosa*) 12 (*T. pyriformis*) and 14 (*P. denitrificans*)

vide the molecular basis for the observed large variations in reactivity referred to the eukaryotic mitochondrial oxidase system. It may be deduced that while there is an essentially continous chain of positive residues surrounding the partially exposed site of electron insertion and withdrawal in all the reactive cytochromes *c* and c_2, no such feature is noted in the unreactive cytochrome *c* from the protozoan, *T. pyriformis*, or the cytochrome "*c*-551" from the pseudomonad, *P. aeruginosa*. Further requirements of surface features among the reactive cytochromes *c*, in which certain particularly essential lysines (those at positions 13 and 27) are displaced (as in *R. vannielii* cytochromes c_2) or missing (as in the eukaryotic *T. pyriformis* cytochrome *c*), can be deduced to be molecular requirements for reactivity.

The quantitative expression of these molecular features in terms of the kinetic constants is given in Table 3. Thus, the fivefold lower value of K_1, the binding constant for the first complex formation in *C. fasciculata* cytochrome *c*, relative to horse heart cytochrome *c*, can be rationalized as arising from the absence of lysines 25 and 27.

However, the value for the second binding equilibrium, K_2, is four times larger for the *C. fasciculata* protein (Errede and Kamen 1978), which suggests that absence of lysines 25 and 27 affects the orientation of the first bound cytochrome c so that the second site in this protein is better exposed for reaction than it is in the horse protein.

Other deductions can be made relating to the relative reactivities of the cytochromes c_2 in terms of lysine placement on the front surface, but enough has been said already to indicate the nature of this approach and the direction it will take as more precise structural data are obtained. Moreover, other test systems are in prospect as eventually useful in the general development of paradigms for chemical recognition of the cytochromes c by their organelles. Thus, there are the soluble eukaryotic redox enzymes, represented in particular by yeast cytochrome c peroxidase (Yonetani and Ray 1965) — which have already been used as kinetic probes of cytochrome c structure (Kang et al. 1977) — and *Vitreoscilla* cytochrome c type oxidase (Liu and Webster 1974). Numerous bacterial redox systems, i.e., the well-studied *Pseudomonas aeruginosa* nitrite reductase (Kamen and Horio 1970), *Thiobacillus novellus* sulfite cytochrome c reductase (Yamanaka 1975), and *P. denitrificans* cytochrome c oxidase (Korszun and Salemme 1977) have been described.

Approaches based on chemically modified cytochromes c (reviewed in Brautigan et al. 1978), in which effects of singly modified lysines or other surface residues on kinetic responses to oxidase have been studied, are currently exhibiting vigorous development. A recent report (Speck et al. 1979) extends this approach to interactions with Complex IV, using a coenzyme Q analog to effect reduction and thereby obviate the complications arising from the use of Complex I and NADH as reductant (Errede and Kamen 1978). It is encouraging that these direct determinations support conclusions reached by the comparative biochemical approach as regards critical cationic residues involved in interactions of cytochromes c with associated mitochondrial redox systems.

Acknowledgments. Financial support provided by grants from the National Institutes of Health (GM-18528) and the National Science Foundation (BMS-75-13608) made possible many of the researches quoted in this presentation.

References

Aasa R, Albracht SPJ, Falk K-E, Lanne B, Vänngård T (1976) EPR signals from cytochrome c oxidase. Biochim Biophys Acta 422: 260–272

Almassy RJ, Dickerson RE (1978) Pseudomonas cytochrome c_{551} at 2.0 Å resolution enlargement of the cytochrome c family. Proc Natl Acad Sci USA 75: 2674–2678

Ambler RP (1978) Cytochrome c and copper protein evolution in prokaryotes. In: Leigh GJ (ed) The evolution of metalloenzymes, metalloproteins and related materials. Sympos Univ Sussex, p 100

Bartsch RG (1968) Bacterial cytochromes. Annu Rev Microbiol 22: 181–200

Bartsch RG (1978) In: Clayton RK, Sistrom WR (eds) The photosynthetic bacteria, Chap 13, p 249. Plenum, New York

Benga G, Borza V (1975) Differences in reactivity of cytochrome oxidase from human liver mitochondria with horse and human cytochrome c. Arch Biochem Biophys 169: 354–357

Brautigan DL, Ferguson-Miller S, Margoliash E (1978) Mitochondrial cytochrome c preparation and activity of native and chemically modified cytochrome c. Methods Enzymol 53: 128–164

Brautigan DL, Feinberg BA, Hoffman BM, Margoliash E, Peisach J, Blumberg WE (1979) Multiple low spin forms of the cytochrome c ferrihemochrome. Epr spectra of various eukaryotic and prokaryotic cytochromes c. J Biol Chem 252: 574–582

Davis KA, Hatefi Y, Salemme FR, Kamen MD (1972) Enzymic redox reactions of cytochromes c. Biochem Biophys Res Commun 49: 1329–1335

Dickerson RE, Timkovich R (1975) Cytochromes c. In: Boyer PD (ed) The enzymes, 3rd ed, vol 11, pp 397–547. Academic Press, New York

Dobson CM, Hoyle NJ, Geraldes CF, Wright PE, Williams RJP, Bruschi M, Le Gall J (1974) Outline structure of cytochrome C_3 and consideration of its properties. Nature 249: 425–429

Errede B, Kamen MD (1978) Comparative kinetic studies of cytochromes c in reactions with mitochondrial cytochrome c oxidase and reductase. Biochemistry 17: 1015–1027

Errede B, Haight GP Jr, Kamen MD (1976) Oxidation of ferrocytochrome c by mitochondrial cytochrome c oxidase. Proc Natl Acad Sci USA 73: 113–117

Ferguson-Miller S, Brautigan DL, Margoliash E (1976) Correlation of the kinetics of electron transfer activity of various eukaryotic cytochromes c with binding to mitochondrial cytochrome c oxidase. J Biol Chem 251: 1104–1115

Florkin M, Stotz FH (eds) (1965) The classification and nomenclature of cytochromes. In: Comprehensive biochemistry, vol XIII, 2nd ed, pp 18–24. Elsevier/North-Holland, Amsterdam New York

Haser R, Pierrot M, Frey M, Payan F, Astier UP, Bruschi M, Legall J (1979) Structure and sequence of the multihaem cytochrome c_3. Nature 282: 806–810

Horio T, Kamen MD (1970) Bacterial cytochromes: II. Functional aspects. Annu Rev Microbiol 24: 399–428

Huynh BH, Emptage MH, Münck E (1978) Mössbauer study of cytochrome c_2 from *Rhodospirillum rubrum.* Sign of the product g_x g_y g_z of some low spin ferric heme proteins. Biochim Biophys Acta 534: 295–306

Kamen MD (1973) Toward a comparative biochemistry of cytochromes. Protein, Nucleic Acid and Enzyme. Kyoritsu Publ. Co (Tokyo) 18, 753–773

Kamen MD, Horio T (1970) Bacterial cytochromes: I. Structural aspects. Annu Rev Biochem 39: 673–700

Kamen MD, Dus KM, Flatmark T, de Klerk H (1972) Cytochromes c. In: King TE, Klingenberg M (eds) Electron and coupled energy transfer in biological systems, vol I/A, pp 243–324. Dekker, New York

Kang CH, Ferguson-Miller S, Margoliash E (1977) Steady state kinetics and binding of eukaryotic cytochromes c with yeast cytochrome c peroxidase. J Biol Chem 252: 919–926

Keller RM, Pettigrew GW, Wüthrich K (1973) Structural studies by proton NMR of cytochrome c-557 from Crithidia oncopelti. FEBS Lett 36: 151–156

Kilpatrick L, Erecinska M (1977) Mitochondrial respiratory chain of Tetrahymena pyriformis. The thermodynamic and spectral properties. Biochim Biophys Acta 460: 346–363

Koppenol WH, Vroonland CAJ, Braams R (1978) The electric potential field around cytochrome c and the effect of ionic strength on reaction rates of horse cytochrome c. Biochim Biophys Acta 503: 499–508

Korszun ZR, Salemme FR (1977) Structure of cytochrome c_{555} of chlorobium thiosulfatophilum: Primitive low-potential cytochrome c. Proc Natl Acad Sci USA 74: 5244–5247

Kowalsky A (1965) Nuclear magnetic resonance studies of cytochrome c. Possible electron delocalization. Biochemistry 4: 2382–2388

Lemberg R, Barrett J (1973) Bacterial cytochromes. Academic Press, New York

Liu CY, Webster DA (1974) Spectral characteristics and interconversions of the reduced, oxidized, and oxygenated forms of purified cytochrome o. J Biol Chem 249: 4261–4266

Margalit R, Schejter A (1970) Thermodynamics of the redox reactions of cytochromes of five different species. FEBS Lett 6: 278–280

Margoliash E, Ferguson-Miller S, Tulloss J, Kang CH, Feinberg BA, Brautigan DL, Morrison M (1973) Separate intramolecular pathways for reduction of oxidation of cytochrome c in electron transport chain reactions. Proc Natl Acad Sci USA 70: 3245–3249

Maurel P, Douzou P, Waldmann J, Yonetani T (1978) Enzyme behaviour and molecular environment. The effects of ionic strength, detergents, linear polyanions and phospholipids on the pH profile of soluble cytochrome oxidase. Biochim Biophys Acta 525: 314–324

Nicholls P (1974) Cytochrome c binding to enzymes and membranes. Biochim Biophys Acta 346: 261–310

Pfennig N, Trüger HG (1973) The rhodospirillales (phototrophic or photosynthetic bacteria). In: Lechevalier HA (ed) Handbook of microbiology, vol I, pp 17–27. CRC Press, Cleveland/Ohio

Redfield AG, Gupta RK (1971) Pulsed NMR study of the structure of cytochrome c. Cold Spring Harbor Symp Quant Biol 36: 405–411

Salemme FR (1977) Structure and function of cytochromes c. Ann Rev Biochem 46: 299–329

Salemme FR, Freer ST, Yuong NGH, Alden RA, Kraut J (1973a) The structure of oxidized cytochrome c_2 of *Rhodospirillum rubrum*. J Biol Chem 248: 3910–3921

Salemme FR, Kraut J, Kamen MD (1973b) Structural bases for function in cytochromes c. An interpretation of comparative X-ray and biochemical data. J Biol Chem 248: 7701–7716

Scholes PB, McLain G, Smith L (1971) Purification and properties of a c-type cytochrome from micrococcus denitrificans. Biochemistry 10: 2072–2076

Smith GM, Kamen MD (1974) Proton magnetic resonance spectra of *Rhodospirillum rubrum* cytochrome c_2. Proc Natl Acad Sci USA 71: 4303–4306

Smith L, Conrad H (1956) A study of the kinetics of the oxidation of cytochrome c by cytochrome c oxidase. Arch Biochem Biophys 63: 403–413

Smith L, Davies HC, Nava ME (1976) Evidence for binding sites on cytochrome c for oxidases and reductases from studies of different cytochromes c of known structure. Biochemistry 15: 5827–5831

Speck SH, Ferguson-Miller S, Osheroff N, Margoliash E (1979) Definition of cytochrome c binding domains by chemical modification: Kinetics of reaction with beef mitochondrial reductase and functional organization of the respiratory chain. Proc Natl Acad Sci USA 76: 155–159

Staudenmayer N, Ng S, Smith MB, Millett F (1977) Effect of specific trifluoroacetylation of individual cytochrome c lysines on the reaction with cytochrome oxidase. Biochemistry 16: 600–604

Swanson R, Trus BL, Mandel G, Kallai OB, Dickerson RE (1977) Tuna cytochrome c at 2.0 Å resolution. I. Ferricytochrome structure analysis. J Biol Chem 252: 759–775

Takano T, Kallai OB, Swanson R, Dickerson RE (1973) The structure of ferrocytochrome c at 2.45 Å resolution. J Biol Chem 248: 5234–5255

Takemori S, Wada K, Ando K, Hosokawa M, Sekuzu I, Okunuki K (1962) Studies of cytochrome A. VIII. Reaction of cytochrome A with chemically modified cytochrome c and basic proteins. J Biochem 52: 28–37

Wada K, Okunuki K (1968) Studies on chemically modified cytochrome c. J Biochem (Japan) 64: 667

Wada K, Okunuki K (1969) Studies on chemically modified cytochrome c. II. The trinitrophenylated cytochrome c. J Biochem (Japan) 66: 249–262

Wood PM (1977) Roles of c-type cytochromes in algal photosynthesis. Eur J Biochem 72: 605–612

Wood PM, Willey DL (1980) Use of fluorescent gelsen characterization of a membrane cytochrome c from *Pseudomonas aeruginosa*. Fems Letters 7: 273–277

Wüthrich K (1970) Structural studies of hemes and hemoproteins by nuclear magnetic resonance spectroscopy. Struct Bonding (Berlin) 8: 53–121

Yamanaka T (1972) Evolution of cytochrome c molecule. Adv Biophys 3: 227

Yamanaka T (1975) Comparative study on the redox reactions of cytochromes c with certain enzymes. J Biochem (Japan) 77: 493–499

Yamanaka T, Okunuki K (1974) Cytochromes. In: Neilands J (ed) Microbiol iron metabolism, pp 349–400. Academic Press, New York

Yonetani T, Ray CS (1965) Studies on cytochrome c peroxidase. I. Purification and some properties. J Biol Chem 240: 4503–4508

3. Recognition of Ligands by Haem Proteins

M.F. PERUTZ

A. Introduction

How does haemoglobin recognise its substrate, oxygen, and discriminate against its toxic inhibitor, CO? What makes haemoglobin an oxygen carrier and cytochrome b_5 an electron transfer enzyme? Why is the peroxidase activity of methaemoglobin so weak, compared with peroxidase enzymes, seeing that their absorption spectra and many of their reactions are so similar? What role does the porphyrin play in all these proteins? The oxygen affinity of haemoglobin is lowered by a variety of physiologically important ligands. How does haemoglobin recognize them?

B. Recognition of Oxygen by Haemoglobin and Discrimination Against CO

In all haem proteins of known structure the fifth coordination position at the iron is occupied by the imidazole of a histidine. Chemists experimenting with model compounds found that haems linked to other bases such as pyridine or piperidine combine with CO but not with oxygen. The N_ϵ of imidazole is the strongest Lewis base Nature has available; it acts as an electron donor to the iron, but why should such electron donation be needed for the binding of oxygen?

In reacting with oxygen the haem iron remains formally ferrous, but in 1964 Joseph Weiss suggested that formation of the bond is accompanied by the transfer of an electron from the iron to the oxygen, forming superoxide ion and leaving the iron formally ferric. Linus Pauling poured scorn on Weiss' suggestion, but experiments have confirmed the transfer of electrons from the iron to the oxygen, which explains why a strong base in the fifth coordination position is essential for its recognition.

How does haemoglobin discriminate against CO? To answer this question we must consider the differences in electronic structure between oxygen and CO. The oxygen molecule has one unpaired electron each in the π_x^* and π_y^* orbitals and therefore has a spin of $S = 1$. On combination with iron, electron donation from the iron d_π orbitals raises the energy of one of the oxygen π^* orbitals, so that the two electrons pair in the

MCR Laboratory of Molecular Biology, Cambridge CB2 2QH, Great Britain

other. Due to that asymmetry the oxygen molecule binds at an angle of $60°$ to the haem axis. In CO, on the other hand, the π_x^* and π_y^* orbitals are both empty and are avid electron acceptors. CO therefore binds parallel to the haem axis and its binding does not depend on the presence of a strong base on the other side of the haem.

Of the various model compounds that combine reversibly with molecular oxygen the one that comes nearest to mimicking haemoglobin is Collman's picket fence complex, yet it combines with CO irreversibly, while in haemoglobin the partition coefficient between oxygen and CO is 250. About 1% of human haemoglobin is constantly blocked by CO produced endogenously in the breakdown of porphyrin. If the partition coefficient were much larger than 250, this endogenously produced CO would have as deleterious an effect on oxygen transport as cigarette smoke has in heavy smokers. Experiments show that the CO affinity of haemoglobin is lowered by the distal haem ligand His E7 which is so placed as to offer no steric hindrance to the inclined oxygen, but does impede axially bound ligands like CO. X-ray analysis of carbonmonoxyhaemoglobin and neutron analysis of carbonmonoxymyoglobin have proved that steric hindrance by these two residues forces the CO to lie inclined to the haem axis, while normally FeCO bonds are linear. The role of the distal histidine in lowering the CO affinity has been confirmed by the unusual properties of haemoglobin Zürich, an abnormal haemoglobin in which the distal histidine E7β is replaced by arginine whose side chain protrudes at the surface and leaves the haem pocket empty.

The carriers of this abnormality are heterozygotes, and about one third of their haemoglobin has the Arg E7β. I found that the abnormal β subunits of these patients were saturated with CO when I received their blood, because the substitution of Arg for His had raised the partition coefficient between oxygen and CO from 250 to 500. I then examined another abnormal haemoglobin in which the distal valine E11β is replaced by alanine (Hb Sydney). The partition coefficient of this mutant haemoglobin was normal, showing that the distal valine is inessential for the discrimination against CO. The distal histidine has a second function. Being a strong base, its free N_ϵ slows down both autoxidation and acid catalysed oxidation of the haem iron and thus helps to keep it ferrous.

C. Role of Iron and Porphyrin

Cobaltous iron is the only metal that can substitute for ferrous iron in haemoglobin and combine reversibly with molecular oxygen, but the association constant of cobalt haemoglobin with oxygen is 400 times smaller than that of iron haemoglobin so that this metal is physiologically useless. For experiments in vitro or with model compounds it has the advantage of not being so easily oxidized.

Cobald chelates have taught us something about the role of the porphyrin in the recognition of oxygen. It seems that any square pyramidal complex of cobaltous iron with an electronegative ligand in trans position to the oxygen is capable of combining with an oxygen molecule in the bent conformation, even $[Co(CN)_5]^{3-}$. Analogous ferrous chelates cannot do so because they become irreversibly oxidized. In haemoglobin, therefore, the porphyrin together with the proximal histidine may act chiefly as a scaffolding to provide square pyramidal coordination for the ferrous iron atom, though the porphyrin may also play a part in the tuning of the oxygen affinity. In electron transfer enzymes, the porphyrin together with the apical ligands provides octahedral coordination for the iron, but in addition the porphyrin acts like a metal plate conducting electrons along its π orbitals from its exposed edge to and from the iron at its centre. It has even been suggested that the surrounding protein may polarise these orbitals so that one pathway becomes favoured for electron donation and another for electron acceptance, but this seems rather fanciful.

D. Protection of Haemoglobin from Autoxidation and Enzymatic Reduction of Methaemoglobin

Free haem is oxidized quickly by oxygen. Kendrew and I used to believe that in myoglobin and haemoglobin oxidation was retarded by the largely nonpolar lining of the haem pockets, but experiments on model compounds have proved us wrong. They have shown that oxidation of haem proceeds via an intermediate bridged μ-oxo or μ-peroxo complex of two adjacent haems. The true function of the haem pockets is to prevent the formation of such complexes by shielding the haems from each other. A nonpolar environment turns out to be inessential; on the contrary, the FeO_2 bond, being polar, is strengthened by polar environments, while the Fe-CO bond is not, and oxygen protects shielded haems from oxidation.

All the same, oxidation of haemoglobin proceeds slowly even in vivo, and red cells have evolved an NADH-dependent reductase system that acts as a scavenger of methaemoglobin. How do haemoglobin and methaemoglobin reductase recognize each other? The outside edges of the haem pockets are flanked by two lysines. We know that these are inessential for oxygen transport, because abnormal haemoglobins in which these lysines are replaced function normally. I suggest, therefore, that they recognize methaemoglobin reductase, just as the lysines surrounding the entrance to the haem pocket of cytochrome c recognize cytochrome oxidase and peroxidase, and probably also cytochrome c reducing enzymes. I suspect that methaemoglobin reductase has acidic groups complementary to these lysines which allow formation of a transient salt bridge complex and transfer of an electron to the edge of the haem.

E. Cytochromes

Cytochrome b_5 has the same protohaem as myoglobin and haemoglobin; why does it act as an electron rather than an oxygen carrier? The essential difference consists of nothing more than a small shift in the position of the distal histidine whose N_ϵ is 4 Å from the centre of the porphyrin in haemoglobin, but only 2 Å in cytochrome b_5. In consequence, the iron atom in cytochrome b_5 is six-coordinated and low spin in both the ferrous and ferric states, whereas in the absence of oxygen the iron in myoglobin and haemoglobin is five-coordinated high spin in the ferrous and six-coordinated low spin in the ferric state. In haemoglobin, the change in spin and coordination number is accompanied by marked changes in the structure of the surrounding protein, so that substantial activation energies must be needed for oxidation and reduction. By contrast, a change of valency of the six-coordinated low spin iron in cytochrome b_5 entails no significant stereochemical changes, so that the activation energy is small and the reactions are correspondingly faster. Moreover, the binding of a second electron-donating histidine with a higher affinity for ferric than for ferrous iron lowers the redox potential by about 100 mV compared to that of haemoglobin.

In most cytochrome c's the haem iron is also six-coordinated, but the sixth ligand is a methionine sulphur instead of a histidine. This may act as an electron acceptor or as a weaker donor than histidine so that the redox potentials of cytochrome c are generally higher than those of b's. The sulphur atoms in the thioester bridges linking the haem to the protein in cytochrome c also seem to make the redox potential more positive. One of them protrudes from the surface of the molecule and may serve as a contact for electron transfer.

F. Peroxidases

The iron in peroxidases is ferric and high spin, as in methaemoglobin, and the two proteins have similar absorption spectra, but while methaemoglobin shows only very slight peroxidatic activities, peroxidases are powerful catalysts. Their haem iron can donate an electron to H_2O_2 with the formation of Fe^{4+}, yet according to spectral and X-ray evidence it is linked to a histidine just like the iron in haemoglobin. What pushes the electron out? The answer was obscure until the electron density map of cytochrome c peroxidase recently obtained by Dr. Joseph Kraut and his colleagues showed that the histidine is polarised by a buried carboxylate rather as the histidine in the charge relay system of the serine proteinases is polarised by a buried aspartate. Haemoglobin and peroxidase also have different residues on the distal side of the haem pocket; the distal histidine is present in both, but while haemoglobin contains a valine which remains inert, peroxidase contains a tryptophan and an arginine which take part in the catalytic activity. Yet the buried carboxylate is likely to be the vital feature that allows peroxidases to recognise H_2O_2 as a substrate for reduction, and is likely to be present also in catalase whose structure is still unknown.

G. Recognition of Other Haemoglobin Ligands

Haemoglobin has four other physiologically important ligands besides oxygen: H^+, CO_2, Cl^-, and D-glycerate-2,3-bis(phosphate) (DPG for short). Their binding is antagonistic to the binding of oxygen. This antagonism arises because haemoglobin is an allosteric protein which alternates between two structures, the deoxy or T structure with a low oxygen affinity and a high affinity for the other four ligands, and the oxy or R structure in which these relative affinities are reversed. The T structure generates specific binding sites where these ligands are recognised, and these sites are absent in the R structure. All the binding sites are designed so that these ligands can form salt bridges, but some of the sites overlap so that the binding of certain pairs of ligands becomes competitive.

4. Influences of Solvent Water on the Transition State Affinity of Enzymes, Protein Folding, and the Composition of the Genetic Code

R. WOLFENDEN

A. Introduction

This symposium is a special delight for one of those who benefited from Dr. Lipmann's kindness in accepting graduate students after his move to New York in 1957. By then he had numerous scientific progeny, and my relationship to him is complicated by the fact that I can also count myself among his scientific grandchildren through apprenticeships with Hans Bomann and Bill Jencks. A good deal of what I am going to say, about enzyme interactions with high energy intermediates in substrate transformation, had its genesis in conversations about group potential (Lipmann 1941) and in some ideas expressed a generation later by Jencks (1966) in a volume commemorating the 1941 review.

Enzymes share with midwives (and other catalysts) a remarkable talent for easing the difficult passage of substances from one metastable condition to another. To explain this ability, it seems necessary to suppose that the "grip" of the enzyme tightens as the substrate is chemically altered, relaxing later as the product is generated and released. This is easiest to visualize (Fig. 1) for one of the many enzymes that transform a single substrate to a single product (Wolfenden 1969a), but it is also true for enzymes that

ANY REACTION : $R \xrightarrow{k} P$

CAN BE CONSIDERED TO INVOLVE AN EQUILIBRIUM

ACTIVATION : $R \xrightleftharpoons{K^{\ddagger}} R^{\ddagger}$

FOLLOWED BY DECOMPOSITION OF THE ACTIVATED

COMPLEX : $R^{\ddagger} \longrightarrow P$

THE RATE OF DECOMPOSITION OF $R^{\ddagger}$ IS ASSUMED

TO BE UNIVERSAL, $\sim 10^{13} s^{-1}$ AT 25° C

AT ROOM TEMPERATURE, $k = 10^{13} \, K^{\ddagger}$

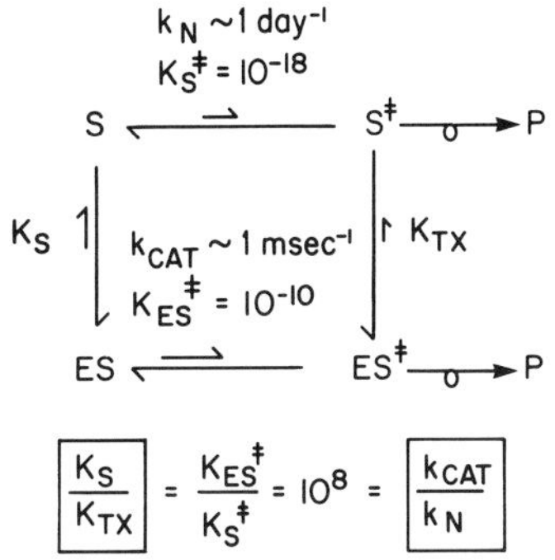

$$k_N \sim 1 \, day^{-1}$$
$$K_S^{\ddagger} = 10^{-18}$$

$$S \xrightleftharpoons{} S^{\ddagger} \underset{0}{\longrightarrow} P$$

$$K_S \Big| \quad k_{CAT} \sim 1 \, msec^{-1} \Big| K_{TX}$$
$$K_{ES}^{\ddagger} = 10^{-10}$$

$$ES \xrightleftharpoons{} ES^{\ddagger} \underset{0}{\longrightarrow} P$$

$$\boxed{\frac{K_S}{K_{TX}}} = \boxed{\frac{K_{ES}^{\ddagger}}{K_S^{\ddagger}}} = 10^8 = \boxed{\frac{k_{CAT}}{k_N}}$$

Fig. 1a, b. a Absolute reaction rate theory. b Equation for transition state binding. K_S is the dissociation constant of the ES complex. K_{TX} is the formal dissociation constant for the enzyme-substrate complex in the transition state. Rate and dissociation constants are in the range commonly encountered

Department of Biochemistry, University of North Carolina, Chapel Hill, North Carolina 27514, USA

must solve the additional problem of gathering several substrates from dilute solution (Wolfenden 1972; Lienhard 1973). If the rate of conversion of a substrate to a product increases on an enzyme, its equilibrium constant for activation increases by the same factor, usually well in excess of 10^8. To explain this enhancement, we have to suppose that the formal affinity of the enzyme for the substrate increases to the same degree as they pass from the ES complex to the transition state. One can ask whether it is necessary for this path to be followed, or whether the enzyme might not in principle simply pluck from aqueous solution activated forms of the substrate for conversion to product? For many, and perhaps the majority, of enzyme reactions, the overall rate of conversion of substrate to product (measured by k_{cat}/K_m) approaches too closely the limits imposed by diffusional encounter to allow mechanisms requiring species that are not reasonably populous (Wolfenden 1974). We can conclude that many enzyme reactions tend to occur by combination of enzyme and substrate in the ground state as shown in Fig. 1.

B. Transition State Analogs

Catalysis can accordingly be said to depend on, and consist in, an enhanced affinity of the enzyme for highly activated intermediates in substrate transformation, as compared with the substrate itself. Transition state analogs are stable compounds, designed to resemble fleeting intermediates in substrate transformation. Because of this resemblance, they are bound by enzymes much more tightly than substrates are bound. Because they are chemically stable and cannot collapse to form products, their affinity for the enzyme persists and they are expected to be strong metabolic inhibitors (for reviews, see Wolfenden 1972; Lienhard 1973; Wolfenden 1976). A complete list of potential transition state analogs would now include over 100 cases, some of which occur naturally in the form of antibiotics. Evidently Nature designs enzymes to bind transition states tightly, and biological weapons to find this weakness.

An antibacterial agent elaborated by *Nocardia interforma* provides a good example. This antibiotic, named conformycin because it was found in culture filtrates along with formycin and showed synergistic activity in inhibiting the growth of bacteria, is bound by mammalian adenosine deaminase 7 orders of magnitude more tightly than adenosine and 8 orders of magnitude more tightly than the substrate for the reverse reaction, inosine (Fig. 2) (Ohno et al. 1974; Cha et al. 1975). Strong evidence for a mechanism involving direct water attack on adenosine is the enzyme's ability to catalyze covalent addition of water to other heterocycles (Evans and Wolfenden 1972), and coformycin bears an obvious resemblance to the intermediate shown in *brackets*. Earlier efforts to produce analogs of this intermediate had resulted in modified purines with tetrahedrally substituted carbon at C-6 that were strongly inhibitory (Evans and Wolfenden 1970). When the structure of coformycin was finally solved by synthesis and by X-ray crystallography, it was startling to learn how much better *Nocardia* had succeeded than ourselves, by the simple expedient of elaborating an analog with a 7-membered ring that, unlike the purine nucleus, can accomodate a hydroxyl group bonded stably to C-6 in an sp_3 configuration.

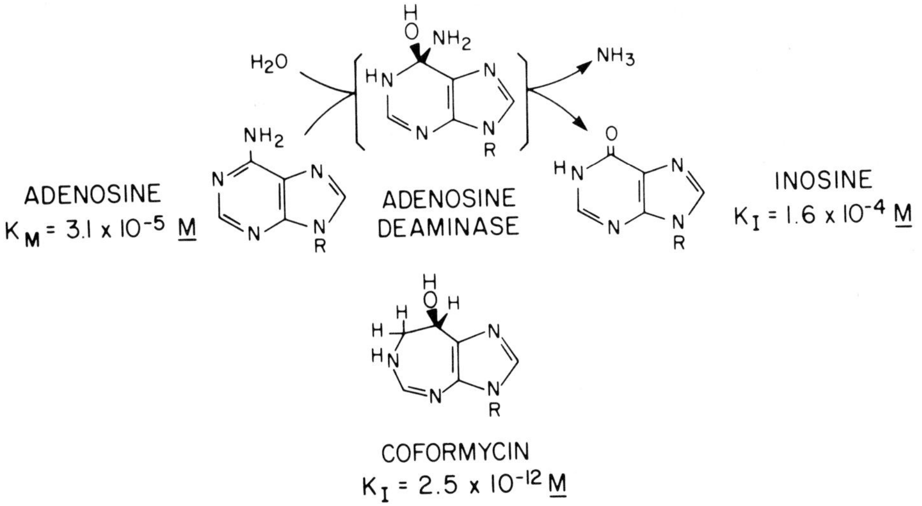

Fig. 2. Action and inhibition of adenosine deaminase

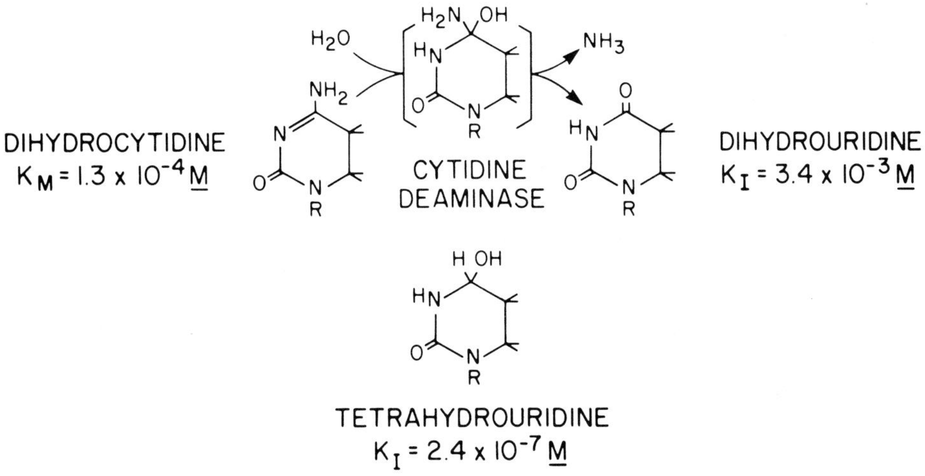

Fig. 3. Action and inhibition of cytidine deaminase

Several other analogs that show really exceptional affinities, and for which new structural information is available, are shown in Figs. 3–7. Cytidine deaminase (Fig. 3) is strongly inhibited by tetrahydro-uridine, an analog of a tetrahedral intermediate in direct attack by water on cytidine, or on the alternative substrate 5,6-dihydrocytidine (Evans et al. 1975), and a number of proteolytic enzymes of the papain-chymotrypsin variety are subject to competitive inhibition by aldehydes (Fig. 4) that resemble the acyl portion of good substrates (Westerik and Wolfenden 1972; Thompson 1973). Recent evidence from isotope effects on binding, cross-saturation NMR, and protein crystallography shows that these aldehydes are bound as hemiacetals resembling the tetrahedral intermediates that might be expected to be formed during a double displacement reaction (Lewis and Wolfenden 1977; Clark et al. 1977; Brayer et al. 1979).

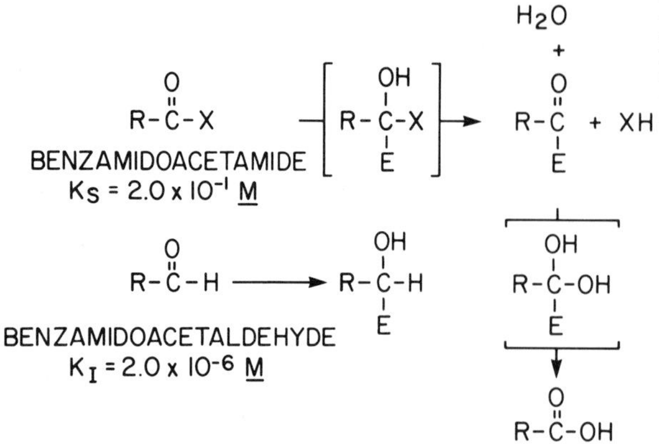

H₂O is shown as H_2O

Fig. 4 region:

$$\underset{\substack{\text{BENZAMIDOACETAMIDE} \\ K_S = 2.0 \times 10^{-1}\ \underline{M}}}{R-\overset{\overset{\text{O}}{\|}}{C}-X} \quad \left[\ R-\overset{\overset{\text{OH}}{|}}{\underset{\underset{\text{E}}{|}}{C}}-X\ \right] \longrightarrow R-\overset{\overset{\text{O}}{\|}}{\underset{\underset{\text{E}}{|}}{C}}\ +\ XH$$

$$\underset{\substack{\text{BENZAMIDOACETALDEHYDE} \\ K_I = 2.0 \times 10^{-6}\ \underline{M}}}{R-\overset{\overset{\text{O}}{\|}}{C}-H} \longrightarrow R-\overset{\overset{\text{OH}}{|}}{\underset{\underset{\text{E}}{|}}{C}}-H \qquad R-\overset{\overset{\text{OH}}{|}}{\underset{\underset{\text{E}}{|}}{C}}-OH$$

$$R-\overset{\overset{\text{O}}{\|}}{C}-OH$$

Fig. 4. Action and inhibition of papain. K_i value shown is for anhydrous benzamidoacetaldehyde, the form of the inhibitor that combines with papain

These inhibitors and intermediates differ from the substrates mainly in their steric configuration. Another group of analogs are distinguished by their electrostatic properties. 2-Phosphoglycollate, designed to resemble an ene-diolate intermediate in the action of triosephosphate isomerase (Wolfenden 1969b), was recently shown to be bound by the enzyme as the species with three negative charges. We also know that a proton is taken up by the enzyme as the inhibitor is bound, probably by glu-165, the essential residue that is thought to serve as a general base catalyst for proton transfer in this reaction (Fig. 5) (Hartman et al. 1975; Cambell et al. 1978). Oxalate is a powerful inhibitor of many enzymes that are believed to generate or utilize pyruvic acid in its enolate form, or as the carbanion that is believed to be generated during the action of lactate oxidase (Fig. 6) (Ghisla and Massey 1977). Nitrate ion is a very strong inhibitor of

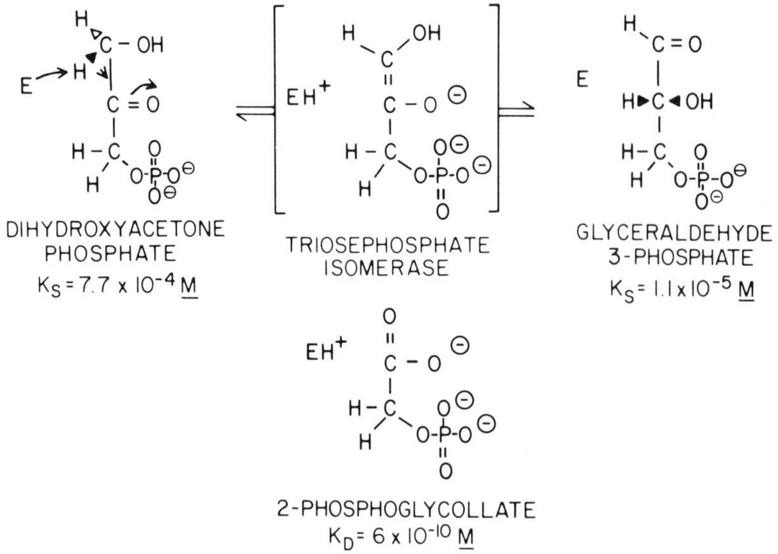

DIHYDROXYACETONE PHOSPHATE
$K_S = 7.7 \times 10^{-4}\ \underline{M}$

TRIOSEPHOSPHATE ISOMERASE

GLYCERALDEHYDE 3-PHOSPHATE
$K_S = 1.1 \times 10^{-5}\ \underline{M}$

2-PHOSPHOGLYCOLLATE
$K_D = 6 \times 10^{-10}\ \underline{M}$

Fig. 5. Action and inhibition of triosephosphate isomerase. K_D value shown is for tri-anionic 2-phosphoglycollate dissociating from the conjugate acid form of the enzyme with $pK_a = 3.9$ (Hartman et al. 1975)

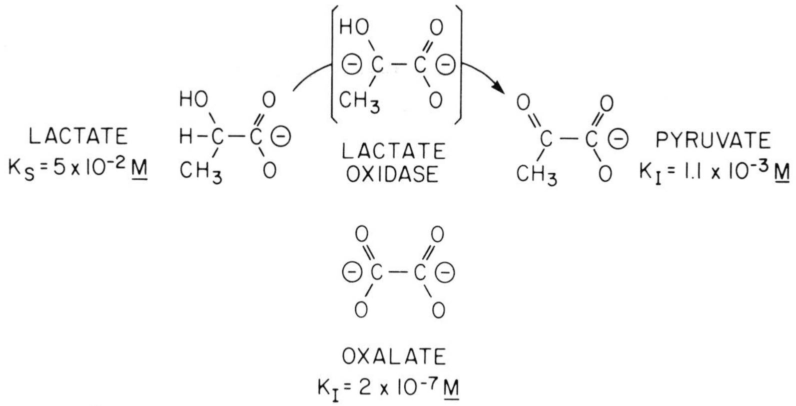

Fig. 6. Action and inhibition of lactate oxidase

some kinases, and infrared spectroscopy (Reed et al. 1978) supports the suggestion (Milner-White and Watts 1971) that this ion forms a complex in which it is sandwiched between the phosphoryl donor and acceptor, forming a dead-end complex that resembles the transition state. The complex resulting from quasi-irreversible inhibition of glutamine synthetase by methionine sulfoximine and ATP (Rowe et al. 1969) has been studied extensively by magnetic resonance spectroscopy, and the results accord with a possible resemblance between the complex (Villafranca et al. 1975) and an intermediate formed

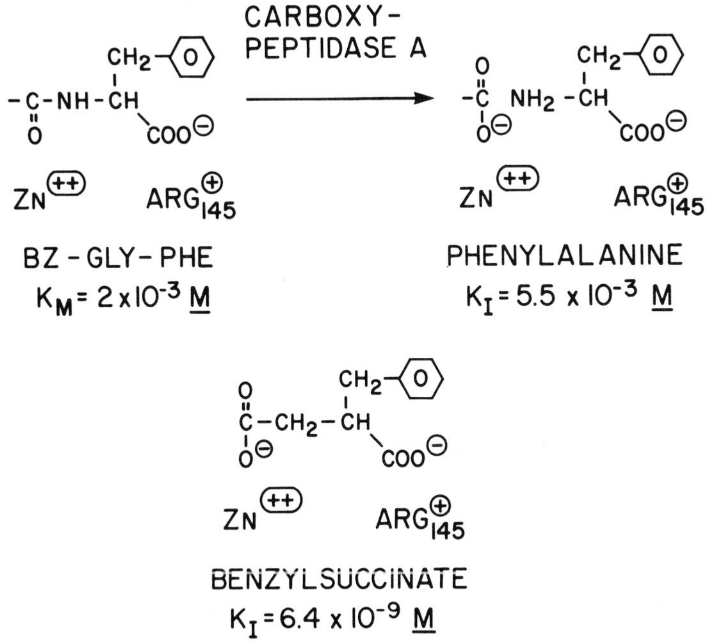

Fig. 7. Action and inhibition of carboxypeptidase A

during ammonia attack on γ-glutamyl phosphate by ammonia as proposed by Meister and his associates. As a final example, carboxypeptidase A is strongly inhibited by benzylsuccinic acid, designed to resemble the combined products of peptide hydrolysis (Byers and Wolfenden 1973). The pH-dependence of K_i suggests binding of the monoanion, but ^{13}C n.m.r. studies show that the inhibitor is actually bound as the dianion. These results appear to be consistent with displacement of a hydroxide ion from the active site zinc by benzylsuccinate, or with the uptake of a proton by the enzyme as the inhibitor is bound (Fig. 7) (Palmer et al. 1980).

C. Shape Changes and the Role of Distant Binding Determinants

Inhibitors of this kind are frequently bound slowly, with evidence of changes in the conformation of the enzyme (Wolfenden 1976, and references cited therein). Under certain conditions crystals of triosephosphate isomerase, for example, undergo contractions of the unit cell (Johnson and Wolfenden 1970) that are so extensive that the structure of the inhibited enzyme remains something of a mystery several years after the structure of the native enzyme was solved by David Phillips and his associates (Banner et al. 1975). Crystals of carboxypeptidase are extensively disordered by benzylsuccinate (Byers and Wolfenden 1973), so that the rough structure indicated for this complex in Fig. 7 can only be inferred from the structure of the weaker complex formed by benzylsuccinate with thermolysin; the image of this latter complex is, however, of such high quality that the location of single water molecule (possibly similar to that of substrate water) can be observed next to the position of the scissile peptide bond (Bolognesi and Matthews 1979). It seems reasonable to suppose that limited conformational changes may also occur during enzyme catalysis, allowing the ES complex to tighten in such a way that the activated complex is momentarily enclosed, just as a jewel is enclosed in its setting; such a program would allow the structural requirements of diffusion-controlled substrate access to be reconciled with the maximization of attractive contacts between the enzyme and the substrate in the transition state (Wolfenden 1974). By accommodating these conflicting requirements, shape changes may contribute to catalytic efficiency in a manner that differs from, but is not inconsistent with, the rationale for induced fit proposed by Koshland (1959).

A general feature of enzyme behavior, reinforced by observations with transition state analogs, is that seemingly irrelevant parts of the substrate may play an important role in transition state binding. Any binding determinant that stabilizes the enzyme-substrate complex in the transition state enhances k_{cat}/K_m. Whether or not the effect of that particular binding determinant is expressed in K_m (rather than in k_{cat}) depends on whether or not its binding influence is also present in the ES complex (Wolfenden 1972), although its detection may sometimes be clouded by compensating effects of other kinds (Jencks 1975). Adenosine deaminase, for example, deaminates adenosine at a limiting rate several orders of magnitude more rapid than its limiting rate for deamination of adenine, whereas it seems to bind both compounds with similar affinity (Wolfenden et al. 1969). Correspondingly, removal of substituent ribose diminishes the enzyme's affinity for a possible transition state analog by several orders of magnitude (Wolfenden et al. 1977). Very clear evidence has also been obtained for the importance of distant binding determinants on transition state stabilization by proteases that are equipped with serine as the active site nucleophile (Thompson 1974; Bauer et al. 1976).

During the evolution of an efficient enzyme, it is helpful to lower K_m only so long as K_m exceeds the concentration of the substrate in vivo (Cleland 1967). However, it is always helpful to increase k_{cat}. The expression of binding interactions in k_{cat}, requiring that their effects appear in the transition state but not the enzyme-substrate complex, is a trick requiring such chemical ingenuity that evolution in this direction might be expected to have proceeded only so far as was physiologically useful. It is sometimes suggested than any enzyme, for which k_{cat}/K_m has attained the limits imposed by diffusional encounter, and for which K_m exceeds the physiological substrate concentration, has attained a state of evolutionary perfection. But an enzyme molecule with even higher values of k_{cat} and K_m would presumably be able to process even larger quantities of substrate per unit time, if higher concentrations were available to it. In terms of the evolution of the organism as a whole, it would be difficult to say whether any existing enzyme has reached the point where it can undergo no further improvement. We may speculate that physiological substrate concentrations in modern organisms have risen through evolution to the point where a balance is just struck between the chemical limitations of enzymes *as they presently exist,* and the energetic cost of synthesizing more of the various enzymes involved in each metabolic pathway. It seems evident that enzymes would be of little use to a primitive cell represented by a mere bag of very dilute substrates: diffusional encounter would be too infrequent for efficient catalysis to occur. Within solubility limitations condensation may represent one of the more significant measures of evolutionary advancement. If advanced forms of life exist in other parts of the universe, we may find that they possess hyperactive enzymes operating rapidly on substrates present at concentrations higher than those in organisms with which we are familiar.

D. Catalysis as a Problem in Extraction

We would like to understand the basis for the extremely high affinity of an enzyme for the altered substrate in the transition state. It is equally important to discover how an enzyme manages to "fend off" the substrate in the ground state, since it is the *increase* in affinity following formation of ES that determines the magnitude of the enzyme's turnover number. We can regard this as a problem in differential extraction from water

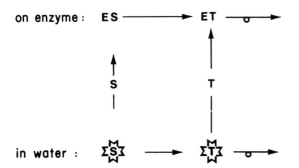

Fig. 8. Enzyme action as a problem in extraction. For catalysis to occur, the enzyme must extract $\overline{T}$, the altered substrate in the transition state, more effectively than it extracts S from solvent water

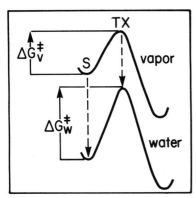

Fig. 9. Solvent retardation

SOLVENT RETARDATION

(Fig. 8) and it is easy to see, following Saul Cohen's (1970) suggestion, how an enzyme might speed up a reaction that proceeds through a transition state that is much less polar than the starting materials. S_N2 reactions, such as the exchange of halide ions into alkyl halides, are known to be subject to "catalysis by desolvation", in the sense that they proceed very much more rapidly in dipolar aprotic solvents than in water (Fig. 9). With the advent of new techniques in mass spectrometry, it has become possible to compare the rates of such reactions in water with the rates of the same reactions in nothing at all, i.e., in the vapor phase. Rate enhancements as large as 17 orders of magnitude are observed for reactions of this kind (Olmstead and Brauman 1977) that involve some delocalization, but no actual loss, of electrostatic charge in the transition state. If an enzyme were to enhance the reactivity of a substrate by removing it from a watery environment, one would suppose that it might be strongly inhibited by a substrate analog that was hydrophobic, and therefore saved the enzyme the trouble of stripping away solvent water. Beautiful cases in point are provided by uncharged analogs of thiamine pyrophosphate devised by Gutowski and Lienhard (1976), that serve as very potent inhibitors of pyruvate dehydrogenase. The nonenzymatic reactions have been shown to proceed very much more rapidly in nonpolar solvents that in water (Crosby and Lienhard 1970). In reactions of this kind, the enzyme presumably drags polar reactants "by the hair" into an environment that destabilizes them (relative to the less polar transition state), and thiamine pyrophosphate appears to be well-equipped with chemical appendages that might serve as the hair.

Much more common are reactions that, in the absence of enzyme, proceed through intermediates that are more polar than the starting materials. Since ionic or polar intermediates are stabilized by watery surroundings, such reactions are understood (although this is difficult to prove experimentally) to be subject to solvent catalysis (Fig. 10), and it is not immediately apparent how catalysis by extraction from water could occur. The alkylation of tertiary amines, and many S_N1 reactions, for example, proceed very much more rapidly in water than in less polar solvents, and nothing useful would seem to be accomplished by moving the reactants into a waterless cavity.

The answer must be that, in some sense, the altered substrate in the transition state is even better stabilized by the enzyme than it is by an aqueous environment. One strat-

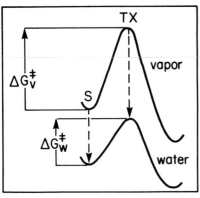

Fig. 10. Solvent catalysis

SOLVENT CATALYSIS

egy for accomplishing this might be to avoid the problem by changing the mechanism so that no intermediate, more polar than the starting material, is generated. β-Galactosidase, for example, appears to follow a mechanism involving formation of a covalent glycosyl-enzyme intermediate, whereas the acid-catalyzed reaction in solution proceeds through a carbonium ion intermediate (Sinnott 1978; Wentworth and Wolfenden 1974). An alternative strategy would be to find a means of stabilizing charges that *are* generated, more effectively than does solvent water (there may, of course, be little real difference between a glycosyl-enzyme intermediate and a tight ion pair). Only a superbly designed chelating agent could extract an ion from aqueous solution, in preference to a neutral molecule closely related in structure to the ion, but we know that metalloproteins have learned to solve this problem. Furthermore, the structural changes that seem to accompany the binding of alkali metal ions by ionophores like valinomycin (Fig. 11) bear a suggestive resemblance to the scenario we have considered above for the enzyme's embrace of the activated substrate, following diffusion-controlled binding. Many of the most effective transition state analogs, such as benzylsuccinate, nitrate or 2-phosphoglycollate, are tightly bound only in their most fully ionized forms, so that the analogy between transition state affinity and metal ion chelation is sometimes rather close. Theroretical calculations of the approximate magnitude of electrostatic effects have been reported (Warshel 1978).

E. The Affinity of Biological Compounds for Solvent Water

When chemical groups interact in a watery environment , water must usually be stripped away (at least in part) from the interacting groups. In attempting to understand the affinities that are observed in watery surroundings, it would be helpful to have a quantitative idea of how easily these groups can be removed from solvent water, since it might then be possible to infer the existence of specific attractive or repulsive forces that may be present. It seemed to us that the affinity of compounds for solvent water is sufficiently important for catalysis, and for chemical recognition in general, to be

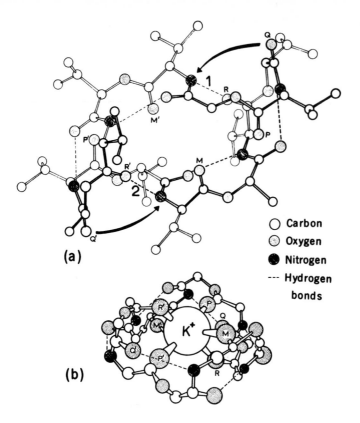

Fig. 11. Uncomplexed (*a*), and potassium-complexed (*b*), conformations of valinomycin, from Duax et al. (1972). Copyright 1972 by the American Association for the Advancement of Science

worth experimental study. Somewhat unexpectedly, our experiments seem to have turned up some information that may have a bearing not only on rates and equilibria, but also on the folding of globular proteins and even the composition of the genetic code.

The hydrophilic character of a molecule can be determined, in an absolute sense, by measuring its equilibrium distribution between the dilute vapor phase and a water solution so dilute that solute-solute interactions can be neglected (Fig. 12). Over the years, a large number of nonpolar compounds have been examined in this way, and results from the literature have recently been correlated (Hine and Mookerje 1975) to

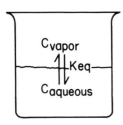

Fig. 12. Equilibrium constant for vapor-water transfer

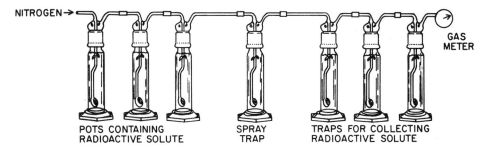

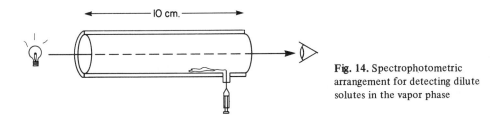

Fig. 13. The "Sniffer". Distribution of solute to vapor is determined from its rate of accumulation in the traps, as a function of the volume of carrier gas transferred (Wolfenden 1978)

Fig. 14. Spectrophotometric arrangement for detecting dilute solutes in the vapor phase

show that affinity for water in complex molecules is usually an additive function of their constituent groups. Until recently, compounds examined have been confined to volatile solutes with substantial vapor pressures. In order to examine more polar compounds of biological interest, we have developed methods of enhanced sensitivity for measuring the partial pressures of compounds over water. These involve purge-and-trap systems (Fig. 13) or direct optical detection of solutes in the vapor phase by infrared or ultraviolet spectroscopy (Fig. 14). Quantities of material in the vapor phase are small, so that it is sometimes necessary to use cuvettes of long light path, and here multiple reflection cells based on the ingenious design of J.U. White (1942) are useful.

Among the neutral organic molecules examined thus far (Fig. 15), the extreme position of acetamide and N-methylacetamide is worth noting (Wolfenden 1978). These simple representatives of the peptide bond are much more "heavily" hydrated than water itself, or than unionized amines and carboxylic acids. It is easy to show that the acylation of amines by carboxylic esters, as in the strongly exergonic biosynthesis of the peptide bond from peptidyl-tRNA, would actually be endergonic in the absence of water (Fig. 16). Some other effects of solvent water on free energies of hydrolysis are shown in Fig. 17. As predicted from theoretical studies (Kalckar 1941; George et al. 1970; Hayes et al. 1978), the high group potential of anhydrides is an intrinsic property of these molecules, whereas group potentials of other carboxylic acid derivatives are overwhelmingly dependent on changing solvation.

We should, I think, be particularly interested in molecules that exhibit a discontinuity or breakdown in free energies of solvation, as calculated from their constituent groups (Fig. 18). These cases are not common. The partial pressures of aqueous *vic*-glycols are larger than expected, consistent with self-hydrogen-bonding in the gas phase (Hine and Mookerjee 1975). On the other hand, P-nitrophenol, imidazole and amides

$$Keq = \frac{MOLES/LITER\,(vapor)}{MOLES/LITER\,(aqueous)}$$

Fig. 15. Water-vapor distribution of simple uncharged organic compounds

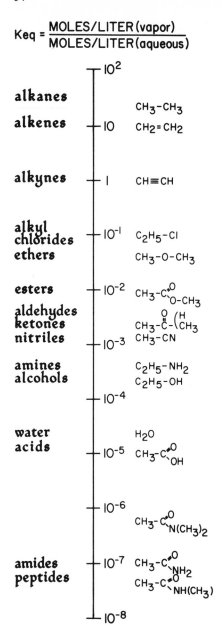

alkanes

alkenes

alkynes

alkyl chlorides
ethers

esters
aldehydes
ketones
nitriles

amines
alcohols

water
acids

amides
peptides

are less volatile than might be expected from their constituent groups; the accepting propensities of one end of these molecules tend to strengthen, through mutually reinforcing inductive effects, the donating properties of the other end of the molecule in hydrogen bonding interactions with water. Apparently small changes in structure, as in these examples, can sometimes lead to large changes in hydration potential. A clever

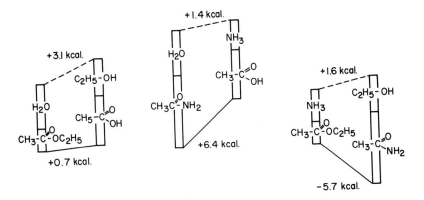

Fig. 16. Free energy changes for reactions occurring in the vapor phase (*broken lines*) and in water (*solid lines*). *Vertical bars* indicate free energies of hydration of reactants and products

		Un-ionized ΔG_{H_2O}	$\Delta G_{solv.}$	ΔG_{vap}
EtOAc	hydrolysis	+ 0.7	−2.4	+ 3.1
EtSAc	hydrolysis	− 2.0	+0.6	− 1.4
AcNH$_2$	hydrolysis	+ 6.4	+4.9	+ 1.5
(Ac)$_2$O	hydrolysis	−15.7	−2.5	−13.2
EtOAc	ammonolysis	− 5.7	−7.3	+ 1.6

Fig. 17. Free energies of reaction in water and in the vapor phase. Changes in free energy of solvation as reactants are converted to products are indicated in the *central column*

catalyst could exploit such differences in ways that we can guess, but do not yet have sufficient information to recognize experimentally.

F. Amino Acid Side Chains, Protein Folding and the Genetic Code

We can also ask about the hydration properties of the amino acid side chains that constitute the active sites of enzymes. This is of considerable interest in relation to protein structure as well, since the early work on hemoglobin suggested that the polar character of the various amino acids tends to determine their relative tendencies to be found at the surface of globular proteins (Perutz 1965). Looking at the atomic coordinates of crystalline proteins, Lee and Richards (1971) developed a model in which solvent water was represented by a rolling sphere with a radius of 1.4 Å, and the exposure of each part of the protein to solvent could be computed. Applying this method to 12 crystalline proteins, Chothia (1976) has calculated the fraction of amino acids of each kind that are inaccessible to solvent over 95% of their surface areas, with the results shown

Group contributions to log $(\frac{V}{W})$ are normally found to be additive.

Exceptions indicate group interaction in the vapor phase and/or in water.

Positive deviants

$$CH_2OH \\ | \\ CHOH \\ | \\ CH_2OH \qquad\qquad CH_2OH \\ | \\ CH_2OH \qquad\qquad CH_2NH_2 \\ | \\ CH_2NH_2$$

Δ log $\frac{V}{W}$ +7.36   +3.02   +1.00

Negative deviants

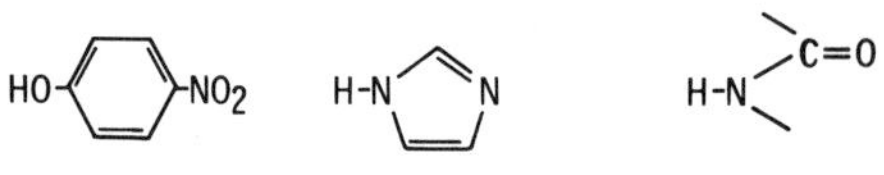

HO—⟨benzene ring⟩—NO$_2$ H–N⟨imidazole ring⟩N   H–N⟩C=O

Fig. 18. Compounds that exhibit nonadditivity of water-leaving tendencies in relation to expectations based on their constituent groups

isoleucine 0.60   serine 0.22
valine 0.54   glutamate 0.18
phenylalanine 0.50   histidine 0.17
cysteine 0.50   proline 0.18
leucine 0.45   tyrosine 0.15
methionine 0.40   aspartate 0.15
alanine 0.38   asparagine 0.12
glycine 0.36   glutamine 0.07
tryptophan 0.27   lysine 0.03
threonine 0.23   arginine 0.01

Fig. 19. Proportion of residues 95% buried in 12 proteins (Chothia 1976)

in Fig. 19. How do these numbers compare with the affinities of amino acid side chains for solvent water? Distribution coefficients from water to alcohol have been determined for some free amino acids (Nozaki and Tanford 1971), and when these are compared with the protein values, the agreement is poor. Since the properties of the free amino acids are doubtless affected to some extent by the presence of the zwitterionic situation of the α-carbon atom, we decided to look at the behavior of the side chains without substituents. Equilibria for their distribution from aqueous solution buffered at pH 7, to the vapor phase, are shown in Fig. 20. The results are very different from those observed with the free amino acids. Here serine and tryptophan are similar, and tryptophan, rather than being the most hydrophobic amino acid, is considerably more hydrophilic than, for example, phenylalanine. This is not a question of the reference phase that is used (similar results are obtained for water-alcohol distribution), but results from

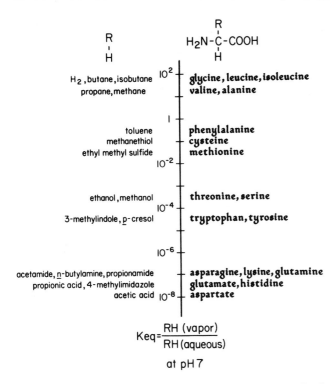

Fig. 20. Dehydration equilibria of amino acid side chains. Distribution coefficients are for the compounds indicated to the *left* of the scale, between buffered solution at pH 7 and the vapor phase

the disruptive local influence of a pair of charged groups on the relative solvation properties of the side chains. To make a semi-quantitative test of Perutz's generalization, we can ask how these figures compare with the distribution of the amino acids in proteins. In Fig. 21 it can be seen that the fraction of each amino acid buried is indeed related to what we may call their "hydration potentials" by analogy with Lipmann's (1941) "group potentials" (Wolfenden et al. 1979). The correlation between the two variables is highly significant, with a probability of less than 1 part in 10^5 of occurring by chance. The only amino acids that are missing are proline (which is not really an amino acid, and whose "side chain" is not directly comparable with the others) and arginine, an extremely hydrophilic amino acid that appears from preliminary data to lie in the expected relationship to the others.

Because of these relationships, it seemed of interest to inquire whether there might be some corresponding pattern in the genetic code. It was early noted that genetic code words specify amino acids that appear to be related in structure or biosynthetic function (Nirenberg et al. 1963), and some attempts at rationalization have been made (Crick 1968; Woese 1969); but the ordering of the side chains in Fig. 20 is too new to have been possible to examine in this way before. The genetic code has been found to be moderately degenerate at the first position of the codon, and highly degenerate at the third position, in the sense that several alternative bases serve to code for the same

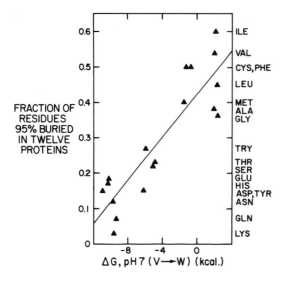

Fig. 21. Exposure of amino acid residues to water in proteins (*ardinate*) plotted as a function of increasing affinity of side chains for water at pH 7, relative to the vapor phase (*abscissa*)

amino acid at these positions; but each amino acid except serine is uniquely associated with the presence of a single base at the second position of the codon. When the 18 amino acids of the present list are arranged in order of increasing hydration potential, a noticeable pattern emerges (Table 1). Adenine, for example, serves as the second code letter for seven amino acids. Every one of them has a hydrophilic side chain. Estimated conservatively, the probability that this would occur by chance is less than 0.0045.

Table 1. Code letters in messenger RNA and DNA

		2nd code letter (mRNA)	2nd anti-code letter (DNA)
	gly	G	C
	leu	U	A
	ile	U	A
	val	U	A
	ala	C	G
	phe	U	A
	cys	G	C
	met	U	A
	thr	C	G
	ser	C(G)	G(C)
	trp	G	C
	tyr	A	T
	gln	A	T
	lys	A	T
	asn	A	T
	glu	A	T
	his	A	T
most hydrophilic	asp	A	T

It seems natural to suppose, with Sonneborn (1965), that coding similarities between amino acids with similar physical properties may have tended to offset the disruptive effects of mutation on the structural stability of proteins during their evolution. Figure 21 provides quantitative support for the view that mutations that would result in the introduction of hydrophilic amino acids, at interior locations that were previously hydrophobic, are expected to be especially damaging. The observed distribution of code letters seems to minimize the likelihood of these events.

G. Prospect

We are encouraged, by these various observations, in supposing that solvent water may exert on organizing influence on biological recognition to a degree that has been only dimly appreciated. The directional preferences of solute-water interactions are considerable, and the stereoenergetic details of solvation are likely to be of special importance in the many cases where solvent water is not fully stripped away from groups as they come together in a watery environment. To describe these details is likely to be much more difficult than to evaluate their overall hydration potentials, and will require the best efforts of organic chemists, spectroscopists, and theoreticians.

References

Banner DW, Bloomer AC, Petsko GA, Phillips DC, Pogson CI, Wilson IA, Corran PH, Furth AJ, Milman JD, Offord RE, Priddle JD, Waley SG (1975) Structure of chicken muscle triose phosphate isomerase determined crystallographically at 2.5 A resolution using amino-acid sequence data. Nature 255: 609–614

Bauer CA, Thompson RC, Blout ER (1976) The active center of *Streptomyces griseus* protease 3 and α-chymotrypsin: Enzyme-substrate interactions remote from the scissile bond. Biochemistry 15: 1291–1295

Bolognesi MC, Matthews BW (1979) Binding of the biproduct analog L-benzylsuccinic acid to thermolysin determined by X-ray crystallography. J Biol Chem 254: 634–639

Brayer GD, Delbaere LTJ, James MNG, Bauer CA, Thompson RC (1979) Crystallographic and kinetic investigations of the covalent complex formed by a specific tetrapeptide aldehyde and serine protease from *Streptomyces griseus.* Proc Natl Acad Sci USA 76: 96–100

Byers LD, Wolfenden R (1973) Binding of the bi-product analog benzylsuccinic acid by carboxypeptidase A. Biochemistry 12: 2070–2078

Campbell ID, Jones RB, Kiener PA, Richards E, Waley SG, Wolfenden R (1978) The form of 2-phosphoglycollic acid bound by triosephosphate isomerase. Biochem Biophys Res Commun 83: 347–352

Cha S, Agarwal RP, Parks PE (1975) Tight-binding inhibitors 2:non-steady state nature of inhibition of milk xanthine-oxidase by allopurinol and alloxanthine and of human erythrocyte adenosine deaminase by coformycin. Biochem Pharmacol 24: 2187–2197

Chothia C (1976) The nature of the accessible and buried sources of proteins. J Mol Biol 105: 1–14

Clark PI, Lowe G, Nurse D (1977) Detection of enzyme-bound intermediates by cross-saturation nuclear magnetic resonance spectroscopy – investigation of papain N-benzoylaminoacetaldehyde complex. J Chem Soc Chem Commun V 1977: 451–453

Cleland WW (1967) Enzyme kinetics. Annu Rev Biochem 36: 77–112

Cohen RM, Wolfenden R (1971) Cytidine deaminase from *Escherichia coli:* Purification, properties, and inhibition by 3,4,5,6-tetrahydrouridine. J Biol Chem 246: 7561–7565

Cohen SG, Vaidya VM, Schultz RM (1970) Active site of α-chymotrypsin. Activation by association-desolvation. Proc Natl Acad Sci USA 66: 249–256

Crick FHC (1968) The origin of the genetic code. J Mol Biol 38: 367–379

Crosby J, Lienhard GE (1970) Mechanisms of thiamine-catalyzed reactions. A kinetic analysis of the decarboxylation of pyruvate by 3,4-dimethylthiazolium ion in water and in ethanol. J Am Chem Soc 92: 5707–5713

Duax WL, Hauptman H, Weeks CM, Norton DA (1972) Valinomycin crystal structure by direct methods. Science 176: 911–914

Evans B, Wolfenden R (1970) A potential transition state analog for adenosine deaminase. J Am Chem Soc 92: 4751–4752

Evans B, Wolfenden R (1972) Hydratase activity of a hydrolase. Adenosine deaminase. J Am Chem Soc 94: 5902–5903

Evans BE, Mitchell G, Wolfenden R (1975) The action of bacterial cytidine deaminase on 5,6-di-hydrocytidine. Biochemistry 14: 621–624

George P, Witonsky RJ, Trachtman M, Wu C, Dorwart W, Richman L, Richman W, Shurayh F, Lentz B (1970) "Squiggle-H_2O": An enquiry into the importance of solvation effects in phosphate ester and anhydride reactions. Biochim Biophys Acta 223: 1–15

Ghisla S, Massey V (1977) Studies on mechanism of action of flavoenzyme lactate oxidase – proton uptake and release during binding of transition state analogs. J Biol Chem 252: 6729–6735

Gutowski JA, Lienhard GE (1976) Transition state analogs for thiamin pyrophosphate-dependent enzymes. J Biol Chem 251: 2863–2866

Hartman FC, Lamuraglia GC, Tomozawa Y, Wolfenden R (1975) The influence of pH on the interaction of inhibitors with triosephosphate isomerase and determination of the pK_a of the active site carboxyl group. Biochemistry 14: 5274–5279

Hayes DM, Kenyon GL, Kollman PA (1978) Theoretical calculations of the hydrolysis energy of some "high-energy" molecules. 2. A survey of some biologically important hydrolytic reactions. J Am Chem Soc 100: 4331–4340

Hine J, Mookerjee PK (1975) The intrinsic hydrophilic character of organic compounds. Correlations in terms of structural contributions. J Org Chem 40: 292–298

Jencks WP (1966) Strain and conformation change in enzyme catalysis. In: Kaplan NO, Kennedy EP (eds) Current aspects of biochemical energetics, pp 273–298. Academic Press, New York

Jencks WP (1975) Binding energy specificity and enzyme catalysis – the circe effect. Adv Enzymol 43: 219–410

Johnson LN, Wolfenden R (1970) Changes in absorption spectrum and crystal structure of triose phosphate isomerase brought about by 2-phosphoglycollate, a potential transition state analog. J Mol Biol 47: 93–100

Kalckar HM (1941) The nature of energetic coupling in biological synthesis. Chem Rev 28: 71–178

Koshland DE Jr (1959) Mechanisms of transfer enzymes. In: Boyer PD, Lardy H, Myrback K (eds) The enzymes, vol I, pp 305–346. Academic Press, New York

Lee B, Richards FM (1971) The interpretation of protein structures: Estimation of static accessibility. J Mol Biol 55: 379–400

Lewis CA Jr, Wolfenden R (1977) Thiohemiacetal formation by inhibitory aldehydes at the active site of papain. Biochemistry 16: 4890–4895

Lienhard GE (1973) Enzymatic catalysis and transition state theory. Science 180: 149–154

Lipmann F (1941) Metabolic generation and utilization of phosphate bond energy. Adv Enzymol 1: 99–162

Milner-White EJ, Watts DC (1971) Inhibition of adenosine 5'-triphosphatecreatine phosphotransferase by substrate-anion complexes. Biochem J 122: 727–741

Nirenberg MW, Jones OW, Leder P, Clark BFC, Sly WS, Pestka S (1963) On the coding of genetic information. Cold Spring Harbor Symp Quant Biol 28: 549–558

Nozaki Y, Tanford C (1971) The solubility of amino acids and two glycine peptides in aqueous ethanol and dioxane solutions. Establishment of a hydrophobicity scale. J Biol Chem 246: 2211–2217

Ohno S, Yagisawa N, Shibahara S, Kondo S, Maeda K, Umezawa H (1974) Synthesis of coformycin. J Am Chem Soc 96: 4326–4327

Olmstead WN, Brauman JI (1977) Gas-phase nucleophilic displacement reactions. J Am Chem Soc 99: 4219–4228

Perutz MF (1965) Structure and function of haemoglobin I. A tentative model of horse oxyhaemoglobin. J Mol Bil 13: 646–668

Reed GH, Barlow CH, Burns RA Jr (1978) Investigations of anion binding sites in transition state analogue complexes of creatine kinase by infrared spectroscopy. J Biol Chem 253: 4153–4158

Rowe WB, Ronzio RA, Meister A (1969) Inhibition of glutamine synthetase by methionine sulfoximine. Studies on methionine sulfoximine phosphate. Biochemistry 8: 2674–2680

Sinnott ML (1978) Ions, ion-pairs and catalysis by LacZ beta-galactosidase of *E. coli.* FEBS Lett 94: 1–9

Sonneborn TM (1965) Degeneracy of the genetic code: Extent, nature and genetic implications. In: Bryson V, Vogel HJ (eds) Evolving genes and proteins, pp 377–397. Academic Press, New York

Palmer AR, Ellis PD, Wolfenden R (1980) The ionization state of benzylsuccinate bound by carboxypeptidase A. Fed Proc 39: in press

Thompson RC (1973) Use of peptide aldehydes to generate transition state analogs of elastase. Biochemistry 12: 47–51

Thompson RC (1974) Binding of peptides to elastase: Implications for the mechanism of substrate hydrolysis. Biochemistry 13: 5495–5501

Villafranca JJ, Ash DE, Wedler FC (1975) Evidence for methionine sulfoximine as a transition state analog for glutamine synthetase from NMR and EPR data. Biochem Biophys Res Commun 66: 1003–1010

Warshel A (1978) Energetics of enzyme catalysis. Proc Natl Acad Sci USA 75: 5250–5254

Wentworth DF, Wolfenden R (1974) Slow binding of D-galactal, a "reversible" inhibitor of bacterial β-galactosidase. Biochemistry 13: 4715–4720

Westerik JO, Wolfenden R (1972) Aldehydes as inhibitors of papain. J Biol Chem 247: 8195–8197

White JU (1942) Long optical paths of large aperture. J Opt Soc Am 32: 285–288

Woese C (1969) Models for the evolution of coding assignments. J Mol Biol 43: 235–240

Wolfenden R (1969a) On the rate-determining step in the action of adenosine deaminase. Biochemistry 8: 2409–2412

Wolfenden R (1969b) Transition state analogs for enzyme catalysis. Nature 223: 704–705

Wolfenden R (1972) Analog approaches to the structure of the transition state in enzyme reactions. Accounts of Chemical Research 5: 10–18

Wolfenden R (1974) Enzyme catalysis: Conflicting requirements of substrate access and transition state affinity. Mol Cell Biochem 3: 207–211

Wolfenden R (1976) Transition state analog inhibitors and enzyme catalysis. Annu Rev Biophys Bioeng 5: 271–306

Wolfenden R (1978) Interaction of the peptide bond with solvent water: A vapor phase analysis. Biochemistry 17: 201–204

Wolfenden R, Kaufman J, Macon JB (1969) Ring-modified substrates of adenosine deaminase. Biochemistry 8: 2312–2415

Wolfenden R, Wentworth DF, Mitchell GN (1977) The influence of substituent ribose on transition state affinity in reactions catalyzed by adenosine deaminase. Biochemistry 16: 5071–5077

Wolfenden R, Cullis P, Southgate C (1979) Water, protein folding and the genetic code. Science 206: 575–577

5. Suicide Substrates: Mechanism-Based Inactivators of Specific Target Enzymes

C. WALSH

A. Introduction

It is a great pleasure to present this paper at a symposium celebrating the 80th birthday of Professor Fritz Lipmann both because of my personal association with him (admittedly stretching back only a mere 15 years) and because he is such a seminal figure in the development of biochemistry and enzymology. I joined the Lipmann laboratory in 1965 as a graduate student to do my doctoral research with Len Spector and initiated mechanistic studies on how citrate was cleaved to acetyl CoA and oxalacetate by the cytoplasmic citrate cleavage enzyme (ATP citrate lyase), discovered by Lipmann and Srere some years earlier in 1953 (Srere and Lipmann 1953). Although (or perhaps because) most other people in the Lipmann laboratory were working on protein biosynthesis, Dr. Lipmann always evinced a lively interest in my experiments and other enzymatic mechanistic studies. Perhaps as a result, I had the pleasure of reading through the autobiographical manuscript of his which became his book *Wanderings of a Biochemist* (Lipmann 1971). That experience increased my appreciation for the breadth of contributions to biochemistry made by the central figure of this meeting.

In this article I shall briefly describe recent studies from my laboratory and several others on mechanism-based enzyme inactivators, also designated as "suicide substrates" (Maycock and Abeles 1076; Rando 1974, 1975, 1978; Walsh 1977). These compounds may be natural or synthetic and are structural analogs to a normal physiological substrate of a target enzyme, with one major difference. They have built into them a latently reactive functional group which becomes activated only during catalytic action by the enzyme. Thus, the target enzyme mistakes the inactivator for a substrate and takes it part way or all the way through the catalytic cycle, unraveling the latent group into a chemically reactive one in the microenvironment of enzyme's active site. If the enzyme molecule becomes covalently modified and loses activity as a catalyst, then it will have committed suicide by processing and activating the suicide substrate. These inactivators differ from other earlier affinity labels or active-site directed reagents, for those previous molecules already have their reactive groups exposed, i.e., α-haloketones or epoxides, and are reactive-free in solution (Shaw 1970). As such they are, in general, too indiscriminate to offer in vivo selectivity. In contrast, the suicide substrates are unreactive when free in solution, since they require catalytic unmasking by the target enzyme.

Departments of Chemistry and Biology, Massachusetts Institute of Technology, Cambridge, Massachusetts 02139, USA

One can expect maximal in vivo selectivity, since at least three factors augur specificity: (1) the normal discrimination in binding steps; (2) the mechanism-based inactivator often requires a specific mechanism for unraveling and may, for example, inactivate flavin-dependent dehydrogenases but not nicotinamide-dependent dehydrogenases, which carry out the same reaction; (3) the reactive species capable of covalent modification is generated in a specified, local microenvironment, the active site of the desired target enzyme.

There is currently a great deal of interest in utilizing these simple concepts for design and testing new therapeutic agents (Walsh and Abeles 1973), and one can see success will require both a knowledge of the synthetic chemistry to build in latently reactive functional groups and also on understanding of the catalytic mechanism of the candidate enzyme to be inactivated. If these in vitro objectives succeed, then one can proceed to the in vivo phase where pharmacokinetics and pharmacodynamics, as always, will condition availability and delivery of the suicide substrate molecules (Walsh and Abeles 1973).

B. Types of Enzymically Activatable Functional Groups

Collected in Table 1 are a representative listing of the types of enzymatically activatable functional groups that are catalytically unmasked by the indicated target enzymes with resultant loss of catalytic activity.

Table 1. Enzymically activatable functional groups and specific enzyme targets

Functional group	Compound	Target enzyme	Nature of covalent adduct
I. *Acetylenes*	1. 3-decynoyl-CoA	β-hydroxydecanoyl thiolester dehydrase	Active site histidine
	2. Pargyline (N-methyl-N-benzyl-propynyl-amine)	Monoamine oxidase	N^5 of bound FAD
	3. Propargylglycine (2-amino-4-pentynoate)	Several PLP-dependent enzymes including: γ-cystathionase, cystathioninine-γ-synthetase, methionine-γ-lyase, L-alanine transaminase, L-aspartate transaminase; also D-amino acid oxidase	Enzymic N,S-atoms
	4. γ-acetylenic GABA	GABA transaminase	Unidentified active site residue
	5. 2-hydroxy-3-butynoate	Several α-hydroxy acid oxidizing flavoenzymes	Flavin modification solely; no apoprotein modification
II. *Olefins*	1. Vinylglycine	L-aspartate transaminase D-amino acid transaminase	Lys 246 in L-Asp. trans.

Table 1. Continued

Functional group	Compound	Target enzyme	Nature of covalent adduct
	2. γ-vinylGABA	GABA transaminase	Unidentified active site residue
	3. Gabaculine	GABA transaminase	Covalent adduct with PLP
	4. 4,5-trans-lysene	L-lysine-ε-transaminase L-ornithine-δ-trans-amianse	Unidentified active residue
	5. Allylisopropyl-acetamide	Liver cytochrome P-450	Heme prosthetic group
III. β-Substituted Amino Acids	1. β-fluoro-D-alanine	E. coli alanine racemase, rabbit serine hydroxy-methylase	Active site cysteine in SHM
	2. α-F-methylDOPA	DOPA decarboxylase	Unidentified residue
	3. β-trifluoroalanine	Several PLP-dependent enzymes	Unidentified residue
	4. O-carbamyl-D-serine	E. coli alanine recemase	Unidentified residue
IV. Miscellaneous	A. Cyclopropyl compounds		
	1. Cyclopropanone	Aldehyde dehydrogenase	–
	2. Cyclopropylamine tranylcypramine (trans-2-phenyl cyclopropylamine)	Monoamine oxidase	–
	B. Penicillin derivatives		
	1. Clavulanate	Plasmid RTEM β-lactamase	serine residue
	2. Penicillin sulfones	Plasmid RTEM β-lactamase	serine residue
	3. 6-β-bromopeni-cillin	Plasmid RTEM β-lactamase	serine residue
	C. Phenylhydrazine	Monoamine oxidase	C_{4a} of bound FAD
	D. Nitrilo compounds		
	1. Cyanoglycine	Plasma amine oxidase	–
	2. β-aminopropio-nitrile	Collagen lysyl-ε-oxidase	–
	E. 5-Fluoro-2′-deo-xyuridylate	Thymidylate synthetase	Enz-cysteine-SH
	5-nitro-2′-deoxy-uridylate	Thymidylate synthetase	Enz-cysteine-SH

References to specific enzymes can be found in the text

I. Acetylenes

The now classical first experiments on acetylenic inactivators were reported some 12 years ago by HelmKamp, Rando, and Bloch (Bloch 1972) who observed time-dependent inactivation of the *E. coli* enzyme β-hydroxydecanoyl thiolester dehydrase by a 3-decynoyl thiolester. Careful experimentation revealed that the enzyme begins catalysis on the suicide substrate by usual generation of a C_2

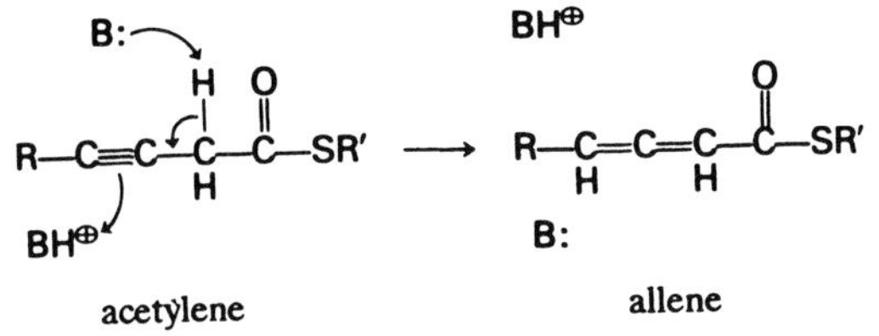

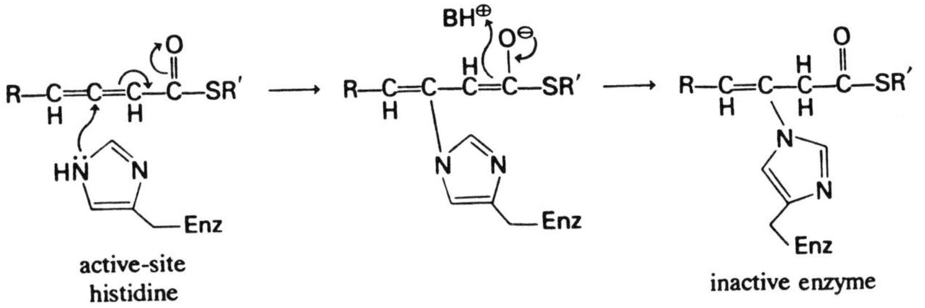

B: BH⊕

$$R-C\!\equiv\!C-\overset{H}{\underset{H}{C}}-\overset{O}{\overset{\|}{C}}-SR' \longrightarrow R-C\!=\!C\!=\!\overset{}{C}-\overset{O}{\overset{\|}{C}}-SR'$$

BH⊕   B:

acetylene   allene

BH⊕

$$R-C\!=\!C\!=\!C-\overset{O}{\overset{\|}{C}}-SR' \longrightarrow R-C\!=\!C-C\!=\!C-SR' \longrightarrow R-C\!=\!C-\overset{H}{\underset{H}{C}}-\overset{O}{\overset{\|}{C}}-SR'$$

HN N   N N   N

—Enz   —Enz   Enz

active-site   inactive enzyme
histidine

carbanion, but this anion can be reprotonated not only at C_2 but also at C_4 in the acetylenic case to effect a net conversion, at the enzyme active site, of an unconjugated acetylene to a conjugated allene (a propargylic rearrangement), which is the uncovered alkylating electrophile that is covalently captured by an enzymic histidine group with suicidal outcome. This result has been generalized for the design of acetylene-containing suicide substrates with an acetylene adjacent to a site of known (or suspected) enzyme-mediated carbanion formation to uncover the proximal electrophile via the propargylic rearrangement (Walsh 1977; Walsh and Abeles 1973). It is possible, for example, that pargyline (and other propynyl amine congeners) inactivates the flavin-dependent mono-amine oxidases by such a rearrangement. Alternatively in this redox enzyme, oxidation at the N-propynyl methylene to the product imine generates a conjugated acetylene imine activated for Michael attack by N^5 of reduced FAD (Maycock et al. 1976).

Propargylglycine is a naturally occurring amino acid (Scannell et al. 1971) from mushrooms and we (Walsh and Abeles 1973) have studied its inactivating properties towards a number of pyridoxal-P dependent enzymes, including transaminases and, particularly, enzymes catalyzing transformations at the γ-carbon of amino acid substrates (Marcotte and Walsh 1975, 1978; Johnston et al. 1979). We have also reported on its oxidation by and "wounding" of the flavoprotein D-amino acid oxidase (Marcotte and Walsh 1978). The interesting feature of propargylglycine is that the 4,5-acetylene is insulated from the C_2 amino group by the C_3 methylene. Enzymes which can generate not only α-carbanion equivalents, but also stabilized β-carbanion equivalents, are the specific targets to generate the 3,4,5-allene in conjugation to a C_2-p-quinoidal PLP adduct as the proximal killing agent. Unraveling of the latent allene also proceeds by propargylic rearrangement here.

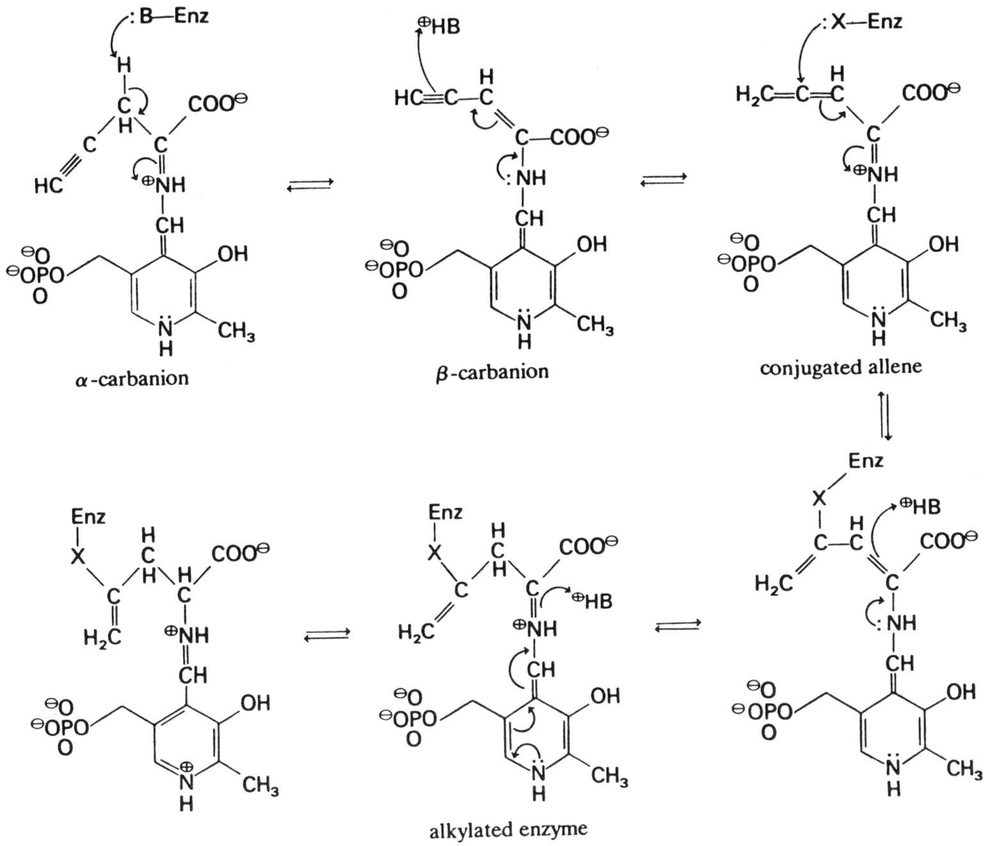

α-carbanion   β-carbanion   conjugated allene

Enz

alkylated enzyme

II. Olefins

The major route for enzymic activation of olifenic double fonds is oxidation at a locus adjacent to them to generate a polarized conjugated system, set up for a 1,4-Michael (conjugate) addition process (Rando 1975; Walsh 1977). Thus a *net* oxidation, independent of specific mechanism, will uncover the reactive functionality, again an electrophile to be captured by nucleophilic amino acid side chains of enzyme (or coenzyme).

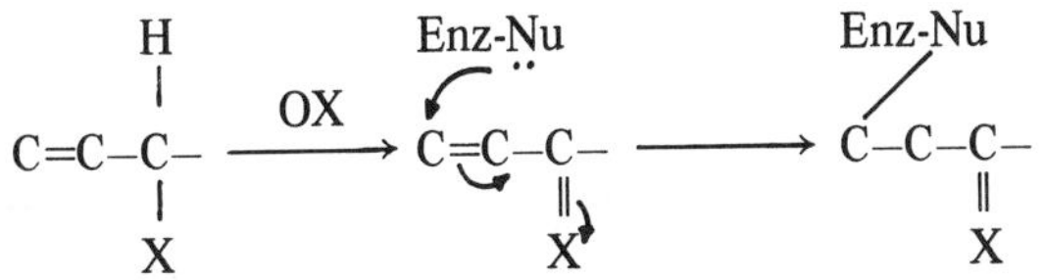

Three specific examples are noted below.

γ-Vinylglycine has been synthesized by the group at the Merrell Research Center in Strasbourg and found to be an effective irreversible inactivator of brain γ-amino-butyrate(GABA) transaminase both in vitro and in vivo and shows promise as an anti-epileptic candidate, rationally designed (Seiler et al. 1978).

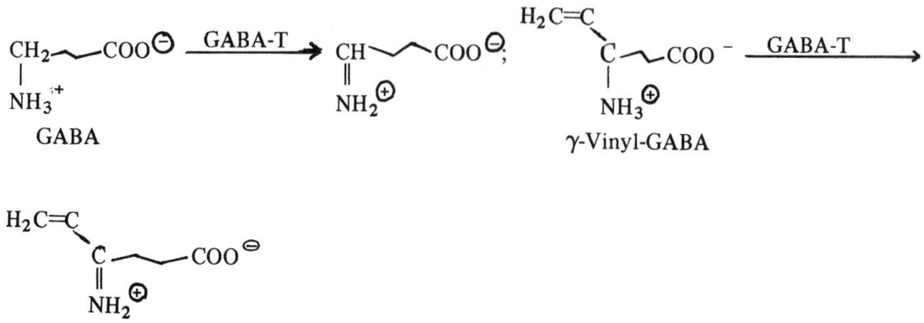

Activated for Michael attack

A naturally occurring cyclohexadienoid amino acid, gabaculine (Mishima et al. 1976), from *S. tocayensis,* also inactivates GABA-transaminase after transamination by the enzyme but not by Michael addition of an enzyme nucleophile. Rather the low energy path for decomposition of the initial exocyclic imine is aromatization to a meta anthranilyl derivative of pyridoxamine-P. This captures the coenzyme in a nonhydrolyzable secondary amino linkage and, since the cofactor does not dissociate readily from the enzyme, causes inactivation by covalent modification of cofactor.

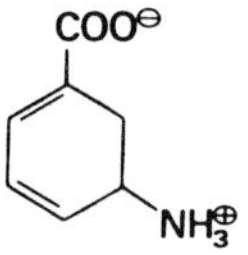

gabaculine

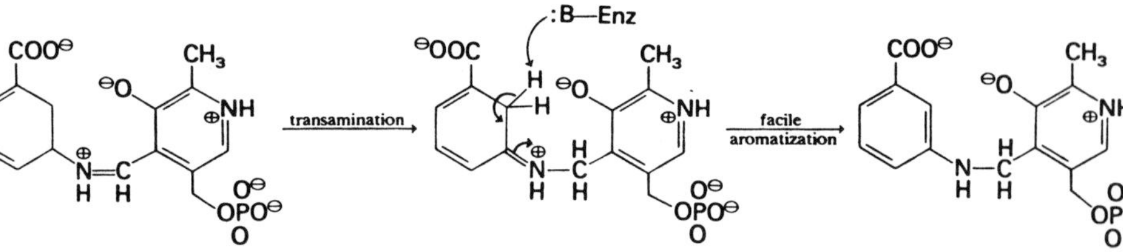

stabile anthranilate
derivative

A third olefinic suicide substrate of note is allylisopropylacetamide, a drug known to induce porphyria from heme destruction in the liver with accumulation of a characteristic green pigment, a heme derivative, in treated livers. The molecular cause of this toxicity may arise from catalytic processing of allylisopropylacetamide by the hepatic microsomal cytochrome P_{450} monooxygenase. Ortiz de Montellano et al. (1978) have recently suggested, from NMR and mass spectroscopic evidence, that one of the venyl groups of the heme moiety of the P_{450} monooxygenase is covalently derivitized by the allylic substrate in a suicidal fashion. Since this monooxygenase normally converts double bonds to epoxides, it is suggested that the heme vinyl group may be modified by an activated electrophile as shown. This represents a different

Heme alkylation

catalytic unmasking from that effected by GABA transaminase above.

III. β-Substituted Amino Acids

A variety of synthetic β-substituted amino acids have recently been prepared and used as suicide substrates for a variety of pyridoxal-P-dependent enzymes; including physiologically and pharmacologically important racemases and amino acid decarboxylases (Walsh et al. 1978; Kollonitsch et al. 1978).

Among the β-substituted amino acids with good leaving groups fluorine has recently found favor because of its small size and ease of intramolecular elimination. A prototypic example is β-fluoro-D-alanine synthesized by Kollonitsch and colleagues at Merck (Kollonitsch and Barash 1976) and shown to have broad spectrum activity in terminating infections in experimental animals. We have validated their suggestion that bacterial alanine racemase is the site of antibiotic action by proving that D-fluoro-alanine (and the L-isomer) is a suicide substrate, partitioning between harmless turnover to pyruvate, F^-, and NH_4^+ on the average 800 times per enzyme inactivation event (Wang and Walsh 1978). This partition ratio is independent of α-carbon chirality and independent of the nature of the leaving group X (e.g., fluoro, chloro, acetyl, carbamyl) consistent with the chemically symmetric aminoacrylate-PLP intermediate as the proximal electrophile partitioning between H_2O attack at C_2 (successful turnover) and addition of an enzymic nucleophile at C_3 (alkylation and inactivation as suggested below). O-Carbamyl-D-serine, a natural antibiotic elaborated by Streptomycetes, is thus a suicide substrate for *E. coli* alanine racemase (Wang and Walsh 1978).

A variety of other fluorinated amino acids have been designed and tested as suicide substrates for PLP-dependent amino acid decarboxylases (Kollonitsch et al. 1978), among them α-fluoromethylDOPA (and α-difluoromethylDOPA; Palfreyman et al. 1978). Since the α-CH bond of DOPA is not broken during catalytic action of this key aromatic amino acid decarboxylase, it will slowly decarboxylate α-methylDOPA (aldomet) to α-methyldopamine and this is the basis of action of the antihypertensive aldomet. When α-CH$_2$F is substituted for α-CH$_3$, a net elimination of CO$_2$ and F$^-$ can occur in a stoichiometric low energy path to generate another eneamino-PLP susceptible to conjugate addition by an enzyme nucleophile.

Py = PYRIDOXAL PHOSPHATE RING SYSTEM·

When one uses β-trifluoromethyl groups rather than β-monofluoromethyl groups, the suicide substrate, e.g., trifluoroalanine (Silverman and Abeles 1976), appears much more efficient at inactivation in any given catalytic cycle. For example, the partition ratio for alanine racemase for trifluoroalanine turnovers per inactivation events is $< 10/1$ compared to the 800/1 for monofluoroalanine. Silverman and Abeles (1977), moreover, have shown that the initial adduct from trifluoroalanine and γ-cystathionase can go on and lose the other two fluorines by intramolecular elimination to yield an acyl oxidation state (an amide in γ-cystathionase) at C_3 of the original trifluoroalanine (Silverman 1979). Finally, difluoromethyl groups have also been used and in particular the Merrell group (Seiler et al. 1978) has found that α-CHF$_2$-ornithine is an effective inactivator or ornithine decarboxylase, the first enzyme in polyamine biosynthesis, and has interesting in vivo effects as well.

It is worth specific note that β-haloamino acids are likely to be specific suicide agents for *PLP-dependent enzymes,* since HX elimination from a typical β-halo amino acid such as fluoroalanine generates a free eneamino acid (e.g., amino acrylate) which is *nucleophilic, not electrophilic,* at C_3, i.e., it is a carbanion equivalent at C_3 and will not add nucleophilic reagents. But in PLP-enzymes HX-loss from an amino acid PLP aldimine yields an eneamino-PLP adduct, where the PLP serves as an electron sink to *functionally reverse the polarity* at C_3 and render it *electrophilic* (Walsh 1979). This is the chemical basis for the β-replacement reactions of normal biosynthesis (e.g., O-acetylserine → cysteine).

A final instance of β-fluorinated amino acids as specific enzyme suicide substrates is our recent demonstration (Wang 1979; Wang et al. 1979) that D-fluoro-alanine, but not L-fluoroalanine, inactivates rabbit liver serine transhydroxymethylase (SHM) in a reaction accelerated 660-fold by the normal reaction cofactor tetrahydrofolate. Although the normal aldol cleavage of L-serine to glycine and $N^{5,10}$-CH_2-THF does not involve an α-carbanion equivalent in the cleavage direction, in the back direction a stabilized glycine α-anion is required for C_α-C_β bond formation in the serine product. The α-H of glycine specifically removed corresponds to the α-H of a D-amino acid. Thus, we predicted HF-elimination from D-fluoroalanine and indeed have observed a 60/1 partition ratio for pyruvate formation per SHM molecule inactivated. The enzymic nucleophile is an active site cysteine sulfhydryl group. However, despite the dramatic acceleration of killing by THF (to a maximal turnover number of 95 min^{-1}), the K_m for D-fluoroalanine is ca. 50 mM and so unlikely to have any in vivo utility.

IV. Suicide Substrates of Miscellaneous Structure

Three additional suicide substrates merit note both because of the chemistry of the suicidal inactivation process and because of the physiological importance of the target enzyme taken out of action.

The first class consists of compounds which can generate cyclopropanone equivalents. The key chemical property utilized is that cyclopropanone structures are strained and destabilized relative to addition products. Thus cyclopropanone is extensively hydrated or capturable rapidly by other nucleophiles. If the nucleophilic group is at an

enzyme active site, the tetrahedral adduct, if stable, may be an inactive species. Thus Wiseman and Abeles (1979) have shown that the mushroom amino acid cuprine acts as a source of cyclopropanone (probably via a glutaminase process), which is an effective suicide substrate for aldehyde dehydrogenase, known to have a reactive SH group at the active site.

An additional example may be provided by tranylcypramine (trans-2-phenylcyclo-propylamine) a monoamine oxidase inhibitor, at one time used clinically. Recent kinetic evidence suggests oxidation by MAO to a cyclopropylimine species which may covalently modify the enzyme-bound FAD.

$$FADH_2\text{-Enz}$$

$$C_6H_5 \overset{H}{\underset{H\ \ NH_2}{\triangle}} \rightarrow C_6H_5 \triangle_{NH_2} + \rightarrow C_6H_5 \underset{NH_2}{\triangle} FAD\text{-Enz}$$

A second recent group of potentially important suicide substrates comprise various penicillin derivatives which inactivate β-lactamases (penicillinase and cephalosporinase activity). The enzyme best studied for inactivation is the plasmid-coated RTEM lactamase responsible for much of the penicillin resistance developed by clinically important pathogenic bacteria (Fisher and Knowles 1978). A specific compound of note is clavulanate, studied in detail by Fisher, Charnas and Knowles (1978). The key features of inactivation involve normal acyl enzyme formation during β-lactam ring opening, but then intramolecular fragmentation competes with the normal hydrolytic breakdown of the acyl enzyme species. Fragmentation leaves the enzyme as a covalent derivative in an inactive state. These kinds of mechanism-based penicillinase inactivators could be very useful alone or in combination with other penicillins since they should act synergistically both to remove the major molecular basis of resistance and deliver more active antibiotic molecules.

The third category involves 5-substituted 2'-deoxyuridylate analogs as suicide substrates for thymidylate synthetase, a key enzyme in DNA biosynthesis. The paradigm has been 5-fluorodeoxyuridylate (FdUMP), a clinically useful anticancer agent (Pogolotti and Santi 1977). It is processed analogously to the normal substrate 2'-deoxy-UMP up to the point of a ternary covalent complex between an enzymic SH group added to C_6 and a $C_5\text{-CH}_2$-tetrahydrofolate linkage. The normal substrate undergoes an apparent

1,3-hydride shift subsequently to yield the C_5-CH_3 group in thymidylate and 7,8-di-hydrofolate. The FdUMP ternary complex does not break down rapidly and the enzyme is thus trapped in mid-catalytic cycle (half time for breakdown is several hours at room temperature).

5-fluoro-2'-deoxyuridylate
(F-dUMP)

A more recent 5'-substituted 2'-deoxyUMP reported by Santi et al. (1978) is the 5-nitro-dUMP. The nitro group is sufficiently good at stabilizing an anion at C_5 that the enz-cysteinyl-5-nitrodUMP adduct accumulates at the active site and generates in-activated enzyme even in the absence of tetrahydrofolate.

C. Criteria for Mechanism-Based Inactivation

A variety of criteria have been elaborated elsewhere (Rando 1975, 1978; Walsh 1977) for determination whether enzyme inactivation has occurred by a "suicidal" or "mech-anism-based" process. These include kinetic and chemical criteria. Since covalent modi-fication of the enzyme catalyst occurs, loss of activity should be time-dependent and irreversible provided the derivitized enzyme is a stable adduct. An additional common kinetic outcome is pseudo first-order loss of activity conforming with a scheme involv-ing a binding step (saturable) that precedes first-order inactivation processes from bound inactivator.

$$E + I \rightleftarrows E \cdot I \xrightarrow{\text{inactivation}} E - I$$

Chemical criteria involve determination of adduct stoichiometry, usually performed with radioactively labeled suicide substrate, labeled in the fragment expected to undergo covalent attachment to the enzyme. For specific mechanism-based inactivation, typically one mol of label is incorporated per mol of active site, although a number of instances of "half-site reactivity" have been seen with propargylglycine killing of pyridoxal-P-dependent enzymes (Johnston et al. 1979; Silverman 1979).

Finally, the *partition ratio* − the number of times a potential suicide substrate is processed to product which diffuses away from the active site and leaves an enzyme molecule unharmed, compared to the frequency of a killing event − can be determined experimentally (Walsh 1977) and specific ratios were noted for D-fluoroalanine killing of alanine racemase (Wang and Walsh 1978) and serine hydroxymethylase (Wang 1979; Wang et al. 1979) earlier in this paper. A generalized partition scheme is indicated below, where S' is the suicide substrate and P' the product. This ratio k_3/k_4 can be diagnostic of mechanism and indicative of the structure of the

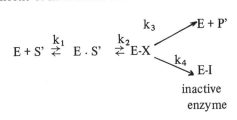

partitioning intermediate E-X. Also, it is an important ratio for potential in vivo evaluation, since P' is a reactive product molecule where the latent functional group has been uncovered. To the extent that P' is released and can react with cellular (or extracellular) molecules other than the target enzyme, specificity will be decreased, side effects will increase. Thus maximal specificity for a therapeutic suicide substrate would occur with as low a k_3/k_4 ratio as possible. In the limit, k_3 would be zero, no product molecules would get away, one equivalent of suicide substrate would be completely efficient in killing an equivalent of enzyme. In practice, gabaculine appears completely efficient in killing various GABA transaminases and vinylglycine nearly so ($k_3/k_4 = 1/9$) in aspartate transaminase inactivation (Rando 1975).

It is worth noting that most of the suicide substrates discussed can be unraveled catalytically to generate a reactive electrophile as the proximal inactivating agent. This chemistry is utilized to trap nucleophilic groups (oxygen, nitrogen, or sulfur) of enzymic amino acid side chains or coenzymes. In the penicillinase and thymidylate synthetase inactivations noted, an initial nucleophilic addition step occurring in normal catalysis is then followed by some aberrant consequence, such as a fragmentation (penicillinase) or an inability to undergo a 1,3-hydride shift (thymidylate synthetase). Most enzymes do not have corresponding electrophilic groups that can be correspondingly trapped by unmasking nucleophilic functionalities on potential suicide substrates, but flavin-dependent and pyridoxal P-dependent enzymes carry electrophilic centers on their coenzymes, and it may be that this reversed chemistry accounts for covalent modification of the flavin in target flavoenzymes by hydroxybutynoate and phenyldiazene inactivations, and of the pyridoxal cofactor in gabaculine-induced GABA transaminase suicide (Rando 1975).

Finally, one can consider categorizing enzymes susceptible to mechanism-based inactivators on the basis of the chemistry the enzymes use in catalysis and the nature of the chemical transformation. Thus enzymes using carbanion chemistry are susceptible to acetylenic inhibitors while redox enzymes are susceptible to acetylenes and olefins at sites adjacent to the oxidizable locus. Hydrolytic enzymes, carrying out group transfer of an electrophilic fragment to water, are more difficult and penicillinases are an important special subset but the fragmentation of acyl enzyme intermediates may be generalizable. To date phosphoryl transfer enzymes have been an undertargeted category but certainly an interesting one (e.g., cyclicAMP-dependent protein kinases could be interesting targets). Ortiz de Montellano et al. (1979) have recently synthesized phosphate esters of α-difluoroalcohols and suggested that phosphoryl transfer would liberate the free α-difluoroalcohol and this is, in aqueous environment, a reactive acyl fluoride

$$\underset{O}{\overset{O}{ROPOCH_2R'}} \xrightarrow{R''X} \underset{O}{\overset{O}{R''XPOR}} + HO\underset{F}{\overset{F}{-C}}-R' \xrightarrow{-HF} O=\underset{F}{\overset{}{C}}-R$$

acyl fluoride

equivalent (Ortiz de Montellano and Vinson 1979). With careful attention to the nature of the latent functionality (and its synthesis) and the catalytic mechanism of the target enzyme, the design of a suicide substrate for almost any category of enzyme may soon become feasible.

References

Bloch K (1969) Enzymatic synthesis of monounsaturated fatty acids. Acc Chem Res 2: 193

Bloch K (1972) In: Boyer P (ed) The enzymes, 3rd ed, vol V, p 441. Academic Press, New York

Fisher J, Knowles J (1978) Bacterial resistance to β-lactams: The β-lactamases. In: Clarke FH (ed) Annual reports of medicinal chemistry, vol XIII, p 239. Academic Press, New York

Fisher J, Charnas R, Knowles J (1978) Kinetic studies on the inactivation of *E. coli* RTEM β-lactamase by clavulanic acid. Chemical studies on the inactivation of *E. coli* RTEM β-lactamase by clavulanic acid. Biochemistry 17: 2180, 2185

Johnston M, Jankowski D, Marcotte P, Soda K, Walsh C (1979) Suicide inactivation of bacterial cystathionine γ-synthetase and methionine γ-lyase during processing of L-propargylglycine. Biochemistry 18: 1729

Kollonitsch J, Barash L (1976) Organofluorine synthesis via photofluorination: 3-fluoro-D-alanine and 2-deuterio analogue, antibacterials related to the bacterial cell wall. J Am Chem Soc 98: 5591

Kollonitsch J, Barash L, Kahan F, Kropp H (1973) New antibacterial agent *via* photofluorination of a bacterial cell wall constituent. Nature 243: 346

Kollonitsch J, Patchett A, Marburg S, Maycock A, Perkins L, Doldouras G, Duggan D, Aster S (1978) Selective inhibitors of biosynthesis of aminergic neurotransmitters. Nature 274: 906

Lipmann F (1971) Wanderings of a biochemist. Wiley, New York

Marcotte P, Walsh C (1975) Active site-directed inactivation of cystathionine γ-synthetase and glutamic pyruvic transaminase by propargylglycine. Biochem Biophys Res Commun 62: 677

Marcotte P, Walsh C (1978) Properties of D-amino acid oxidase covalently modified upon its oxidation of D-propargylglycine. Biochemistry 17: 2864

Maycock A, Abeles R (1976) Suicide enzyme inactivators. Acc Chem Res 9: 313

Maycock A, Abeles R, Salach J, Singer T (1976) The structure of the covalent adduct formed by the interaction of 3-dimethylamino-1-propyne and the flavine of mitochondrial amine oxidase. Biochemistry 15: 114

Mishima H, Kurihara H, Kobayashi K, Miyazawa S, Terahara A (1976) γ-aminobutyrate aminotrans-
ferase inhibitor of microbial origin (enzyme inhibitor; cyclohexa-1,3-diene amino acid). Tetra-
hedron Lett 7: 537
Ortiz de Montellano PR, Vinson W (1979) Carboxylic and phosphate esters of α-fluoro alcohols. J
Am Chem Soc 101: 2222
Ortiz de Montellano PR, Mico BA, Yost GS, Correra MA (1978) Suicidal inactivation of cytochrome
P-450: Covalent binding of allylisopropylacetamide to the heme prosthetic groups. p 337 in:
see ref. Seiler et al. 1978
Palfreyman MG, Danzin C, Jung MJ, Fozard JR, Wagner J, Woodward JK, Aubry M, Dage RC, Koch-
Weser J (1978) Substrate-induced irreversible inhibition of aromatic L-amino acid decarboxylase
by α-difluoromethyl-DOPA. p 221 in: see ref. Seiler et al. 1978
Pogolotti AL Jr, Santi DV (1977) The catalytic mechanism of thymidylate synthetase. In: van Ta-
melen (ed) Bioorganic chemistry, vol 1, Enzyme action, p 277. Academic Press, New York
Rando R (1974) Chemistry and enzymology of k_{cat} inhibitors. Science 185: 320
Rando R (1975) Mechanisms of action of naturally occurring irreversible enzyme inhibitors. Acc
Chem Res 8: 281
Rando R (1978) Principles of catalytic enzyme inhibition. p 13 in: see ref. Seiler et al. 1978
Santi DV, Wataya Y, Matsuda A (1978) Approaches to the design of mechanism-based inhibitors
of pyrimidine metabolism. p 291 in: see ref. Seiler et al. 1978
Scannell J, Preuss D, Demny T, Weiss F, Williams J, Stempel A (1971) Antimetabolites produced
by microorganisms. II L-2-amino-4-pentynoic acid. J Antibiot 24: 239
Seiler N, Jung MJ, Koch-Weser J (eds) (1978) Enzyme-activated irreversible inhibitors. Elsevier/
North-Holland, New York
Shaw E (1970) Chemical modification by active-site-directed reagents. In: Boyer P (ed) The enzymes,
3rd ed, vol I, p 91. Academic Press, New York
Silverman R (1979) In: Kalman T (ed) Medicinal chemistry VI. Elsevier/North-Holland, New York
(in press)
Silverman R, Abeles R (1976) Inactivation of pyridoxal phosphate dependent enzymes by mono-
and polyhaloalanines. Biochemistry 15: 4718
Silverman R, Abeles R (1977) Mechanism of inactivation of γ-cystathionase by β,β,β-trifluoroalanine.
Biochemistry 16: 5515
Srere P, Lipmann F (1953) An enzymatic reaction between citrate, adenosine triphosphate and
coenzyme A. J Am Chem Soc 75: 4874
Walsh C (1977) Recent developments in suicide substrates and other active site-directed inactivating
agents of specific target enzymes. In: Horizons in biochemistry and biophysics, vol III, p 36.
Addison-Wesley, Reading, MA
Walsh C (1978) Chemical approaches to the study of enzymes catalyzing redox transformations.
Ann Rev Biochem 47: 881
Walsh C (1979) Enzymatic reaction mechanisms (see Chapter 4). Freeman, San Francisco
Walsh C, Abeles R (1973) Acetylenic enzyme inhibitors. Inactivation of γ-cystathionase, in vitro
and in vivo, by propargylglycine. J Am Chem Soc 95: 6125
Walsh C, Johnston M, Marcotte P, Wang E (1978) Studies on suicide substrates for pyridoxal-P
linked enzymes. p 177 in: see ref Seiler et al. 1978
Wang E (1979) Mechanistic studies on alanine racemase and serine transhydroxymethylase. PhD
diss, Biol Dep. M.I.T.
Wang E, Walsh C (1978) Suicide substrates for the alanine racemase of E. coli B. Biochemistry 17:
1313
Wang E, Kallen R, Walsh C (1979) D-fluoroalanine: A suicide substrate for serine transhydroxy-
methylase. In: Kisluik R, Brown G (eds) Chemistry and biology of pteridines, p 507. Elsevier/
North-Holland, New York
Wiseman J, Abeles R (1979) Mechanism of inhibition of aldehyde dehydrogenase by cyclopropanone
hydrate and the mushroom toxin coprine. Biochemistry 18: 427

6. Recognition: the Kinetic Concepts

J. NINIO and F. CHAPEVILLE

A. Limitations of the Lock and Key Concept

Following Pauling's analysis (Pauling 1948, 1958), most enzymologists relate enzyme specificity to the supposed geometric complementarity between enzyme and substrate, the structural match being realized either at the initial stage of binding or later during the formation of a transition complex (Koshland 1958; Wolfenden 1972; Fersht 1974; Jencks 1975). The concept of stereochemical complementarity (Fischer 1894), which is fundamental to our understanding of the replication of the genetic material (Pauling and Delbrück 1940; Watson and Crick 1953), was invoked whenever there was a problem of molecular specificity, from immunology (Ehrlich 1900; Haurowitz 1967) to molecular evolution (Woese 1967). Yet there is now a rapidly growing body of knowledge which is being produced and organized along a quite different mode of thinking. The following are a few examples of recent experimental findings which created difficulties for the geometric viewpoint and contributed to the spread of the kinetic ideas.

I. tRNA and Amino Acyl-tRNA Ligase Interactions

When a mixture of tRNA's from yeast is acylated by the yeast valyl-tRNA ligase, there is an almost exclusive formation of valyl-tRNA^{Val}. Thus one may think that the tRNA ligase is recognizing specifically the tRNA^{Val} isoacceptors, that its binding site is somewhat complementary in shape to the surface of the yeast tRNA's for valine. Since the ligase is highly specific for its cognate tRNA's, one would expect that, when presented with a set of tRNA's from a distant species, *B. subtilis* for instance, it will fail to recognize them. Quite to the contrary, the yeast valyl-tRNA ligase does acylate nearly all tRNA's of *B. subtilis* (Giegé et al. 1974). Eventually, a ligase will even acylate a foreign RNA molecule having only a vague resemblance to a tRNA (Pinck et al. 1970; Siberklang et al. 1977).

The ligase does not recognize in an absolute manner its cognate tRNA's but has a strategy for choosing the good tRNA's among the usual set of competing tRNA's of the cell. It has a certain general way of interacting with tRNA's and specific ways of rejecting the undesirable tRNA's of the cell. Since there is no need to develop defences against

Institut de Recherche en Biologie Moleculaire, Universite de Paris, Tour 43, 2 Place Jussieu, 75221 Paris Cedex 05, France

the foreign tRNA's of distantly related species, the ligase does often acylate them. Recognition is essentially negative. It is easier to find common features to the tRNA's that
are rejected by a ligase, than to specify the characteristics of a recognition site of all
tRNA's acylated by a same ligase (Yarus et al. 1977). One of the most common ways
of rejecting undesirable substrates is by proofreading, i.e., destruction of the erroneous
products.

II. Repressor-Operator Interactions

Since the repressor has to recognize a very small stretch of DNA, representing less than
one part in 10^5 of the *E. coli* chromosome, it was thought at the beginning that the
repressor would only bind to the operator, or to closely related sequences. Quite to the
contrary, the *Lac* repressor is able to bind to almost any stretch of natural or artificial
DNA like Poly (dA-dT). Poly (dA-dT) or Poly (dT-dT-dG).Poly (dC-dA-dA) (Riggs et
al. 1972). To make the situation worse, a repressor mutant named X86 was found,
presenting a 50-fold increase in affinity for the operator DNA; yet, the bacteria producing it are constitutive, i.e., they behave as though the repressor was absent or inactive,
which is not the case (Pfahl 1976). This is an instance of a rather frequent phenomenon.
By increasing the fit between the recognizing molecule and the recognized substrate,
one increases correlatively the fit between the former and several competitors of the
latter, and the overall specificity decreases. A detailed quantitative treatment of the
binding of the repressor to the various sites of DNA would show that there is an increased probability for the X86 mutant repressor to bind at the nonoperator sites and
thus an increased probability for the operator to be free of repressor (von Hippel et al.
1974). Similarly, it is known that the antibodies which have the greatest affinities for
their antigens are also those which cross-react the most easily with other antigens (Kharush 1978).

III. Ribosome Mutants

The *ram* mutants of *E. coli* contain ribosomes that make highly erroneous proteins. For
instance, in a *ramI* mutant, the β-galactosidase (measured by precipitation with a specific antibody) is only 30% active, implying that more than 70% of the synthesized
chains have abnormal sequences (Rosset and Gorini 1969). The geometric interpretation
is that the codon-anticodon recognition site is distorted in the mutant ribosome, making noncomplementary associations easier, and complementary associations more difficult than in wild type. Thus, one expected the complementary associations between
the amber codon UAG and the strictly complementary anticodon of su_3^+ tRNA^{Tyr} to
be disadvantaged on *ram* ribosomes (Apirion et al. 1969; Gorini 1971). The opposite
was observed. The translation of codons by complementary anticodons (as measured
by levels of nonsense and missense suppression at specific sites) was improved with am-
biguous ribosomes and depressed with the high-fidelity mutants (Rosset and Gorini
1969; Strigini and Gorini 1970).

IV. DNA Polymerase Mutants

DNA polymerases from phage, bacteria, and some lower eukaryotes are able to excise from the growing chain the last nucleotide that had been incorporated (review in Bernardi and Ninio 1978). Excision is particularly efficient when the terminal nucleotide is mismatched. Such DNA polymerases combine two specificities: one at the incorporation level in the primary choice between correct and incorrect substrates and the other at the excision (proofreading) level in the relative efficiencies of elimination of correctly and incorrectly matched nucleotides. As with ribosomes, there are accuracy mutants of DNA polymerases. Bessman et al. (1974) studied several mutant DNA polymerases from phage T4 and obtained with the L141 mutant quite remarkable results. This DNA polymerase was shown to be less specific than wild type at the incroporation level. At the proofreading level, the ratio of the efficiencies of excision toward correct and incorrect bases was also indicating a poorer specificity in L141. With the geometric viewpoint in mind, one would deduce that both incorporation and excision sites of the wild-type DNA polymerase have a better design than the corresponding sites in L141, and that the wild-type polymerase must be more accurate than the mutant. The opposite is true: L141 is an antimutator DNA polymerase, making much less mistakes than wild type.

V. Immunology

According to the lock and key concept, since every antibody matches exactly its substrate, and since there are millions of potential antigens, many of them never experienced before in evolution, the immune system must be able to generate millions of unique antibody sequences . . . (see Ninio 1979).

B. Kinetic Concepts

I. Proofreading

At many stages in the processes of information transfer in the cell, erroneous products are made, then destroyed. Aminoacyl-tRNA ligases can proofread at two stages. Some of the ligases are able to destroy the aminoacyladenylate (Baldwin and Berg 1966) or capable of hydrolyzing the aminoacyl-tRNA (Yarus 1972; Eldred and Schimmel 1972; Bonnet et al. 1972; Hopfield et al. 1976). DNA polymerases from a variety of sources are able to proofread by excision of the terminal nucleotide (Brutlag and Kornberg 1972; Muzyczka et al. 1972; Hershfield 1973; review in Bernardi and Ninio 1978). On the ribosome, after the formation of the peptide bond, the peptidyl-tRNA may fall off and will do so with appreciable chances if the codon-anticodon association is particularly weak (Menninger 1978; Caplan and Menninger 1979). Then, the peptidyl-tRNA will be hydrolyzed by one of the numerous scavenging enzymes of the cell, the peptidyl-tRNA hydrolase (Chapeville et al. 1969). While little is known of the physico-chemical basis of the DNA polymerase proofreading activity, the events on the aminoacyl-tRNA ligase have been investigated in considerable detail (v.d. Haar and Cramer 1976; Fersht and Kaethner 1976; Igloi et al. 1977; Fersht 1977).

II. Kinetic Modulation

Consider that when the correct substrate binds to an enzyme, the two molecules stick together an average time t_c, t_i being the average sticking time characterizing the association of the enzyme with an "incorrect" substrate. Assume that on the average the enzyme takes a time T to perform its catalytic act. If T is very short compared to t_c and t_i, whenever a substrate, correct or incorrect, binds to the enzyme, it will be transformed into product. In this case, there is no discrimination between correct and incorrect substrates, based on the different values of the sticking times t_c and t_i. If, on the contrary, T is not too small compared to t_c and t_i, the smaller the sticking time, the higher the chances that the substrate will be dissociated before the catalytic event has had a chance to occur. Thus, the enzyme will show a preference for the substrate having the larger sticking time. By changing in a totally nonspecific way the value of T, one may move from a situation in which the enzyme does not discriminate between the substrates, to a situation in which the substrates are discriminated according to their sticking times (Ninio 1974). This notion can be generalized. The overall specificity in any reaction depends upon the interplay between the specific and the nonspecific kinetic constants. The L141 DNA polymerase is less specific than wild type at the level of the "specific" kinetic constants, but shows a much more favorable balance between the kinetic constants of polymeraization and excision, thus taking a better advantage of the discriminatory capacities of the individual acts.

The fidelity of translation has been shown to be influenced by factors that push or pull nonspecifically certain steps of protein synthesis. It varies with the concentration of EF-G (Gavrilova et al. 1976) and is very sensitive to the GTP/GDP ratio (Jelenc and Kurland 1979). Correlations between accuracy and speed of ribosomes have been described by several authors (Galas and Branscomb 1976; Zengel et al. 1977; Twilt et al. 1979; Piepersberg et al. 1979). The accuracy of mutant T4 DNA polymerases correlates with the kinetic balance between polymerization and excision (Muzyczka et al. 1972; Bessman et al. 1974).

A new branch of enzymology has developped, dealing with the fidelity of enzyme systems in relation to the mechanism of the reaction (Ninio 1974; Hopfield 1974; Ninio 1975; Kurland et al. 1975; Blomberg 1977; Ninio 1977; Kurland 1978; Galas and Branscomb 1978; Bernardi et al. 1979). All the effects that seemed paradoxical from the geometric viewpoint are easily interpreted with the kinetic concepts.

III. Kinetic Amplification

There are ways to construct highly reliable computers from non reliable components, provided the components are wired intelligently (von Neumann 1956; McCulloch 1960; Winograd and Cowan 1963). Similarly, there are enzyme mechanisms which permit to achieve a higher accuracy than would result from the simple combination of the specificities of the individual steps (Hopfield 1974; Ninio 1975, 1977; Blomberg 1977; readable account in Guéron 1978). These schemes require nonspecific destructive reactions like apparently futile ATPase or GTPase activities. Some "wiring" details of the reaction scheme are given a particular significance. For instance, the enhancement of the specificity of the proofreading activity of a tRNA ligase requires a delayed departure of the acylated tRNA from the enzyme (Ninio 1975), i.e., the acylated tRNA will preferentially leave after a new reaction cycle has started. Whether or not the cell is taking advantage of the kinetic amplifivation possibilities is not known at present, although there are some encouraging results in this direction (Thompson and Stone 1977; Bernardi et al. 1979).

C. Extensions

The kinetic concepts were developped in the analysis of the fidelity of particular enzyme systems, mainly in relation to protein synthesis and DNA replication. Other areas of biology may benefit from the introduction of kinetic thinking. In immunology, attempts have been made to deal with antibody specificity in a quantitative manner (Bell 1974; Inman 1978). With the development of the network theory (Jerne 1976; Hoffmann 1975; Richter 1975), specificity is no more considered at the individual level but is seen to arise as a result of a complex set of interactions within a wide family of molecules. Furthermore, there is now evidence that the IgG-IgM switch does not involve changes in the variable region (the region which recognizes the antigen), but changes in the constant region (Wang et al. 1977). Thus the immune system would refine its response through a kind of kinetic modulation (see also Ninio 1979).

The kinetic concepts may turn out to be essential for an understanding of regulation in eukaryotes. The state of chemical modification (phosphorylation or acetylation) of the proteins that bind to DNA in chromatin seems to determine whether or not transcription will occur (Allfrey et al. 1964; Kleinsmith et al. 1976; Marushije 1976). Yet, if phosphorylases and kinases are taken individually, they do not display a very high degree of specificity. But it may well be that the system taken as a whole and dynamically is very specific. Since the phosphorylase performs a destructive reaction and the kinase degrades ATP, both classes of enzymes may in principle use kinetic amplification mechanisms. Furthermore, the simultaneous presence of two antagonistic series of reactions (addition and removal of phosphate groups) may result in a pattern of phosphorylation that could be much more specific than expected from the study of any individual phosphorylase or kinase. It is worth to examine in this context the turnover of phosphate and acetyl groups on the DNA binding proteins (e.g., as done by Chestier and Yaniv 1979).

References

Allfrey VG, Faulkner R, Mirsky AE (1964) Acetylation and methylation of histones and their possible role in the regulation of RNA synthesis. Proc Natl Acad Sci USA 51: 786–794

Apirion D, Phillips SL, Schlessinger D (1969) Approaches to the genetics of *Escherichia coli* ribosomes. Cold Spring Harb Symp Quant Biol 34: 117–128

Baldwin AN, Berg P (1966) Transfer ribonucleic acid induced hydrolysis of valyladenylate bound to isoleucyl ribonucleic acid synthetase. J Biol Chem 241: 839–845

Bell GI (1974) Model for the binding of multivalent antigen to cells. Nature 248: 430–431

Bernardi F, Ninio J (1978) The accuracy of DNA replication. Biochimie 60: 1083–1095

Bernardi F, Saghi M, Dorizzi M, Ninio J (1979) A new approach to DNA polymerase kinetics. J Mol Biol 129: 93–112

Bessman MJ, Muzyczka N, Goodman MF, Schnaar RL (1974) Studies on the biochemical basis of spontaneous mutation. III. The incorporation of a base and its analogue into DNA by wild-type, mutator and antimutator DNA polymerases. J Mol Biol 88: 409–421

Blomberg C (1977) A kinetic recognition process for tRNA at the ribosome. J Theor Biol 66: 307–325

Bonnet J, Giege R, Ebel J-P (1972) Lack of specificity in the amino-acyl-tRNA synthetase-catalysed deacylation of amino acyl-tRNA. FEBS Lett 27: 139–144

Brutlag D, Kornberg A (1972) Enzymatic synthesis of deoxyribonucleic acid. XXXVI. A proofreading function for the $3' \rightarrow 5'$ exonuclease activity in deoxyribonucleic acid polymerases. J Biol Chem 247: 241–248

Caplan AB, Menninger JR (1979) Tests of the ribosomal editing hypothesis: amino acid starvation differentially enhances the dissociation of peptidyl-tRNAs from the ribosome. J Mol Biol 134: 621–637

Chapeville F, Yot P, Paulin D (1969) Enzymatic hydrolysis of N-acyl-aminoacyl transfer RNAs. Cold Spring Harb Symp Quant Biol 34: 493–498

Chestier A, Yaniv M (1979) Rapid turnover of acetyl groups in the four core histones of simian virus 40 minichromosomes. Proc Natl Acad Sci USA 76: 46–50

Ehrlich P (1900) On immunity with special reference to cell life. Proc Roy Soc B 66: 424–448

Eldred EW, Schimmel PR (1972) Rapid deacylation by isoleucyl transfer ribonucleic acid synthetase of isoleucine specific transfer ribonucleic acid aminoacylated with valine. J Biol Chem 247: 2961–2964

Fersht AR (1974) Catalysis, binding and enzyme-substrate complementary. Proc R Soc London Ser B 187: 397–407

Fersht AR (1977) Enzyme structure and mechanism. Freeman, San Francisco

Fersht AR, Kaethner MM (1976) Enzyme hyperspecificity. Rejection of threonine by the valyl-tRNA synthetase by misacylation and hydrolytic editing. Biochemistry 15: 3342–3346

Fischer E (1894) Einfluß der Configuration auf die Wirkung der Enzyme. Chem Ber 27: 2985–2993

Galas DJ, Branscomb EW (1976) Ribosome slowed by mutation to streptomycin resistance. Nature 262: 617–619

Galas DJ, Branscomb EW (1978) Enzymatic determinants of DNA polymerase accuracy. Theory of coliphage T4 polymerase mechanisms. J Mol Biol 124: 653–687

Gavrilova LP, Kostiashkina OE, Koteliansky VE, Rutkevitch NM, Spirin AS (1976) Factor-free ("non-enzymic") and factor-dependent systems of translation of polyuridylic acid by *Escherichia coli* ribosomes. J Mol Biol 101: 537–552

Giegé R, Kern D, Ebel J-P, Grosjean H, de Henau S, Chantrenne H (1974) Incorrect aminoacylations involving tRNAs or valyl-tRNA synthetase from Bacillus stearothermophilus. Eur J Biochem 45: 351–362

Gorini L (1971) Ribosomal discrimination of tRNAs. Nature New Biol 234: 261–264

Guéron M (1978) Enhanced selectivity of enzymes by kinetic proofreading. Am Sci 66: 202–208

Haar F von der, Cramer F (1976) Hydrolytic action of aminoacyl-tRNA synthetases from baker's yeast: "chemical proofreading" preventing acylation of tRNA^{ile} with misactivated valine. Biochemistry 15: 4131–4138

Haurowitz F (1967) The evolution of selective and instructive theories of antibody formation. Cold Spring Harb Symp Quant Biol 32: 559–567

Hershfield MS (1973) On the role of deoxyribonucleic acid polymerase in determining mutation rates. Characterization of the defect in the T4 deoxyribonucleic acid polymerase caused by the TS L88 mutation. J Biol Chem 248: 1417–1423

Hippel PH von, Rezvin A, Gross CA, Wang AC (1974) Non-sepcific DNA binding of genome regulating proteins as a biological control mechanism: I. The lac operon: equilibrium aspects. Proc Natl Acad Sci USA 71: 4808–4812

Hoffmann GW (1975) A theory of regulation and self-nonself discrimination in an immune network. Eur J Immunol 5: 638–647

Hopfield JJ (1974) Kinetic proofreading: a new mechanism for reducing errors in biosynthetic processes requiring high specificity. Proc Natl Acad Sci USA 71: 4135–4139

Hopfield JJ, Yamane T, Yue V, Coutts SM (1976) Direct experimental evidence for kinetic proofreading in amino acylation of tRNA^{Ile}. Proc Natl Acad Sci USA 73: 1164–1168

Igloi GL, von der Haar F, Cramer F (1977) Hydrolytic action of aminoacyl-tRNA synthetases from baker's yeast: "chemical proofreading" of thr-tRNA^{Val} by valyl-tRNA synthetase studied with modified tRNA^{Val} and amino acid analogues. Biochemistry 16: 1696–1702

Inman JK (1978) The antibody combining region: speculations on the hypothesis of general multispecificity. In: Bell GI, Perelson AJ, Good RA (eds) Theoretical immunology, pp 243:278. Dekker, New York

Jelenc PC, Kurland CG (1979) Nucleoside triphosphate regeneration decreases the frequency of translation errors. Proc Natl Acad Sci USA 76: 3174–3178

Jencks WP (1975) Binding energy, specificity and enzyme catalysis: the Circe effect. Adv Enzymol 43: 219–410

Jerne NK (1976) The immune system: a web of V-domains. The Harvey Lectures, Ser 70, pp 93–110. Academic Press, New York

Kharush F (1978) The affinity of antibody: range, variability and the role of multivalence. In: Litman GW, Good RA (eds) Comprehensive immunology, vol V, immunoglobulins, pp 85–116. Plenum, New York

Kleinsmith LJ, Stein J, Stein G (1976) Dephosphorylation of nonhistone proteins specifically alters the pattern of gene transcription in reconstituted chromatin. Proc Natl Acad Sci USA 73: 1174–1178

Koshland DE Jr (1958) Application of a theory of enzyme specificity to protein synthesis. Proc Natl Acad Sci USA 44: 98–104

Kurland CG (1978) The role of guanine nucleotides in protein biosynthesis. Biophys J 22: 373–392

Kurland CG, Rigler R, Ehrenberg M, Blomberg C (1975) Allosteric mechanism for codon-dependent tRNA selection on ribosomes. Proc Natl Acad Sci USA 72: 4248–4251

Marushije K (1976) Activation of chromatin by acetylation of histone side chains. Proc Natl Acad Sci USA 73: 3937–3941

McCulloch WS (1960) The reliability of biological systems. In: Yovits MC, Cameron S (eds) Self-organizing systems, pp 264–281. Pergamon, New York

Menninger JR (1978) The accumulation as peptidyl-transfer RNA of isoaccepting transfer RNA families in *Escherichia coli* with temperature-sensitive peptidyl-transfer RNA hydrolase. J Biol Chem 253: 6808–6813

Muzyczka N, Poland RL, Bessman MJ (1972) Studies on the biochemical basis of spontaneous mutation. 1. A comparison of the deoxyribonucleic acid polymerases of mutator, antimutator, and wild type strains of bacteriophage T4. J Biol Chem 247: 7116–7122

Neumann J von (1956) Probabilistic logics and the synthesis of reliable organisms from unreliable components. In: Shannon CE, McCarthy J (eds) Automata studies, pp 43–98. Princeton Univ Press, Princeton

Ninio J (1974) A semi-quantitative treatment of missense and nonsense suppression in the *str*A and *ram* ribosomal mutants of *Escherichia coli*. Evaluation of some molecular parameters of translation *in vivo*. J Mol Biol 84: 297–313

Ninio J (1975) Kinetic amplification of enzyme discrimination. Biochimie 57: 587–595

Ninio J (1977) Are further kinetic amplification schemes possible? Biochimie 59: 759–760

Ninio J (1979) Approches moléculaires de l'evolution. Masson, Paris

Pauling L (1948) The nature of forces between large molecules of biological interest. Nature 161: 707–709

Pauling L (1958) The probability of errors in the process of synthesis of protein molecules. In: Arbeiten aus dem Gebiet der Naturstoffchemie (Festschrift Prof Dr Arthur Stoll. Zum Siebzigsten Geburtstag, 8 Januar 1957), pp 597–602. Birkhäuser, Basel

Pauling L, Delbrück M (1940) The nature of the intermolecular forces operative in biological processes. Science 92: 77–79

Pfahl M (1976) Lac repressor-operator interaction. Analysis of the X86 repressor mutant. J Mol Biol 106: 857–869

Piepersberg W, Noseda V, Böck A (1979) Bacterial ribosomes with two ambiguity mutations: effects on translational fidelity, on the response to aminoglycosides and on the rate of ptotein synthesis. Mol Gen Genet 171: 23–34

Pinck M, Yot P, Chapeville F, Duranton HM (1970) Enzymatic binding of valine to the 3' end of TYMV-RNA. Nature 226: 954–956

Richter PH (1975) A network theory of the immune system. Eur J Immunol 5: 350–354

Riggs AD, Lin S, Wells RD (1972) Lac repressor binding to synthetic DNAs of defined nucleotide sequence. Proc Natl Acad Sci USA 69: 761–764

Rosset R, Gorini L (1969) A ribosomal ambiguity mutation. J Mol Biol 39: 95–112

Silberklang M, Prochiantz A, Haenni A-L, Rajbhandary UL (1977) Studies on the sequence of the 3'-terminal region of turnip-yellow-mosaic virus RNA. Eur J Biochem 72: 465–478

Strigini P, Gorini L (1970) Ribosomal mutations affecting efficiency of amber suppression. J Mol Biol 47: 517–530

Thompson RC, Stone PJ (1977) Proofreading of the codon-anticodon interaction on ribosomes. Proc Natl Acad Sci USA 74: 198–202

Twilt JC, Overbeek GP, van Duin J (1979) Translational fidelity and specificity of ribosomes cleaved by cloacin DF13. Eur J Biochem 94: 477–484

Wang AC, Wang IY, Fudenberg HH (1977) Immunoglobulin structure and genetics. Identity between variable regions of a μ and a γ^2 chain. J Biol Chem 252: 7192–7199

Watson JD, Crick FHC (1953) Molecular structure of nucleic acids. Nature 171: 738–740

Winograd S, Cowan JD (1963) Reliable computation in the presence of noise. MIT Press, Cambridge, Mass

Woese CR (1967) The genetic code. The molecular basis for genetic expression. Harper, New York

Wolfenden R (1972) Analog approaches to the structure of the transition state in enzyme reactions. Acc Chem Res 5: 10–18

Yarus M (1972) Phenylalanyl-tRNA synthetase and isoleucyl-tRNA^{phe}: a possible verification mehcanism for aminoacyl-tRNA. Proc Natl Acad Sci USA 69: 1915–1919

Yarus M, Knowlton R, Soll L (1977) Aminoacylation of the ambivalent Su$^+$7 amber suppressor tRNA. In: Vogel HJ (ed) Nucleic acid protein recognition, pp 391–408. Academic Press, New York

Zengel JM, Young R, Dennis PP, Nomura M (1977) Role of ribosomal protein S12 in peptide chain elongation: analysis of pleiotropic, streptomycin-resistant mutants of *Escherichia coli.* J Bacteriol 129: 1320–1329

7. Coupled Oscillator Theory of Enzyme Action

M. D. WILLIAMS[1], and J. L. FOX[2]

A. Introduction

I would like to discuss enzyme-substrate interactions today, although the context of my discussion can easily be expanded to include many levels of biological structure. A more general discussion of this approach to the origin and evolution of enzyme catalysis has been presented earlier (Williams and Fox 1974).

Let me begin by posing several experimental observations which are difficult to explain using current conceptualizations. The first of these is the finding of Riggs that the binding of the Lac repressor protein to a section of DNA proceeds at rates which are faster than diffusion limited (Riggs et al. 1970). The second is a similar story. Neumann has argued that the binding of acetyl choline to the postsynaptic receptors also proceeds at a rate which can be explained by either its being faster than diffusion limited or requiring six identical binding sites per peptide chain (Neumann, pers. comm. 1978; Bernhardt and Neumann 1978).

Lastly, I would like to pose a much different and stimulating observation. Figure 1 shows the amino acid sequence of α-chymotrypsin using the numbering for the chymotrypsinogen A sequence. Activation of chymotrypsin proceeds by deletion of two dipeptide sequences which then leave three chains labeled A, B, and C. The disulfide bond from the C terminal region of chain B nearly bisects chain C. The catalytic sites are primarily located on chain C. The bisected halves of chain C are approximately one-third the length of chain B. This might yield a third overtone relationship between the B and C chains. Biscar has empirically correlated the maxima for laser stimulated light absorption of polypeptides by a phenomenological equation (Biscar 1976; Biscar and Kollias 1973a, b). Use of this equation suggests an absorption maximum for the B chain at roughly 850 nm, hardly a chromophorically active region for a colorless protein. If the hydrolysis of the synthetic substrate, benzoyl-L-tyrosine ethyl ester is conducted at 15°C, an activation by laser stimulation of over two-fold is observed as shown in Fig. 2. The half-height bandwidth is about 1 nm, a very narrow bandwidth.

We believe that by shifting our traditional mode of thinking of enzyme-substrate interactions from the particulate mode so overwhelmingly used by biochemists, biologists, and chemists to the wave representation of matter as utilized by physicists for electrons

1 Department of Pharmacology, Division of Biology and Medicine, Brown University, Providence, Rhode Island 02912, USA
2 Department of Zoology, University of Texas at Austin, Austin, Texas 78712, USA

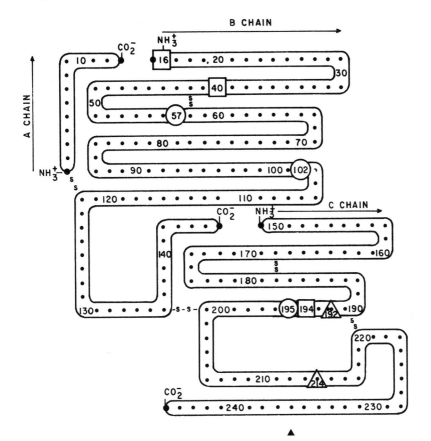

Fig. 1. A schematic representation of the amino acid sequence for bovine α-chymotrypsin (after Hartley). The *numbering* refers to residues in the inactive chymotrypsinogen A precursor molecule. Activation requires the excision of two dipeptides. A disulfide bond from the B chain to the C chain nearly bisects the C chain creating a length roughly one-third that of the B chain. *Triangles* substrate binding residues; *black spots* carboxyl and amino terminal residues; *rings* catalytic residues; and *squares* residues involved in activation

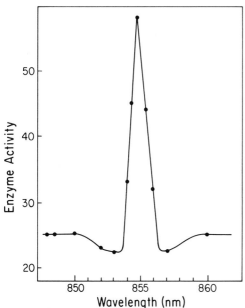

Fig. 2. Activity activation spectrum for hydrolytic reaction of α-chymotrypsin (after Biscar). The experiment was conducted at 15°C and pH 7.1

and oscillating bridges, etc. we can gain new insight into enzyme-substrate interactions. We can pose new experimentally accessible characteristics of the system, but, perhaps more importantly, we can provide a framework for investigation in which an evolutionary continuum can be more easily described. The particulate representation of biological structure is inherently reductionistic. The wave representation of structure is inherently constructionistic and so we can attempt to follow the paths taken by evolution in developing higher levels of complexity. I realize that for scientists who have spent their careers thinking in the particle representation mode a short presentation such as this will be far from convincing, but I will proceed nonetheless.

Fisher's lock and key hypothesis of enzyme function (1894) initiated an unprecedented era of biological thought. Parts of an organism, its enzymes, were providing insight into a technical problem. The mechanism of catalysis is now understood to the extent where its connections to other phenomena are being deduced. The Michaelis-Menten (1913) and Briggs-Haldane (1925) models have related enzyme and ordinary chemical reaction kinetics. The stable intermediate, the enzyme-substrate complex (ES) of Michaelis-Menten theory, serves as a focus of the enzyme catalysis problem. Thermodynamic aspects of enzymic reactions have led to the proposal by Page and Jencks (1971), as well as by Bruice, that enzyme catalysis depends very heavily on changes in entropy during substrate binding. The application of molecular orbital theory in Storm and Koshland's orbital steering model of catalysis (1970) brought the concepts of quantum mechanics to bear on the problem. Discussions by Henderson and Wang (1972), Gold (1971), and Frereira (1973) have amplified this approach to include the concepts of tunneling and orbital symmetry conservation. Koshland (1973) then proposed a possible connection between the Woodward-Hoffman theory (1969) of chemical reactions and enzyme catalysis.

The present authors suggested that the evolution of enzyme catalysis, by introducing time dependence in the catalytic act, was dependent on microscopic and macroscopic phase correlation of enzyme and substrate oscillations and was therefore interpretable with theories such as that of Woodward-Hoffman (Williams and Fox 1974). Enzymes qualify as molecular crystals, since they are at least partly dependent on van der Waals forces for their structure. This has led to a model of enzyme action that originated in the van der Waals force theories of Lifshitz (1955) and Dzyaloshinskii et al. (1961), and the polaron concept of Fano (1956). These concepts, although originally applied to problems in solid-state physics, have provided some especially keen analysis of enzyme catalysis by Kemeny (1974), Kemeny and Goklany (1974), Cope and Straub (1977); Fröhlich (1975), Hoppfield (1974), Bykhovskii (1973), and Biscar (1976). This approach is especially useful in enzyme evolution studies, because it is more macroscopic and can lead to inclusion of the entire structure of the enzyme in considerations of catalysis. Evolution theory of enzyme sequence may thus be someday quantitatively related to the evolution of catalysis. Recent reviews of enzyme catalysis include those of Kirsch (1973), Milligan (1973), and Lienhard (1973).

There are two outstanding features of enzyme catalysis. Enzyme catalyzed reactions proceed at rates up to 10^{14} times as great as those of the uncatalyzed reaction, and the specificity of enzyme catalysis is generally much greater than that of other processes. The evolution of enzyme catalysis presupposes that enzymes at earlier stages of evolution were less specific and slower than those of modern organisms.

B. Enzyme Model

The mechanism of action of enzymes can be comprehensively described by examining an enzyme as a complex system of electromagnetic oscillators. Inherent in this treatment is the quantum mechanical description of the enzyme at a hierachy of structural levels starting at a subatomic level and proceeding through increasingly complex levels of structures to holoenzymes, for this discussion, but on to macromolecular complexes, organelles, cells, organisms and the biosphere.

The quantum mechanical description of matter is based on a treatment of electromagnetic phenomena as wave functions. The extension of this is the development of bounded oscillators with their inherent properties of resonance by energy coupling and, hence, parametric amplification. By analogy to the oriented dipoles of optical absorption and emission processes, these oscillators would be expected to radiate in highly oriented fashions with restricted "beam" diameters. In such a system there would be little loss in amplitude as a function of distance, so that long-range force magnitude would be similar to that at short range.

Let us begin our discussion by examining the simpler system, the substrate, a small molecule generally. A quantum mechanical description of the molecule as a whole is normally obtained by assuming wave functions which are linear combinations of atomic orbitals. Different electron orbitals may be localized in different domains in the molecule. Each wave function will have characteristic modes of oscillation. Coupling of these wave functions may occur, giving rise to new wave functions. Such coupled oscillations depend upon similar frequencies or frequency overtones and proper phase correlations. The magnitudes of such oscillations are a function of the number of coupled wave functions.

Interaction with other oscillators such as water (or other solvent) molecules, may attenuate, enhance, or leave unchanged this propagation. However, in the case of water it is fashionable to discuss ice-like flickering structures of very short time duration. Water is transparent to a wide range of energies, especially in the visible spectrum, but it may also serve as a conductor through tunneling or semiconduction type mechanisms as a medium for propagating the oscillations. This could allow substrate and enzyme to interact and serve to assist in attraction and alignment of the enzyme-substrate complex over greater distances than are commonly discussed.

The requirements of enzyme-substrate complex formation are a coupling of the electromagnetic oscillations of these two components. Thus, the enzyme must possess oscillations which can interact with those of substrate. This interaction may be constructed by oscillations restricted to single or a small set of residues (Leinhard 1973). The sum total of the residues in the "active" binding site must correlate with at least part of the oscillator manifold of the substrate. In most cases, the number of oscillations which may couple between substrate and enzyme will be low and only at short range will coupling be of large magnitude, reflected in turn by significant orientation or binding effects for E-S complex formation. Even in cases where certain types of energy are not diminished by distance, other energies, such as mechanical, will be diminished by distance. If a large region of enzyme (or protein) and substrate can couple energetically, even at longer range, large-scale orientation and binding attraction effects might be observed.

If this were the case, one would expect the binding process to occur at a rate greater than that limited by diffusion. An example of this may be the highly specific binding of the β-lactose regulatory protein to a large region of a DNA double helix, pointed out earlier (Riggs et al. 1970).

Once the substrate approaches a short-range interaction with the enzyme, significant coupling (phasing, etc.) between the oscillations of the enzyme and the substrate may occur. In fact, this coupling will lead to the formation of a new oscillation system with altered center of mass. For substrate approach, binding, and orientation, a multidimensional energy profile can be calculated. Due to the number of oscillators present, a variety of geometric axes will determine the exact shape of this energy profile. The process will seek the lowest energy path, being forced by thermal energy to fluctuate around the electromagnetic energy minimum. These are reflected thermodynamically by the magnitude of the entropy term for the binding.

When the E-S complex forms, a trivial phase change occurs because of the introduction of the new E-S complex species in the cell. The Michaelis-Menton theory draws attention to the importance of the number of particles in the reaction system by emphasizing the union of enzyme and substrate. The theory of superconductors was developed around a notion that in such systems the dominant eigenfunction changes from an intensive variable to an extensive one, such as the number of particles (Penrose and Onsager 1956). The entropy of phase changes in solids fluctuates more and more slowly as the critical point is approached, and the degrees of freedom become successively limited. The formation of the enzyme-substrate complex represents a phase transition. Several authors have produced evidence for superconductivity in enzymes (Cope and Straub 1977; Fröhlich 1975; Hoppfield, 1974).

The alteration of the center of mass would be expected to be large for enzymes requiring a specific addition order of substrate. Alteration of the first binding site could alter the second binding site to promote binding of the second substrate. The specificity of an enzyme, such as prokaryotic transhydrogenase for TPNH as its reductant and DPN+ as its oxidant, may illustrate this point (Kaplan et al. 1952; Fox 1978).

For enzymes which do not possess an obligatory, but rather a random substrate addition order, only a small geometric alteration of the alternative substrate binding site could be tolerated. One would expect the majority of random addition enzymes to possess rather low levels of substrate specificity with a minimal level of oscillator coupling necessary to affect binding. Binding constants would be predicted to be lower than for sequential addition enzymes.

This discussion has not yet touched upon the function of the catalytic site. The binding of substrates is usually discussed as approximating an equilibrium with only a small percentage of E-S complex converting to product. The binding process is considered by us to account also largely for the orientation in enzymic catalysis. Thus, the energetics are described by a series of small enthalpic terms. The binding (and associated orientation) will contribute to the entropy to a much greater extent than will the chemical step.

The orientation of substrates by the binding sites leads to a reduced number of degrees of freedom for the substrates. Geometric changes induced by the substrate binding assist in the orientation of the chemical sites and reactants, as do direct oscillator coupling. The energetics of the chemical reaction are to be largely enthalpic in nature, at this point, with a relatively large energy barrier to be crossed by a small number of processes, as compared with the binding. The greatly restricted degrees of freedom for substrate lower the reaction free energy barrier of the classical transition state.

The reaction itself requires the participation of several clusters of atoms involved in various forms of proton transfer, bond breaking and forming, etc. Each alteration in one location induces alterations in the potential pathway followed by subsequent steps. Minimal geometries expressed on a potential energy diagram define relatively more stable states which may, in some cases, be observed as discrete meta-stable intermediates. The reaction pathway is defined by the maximum overlap of oscillator manifolds for the regions of the reactants and products which underwent modification, and the participating catalytic centers. Oscillations from several parts of the catalytic site combine to yield a manifold that approximates that of the substrates. The greater this similarity, the greater the coupling, the more rapid the reaction. This translates to a requirement for minimizing the symmetry changes, which translates to maximizing selection rules, which are basically the popularly known Woodward-Hoffman Rules. These evolved from theory presented by physicists and physical chemists in the 1920's and 1930's (Silver 1974). In short, the manifolds from substrate and parts of the protein undergo parametric amplification by the phase-coupled resonance of their electromagnetic oscillations. This provides the energy necessary for the chemical step and is generally considered to occur within a single vibration for a simple system. This will require the composite vibration of a "macro" oscillator, which is a linear combination of a number of interactions for an enzyme.

Rate processes are accelerated when resonance conditions apply. Three considerations apply in the collisional interactions of molecules with resonant or near-resonant modes (Rhodes and Szoeke 1972). The colliding molecules must have a minimum kinetic energy to overcome the threshold created by any difference, ΔE, in their near-resonant energy levels. These collisions selectively populate states; a force which is time varying is generated by the interacting molecules' relative motion coupling their internal and translational modes. The effective Fourier component of this force is $E/\hbar$ where E is the resonant energy level. The resonance transfer probability decreases approximately as E^{-x} where $x = \Delta E t_c/\hbar$ and $t_c = a/v$ (t_c is the collision time, a is the range of the molecular force, and v is the relative velocity of the two molecules). In gases t_c is about equal to 10^{-10} s, thus $\Delta E \leqslant 10^{-15}$ ergs, or ~ 0.1 KT. The actual intermolecular force coupling to internal motion is the third consideration. Van der Waals forces dominate with decreasing ΔE. Similar accelerating effects apply for electronic resonances, as for instance in Forster energy transfer.

Enzymes introduce a degree of irreversibility by dissipating the energy from the chemical "transition state" in a variety of ways to prevent decay to reactants, rather than products. In the mechanism called Ping Pong by Cleland (1970), where the first substrate reacts with enzyme completely and its product is released leaving a modified enzyme prior to the second substrate addition, this process has been carried to an extreme. In a sequential type mechanism with ordered and nonordered substrate additions, the dissipation of the energy released by reactions may couple with other parts of the protein to rapidly move from the active site. In a few special cases, the active site may possess components capable of energy depopulation by radiative processes giving rise to ultraweak, or perhaps, high-intensity luminescence. This polarization of reaction direction has been well explored in hysteresis models and experiments (Frieden 1970; Gutfreund 1971).

This energy dissipation serves several functions. By depopulating upper levels of an energy manifold, altered geometries may occur. This could enhance the departure of product. Since product normally closely resembles reactant, this will not be a terribly large effect, but will vary according to the specific example under consideration. The population of higher energy levels of parts of the protein away from the active site will probably temporarily destabilize the protein, but periods of time in which a functioning enzyme is bound in E-S complex will offset this effect by restabilizing the enzyme and, thus, lead to longer average lifetimes.

C. Raman Scattering

Raman spectra give information about molecular structure. When ordinary light sources are employed, Raman spectra in the visible region are very weak and are usually ignored. Occasionally a molecule is encountered that produces a fluorescence spectrum with a lot of sharp detail superimposed on the broad fluorescence peak. These are frequently cases of "hot" fluorescence, also called resonance Raman scattering. When the exciting beam wavelength approaches a conventional fluorescence transition, the Raman effect is enhanced. If the exciting beam includes wavelengths that are multiples of vibrational transitions that are Raman active, then these transitions are found superimposed on the fluorescence emissions. Fluorescence spectra of acetaldehyde (Longin 1960) and chlorophyll-a (Menzel 1974) demonstrate such effects. Resonance Raman spectra have been obtained for many proteins including cytochrome-c oxidase, α-chymotrypsin, hemoglobin, and cytochrome-c. Ordinary, as opposed to resonance, Raman spectra have been obtained for numerous polypeptides, proteins, enzymes, nucleosides, nucleotides, nucleic acids, carbohydrates, steroids, membranes, lipids, and calcified tissue (Parker 1975).

The feature of molecules that is particularly important for observation of the Raman effect is their molecular polarizability or, in the case of crystals, the susceptibility. Polarization fields associated with excited states generate polarization waves. The electromagnetic waves described by Maxwell's equations couple with these polarization waves. These may result from the coupling of a photon and exciton (as in semiconduction), the electron gas of a conductor, a lattice wave, or a wave of spin transitions. The regenerative coupling of polarization waves and electromagnetic fields provides exactly the type of time dependent feedback necessary for evolutional selection.

In the case of biological macromolecules only isolated chains usually have a center of (large scale) symmetry. Molecules which are noncentrosymmetric may exhibit piezoelectric properties. This may even apply to individual chains if their longitudinal symmetry is broken by selective excitation of one side. The result is a polariton in the LO (longitudinal-optical) model. Activation of these modes is related to the intensity of infrared transitions and is Raman forbidden. The ground state (thermal) reaction path is thus coupled with LO phonon generation which is allowed to the same extent that an excited state path of the same symmetry is forbidden. The transition to the excited state is forbidden unless symmetry is broken by the input energy promoting the transition. Polarizability and electric fields including those of enzyme, substrate, cell, and environment interact in a manner that results in their mutual enhancement in a successful reaction. Enzyme catalysis thus employs evolution on a microscopic scale. If a polarization or electric field grows, it promotes its survival by enhancing the (other) field that generated it. Thus fields and polarizations in cells and enzymes interact in successful catalysis.

It has recently been experimentally demonstrated that enzymes, in addition to many other biological materials, exhibit piezoelectric properties (Vasileva-Popova et al. 1975). A theory of piezoelectric enzyme function has been proposed (Caserta and Gerigni 1974). An outstanding feature of LO mode contribution to enzyme function is the development of long range electric fields in the direction of the LO mode. These fields exist because the LO mode develops periodic concentrations of positive and negative charge. These long range fields were investigated by Fröhlich, who has suggested their importance in enzyme catalysis and other biological processes.

D. Discussion

A model of enzyme catalysis has been presented. This model is based upon the coupling of oscillations inherent in the substrate(s) and the enzyme.

The coupling of oscillations leads to higher and higher orders of complexity. The addition of coupled oscillations holds not only for substrates and enzymes but for enzyme complexes, membranes, cellular organelles, cells, etc. There is a natural tendency for simple systems to evolve into more complex systems, because the process of growth of complexity results from the mutually enhanced polarization and field strength obtained from this increase in complexity.

Each individual in a system generates electromagnetic fields. These fields produce polarizations in neighboring individuals. The polarizations, in turn, enhance the original fields. This is a form of feedback mechanism. Feedback mechanisms create stability in a system and provide a means for amplification.

The chemical event catalyzed by an enzyme represents a microscopic form of evolution dependent upon polarizabilities and field strengths. These, in turn, have to be related to the electromagnetic features of the enzyme active site, the holoenzyme, enzyme complexes, membranes, organelles, cells, etc. The result is an increase in the complexity of the cellular environment by the generation of a new molecule.

We may see support for the hypothesis of the increase of complexity by the evolution of oscillator systems in the following examples. Analysis of the amino acid sequence for the flavodoxin from *Clostridium pasteurianum* reveals strong evidence for a basic, repeating pentapeptide sequence (Kobayashi and Fox 1976, 1978) which duplicated 22 times. At a higher level of internal structure, we may examine the case of c-phycocyanin, a photon transferring protein found in the light harvesting step of photosynthesis in blue-green algae. This protein consists of two polypeptide chains with roughly 18,000 Dalton molecular weight. The β-chain possesses two linear tetrapyrrole chromophores while the α-chain only possesses one chromophore (Glazer and Fang 1973). The α-chain transfers its energy to the β-chain (Glazer et al. 1973). Thus, we see one half of the oscillator system per chromophore in the β-chain of what we see in the α-chain, in a manner analogous to that related before for the B and C regions of chymotrypsin.

The harmonic relationships in cells are suggested by examination of a histogram of enzyme-complex molecular weights (see Fig. 3) where it is apparent that multiples and subharmonics of 35,000 daltons are dominant. This idea is not new; it was suggested by Svedberg in 1929, but has been largely discounted in recent years.

The coupled oscillator theory has been primarily applied to a discussion of enzymic catalysis in this paper. However, it may be simply extrapolated to higher levels of organization. This theory suggests a variety of experiments to examine the function and evolution of organisms and their components.

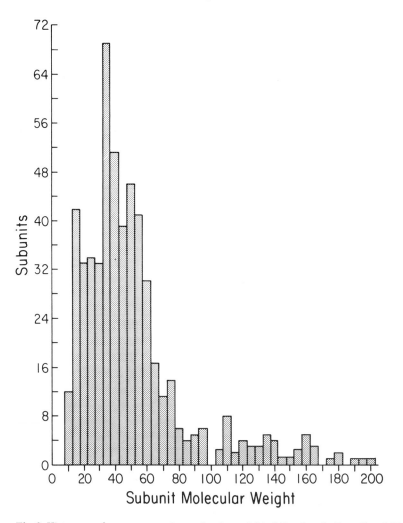

Fig. 3. Histogram of enzyme complex molecular weights (after data in Darnall and Klotz). A fundamental appears at 35,000 daltons with possibly harmonically related peaks at 17,000, 50,000, 75,000, 110,000, 135,000 and 160,000 daltons

Acknowledgments. We wish to thank the N.I.H. for a Training Grant Predoctoral Fellowship (MDW) and for research support (GM 21688) to JLF. We also thank the NSF for research support (GB 43094) and Terry L. Welsher for discussion.

References

Berhardt J, Neumann E (1978) Kinetic analysis of receptor-controlled tracer efflux from sealed membrane fragments. Proc Natl Acad Sci USA 75: 3756–3760

Biscar JP (1976) Photon enzyme activation. Bull Math Biol 38: 29–38

Biscar JP, Kollias N (1973a) The non-fluorescent broad band of poly L-glutamic acid. Phys Lett 44A: 373–374

Biscar JP, Kollias N (1973b) Resolved pseudo-Raman band of pga. Phys Lett 45A: 189–190

Briggs GE, Haldane JBS (1925) A note on the kinetics of enzyme action. Biochem J 19: 338–339

Bruice TC (1976) Some pertinent aspects of mechanism as determined with small molecules. Ann Rev Biochem 45: 331–374

Bykhovskii VK (1973) Transmission of coherence to the electron shell of biological macromolecules and their complexes. Biophysics 18: 199–201

Caserta G, Cervigni T (1974) Piezoelectric theory of enzymic catalysis as inferred from the electro-mechanochemical principles of bioenergetics. Proc Natl Acad Sci USA 71: 4421–4424

Cleland WW (1970) Steady state kinetics. In: Boyer PD (ed) The enzymes, vol II, 3rd ed. Academic Press, New York

Cope FW, Straub KD (1977) Kinetics of the light flash of the living firefly: a supramolecular process. Physiol Chem Phys 8: 343–347

Darnall DW, Klotz IM (1975) Subunit constitution of proteins: a table. Arch Biochem Biophys 166: 651–682

Dewar MJS (1974) Comments on two early papers on the theory of pericycle reactions. In: Orbital symmetry papers. American Chemical Society, Washington

Dyzaloshinskii IE, Pitaevskii LP (1961) The general theory of van der Waals forces. Adv Phys 10: 165–209

Evans MG (1939) The activation energies of reactions involving conjugated systems. Trans Faraday Soc 35: 824–834

Evans MG, Warhurst E (1938) The activation energy of diene association reactions. Trans Faraday Soc 34: 614–624

Fano U (1956) Atomic theory of electromagnetic interactions in dense materials. Phys Rev 103: 1202–1218

Fisher E (1894) Einfluß der Configation auf die Wirkung der Enzyme, II. Berichte 27: 3479–3483

Fox JL (1978) A model for transhydrogenase. J Theor Biol 70: 335–337

Frereira R (1973) Enzymes as orbital symmetry adaptors. J Theor Biol 39: 665–668

Frieden C (1970) Kinetic aspects of regulation of metabolic processes. J Biol Chem 245: 5788–5799

Fröhlich H (1975) The extraordinary dielectric properties of biological materials and the action of enzymes. Proc Natl Acad Sci USA 72: 4211–4215

Glazer AN, Fang S (1973) Chromophore content of blue-green algal phycobiliproteins. J Biol Chem 248: 659–662

Glazer AN, Fang S, Brown DM (1973) Spectroscopic properties of C-phycocyanin and of its α and β subunits. J Biol Chem 248: 5679–5685

Gold HG (1971) Proton tunneling and enzyme catalysis. Acta Biotheor XX: 29–40

Gutfreund H (1971) Transients and relaxation kinetics of enzyme reactions. Ann Rev Biochem 40: 315–344

Hartley B (1964) Amino acid sequence of bovine chymotrypsinogen A. Nature 201: 1284–1287

Henderson R, Wang JH (1972) Catalytic configurations. Annu Rev Biophys Eng 1: 1–25

Hoppfield JJ (1974) Electron transfer between biological molecules by thermally activated tunneling. Proc Natl Acad Sci USA 71: 3640–3644

Kaplan NO, Colowick SP, Neufield EF (1952) Pyridine nucleotide transhydrogenase. J Biol Chem 195: 107–119

Kemeny G (1974) Collective aspects of conformons and the electron transport chain. J Theor Biol 48: 231–241

Kemeny G, Goklany IM (1974) Quantum mechanical model for conformons. J Theor Biol 48: 23–38

Kirsch JF (1973) Mechanism of enzyme action. Annu Rev Biochem 42: 105–234

Kobayshi K, Fox JL (1976) Molecular evolution of flavodoxin. A hypothesis for the genetically coded evolutionary development of a functional protein from a primitive pentapeptide. Viva Origino 5: 1–9

Kobayashi K, Fox JL (1978) The evolution of protein sequences by repetitious gene duplication: clostridial flavodoxin. J Mol Evol 11: 233–243

Koshland DE Jr (1973) The contribution of orientation to the catalytic power of enzymes. In: From theoretical physics to biology (Proc 3rd Int Conf), pp 286–302. Karger, Basel

Leinhard GE (1973) Enzymatic catalysis and transition-state theory. Science 180: 149–154

Lifshitz EM (1955) The theory of molecular attractive forces between solids. Sov Phys JETP 2: 73–83

Longin P (1960) Luminescence en phase vapour et en solution etendue cristallissee a 77°K de quelques aldehydes et cetones eliphatiques. CR Acad Sci Paris 251: 2499–2501

Menzel ER (1974) Vibrationally hot fluorescence in chlorophyll-a. Chem Phys Lett 24: 545–548

Michaelis L, Menten ML (1913) Die Kinetik der Invertwirkung. Biochem Z 49: 333–369

Milligan WO (1973) The Robert A. Welch foundation conference on chemical research XV. Bio-organic chemistry and mechanisms. R.A. Welch Foundation, Houston

Page MI, Jencks WP (1971) Entropic contributions to rate accelerations in enzymic and intramolecular reactions and the chelate effect. Proc Natl Acad Sci USA 68: 1678–1683

Parker F (1975) Biochemical applications of infrared and Raman spectroscopy. Appl Spectrosc 29: 129–147

Penrose O, Onsager L ((1956) Bose-Einstein condensation and liquid helium. Phys Rev 104: 576–596 596

Rhodes CK, Szoeke A (1972) Gaseous lasers: atomic molecular and ionic. In: Arecchi FT, Schulz-Dubois EO (eds) Laser handbook. Elsevier, New York

Riggs AD, Suzuki H, Bourgeois S (1970) *lac* repressor-operator interaction. J Mol Biol 48: 67–83

Silver DM (1974) Hierarchy of symmetry conservation rules governing chemical reaction systems. J Am Chem Soc 96: 5959–5967

Strom DR, Koshland DE (1970) A source for the special catalytic power of enzymes: orbital steering. Proc Natl Acad Sci USA 66: 445–452

Svedberg T (1929) Mass and size of protein molecules. Nature 123: 871

Vassileva-Popova JG, Vassilev NN, Iliev R (1975) A possible transfer of physical information through the conversion of mechanical energy into electrical: a piezoelectric effect. In: Vassileva-Popova JG (ed) Proteins, physical and chemical bases of biological information transfer, pp 137–146. Plenum Press, New York

Williams MD, Fox JL (1974) The origin and evolution of enzyme catalysis. In: Dose K, Fox SW, Deborin GA, Pavloskaya TE (eds) The origin of life and evolutionary biochemistry, pp 461–468. Plenum Press, New York

Woodward RB, Hoffman R (1969) The conservation of orbital symmetry. Angew Chem Int Ed Engl 8: 781–853

8. Stereochemical Aspects of Chain Lengthening and Cyclization Processes in Terpenoid Biosynthesis

O. CORI[1], L. CHAYET[1], M. DE LA FUENTE[1], L. A. FERNANDEZ[1],
U. HASHAGEN[1], L. PEREZ[1], G. PORTILLA[2], C. ROJAS[1], G. SANCHEZ[1],
and M. V. VIAL[1]

Abbreviations

IPP	Isopentenyl pyrophosphate (C_5)
DMAPP	Dimethylallylpyrophosphate (C_5)
GPP	Geranylpyrophosphate (E-C_{10})
NPP	Nerylpyrophosphate (Z-C_{10})
2,6 di E FPP	2E,6E Farnesylpyrophosphate (C_{15})
2Z,6E FPP	2Z,6E Farnesylpyrophosphate (C_{15})
DTNB	5,5' dithio bis-(nitrobenzoic) acid
TLC	Thin layer chromatography
GLC	Gas chromatography.

The terms E and Z are used instead of "trans" and "cis". The prochirality of protons is referred to IPP and not to mevalonic acid, which changes the precedence rule for the ligands.

A. Introduction

Whether it is due to the influence of Professor Fritz Lipmann or to the essence of Biochemistry, much of the work of his former students has been concerned with phosphorylated compounds. Perhaps it would be better to conclude that Dr. Lipmann and the essence of Biochemistry are one and the same. . . .

Terpenoids or isoprenoids, which have been for more than a century a favorite topic of organic chemists, do not escape this general pattern. They are more or less complex molecules derived from isoprene units (C_5H_8). Very often end products are chiral, although their pyrophosphorylated immediate precursors are not, and this involves stereochemical problems in their biosynthesis.

The chain length of terpenoid components of plant tissues ranges from one to several thousands of isoprene units (Hemming 1977). The conformation around double bonds may be E as in squalene, or carotenes or Z as in nerol or rubber. In polyprenols both conformations are present in the same molecule.

1 Universidad de Chile, Faculty of Chemical Sciences, Casilla 233, Santiago 1, Chile
2 Universidad de Chile, Dep. of Chemistry, Faculty of Sciences, Casilla 233, Santiago 1, Chile

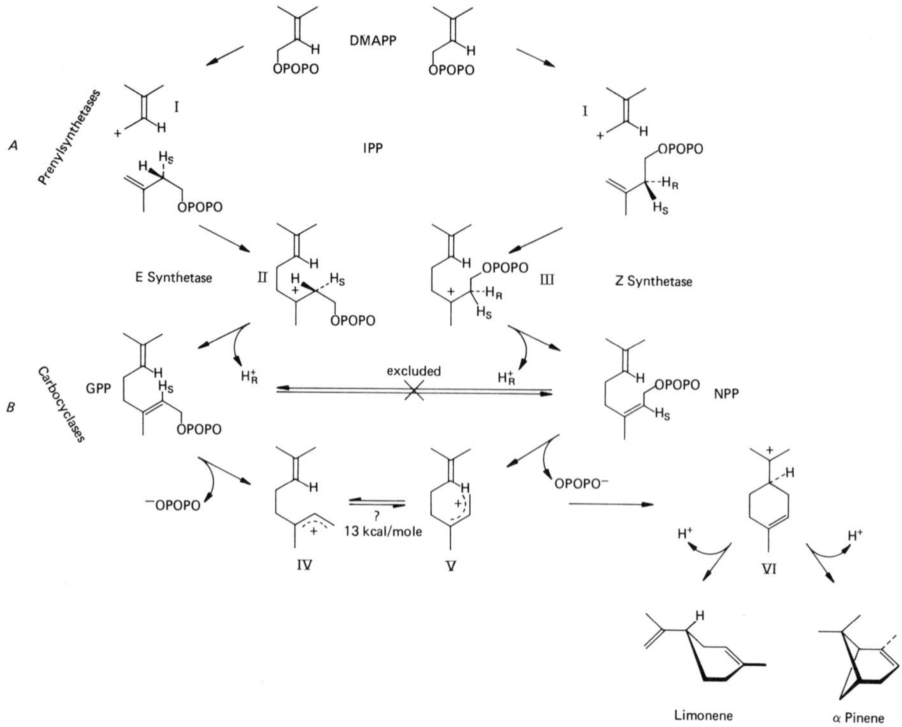

Fig. 1. Probable reaction mechanisms of stereospecific prenylsynthetases and of carbocyclases

The biosynthesis of terpenoids in animal tissues is confined to a few triterpenoids and dolichol and the E conformation is prevalent.

In plant tissues, on the other hand, the number of isoprenoids of varying structure, conformation and configuration ranges in the thousands. Thus it seems at this moment more challenging to explore biosynthesis of terpenoids by plant enzymes in spite of the inherent difficulties involved (Croteau 1975).

In the majority of cases, chain length and conformation of end products are defined by the distribution and characteristics of prenylsynthetases. These enzymes catalyze the addition of the olefinic double bond of isopentenylpyrophosphate (IPP) to a carbocation derived from an allylic pyrophosphate of 5, 10, 15 or more carbon atoms, such as dimethylallylpyrophosphate (DMAPP), geranylpyrophosphate (GPP), farnesyl-pyrophosphate (FPP), etc. (Fig. 1A).

The molecular geometry of these allylic pyrophosphates presents alternatives of E or Z conformation around the double bonds.

These diastereomers could arise either from stereoselective biosynthesis or through an E − Z isomerization process. Furthermore, the prochirality of C_2 in IPP as well as in the intermediate carbocation resulting from condensation (Fig. 1A, II or III) poses the question as to which of both protons is eliminated in the condensation process. The elegant experiments of Popják and Cornforth (1966) have established that in several cases the formation of an E product implies the elimination of the pro-R proton of IPP and that the proton of the opposite prochirality was eliminated when Z double bonds

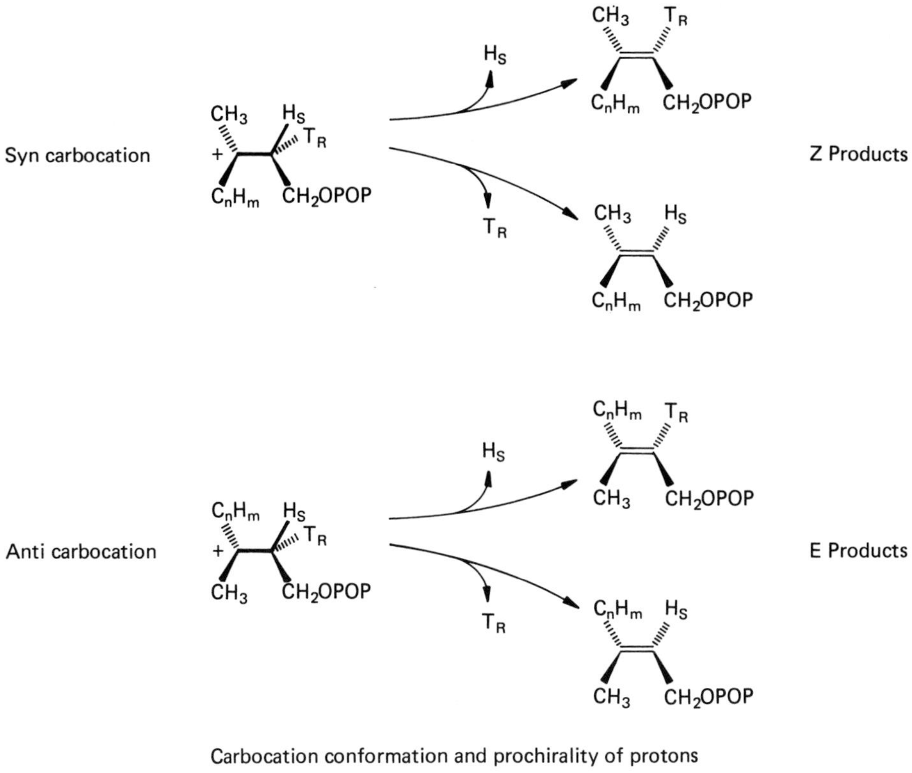

Carbocation conformation and prochirality of protons

Fig. 2. Conformation of carbocation, chirality of the eliminated proton and product conformation

were formed. However, it may be visualized (Fig. 2) that the prochirality of proton elimination is not chemically coupled to a given conformation of the products (Bunton and Cori 1978) and that this relationship is established only through the stereospecificity of the prenylsynthetases involved, which may exhibit species differences.

The distance or proximity between certain carbon atoms in a precursor molecule is a determinant condition for further transformation of an allylic pyrophosphate into cyclic end products. The smaller the ring formed, the more stringent are the steric requirements. Thus it was anticipated from the analysis of models (Valenzuela et al. 1966) that nerylpyrophosphate (NPP), the Z isomer and not GPP should be the obligatory precursor of cyclic monoterpenes such as limonene or the pinenes (Fig. 1B).

Our research group in Santiago has been interested for several years in two aspects of terpenoid biosynthesis in plants: The conformation of the immediate precursors of cyclic monoterpene hydrocarbons in the carbocyclase reaction and the mechanism of formation of E and Z allylic pyrophosphates by prenylsynthetases. We have attempted to approach these two stereochemical problems using partially purified enzymes from Pinus and Citrus and comparing them with nonenzymic solvolysis reactions of C_{10} allylic pyrophosphates mediated by bivalent metals.

B. Stereochemistry of Chain Lengthening in Plant Tissues

Incubation of crude extracts either from *Pinus radiata* seedlings or from the flavedo of *Citrus sinensis* with precursors such as $2\text{-}^{14}C$ mevalonic acid or $1\text{-}^{14}C$ IPP results in the formation of four pyrophosphorylated allylic products: GPP, NPP, 2,6 di E FPP and 2Z,6E FPP. They were identified by GLC or TLC of the corresponding alcohols or of their trimethylsylyl ethers after enzymic hydrolysis with potato apyrase (Del Campo et al. 1977) plus phosphomonoesterase from *E. coli*. It becomes thus evident that these plant tissues generate both E and Z double bonds in the chain lengthening process. For convenience, we will refer to these two activities as "E synthetase" and "Z synthetase".

It has been assumed that the Z isomers of allylic pyrophosphates could be formed from the E compounds by an isomerase, but evidence in the literature for the isomerization of allylic pyrophosphates is indirect (Shine and Loomis 1974; Evans and Hanson 1976). As opposed to this assumption, a crude extract from orange flavedo, which forms both E and Z C_{10} or C_{15} allylic pyrophosphates, does not interconvert them (Fig. 3). The method used to exclude an E-Z isomerization of pyrophosphates would have detected an interconversion of the order of 0.5%.

A redox isomerization system which requires NAD^+ has been described in extracts from fresh flavedo of *Citrus sinensis* (Chayet et al. 1973). This enzyme system transforms the condensation products of $1\text{-}^{14}C\text{-}IPP$ plus GPP into 2,6 di E and 2Z,6E farnesols, although it does not interconvert free alcohols. Intermediate aldehydes have been identified, and the reaction sequence has been shown to be

$$2,6 \text{ di E FPP} \xrightarrow{H_2O} 2,6 \text{ di E farnesol} \xrightarrow{NAD^+} 2,6 \text{ di E farnesal} \longrightarrow 2Z, 6E \text{ farnesal}$$

$$\xrightarrow{NADH} 2Z, 6E \text{ farnesol}$$

Trapping aldehydes with 8 mM aniline blocks this E-Z interconversion. Prenylsynthetase from an acetone powder of orange flavedo forms E and Z FPP both in the absence and in the presence of aniline. This excludes the redox isomerization pathway which may be important in the whole plant, where there is no shortage of ATP. The isomerization of prenylpyrophosphates through this route would require a dephosphorylation of the pyrophosphates and two subsequent phosphorylations by not well characterized prenylkinases (Madyastha and Loomis 1969) with the inherent expenditure of ATP.

The stereochemistry of proton elimination in the condensation process has been established by the use of doubly labeled mevalonic acid or IPP. Prenylsynthetases from *Citrus* or *Pinus* eliminate tritium from the 2-pro-R position of IPP independently of the conformation of the double bond formed (Jedlicki et al. 1972). This agrees with the results of other research groups in *Rosa* (Banthorpe et al. 1972), in *Pinus* (Banthorpe and Le Patourel 1972), and in *Persea* (Robinson and Ryback 1969). It may be concluded that product conformation and the configuration of the proton eliminated by prenylsynthetases are linked through the configuration of the active site of the enzyme and not through a chemical process (Fig. 2).

Crude extracts from acetone powders from the flavedo of *Citrus sinensis* form 2,6 di E FPP from $1\text{-}^{14}C$ IPP plus GPP at a rate of more than ten times the rate of forma-

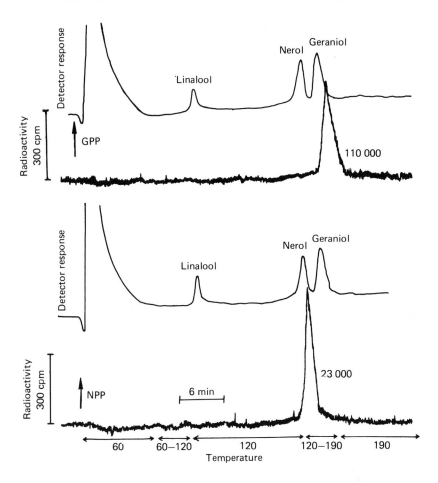

Fig. 3. *Absence of isomerization of GPP or NPP by an extract from fresh flavedo or Citrus sinensis.* 40 μM 2-^{14}C GPP (120,000 cpm) or NPP (56,000 cpm) were incubated for 2 h at 30°C with an extract from orange flavedo (0.8 mg/ml of protein), 2 mM 2-mercaptoethanol, 30 mM KF, 2 mM $MgCl_2$, 2 mM $MnCl_2$ in 33 mM TRIS-HCl buffer pH 7.4. The resulting alcohols and hydrocarbons were extracted with hexane and the remaining pyrophosphorylated substrates were hydrolyzed for 2 h with potato apyrase (del Campo et al. 1977) plus phosphomonoesterase from *E. coli.* The alcohols thus resulting from the hydrolysis of NPP or GPP were extracted with hexane and analyzed by GLC on a 2% polyethylene glycol adipate column at 110°C with added carrier alcohols. Radioactivity was monitored by means of a gas-phase Geiger counter attached to the gas chromatograph. The *upper tracings* are the carrier peaks, the *lower ones* the monitored radioactivity. The *figures* indicate the total radioactivity found in the hexane extract

tion of 2Z, 6E FPP. After a 40-fold purification of the enzyme through ammonium sulfate precipitation, intervent dilution chromatography, and ion exchange chromatography, the E/Z ratio approaches 1.0. Heating this enzyme preparation in the presence of GPP-Mg completely inactivates the Z synthetase while the E activity is fully protected (Fig. 4). GPP or Mg^{2+} do not protect separately the E or the Z synthetase, 2,6E FPP has comparable effects to GPP, whereas Mg-dimethylallylpyrophosphate is almost ineffec-

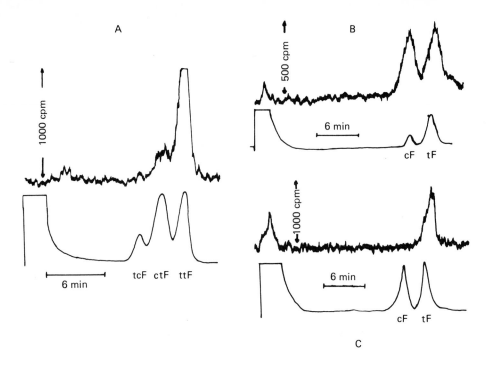

Fig. 4. *Dissociation of E and Z prenylsynthetase activity through purification procedures or heat inactivation.* **A** Formation of 2,6 di E FPP (ttF) and 2Z,6E FPP (ctF) by an extract from an acetone powder of *C. sinensis* flavedo. **B** Products formed by a purified prenylsynthetase preparation. **C** Products formed by the same purified preparation after heating it for 10 min at 60°C in 100 mM KHPO$_4$-buffer pH 7.4 under the following conditions: Protein concentration 0.08 mg/ml; 7.5 mM MgCl$_2$ and 19 μM GPP. The enzyme was assayed with 1-^{14}C IPP plus GPP and the resulting products were analyzed as described in Fig. 3. *Upper tracings:* Radioactivity, *lower tracings:* Carrier peaks

tive. Aging for 6 days at 4°C also causes inactivation of the Z and not of the E synthetase activity.

The effect of bivalent metals on the synthesis of the E or Z diastereomers is different. A partially purified preparation of *C. sinensis* prenylsynthetase forms both 2,6 di E and 2Z, 6E FPP in a 4:1 ratio in the presence of 10 mM Mg^{2+}. The ratio is the same in the presence of Mn^{2+}. In the presence of Co^{2+} total prenylsynthetase activity is 25% higher than with Mg^{2+} and it is 90% lower in the presence of Ni^{2+}, but in the last two cases the formation of 2Z, 6E FPP is completely abolished.

Another observation pointing in the same direction is that some acetone powder preparations are devoid of Z synthetase activity. All this furnishes preliminary evidence for the existence of two stereospecific C$_{15}$ prenylsynthetases in *C. sinensis*.

E prenylsynthetase from hog liver and pumpkin is specific for the E conformation of the allylic substrate (Nishino et al. 1973). The enzymes from *Pinus* and *Citrus* are also stereospecific: NPP, although a product of prenylsynthetase can not act as an IPP acceptor (Jacob et al. 1972) and the C$_2$-C$_3$ satured compound citronellylpyrophosphate is not a substrate but inhibits prenylsynthetase (Table 1).

Table 1. Effects of substrates and inhibitors on the rate of inactivation of E-prenylsynthetase from *Citrus sinensis* by DTNB

Ligand (10 μM)	$\dfrac{I_{50}{}^{a}}{K_m}$	No metal	k (sec^{-1} M^{-1}) 10 mM Mg^{2+}	10 mM Mn^{2+}
Geranylpyrophosphate (K_m = 0.7 μM)		22.0 ± 2.05	2.7 ± 0.25	
Dimethylallylpyrophosphate		19.8 ± 1.6	4.5 ± 0.6	
Citronellylpyrophosphate	11	16.0 ± 1.5	4.5 ± 0.8	
Farnesylpyrophosphate	10.0	10.0 ± 1.0	2.8 ± 0.6	
Nerylpyrophosphate	25	10.0 ± 0.9	6.3 ± 0.5	
No addition		10.6 ± 1.0	7.0 ± 0.4	11.0 ± 1.1
Isopentenylpyrophosphate (K_m = 4.9 μM)		28.6 ± 1.6	9.8 ± 1.1	
Geranylmonophosphate	50	27.6 ± 2.3	9.4 ± 1.6	
Nerylmonophosphate	70	10.3 ± 0.9	13.5 ± 1.5	
Inorganic pyrophosphate	250	8.6 ± 0.7	11.5 ± 1.1	

0.5–1.5 mg of prenylsynthetase were incubated at 0°C for varying periods of time with 0.8–1.0 mM of DTNB in 100 mM K$_2$HPO$_4$ buffer at pH 7.4. Inactivation was stopped by dilution and the residual enzyme was assayed by the formation of acid labile prenols (Beytia et al. 1969) from 1-^{14}C IPP plus unlabeled GPP

[a] The term I_{50} is the concentration of inhibitor necessary to reduce the reaction rate by 50% at a given substrate concentration. It will be used throughout this communication as a measure of inhibitor effectiveness, since it equals K_i for a noncompetitive inhibitor and $K_i (1 + S/K_m)$ for a competitive inhibitor. For comparison purposes, the ratio $I_{50}K_m$ will be used

Prenylsynthetases exhibit an absolute requirement for bivalent metals. The hog liver enzyme requires the presence of substrate for the binding of metals (King and Rilling 1977).

Inactivation by concentrations from 4 μM to 1 mM of the sulfhydryl group reagent DTNB of an E prenylsynthetase preparation from *C. sinensis* and protection experiments with substrates, inhibitors, and metals agree with this: Metal complexes of allylic substrates or inhibitors generally protect the enzyme. whereas the same compounds in the absence of Mg do not affect or tend to increase the rate of inactivation (Table 1). It may be thus concluded that the free ligands are not substrates, but they interact with the enzyme, changing its conformation and exposing some essential SH groups to DTNB, whereas Mg^{2+} bound ligands cause the opposite conformational change, which results in a lower rate of inactivation. Protection from heat inactivation or from aging by GPP-Mg or FPP-Mg and not by the uncomplexed ligands also indicates that substrate or product and metal-substrate or metal-product bind in a different way. Experiments are in progress to test the effect of DTNB on product specificity of prenylsynthetases.

C. Biosynthesis of Cyclic Monoterpene Hydrocarbons

Inspection of molecular models shows that the distance between C_1 and C_6 of the E compound GPP, previously assumed to be the precursor of these terpenoids, is too large to allow for the formation of a six carbon ring (Fig. 1B).

The formation of the Z diastereomer, NPP from mevalonic acid (Beytía et al. 1968) and its incorporation into cyclic monoterpenes by crude extracts from *P. radiata* seedlings confirmed its biosynthetic role (Cori 1969). The E diastereomer GPP is not a substrate for the Pinus enzyme, nor is it for the synthesis of α terpineol by enzymes from *Mentha piperita* (Croteau et al. 1973). Both GPP and NPP are substrates for γ terpinene synthesis by enzymes from *Salvia officinalis* (Poulose and Croteau 1978).

Partially purified enzymes from the flavedo of *Citrus limonum* form α pinene, β pinene, sabinene, and limonene from NPP or GPP with practically the same values of V_{max}/K_m and with an absolute requirement for Mn^{2++} although Co^{2+} can replace it partially (Chayet et al. 1977). Since isomerization of GPP to NPP has been carefully ruled out in our system (Chayet et al. 1977), we assume that either GPP or NPP may be bound by carbocyclase eliminating pyrophosphate and that the resulting carbocation IV (Fig. 1B) may rotate to take the *syn* conformation (Fig. 1B, V). This would be followed by cyclization and regiospecific proton elimination to form the final cyclic hydrocarbons (Fig. 1B).

It may be objected that this anti-syn conformational change is limited, since nonenzymic acid solvolysis produces 5 to 70 times more cyclic products from NPP than from GPP (Valenzuela and Cori 1967; Bunton et al. 1972), and that there is a rotation barrier of 12–14 kcal/mol for allylic carbocations (Deno et al. 1970; Bollinger et al. 1970; Allinger and Siefert 1975). The activation energy values for the carbocyclase reaction at pH 7.0 and in the presence of 3 mM Mn^{2+} are 12.6 ± 2 kcal/mol for NPP, which does not require rotation, and 15.8 ± 2 kcal/mol for GPP. Since rate constants of hydrocarbon formation from GPP and NPP are essentially the same, it may be concluded that factors different from rotation barriers of the intermediate allylic carbocations are the rate limiting steps.

An alternative hypothesis for cyclic hydrocarbon formation from both the E and the Z substrates would be the participation of a common intermediate. The tertiary ester linalylpyrophosphate (Fig. 5) would be a stereochemically plausible intermediate (Potty et al. 1970; George-Nascimento and Cori 1971; Bunton and Cori 1978) since the migration of P-O-P from C_1 to C_3 would occur perpendiculary to the double bond. However, there is no direct evidence of the existence or formation of linalylpyrophosphate. Linalool is formed from mevalonic acid or from isopentenylpyrophosphate (George-Nascimento and Cori 1971), but this is probably an artifact, due to the nonenzymic solvolysis of allylic pyrophosphates mediated by Mn^{2+} ions (George-Nascimento et al. 1971). When tested as a substrate for cineole formation by *Salvia* enzymes (Croteau and Karp 1977), it proved to be a very poor substrate, although a 0.13% incorporation in 3 days into monoterpenes of the peel of *Citrus hassaku,* as compared with 0.007 and 0.004 for NPP and GPP, has been reported (Suga et al. 1977). Our preliminary results with a crude preparation of chemically synthetized 1-^{3}H linalylpyrophosphate suggest that it is not a substrate for Citrus carbocyclase.

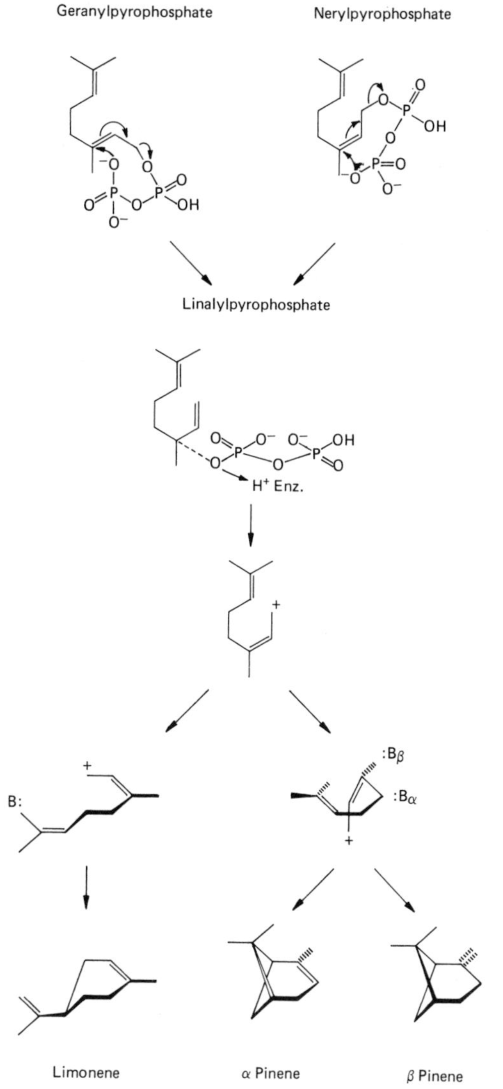

Geranylpyrophosphate   Nerylpyrophosphate

Fig. 5. Linalyl pyrophosphate as a possible intermediate in cyclic terpene hydrocarbon formation

Linalylpyrophosphate

H$^+$ Enz.

:Bβ

:Bα

B:

Limonene α Pinene β Pinene

In order to test a possible binding of a linalyl ester to carbocyclase, we tried to inhibit the enzyme with linalylmonophosphate, taking into account the fact that geranyl and neryl monophosphate, which are not substrates, inhibit the enzyme (Table 2). Linalylmonophosphate has no effect at concentrations below 110 μM, and it exhibits the highest I_{50}/K_m ratio of all the inhibitors tested. Thus K_i may be estimated between 100 and 430 μM, depending whether the inhibition is assumed to be competitive or noncompetitive (see Table 1). The satured analog, citronellylpyrophosphate is, on the other hand, the best inhibitor tested.

Table 2. Effect of inhibitors on the synthesis of terpene hydrocarbons by carbocyclase from *Citrus limonum*

Inhibitors	$\dfrac{I_{50}}{K_m}$	
	With NPP 3.9 μM (K_m = 1.3 μM)	With GPP 7.5 μM (K_m = 2.5 μM)
Citronellylpyrophosphate	4.6	7.6
Geranylmonophosphate	25	13
Citronellylmonophosphate	58	40
Nerylmonophosphate	138	48
Inorganic pyrophosphate	100	122
Linalylmonophosphate	260	172

Citrus carboxyclase has very stringent requirements for the polar moiety of the substrate: Monophosphates are not substrates, although they may bind as inhibitors. Structural requirements for the organic moiety are less stringent, since the enzyme binds E or Z substrates or inhibitors as well as satured analogs. Steric factors may hinder the binding of a tertiary ester, but this point requires further research.

The results may also be interpreted in terms of two carbocyclases, specific for either NPP or GPP.

The latter assumption is supported by several findings. Some aged enzyme preparations lose their ability to utilize NPP. This hints at the existence of two enzymes and is one more argument against the E-Z isomerization hypothesis: A system forming cyclic hydrocarbons sequentially by E-Z isomerization of GPP followed by cyclization of NPP would rather lose its ability to utilize GPP but not NPP. Actual experiment is showing exactly the opposite situation.

The formation of three or four different hydrocarbons by our lemon carbocyclase may be probably due to the presence of several carbocyclases. On the one hand, many of more recent preparations do not form sabinene. Furthermore, Fig. 6 shows that through purification or aging a carbocyclase preparation may change its product pattern, either increasing limonene formation or losing its ability to form α pinene.

Hydrocarbon formation by carbocyclase requires the presence of Mn^{2+}. Other ions such as Mg^{2+}, Cu^{2+} or Zn^{2+} cannot replace it, but Co^{2+} has 16% of the effectivity of Mn^{2+} with GPP and 33% with NPP. The optimum Mn^{2+} concentration is 3 mM; under these circumstances it exceeds the substrate concentration by a factor of 23 or more and practically all substrate would be in the form of Mn complex. The requirement of this high metal/substrate ratio suggests that it may have a further function, such as inducing the adequate conformation of substrate and enzyme to allow cyclization to occur. This is analogous to the mechanism proposed for Mg in prenylsynthetase from avian liver (King and Rilling 1977).

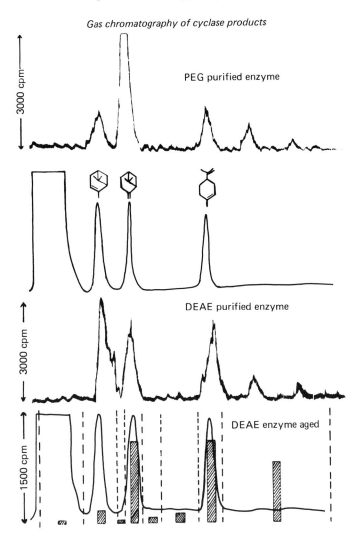

Fig. 6. *Hydrocarbon formation from 1-³H GPP by different preparations of carbocyclase from Citrus limonum.* The enzyme preparation (0.2–0.4 mg/ml) was incubated in 100 mM TRIS buffer as described (Chayet et al. 1977). The resulting radioactive hydrocarbons were added adequate carriers and they were identified by GLC as described. Radioactivity was either monitored by means of an attached gas-phase Geiger counter (*First* and *third tracings*) or by collection of the effluent in cooled U tubes and subsequent scintillation counting (*Hatched bars under bottom tracings*)

D. Nonenzymic Reactions of Allylic Pyrophosphates

The first evidence of stereochemical selectivity in the cyclization of allylic compounds was obtained in an acid medium (Zeitschel 1906). This was also our basis for postulating NPP as a precursor of cyclic monoterpenes (Valenzuela and Cori 1967). Time course

Table 3. Solvolysis products of NPP and GPP in the presence of Mn^{2+}

	from NPP	from GPP
Linalool	17	73
αterpineol	68	6
Nerol	8.5	0
Geraniol	0	14
Limonene	5	2
Myrcene	1	3
Unidentified	0.5	2
Cyclic/non ciclic products ratio	2.8	0.08

$1-^3H$ GPP or 1^3H NPP were prepared by reduction of the aldehydes with $NaB-^3H_4$ and subsequent phosphorylation. 100 μM substrates were incubated in the presence of (5 mM) Mn^{2+} in 0.1 M N-tris (Hydroxymethyl) methyl 2-amino-ethane sulfonate buffer (TES) pH 7.0 for varying periods of time. Reaction was stopped by cooling to $0°C$, extracting the products with hexane and counting the total radioactivity in the hexane phase by conventional beta scintillation spectrometry. Individual products were identified by GLC or TLC and comparison with authentic standards

curves point to linalool as an intermediate in acid catalyzed cyclization of geraniol (Baxter et al. 1978). Equilibrium studies in a medium of low nucleophyllicity such as 85% phosphoric acid forms 72% of cyclic products from geraniol, 91% from nerol, and 79% from linalool (McCormick and Barton 1978), although this may reflect thermodynamic rather than kinetic control.

It was mentioned above that carbocyclase requires Mn^{2+}. This, and the observation that this ion promotes the splitting of sesquiterpenyl pyrophosphates (George-Nascimento et al. 1971), led us to explore its effect on NPP and GPP. These substrates are transformed nonenzymically into rearranged tertiary allylic alcohols and terpene hydrocarbons in the presence of 3 mM Mn^{2+} at pH 7.0, ionic strength of 250 mM, and temperatures between 20° and 50°C (Table 3). The overall pattern shows a predominant formation of cyclic products in the solvolysis of NPP and noncyclic for GPP. The ratio of cyclic to noncyclic products formed from NPP is the same in acid and in Mn^{2+}-mediated solvolysis (Valenzuela and Cori 1967). For GPP, on the other hand, this ratio is of 0.02 in acid solvolysis and four times higher in the Mn^{2+} mediated reaction (Table 3). This could point to a difference between hydronium and manganese ions in the occurrence of conformational changes of carbocation IV (Fig. 1B), as we assume to be the case in the reaction catalyzed by carbocyclase. The fact that Mn^{2+} is also more effective than Cu^{2+}, Zn^{2+} or Mg^{2+} in hydrocarbon formation from both substrates further contributes to the validity of this analogy.

The overall energy of activation of these metal-mediated solvolyses is of the order of 30 kcal/mol for both GPP and NPP. The activation energy for total hydrocarbon formation is about 2 kcal per mol higher than for alcohols, which is in agreement with the differences observed between elimination and substitution in carbocation-mediated reactions (Ingold 1953).

One may visualize chain lengthening by prenylsynthetases and cyclization by carbocyclases as mechanistically similar processes:

An allylic carbocation is generated by pyrophosphate elimination assisted by metal ions. It adds to an olefinic double bond, intramolecularly in the case of carbocyclases

(Fig. 1B), or to a second molecule (IPP) in prenylsynthetases (Fig. 1A). In both cases the reaction ends with the elimination of a proton from a carbocation (II, III, IV or V) to give the stable products. Their stereochemistry is defined by the molecular geometry of the enzymes and not by any foreseeable coupling between product conformation and chirality of proton elimination.

E. Conclusions

We know a good deal more about the steric requirements of prenylsynthetases than those of carbocyclases but most of our knowledge stems from the biochemistry of animal tissues. Since carbocyclases are not found in animals, and Z prenylsynthetases are probably far less abundant, we think that we must be prepared to face the difficulties encountered in the study of plant enzymes if we want to obtain a coherent picture of these two variants of addition mechanisms.

Acknowledgments. This work was supported by grant B 005-781 from the Servicio de Desarrollo Cientifico, Creacion Artistica y Cooperacion Internacional, Universidad de Chile, by grant 79-19 from PNUD/UNESCO, by Fundacion H. Otero V., and by a Cooperative Program sponsored by CONICYT (Chile) and N.S.F. (U.S.A.). The latter allowed the participation of Professor C.A. Bunton, from the University of California, Santa Barbara.

Mr. Gonzalo Perez generously supplied Citrus fruits from Huertos de Betania, Mallarauco.

References

Allinger NL, Siefert JH (1975) Organic quantum chemistry XXXIII. Electronic spectra and rotation barriers of vinyl borane, allyl cations and related compounds. J Am Chem Soc 97: 752–760

Banthorpe DV, Le Patourel NJ (1972) The biosynthesis of (+) α pinene in *Pinus* species. Biochem J 130: 1055–1061

Banthorpe DV, Le Patourel NJ, Francis MJO (1972) Biosyntehsis of geraniol and nerol and their β-D-glucosides in *Pelargonium graveolens* and *Rosa dilecta.* Biochem J 130: 1045–1054

Baxter RL, Laurie WA, MacHale D (1978) Transformation of monoterpenoids in aqueous acid. The reaction of linalool, geraniol, nerol and their acetates in aqueous citric acid. Tetrahedron 34: 2195–2199

Beytía E, Valenzuela P, Cori O (1968) Terpene biosynthesis: Formation of nerol, geraniol and other prenols by an enzyme system from *Pinus radiata.* Arch Biochem Biophys 129: 346–356

Bollinger JM, Brinich JM, Olah GA (1970) Stable carbonium ions XCVI. Propadienylhalonium ions and 2-haloallyl cations. J Am Chem Soc 92: 4025–4033

Bunton CA, Cori O (1978) Mechanistic aspects of terpenoid biochemistry. Interciencia (Caracas) 3: 291–297

Bunton CA, Leresche JP, Hachey D (1972) Deuterium isotope effect in cyclization of monoterpenoids. Tetrahedron Lett 2431–2434

Bunton CA, Cori O, Hachey D, Leresche JP (1979) Cyclization and allylic rearrangement in solvolyses of monoterpenoids. J Org Chem 44: 3238–3244

Chayet L, Pont-Lezica R, George-Nascimento C, Cori O (1973) Biosynthesis of isomeric sesquiterpene aldehydes by soluble enzymes from orange flavedo. Phytochemistry 12: 95–101

Chayet L, Rojas C, Cardemil E, Jabalquinto AM, Vicuña JR, Cori O (1977) Biosynthesis of mono-terpene hydrocarbons from 1-^{3}H nerylpyrophosphate and geranylpyrophosphate by soluble enzymes from *Citrus limonum*. Arch Biochem Biophys 180: 318–327

Cori O (1969) Terpene biosyntehsis: Utilization of neryl pyrophosphate by an enzyme system from *P. radiata* seedlings. Arch Biochem Biophys 135: 416–418

Croteau R (1975) Biosynthesis of monoterpenes and sesquiterpenes. In: Drauert F (ed) Odour and taste substances, pp 153–165. Carl, Nürnberg

Croteau R, Karp F (1977) Biosynthesis of monoterpenes: Partial purification and characterization of 1,8-cineole synthetase from *Salvia officinalis*. Arch Biochem Biophys 179: 257–265

Croteau R, Burbott AJ, Loomis WD (1973) Enzymatic cyclization of nerylpyrophosphate to α terpineol by cell free extracts from peppermint. Biochem Biophys Res Commun 50: 1006–1012

Del Campo G, Puente J, Valenuela MA, Traverso-Cori A, Cori O (1977) Hydrolysis of synthetic pyrophosphoric esters by an isoenzyme from *Solanum tuberesum*. Biochem J 167: 525–529

Deno NC, Haddon RC, Nowak EN (1970) Energy barriers for rotation about carbon-carbon bonds in allyl cations. J Am Chem Soc 92: 6691–6693

Evans R, Hanson JR (1976) Studies in terpenoid biosynthesis. Part XIV. Formation of the sesqui-terpene trichodiene. J Chem Soc Perkins Trans 1: 326–329

George-Nascimento C, Cori O (1971) Terpene biosynthesis from geranyl and neryl pyrophosphates by enzymes from orange flavedo. Phytochemistry 10: 1803–1810

George-Nascimento C, Pont-Lezica R, Cori O (1971) Non enzymic formation of nerolydol from Farnesyl pyrophosphate in the presence of bivalent cations. Biochem Biophys Res Commun 45: 119–124

Hemming FW (1977) The biosynthesis and biological significance of prenols and their phosphorylat-ed derivatives. In: Tevini M, Lichtenthaler HK (eds) Lipids and lipid polymers in higher plants, pp 183–198. Springer, Berlin Heidelberg New York

Ingold CK (1953) Structure and mechanism in organic chemistry, p 460. Cornell Univ Press, Ithaca

Jacob G, Cardemil E, Chayet L, Téllez R, Pont-Lezica R, Cori O (1972) Synthesis of isomeric far-nesols by soluble enzymes from *Pinus radiata* seedlings. Phytochemistry 11: 1683–1688

Jedlicki E, Jacob G, Faini F, Cori O, Bunton CA (1972) Stereospecificity of isopentenylpyrophos-phate isomerase and prenyltransferase from *Pinus* and *Citrus*. Arch Biochem Biophys 152: 590–596

King HL, Rilling HC (1977) Avian liver prenyltransferase. The role of metal in substrate binding and the orientation of substrates during catalysis. Biochemistry 16: 3815–3819

Madyastha KM, Loomis WD (1969) Phosphorylation of geraniol by cell free enzymes from *Mentha piperita*. Fed Proc 28: 665

McCormick JP, Barton DL (1978) Studies in 85% H_3PO_4. II. On the role of the α-terpenyl cation in cyclic monoterpene genesis. Tetrahedron 34: 325–330

Nishino T, Ogura K, Seto S (1973) Comparative specificity of prenyltransferase of pig liver and pumpkin with respect to artificial substrates. Biochim Biophys Acta 302: 33–37

Popják G, Cornforth JW (1966) Substrate stereochemistry in squalene biosynthesis. Biochem J 101: 553–568

Potty VH, Moshonas MG, Bruemmer JH (1970) Cyclization of linalool by enzyme preparations from orange. Arch Biochem Biophys 138: 350–352

Poulose AJ, Croteau R (1978) γ-terpinene synthetase: A key enzyme in the biosynthesis of aromatic monoterpenes. Arch Biochem Biophys 191: 400–411

Robinson DR, Ryback G (1969) Incorporation of tritium from 4-R-(4-^{3}H) mevalonate into abscisic acid. Biochem J 113: 895–896

Shine WE, Loomis WD (1974) Isomerization of geraniol and geranylphosphate by enzymes from carrot and peppermint. Phytochemistry 13: 2095–2101

Suga T, Shishibori T, Morinaka H (1977) The intermediacy of linaloylpyrophosphate in biosyn-thesis of cyclic monoterpenes by higher plants. Proc Annu Meet Japan Biochem Soc, p 1069. Tokyo

Valenzuela P, Cori O (1967) Acid catalyzed hydrolisis of neryl pyrophospahte and geranyl pyro-phosphate. Tetrahedron Lett 32: 3089–3094

Valenzuela P, Beytia E, Cori O, Yudelevich A (1966) Phosphorylated intermediates of terpene biosynthesis in Pinus radiata. Arch Biochem Biophys 113: 536–539

Zeitschel O (1906) Über das Nerol und seine Darstellung aus Linalool. Ber 39: 1780–1793

B. Enzyme Regulation

1. Three Multifunctional Protein Kinase Systems in Transmembrane Control

Y. NISHIZUKA

Abbreviations

EGTA, ethylene glycol bis (β-aminoethyl ether); N,N,N′,N′-tetraacetic acid; SDS, sodium dodecyl sulfate.

A. Introduction

It is my great pleasure, and certainly a great honor, to be invited to present the annual Fritz Lipmann Lecture, which was established in 1974 by the Gesellschaft für Biologische Chemie. I am particularly pleased to be able to dedicate this lecture to Dr. Fritz Lipmann on this occasion to celebrate his 80th birthday.

Our knowledge of enzymatic phosphorylation and dephosphorylation of proteins has been expanding rapidly, and biological roles of such covalent modification of proteins were bound to increase in recent years. The study in diverse fields of this type of biological regulation was initiated as early as in the 1930's by a pioneering work done by Lipmann, who identified first the phosphate that is covalently linked to the hydroxyl group of seryl residue in phosvitin (Lipmann and Levene 1932) and casein (Lipmann 1933), as shown in Fig. 1. The enzyme which transfers the terminal phosphate of ATP to either seryl or threonyl residue in casein, phosvitin, and endogenous phosphoproteins, that we call casein kinase, has later been found in many mammalian tissues by Burnett and Kennedy (1954), and by Rabinowitz and Lipmann (1960). Subsequent to the discovery of cyclic AMP by Sutherland and his co-workers (Sutherland and Rall 1960), the report by Walsh, Perkins, and Krebs (1968) on cyclic AMP-dependent protein kinase

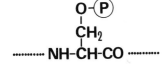

Lipmann, F. & Levene, P.A., 1932   **Fig. 1.** Discovery of serine-phosphate in protein

Department of Biochemistry, Kobe University School of Medicine, Kobe 650, Japan

has shed light on the role of protein phosphorylation reactions in the hormonal control of various cellular activities. In fact, it has been firmly established that cyclic AMP-dependent protein kinase plays roles of crucial importance in various fields of biological regulation. On analogy to this enzyme, cyclic GMP-dependent protein kinase which has been uncovered by Greengard and his co-workers first in lobster tail muscle (Kuo and Greengard 1970a) and later in a variety of mammalian tissues (for review, Nishizuka et al. 1979) has also attracted great attention. During the analysis of various classes of protein kinases we have found recently an additional species of enzyme which is selectively activated by Ca^{2+}, and evidence is now available suggesting that this protein kinase may also be involved in transmembrane control of various cellular activities.

In this lecture I would like to describe the properties, mode of activation, and possible functional specificities of the three sets of protein kinases mentioned above, with special emphasis on the newly found Ca^{2+}-dependent enzyme system. In particular, the coupling of this enzyme activation with phospholipid turnover in membranes will now be proposed and, in the latter part of this lecture, topographical arrangement and compartment in the enzyme substrate interaction will also be proposed to be important factors for the complete understanding of each specific physiological role of these closely similar protein kinase systems. Cyclic AMP-dependent, cyclic GMP-dependent and Ca^{2+}-dependent protein kinases will be tentatively referred to hereafter as simply A-Kinase, G-Kinase, and C-Kinase, respectively.

B. Ca^{2+}-dependent Protein Kinase and Role of Phospholipid

I. Outline of Enzyme

The newly found protein kinase system is schematically outlined in Fig. 2. The enzyme is normally present as an inactive form in soluble fractions in many mammalian tissues. In the presence of Ca^{2+}, the enzyme attaches to membranes to exhibit full catalytic activity. The active factor associated with membranes has been identified as phospholipid. The protein kinase that is activated in this way is able to phosphorylate histone and protamine, and appears to show pleiotropic activities as well described for A-Kinase. This enzyme clearly differs from A- and G-Kinases in the physical and kinetic properties.

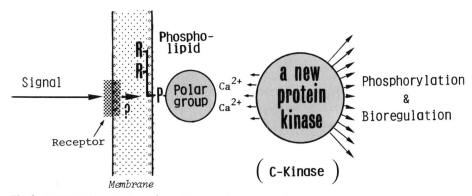

Fig. 2. A proposed mechanism of new transmembrane control

Tissue	C-Kinase	A-Kinase	G-Kinase
	(units/mg protein)*		
Brain	3,270	250	6
Lymphocytes	1,060	320	◁1
Lung	360	290	13
Kidney	280	150	–
Liver	180	130	2
Adipocytes	170	270	2
Heart	110	230	7
Skeletal muscle	80	110	–

* One unit: one pmol Pi transferred into histone per min

Fig. 3. Tissue distribution and activities of three protein kinases. Rat tissues were employed. The homogenates were prepared in the presence of EGTA to remove Ca^{2+}, and centrifuged for 60 min at 100,000 g. Most of enzymatic activities were recovered in the supernatant. Hl histone was used as substrate, and other detailed conditions were described elsewhere (Takai et al. 1975, 1979a)

It has been well known that both A- and G-Kinases are distributed widely in eukaryotic organisms including a variety of vertebrates and invertebrates (for review, Nimmo and Cohen 1977). In comparison with these two previously known enzymes, Fig. 3 shows the tissue distribution and relative activity of C-Kinase in some mammalian tissues. The three classes of enzymes have been assayed under comparable conditions with calf thymus histone as phosphate acceptor and, therefore, the enzymatic activities may be directly compared with each other. It is evident that C-Kinase is distributed in many tissues and organs in amounts that are comparable to A-Kinase, and that the enzyme is present in large quantities particularly in brain and lymphocytes. G-Kinase, in contrast, is very low in enzymatic activity. Most of the work that is described below will deal with the enzymes which were partially purified from rat brain soluble fractions.

II. Relation to Membrane

C-Kinase per se was inactive, and the activation of enzyme was dependent upon the simultaneous presence of membrane and Ca^{2+} as shown in Fig. 4. Either one of these factors alone could not support the enzymatic activity. The membrane employed in this experiment was prepared from rat brain synaptosomes. The enzymatic activity was routinely assayed by measuring the incorporation of the radioactive phosphate of $[\gamma\text{-}^{32}P]ATP$ into Hl histone. The active membrane factor was found in many other tissues and organs including liver, kidney, skeletal muscle, lymphocytes, red blood cells, and adipocytes. This factor was exclusively localized in membranes, and was extracted quantitatively with chloroform-methanol. A series of analyses resulted in the identification of this factor as indeed phospholipid. The result given in Fig. 5 shows the relative activities of various phospholipids. The phospholipids used for this study were of mammalian origins, and all chromatographically pure samples. It was noted that phosphatidylinositol and phosphatidylserine were most effective in supporting the enzymatic

		Kinase activity* (cpm/tube)
Complete system		16,900
Complete – C-Kinase	(0.35 µg)	0
Complete – Membrane**	(0.7 µg)	640
Complete – Ca^{2+}	(1 × 10^{-4} M)	990

* With Hl histone as phosphate acceptor
** Synaptic membranes from rat brain

Fig. 4. Activation of C-Kinase by Ca^{2+} and membrane. C-Kinase employed was purified partially from rat brain as described (Inoue et al. 1977), and the detailed assay conditions were described elsewhere (Takai et al. 1979a)

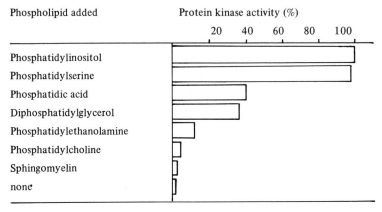

Fig. 5. Relative activities of various phospholipids for activation of C-Kinase. Various phospholipids (6 ng as phosphorus/ml each) were employed. Phosphatidylinositol and phosphatidylserine were obtained from pig liver and bovine brain, respectively, and other phospholipids were from human erythrocytes. Other conditions were identical with those given in Fig. 4

activity. Phosphatidic acid and diphosphatidylglycerol were effective to some extent, whereas neutral phospholipids such as phosphatidylethanolamine, phosphatidylcholine, and sphingomyelin were practically inactive in supporting the enzymatic activity. Although the enzyme showed a marked preference for phosphatidylinositol and phosphatidylserine, unfractionated phospholipid was almost equally as active as pure phosphatidylinositol. Possibly, exact descriptions of specificity for pure phospholipids are of doubtful significance, and a better physiological picture will be clarified by further investigations. Nevertheless, Ca^{2+} was always absolutely needed for the reaction irrespective of the species of phospholipid employed.

In order to look into the mode of activation of this enzyme, the next experiment was performed with synaptic membranes. The enzyme was normally present as an inactive form in soluble fractions, and was activated in such a way that the enzyme, membrane, and Ca^{2+} were associated together to produce a ternary complex which was active to phosphorylate histone and many other proteins. This activation process was freely reversible, and the enzyme, once activated in this way, converted again to an inactive

$$\boxed{\text{Enzyme}} \;+ Ca^{2+} + \text{Membrane} \rightleftharpoons \boxed{\text{Enzyme} \cdot Ca^{2+} \cdot \text{Membrane}}$$

Inactive form   Active form

	C-Kinase in soluble fraction	C-Kinase bound to membranes
$- Ca^{2+}$ (+ EGTA)	92%	8%
$+ Ca^{2+}$ (10^{-5} M)	7%	93%

Fig. 6. Reversible attachment of C-Kinase to membrane in the presence of Ca^{2+}. The enzyme and synaptic membranes were mixed in the presence and absence of Ca^{2+}, and the mixture was centrifuged for 60 min at 100,000 g. The supernatant and precipitates were separately assayed for enzymatic activity. Detailed conditions will be described elsewhere

soluble form when Ca^{2+} was removed with EGTA. This activation and deactivation cycle could be experimentally repeated without loss of the total enzymatic activity. The result shown in Fig. 6 is a typical example of such experiments. A purified preparation of soluble inactive enzyme was first mixed with synaptic membranes, and the mixture was then centrifuged to separate the enzyme from membranes again. In the absence of Ca^{2+}, C-Kinase remained soluble as an inactive form, whereas in the presence of Ca^{2+}, the enzyme was precipitated with membranes as an active form. With phospholipid instead of membranes such a ternary complex was also demonstrated by gel filtration analysis.

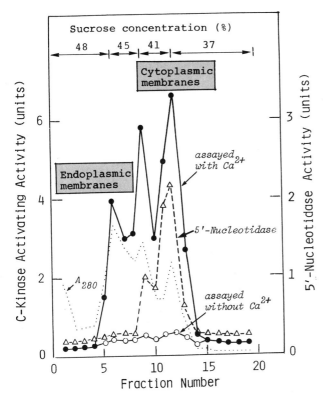

Fig. 7. Specificity of membrane for activation of C-Kinase. A crude membrane fraction was subjected to discontinuous sucrose density gradient analysis by the method of Ray (1970). Each fraction was then assayed for the activity to support C-Kinase in the presence and absence of Ca^{2+}. Detailed conditions were described elsewhere (Takai et al. 1979b)

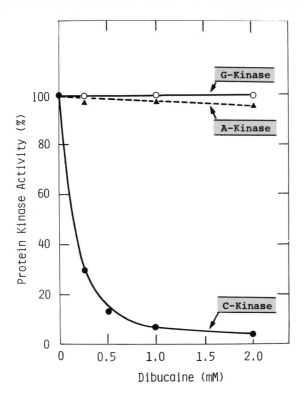

Fig. 8. Selective inhibition of C-Kinase by dibucaine. The three enzymes were assayed under the respective standard conditions in the presence of dibucaine as indicated. Detailed conditions will be described elsewhere

The experiment given in Fig. 7 was designed to show the specificity of membranes. As expected from the foregoing discussions, the C-Kinase activating factor was found not only in cytoplasmic membranes but in endoplasmic membranes as well. In this experiment, a crude membrane fraction was subjected to discontinuous sucrose density gradient analysis. The cytoplasmic membrane was located by measuring 5'-nucleotidase. This result reveals that cytoplasmic membranes as well as endoplasmic membranes show capacity to activate the enzyme in the presence of Ca^{2+}. An additional result shown in Fig. 8 possibly suggests a role of phospholipid in the activation of C-Kinase. Dibucaine is a well-known drug for local anesthesia, and an increasing body of evidence is now available indicating that this drug interacts with membrane phospholipids (for review, Seeman 1972). It was noted that dibucaine selectively inhibited C-Kinase, but did not interfere with A-Kinase, nor with G-Kinase. A preliminary analysis on the mode of this action suggested that this drug did not interact with the active center of enzyme, but blocked the association of enzyme with membranes. In fact, a catalytic fragment of C-Kinase, which was obtained by limited proteolysis with trypsin as described below, was not susceptible to dibucaine over a wide range of concentrations.

C-Kinase was entirely independent of cyclic nucleotides, but absolutely required Ca^{2+} in addition to phospholipid. This effect was very specific for Ca^{2+}, and other divalent cations were totally inactive except for Sr^{2+} which was less than 10% as active as Ca^{2+} as shown in Fig. 9. Analysis on such action of Ca^{2+} indicated that this divalent cation was not necessary for the enzymatic catalysis, but was required for the activation of

Cation at 5×10^{-5} M	Kinase activity (%)
Ca^{2+}	100
Sr^{2+}	8
Mg^{2+}	0
Mn^{2+}	0
Co^{2+}	0
Zn^{2+}	0

Fig. 9. Specificity of divalent cation for activation of C-Kinase. C-Kinase was assayed under standard conditions except that Ca^{2+} was replaced by various cations indicated

enzyme; the best evidence was provided by the fact that again the catalytic fragment of C-Kinase obtained by limited proteolysis was fully active without Ca^{2+} and phospholipid. It is most likely, therefore, that Ca^{2+} interacts with membrane phospholipid rather than with enzyme.

The response of C-Kinase to Ca^{2+} concentrations was markedly dependent on the membranes employed. In Fig. 10 two typical examples are presented. When the enzymatic activity was plotted as a function of Ca^{2+} concentration, a significant difference was observed between synaptic and erythrocyte membranes. An approximate Ka value for Ca^{2+}, the concentration needed for the half maximum activity, is given for each membrane fraction. A series of experiments to clarify the reason for this difference have revealed that the lipid composition of membranes seriously influences the response of enzyme to this divalent cation.

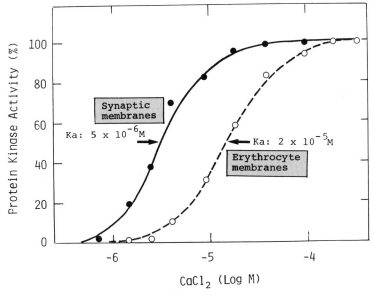

Fig. 10. Response of C-Kinase to Ca^{2+} concentrations. The activation of C-Kinase was assayed in the presence of either synaptic or erythrocyte membranes at various concentrations of Ca^{2+}. Detailed conditions will be described elsewhere

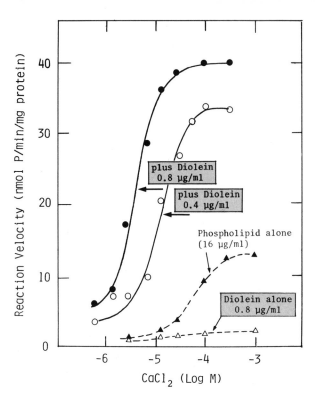

Fig. 11. Effect of diolein on response of C-Kinase to Ca^{2+}. A purified preparation of C-Kinase was assayed in the presence of Ca^{2+} as indicated. Where indicated, phospholipid and diolein were added. In the experiments shown by *solid lines,* diolein (0.4 or 0.8 $\mu g/ml$) was added in addition to phospholipid (16 $\mu g/ml$). Lipid was fractionated by the method of Folch et al. (1957), and a phospholipid fraction obtained by partition chromatography was employed in this study. Detailed conditions will be described elsewhere

III. Role of Phospholipid and Coupling to its Turnover

In the experiments shown in Fig. 11, the reaction velocity of C-Kinase was plotted against Ca^{2+} concentrations. In the presence of phospholipid alone relatively higher concentrations of Ca^{2+} were needed. However, the reaction velocity was greatly enhanced by a very small amount of diolein which was supplemented to the reaction mixture in addition to phospholipid. Concomitantly, the response of enzyme to Ca^{2+} shifted markedly, and full enzymatic activity was obtained at very low Ca^{2+} concentrations. In this experiment, the response of enzyme to Ca^{2+} was examined at two different concentrations of diolein, and it may be emphasized that diolein in an amount of less than 5% of that of phospholipid used greatly enhanced the reaction velocity with the increase in the affinity for Ca^{2+}. Diolein alone showed very little or practically no effect. Such a unique effect of neutral lipid appeared to be very specific for unsaturated diacylglycerol. In the experiment shown in Fig. 12, various synthetic neutral lipids, namely tri-, di-, and monoacylglycerols with saturated and unsaturared fatty acids were examined. In the presence of phospholipid alone, the phosphorylation reaction proceeded slowly with a relatively low affinity for Ca^{2+} as described above. The addition of diolein in an amount of 5% of that of phospholipid sharply decreased the Ka value for Ca^{2+} and increased the reaction velocity. It was evident that none of the other neutral lipids tested could substitute for diolein. Cholesterol was ineffective. It may be noted that phosphatidylinositol from mammalian tissues has been known to be composed of largely

Neutral lipid added		Ka for Ca^{2+} (10^{-6} M)	V$_{max}$*
Phospholipid alone	(16 µg/ml)	70	16
Phospholipid + Triolein	(0.8 µg/ml)	70	18
Phospholipid + Diolein	(0.8 µg/ml)	4	40
Phospholipid + Monoolein	(0.8 µg/ml)	70	18
Phospholipid + Tripalmitin	(0.8 µg/ml)	60	11
Phospholipid + Dipalmitin	(0.8 µg/ml)	60	12
Phospholipid + Monopalmitin	(0.8 µg/ml)	60	12

* nanomoles/min/mg protein

Fig. 12. Specificity of neutral lipid for enhancement of phospholipid-dependent activation of C-Kinase. C-Kinase was assayed in the reaction mixture which contained a fixed amount of phospholipid (16 µg/ml) and various neutral lipids (0.8 µg/ml each) as indicated. The lipids were first mixed in chloroform, dried under nitrogen, and then dispersed by sonication. Detailed conditions will be described elsewhere

unsaturated fatty acids such as arachidonic and oleic acids (Holub et al. 1970). Unsaturated fatty acids are the well-established precursors to prostaglandins.

Hokin and Hokin (1953) first presented evidence two decades ago that phospholipid, particularly phosphatidylinositol, turns over very rapidly in response to various extracellular messengers. Early work on such a phosphatidylinositol turnover response was carried out with various types of exocytotic tissues, nervous tissues, and with salt-excreting glands of seabirds. Subsequent studies developed by a number of investigators, such as by Michell and his co-workers (for review, Michell 1975), have shown that the phosphatidylinositol response can be provoked in virtually any type of tissues which are stimulated by various extracellular messengers including hormones, neurotransmitters, and polyclonal plant mitogens for lymphocytes. Figure 13 shows the pathway for phosphatidylinositol turnover, which was described by Dawson and his co-workers (Dawson et al. 1971). It may be pointed out that the first product of phosphatidylinositol breakdown must be unsaturated diacylglycerol which is very effective to fortify the phos-

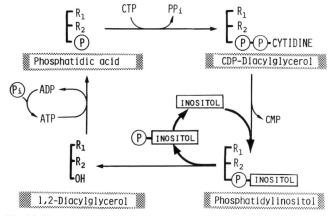

Fig. 13. Phosphatidylinositol turnover

pholipid-dependent activation of C-Kinase, as described above. It is plausible, therefore, that an enzyme which cleaves phosphatidylinositol in a phospholipase C-like manner seems to be of crucial importance for initiating the activation of newly found protein kinase system.

IV. A Possible Mechanism of Activation

Based on the foregoing discussions, it seems to be reasonable to propose a mechanism of C-Kinase activation. It is still possible that C-Kinase may be directly activated simply by the increase in intracellular Ca^{2+} concentrations, since many workers in this field have repeatedly documented that a variety of hormones and other messengers induce the Ca^{2+} influx into cells or the movement of this cation from intracellular pools. However, an alternative mechanism shown in Fig. 14 appears to be more plausible and attractive. A signal of messenger may induce the activation of phospholipase C-like enzyme, which may be associated with membranes. This enzyme causes a partrial hydrolysis of phosphatidylinositol, and then initiates the turnover of this phospholipid on the one hand. On the other hand, the resulting diacylglycerol, particularly which contains unsaturated fatty acids, would greatly enhance or fortify the role of phospholipid in the activation of C-Kinase. It is most likely, therefore, that in this mechanism the activation of C-Kinase may be tightly coupled with phosphatidylinositol turnover, which has remained mysterious for many years. Thus, the C-Kinase activation may possibly be linked to energy expenses, such as CTP breakdown. The enzymes which cleave phosphatidylinositol in a rather specific manner (phosphatidylinositol:inositol phosphotransferases, cyclizing and noncyclizing) have been found in various subcellular fractions such as cytosol and lysosomal fractions of mammalian tissues (Kemp et al. 1961; Irvine et al. 1978). However, the precise nature as well as the mechanism of activation of the enzyme which may be ipso facto involved in the phosphatidylinositol turnover in membranes has remained unexplored. Possibly, it would be one of the most important problems for the next few years. Nevertheless, it is emphasized that in this mechanism C-Kinase may be activated without net increase in the intracellular Ca^{2+} concentration since, when the receptor is stimulated, a small amount of the resulting

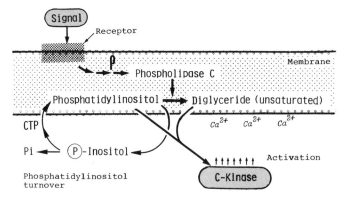

Fig. 14. A possible mechanism of C-Kinase activation

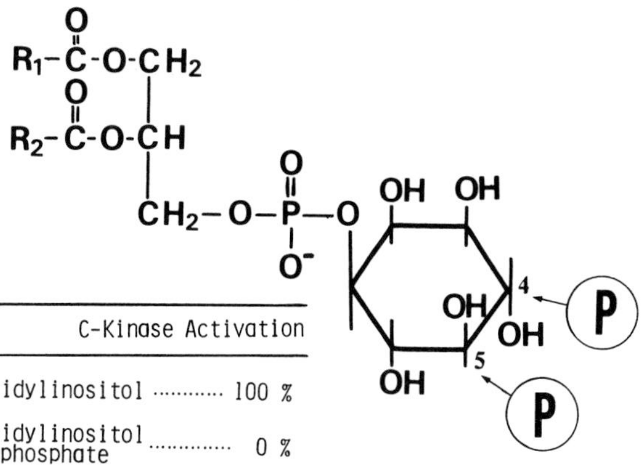

	C-Kinase Activation
Phosphatidylinositol	100 %
Phosphatidylinositol monophosphate	0 %
Phosphatidylinositol diphosphate	0 %

Fig. 15. Structure and activity of phosphatidylinositol phosphates

diacylglycerol would sharply increase the affinity for this divalent cation. In fact, from several lines of evidence it has also been claimed that neither the phosphatidylinositol turnover nor the response to hormones necessarily involves the entrance of Ca^{2+} into the cell.

Another line of evidence, which may be found in the literature, has shown that in plasma membranes of most mammalian cells, particularly of brain tissues, two additional inositol phospholipids are present, which contained one or two mols of phosphate that is esterified at positions 4 and 5 of phosphatidylinositol, as shown in Fig. 15 (Brockerhoff and Ballou 1961: Dawson and Dittmer 1961). Such phosphate groups have also been proposed to turn over rapidly, and possible roles of these phospholipids in neuronal functions, particularly in the cholinergic and α-adrenergic receptors, have been postulated (for reviews, Hawthorne and White 1975; Michell 1975, 1979). Nevertheless, in sharp contrast to phosphatidylinositol, these two phospholipids appeared to be totally inactive, and a preliminary analysis suggested that these phospholipids acted as potent inhibitors for the activation of C-Kinase. The significance of phosphorylation and dephosphorylation of this type of phospholipid has remained to be investigated, although it is of interest to speculate that the reactions might be related to a turn on–turn off mechanism of the C-Kinase activation.

V. Relation to Other Receptors

Most part of Fig. 16 is taken from the Michell's Review, which appeared in 1975 (Michell 1975). Since the first discovery by Hokin and Hokin (1953), numerous studies have established clearly that various hormones, neurotransmitters, and other extracellular messengers provoke rapid turnover of phosphatidylinositol in many types of target cells, as already discussed above. At the same time, it is also clear that several of the receptors that are accepted to produce cyclic AMP do not induce such a phosphatidylino-

Stimulus	Tissues	Phospholipid turnover	Cyclic GMP	Cyclic AMP
Glucagon	Liver, fat cell	⟶	⟶	↑
ACTH	Adrenal gland	⟶		↑
Adrenergic (β)		⟶	→↓	↑
Adrenergic (α)	Various	↑	↑	→↓
Cholinergic (m)	Various	↑	↑	→↓
Vasopressin	Liver	↑		⟶
Glucose	Langerhans	↑	→	⟶
Plant lectin	Lymphocyte	↑	↑	⟶

* Mostly taken from Michell's Review in 1975

Fig. 16. Relation of phosphatidylinositol turnover and cyclic nucleotides. This schedule is mostly taken from a review described by Michell (1975)

sitol response. Inversely, α-adrenergic as well as muscarinic cholinergic stimulation, for instance, which induce the rapid turnover of phosphatidylinositol in various tissues, do not appear to enhance the cyclic AMP production. Thus, a possibility may be raised that the newly found protein kinase system is related to the α-adrenergic and muscarinic cholinergic receptors. A series of studies by many investigators have suggested that the receptor mechanism, which produces phosphatidylinositol turnover, appears to be intimately related to the increase in cyclic GMP concentrations. However, several pieces of information are available indicating that the extracellular messengers that induce phosphatidylinositol turnover do not necessarily elevate the cyclic GMP level. Although the causal relationship between these two observations has not yet been clarified, it is possible that guanylate cyclase in some tissues may be activated by derivatives of unsaturated fatty acids or prostaglandins, both of which could be resulted from the increased turnover of phosphatidylinositol (Graff et al. 1978).

C. Regulatory and Functional Compartment of Three Protein Kinases

I. Enzymology and Mode of Activation

Before discussing the functional specificities, some of the properties of the three species of protein kinases, which are responsible for cyclic AMP, cyclic GMP, and Ca^{2+}, will be briefly described for comparison. As shown in Fig. 17, these enzymes are selectively activated by each specific activating factor. A-Kinase in most tissues shows a tetrameric form with a molecular weight of about 190,000 (for reviews, Rubin and Rosen 1975; Nimmo and Cohen 1977; Nishizuka et al. 1979). G-Kinase obtained from most tissues shows a molecular weight of around 140,000 and consists of two identical subunits (for review, Nishizuka et al. 1979). C-Kinase in its inactive form shows a molecular weight of about 77,000, as estimated by its S value and Stokes radius, but the subunit structure of this enzyme is unknown at present (Takai et al. 1977a, b; Inoue et al. 1977).

	A-Kinase	G-Kinase	C-Kinase
Activating factor	Cyclic AMP	Cyclic GMP	Ca^{2+}, P-lipid
Molecular weight	190×10^3	140×10^3	77×10^3
Subunit structure	R_2C_2 (tetramer)	E_2 (dimer)	?
Trypsin treatment	Activated	Activated	Activated
Catalytic fragment			
(Molecular weight)	40×10^3	34×10^3	51×10^3
(Activating factor)	Independent	Independent	Independent

Fig. 17. Properties of three protein kinases

$$\boxed{C_2 \cdot R_2} \; + 2 \times \text{Cyclic AMP} \; \rightleftharpoons \; \boxed{2 \times C} \; + R_2 \cdot (\text{Cyclic AMP})_2$$

$$\boxed{E_2} \; + 2 \times \text{Cyclic GMP} \; \rightleftharpoons \; \boxed{(E \cdot \text{Cyclic GMP})_2}$$

$$\boxed{E} \; + Ca^{2+} + \text{Membrane} \; \rightleftharpoons \; \boxed{E \cdot Ca^{2+} \cdot \text{Membrane}}$$

Fig. 18. Mode of activation of three protein kinases

Although the three classes of enzymes differ from each other in their physical and kinetic properties, all three enzymes could be irreversibly activated by limited proteolysis (Huang and Huang 1975; Inoue et al. 1976, 1977; Nishizuka et al. 1976; Takai et al. 1977a, b). The catalytic fragments obtained in this way were enzymatically fully active without addition of the respective activating factors, and showed molecular weights of about 40,000, 34,000, and 51,000, respectively (Nishizuka et al. 1976, 1978; Kishimoto et al. 1977; Takai et al. 1977a, b). Nevertheless, such proteolytic activation of enzymes appears to be physiologically less significant, and all three species of protein kinases are activated in reversible manners as shown in Fig. 18.

A-Kinase was first described by Walsh, Perkins, and Krebs (1968) in rabbit skeletal muscle, and shortly afterwards by Langan (1968) in rat liver. Since then, extensive studies were made by many investigators, and Tao, Salas, and Lipmann (1970) were able to show that A-Kinase is composed of two different subunits, that are well-known catalytic and regulatory subunits, and that cyclic AMP is bound to the regulatory subunit, and dissociates enzyme to relieve the fully active catalytic subunit. This mode of activation of A-Kinase was concurrently clarified in this laboratory (Yamamura et al. 1970; Kumon et al. 1970, 1972), and also reported around the same time from the laboratories of Krebs (Reimann et al. 1971), Garren (Gill and Garren 1971) and of Rosen (Erlichman et al. 1971).

In contrast, G-Kinase does not consist of two such distinctly different subunits, but is composed of two identical subunits. This enzyme was first described by Kuo and Greengard (1970a), and subsequent analysis in our laboratory has reached the conclusion that G-Kinase may be activated simply in an allosteric manner without dissociation of the subunits (Takai et al. 1975, 1976; Nishizuka et al. 1976). Such mode of activation of G-Kinase has been recently confirmed by many other investigators by using highly purified enzyme preparations (De Jonge and Rosen 1977; Gill et al. 1977; Flockerzi et al. 1978; Lincoln et al. 1978).

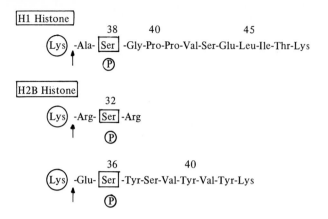

Fig. 19. Tryptic phosphopeptides obtained from H1 and H2B histone fractions phosphorylated by three protein kinases. The three seryl residues were shown to be phosphorylated by either one of these kinases (see text)

II. Substrate Recognition, Primary Sequence

Evidence accumulating in this laboratory during the last several years has indicated that, as far as tested in purified in vitro systems, the three species of enzyme seem to recognize exactly the same seryl and threonyl residues in various phosphate acceptor proteins. For instance, when H1 and H2B histone fractions were extensively phosphorylated separately by these three enzymes, and then subjected to tryptic digestions, the three phosphopeptides listed in Fig. 19 were obtained from the radioactive histone preparations phosphorylated by any one of the three enzymes (Hashimoto et al. 1975, 1976; Kuroda et al. 1976; Takai et al. 1977a). The phosphorylation of serine-38 in H1 histone was reported earlier by Langan (1971) for A-Kinase in both in vitro and in vivo systems. The results indicate that the three seryl residues are equally recognized by the three species of enzymes. Namely, the three protein kinases may not be distinguished by such a sequence analysis with histone as phosphate acceptor. In addition, the amino acid sequences found in the vicinity of the serine differ from each other, and the three enzymes do not appear to recognize a common specific linear sequence of amino acids. However, the three tryptic phosphopeptides all contain at least one basic amino acid, that is lysine in this case, close to the serine phosphorylated. The peptide bonds indicated by *arrows* resisted the hydrolysis with trypsin presumably due to steric hindrance by the phosphate attached. Such common substrate specificities and characteristics of amino acid sequences have also been reported by many investigators in recent years for A- and G-Kinases by using synthetic peptides and other substrate proteins (for reviews, Nimmo and Cohen 1977; Nishizuka et al. 1979; Krebs and Beavo 1979). Although the precise mechanism of substrate recognition is unknown, it is suggestive that the three species of protein kinases are most likely to possess similar active centers and that a basic amino acid seems to play some role in the recognition of substrate proteins.

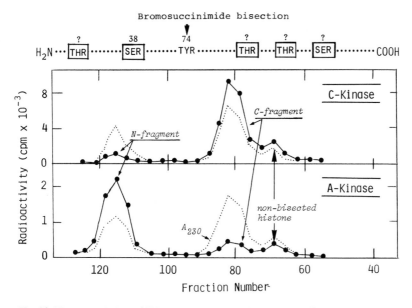

Fig. 20. Phosphorylation of H1 histone by A- and C-Kinases. Calf thymus H1 histone was phosphorylated separately by A- and C-Kinases, and then bisected at the tyrosine-74 by bromosuccinimide (Bustin and Cole 1969). The resulting two fragments were separated by gel filtration on a Sephadex G-100 column, and the radioactivity was determined. Detailed conditions will be described elsewhere

III. Substrate Recognition, Macromolecule Interaction

Although the three classes of enzymes appear to recognize exactly the same serine and threonine in various proteins, the relative reaction velocities greatly differ among the enzyme species as well as the substrate proteins employed. Namely, relative activities of each protein kinase towards various substrates largely depend on the macromolecule interaction between the enzyme and phosphate acceptor protein. A typical example is shown in Fig. 20. In this experiment H1 histone was used as a model substrate, since at different stages of cell cycle this histone has been shown to be phosphorylated at multiple sites in addition to serine-38 (Hohmann et al. 1976). Both serine and threonine near the C-terminal may be heavily phosphorylated in late G1 through M phase (Gurley et al. 1975), although the exact sites of this phosphorylation have not yet been definitely identified. Recent analysis in this laboratory has shown that both A- and C-Kinases are able to phosphorylate multiple hydroxyl groups in H1 histone as judged by fingerprint techniques. Further analysis has revealed that A-Kinase greatly favors the serine-38 which is located near the N-terminal as first reported by Langan (1971), whereas C-Kinase phosphorylates more rapidly both serine and threonine which are located near the C-terminal. This experiment was designed to show such selectivity of the enzymes. The radioactive H1-histone preparation that was phosphorylated separately by A- and C-Kinases was bisected with bromosuccinimide at the tyrosine indicated by an *arrow*. Then, the resulting two fragments thus obtained were separated by gel filtration. The results clearly indicate that the radioactive phosphate incorporated by C-Kinase was primarily localized in the C-terminal fragment, whereas that incorporated by A-Kinase was found mainly in the N-terminal fragment. Acid hydrolysis of each radioactive fragment indicated that in the C-terminal fragment both seryl and threonyl residues were equally phosphorylated, whereas in the N-terminal fragment seryl residue was more predominantly labeled. A small peak of nonbisected histone was probably due to the microheterogeneity of H1 histone (Sherod et al. 1974). G-Kinase showed a similar activity to that of A-Kinase, but, this activity for H1 histone was very slow (see below).

From the foregoing discussions it may be clear that, although the primary sequence around the serine and threonine seems to be important in some way for the recognition of substrate, the shape or size of enzyme and substrate protein, namely the macromolecule interaction between enzyme and substrate, is equally or even more important for the selectivity of reactions.

IV. Substrate Recognition, Topographical Arrangement

In order to demonstrate an additional factor which may also be important for understanding functional specificities of the three protein kinases, the experiment given in Fig. 21 was carried out again with H2B histone as a model substrate, and showed the relative reaction rates of the three enzymes in the presence and absence of DNA in the reaction mixture. With H2B histone alone, A-Kinase rapidly phosphorylated this histone under the given condition, and the reaction of G-Kinase was very slow. If, however, DNA was supplemented to the reaction mixture, the relative reaction velocities were completely reversed. Under the same conditions, except for the presence of DNA, G-Kinase was more active than any of the other kinases, and rapidly phosphorylated the histone

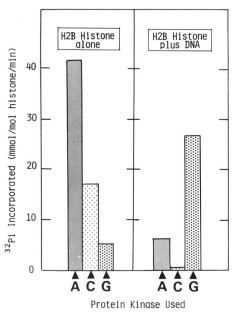

Fig. 21. Relative reaction rates of three protein kinases for H2B histone in the presence and absence of DNA. Each protein kinase was assayed with H2B histone as phosphate acceptor. Where indicated, an equal amount (w/w) of polydeoxyribonucleotide was added to the reaction mixture. Detailed conditions were described elsewhere (Hashimoto et al. 1979)

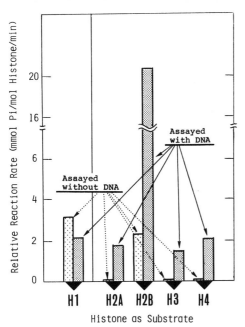

Fig. 22. Effect of DNA on histone phosphorylation by G-Kinase. G-Kinase was assayed in the presence and absence of polydeoxyribonucleotide with various histone fractions as substrates. Equal amounts (w/w) of histone and polydeoxyribonucleotide were added. Detailed conditions were described elsewhere (Hashimoto et al. 1979)

that was associated with DNA. Fingerprint analysis indicated that exactly the same seryl residues were phosphorylated by these kinases both in the presence and absence of DNA.

It was also noted that such a unique effect of DNA observed for G-Kinase was specific for the four histone fractions which consist of nucleosome core. As shown in Fig. 22, in purified in vitro systems G-Kinase reacted with only H1 and H2B histones in the absence of DNA, whereas the reactions with H2A, H3, and H4 histone fractions were extremely slow. If, however, DNA was added to each reaction mixture, the reaction velocities for H2A through H4 histones were greatly enhanced. The phosphorylation of H1-histone was always inhibited and, therefore, it is suggestive that G-Kinase favors histone fractions which are integrated in the nucleosome structure rather than free histone. Such enhancement of reaction by DNA was specific for G-Kinase, and the reactions of A- and C-Kinases were rather inhibited. It was clear that DNA interacted with histone but not with enzyme. This model reaction appears to provide suggestive evidence that the topographical arrangement or subcellular compartment is another important determinant factor for the functions and specificities of protein kinase reactions.

Along the same line, C-Kinase appears to favor greatly the proteins which are associated with membranes. Figure 23 shows autoradiography of SDS-polyacrylamide gel electropherogram which illustrates endogenous substrate proteins of A- and C-Kinases in brain. Both A- and C-Kinases reacted preferentially with the particulate fraction which contained synaptosomes and myelin particles. It was very characteristic that two proteins with a molecular weight of around 20,000, most likely associated with membranes, were very heavily phosphorylated by C-Kinase. In addition to these heavy bands, there appear to be multiple faint bands as indicated by *small arrows*. In this experiment endogenous kinase was inactivated by brief heat treatment. Both A- and C-Kinases ap-

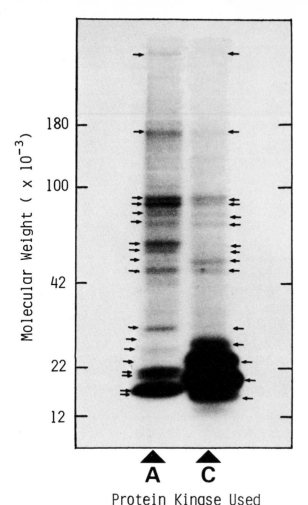

Fig. 23. Autoradiography of electropherogram of brain endogenous proteins phosphorylated by A- and C-Kinases. Rat brain was homogenized, and crude particulate fraction was prepared. This particulate fraction contained mostly synaptosomes and myelin particles. The preparation was phosphorylated separately by A- and C-Kinases, and then subjected to SDS-polyacrylamide gel electrophoresis. The endogenous kinases were inactivated by brief treatment with heat before being employed and, therefore, no radioactive bands were detected unless exogenous kinase was added. Detailed conditions will be described elsewhere

peared to utilize the same spectrum of phosphate acceptors, although the relative activities toward various proteins were again greatly different. The substrate proteins of G-Kinase in this particulate fraction appeared to be very small. Also, soluble cytosol proteins in brain served as poor substrates for both A-, G-, and C-Kinases. Although A-Kinase phosphorylates membrane proteins with considerably high reaction rates, better substrates for A-Kinase are localized in cytosol in other tissues.

Figure 24 shows relative activities of the three kinases for the well-known substrates, namely phosphorylase kinase and glycogen synthetase which were obtained from rabbit skeletal muscle. As well established for A-Kinase (Walsh et al. 1968; Yamamura et al. 1971, 1973), both C- and G-Kinases could phosphorylate and indeed regulate these muscle enzymes. However, the relative reaction velocities of C- and G-Kinases were far less than that of A-Kinase (Kishimoto et al. 1977, 1978; Yamamoto et al. 1978; Takai et al. 1979a).

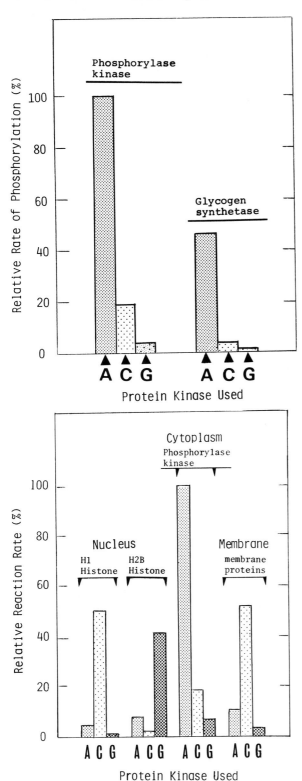

Fig. 24. Relative reaction rates of three protein kinases for muscle phosphorylase kinase and glycogen synthetase. Detailed conditions were described elsewhere (Kishimoto et al. 1977, 1978)

Fig. 25. Relative reaction rates of three protein kinases for four representative substrate proteins in different subcellular fractions. Detailed conditions will be described elsewhere. The reaction rates for H2B histone were estimated in the presence of DNA in the assay mixture

In Sum, Fig. 25 shows the relative reaction velocities measured in vitro with some representative substrates which are located in different subcellular compartments. As already mentioned, the three protein kinases appear to recognize the same spectrum of substrate proteins. However, their relative activities towards various proteins greatly differ from each other. A line of evidence presented above seems to indicate that the three sets of protein kinase systems may each play a specific physiological role, presumably owing to their own specific subcellular compartments.

D. Coda

Although we are still far from a full understanding of this large field of biological regulation, I would like to show two prospects based on the results presented above. The first one possibly has to do with the lipid-protein interaction which I have described. Such lipid-protein interaction in membranes appears to be very important for receptor mechanism. It seems to be very dynamic and tightly coupled with energy expenses. Although other macromolecule interactions, such as those between nucleic acid and protein, have been well clarified and discussed, the precise nature and mechanism of such lipid-protein interaction are scarcely understood. Analysis of this lipid-protein interaction in molecular terms will open a new way to various fields of chemical recognition in biology.

Secondly, there appears to be three species of closely similar protein kinases which may be responsible for the three intracellular messengers, namely Ca^{2+}, cyclic AMP, and cyclic GMP. Although our knowledge of protein kinases is expanding rapidly, the precise roles of such covalent modification reactions in biological regulation have largely remained unexplored. Presumably, the topographical arrangement of compartment in the enzyme and substrate interaction is an important factor for a complete understanding of the chemical basis of transmembrane control.

Before concluding my lecture, I wish to express my sincere congratulations to Dr. Fritz Lipmann for his 80th birthday, and also my deep gratitude to Dr. Francois Chapeville and Dr. Ann-Lise Haenni for the invitation that made this lecture possible.

Acknowledgments. I would like to express indebtedness to my collaborators in the present and the past, who have so efficiently carried out the experiments presented in this lecture. I have been especially grateful in the last few years to my colleagues Drs. H. Yamamura, Y. Takai, S. Matsumura, E. Hashimoto, A. Kishimoto, M. Inoue, Y. Kuroda, K. Sakai, M. Yamamoto, Y. Kawahara, Y. Iwasa, T. Mori, W.E. Criss, R. Minakuchi, U. Kikkawa, Y. Ku, and R. Kii. I am also grateful to Drs. K. Iwai and K. Hamana, of the Institute of Endocrinology, Gumma University, to Drs. T. Fujii and A. Tamura, of Kyoto College of Pharmacy, and to Dr. T. Yamakawa, of the Faculty of Medicine, University of Tokyo, for assistance and advice in performing the experiments for identifying amino acid sequences and phospholipids. My thanks are due also to Dr. K. Hayashi, of the Gunma University School of Medicine, for generous gifts of phosphatidylinositol polyphosphates. Skillful secretarial assistance on the part of Mrs. S. Nishiyama, Miss K. Yamasaki, and Mrs. M. Furuta is gratefully acknowledged.

This investigation has been supported in part by research grants from the Scientific Research Fund of the Ministry of Education, Science, and Culture, Japan.

References

Brockerhoff H, Ballou CE (1961) The structure of the phosphoinositide complex of beef brain. J Biol Chem 236: 1907–1911

Burnett G, Kennedy EP (1954) The enzymatic phosphorylation of proteins. J Biol Chem 211: 969–980

Bustin M, Cole RD (1969) Bisection of a lysine-rich histone by N-bromosuccinimide. J Biol Chem 244: 5291–5294

Dawson RMC, Dittmer JC (1961) Evidence for the structure of brain triphosphoinositide from hydrolytic degradation studies. Biochem J 81: 540–545

Dawson RMC, Freinkel N, Jungalwala FB, Clarke N (1971) The enzymatic formation of myoinositol 1:2-cyclic phosphate from phosphatidylinositol. Biochem J 122: 605–607

Erlichman J, Hirsch AH, Rosen OM (1971) Interconversion of cyclic nucleotide-activated and cyclic nucleotide-independent forms of a protein kinase from beef heart. Proc Natl Acad Sci USA 68: 731–735

Flockerzi V, Speichermann N, Hofmann F (1978) A guanosine 3':5'-monophosphate-dependent protein kinase from bovine heart muscle. J Biol Chem 253: 3395–3399

Folch J, Lees M, Sloane Stanley GH (1957) A simple method for the isolation and purification of total lipids from animal tissues. J Biol Chem 226: 497–509

Gill GN, Garren LD (1971) Role of the receptor in the mechanism of action of adenosine 3':5'-cyclic monophosphate. Proc Natl Acad Sci USA 68: 786–790

Gill GN, Walton GM, Sperry PJ (1977) Guanosine 3':5:-monophosphate-dependent protein kinase from bovine lung. J Biol Chem 252: 6443–6449

Graff G, Stephenson JH, Glass DB, Haddox MK, Goldberg ND (1978) Activation of soluble splenic cell guanylate cyclase by prostaglandin endoperoxides and fatty acid hydroperoxides. J Biol Chem 253: 7662–7676

Gurley LR, Walters RA, Tobey RA (1975) Sequential phosphorylation of H1 histone subfractions in the chinese hamster cell cycle. J Biol Chem 250: 3936–3944

Hashimoto E, Takeda M, Nishizuka Y, Hamana K, Iwai K (1975) Phosphorylated sites of calf thymus histone H2B by adenosine 3':5'-monophosphate-dependent protein kinase from silkworm. Biochem Biophys Res Commun 66: 547–555

Hashimoto E, Takeda M, Nishizuka Y, Hamana K, Iwai K (1976) Studies on the sites in histones phosphorylated by adenosine 3':5'-monophosphate-dependent and guanosine 3':5'-monophosphate-dependent protein kinases. J Biol Chem 251: 6287–6293

Hashimoto E, Kuroda Y, Ku Y, Nishizuka Y (1979) Stimulation by polydeoxyribonucleotide of histone phosphorylation by guanosine 3':5'-monophosphate-dependent protein kinase. Biochem Biophys Res Commun 87: 200–206

Hawthorne JN, White DA (1975) Myo-inositol lipids. Vitam Horm 33: 529–573

Hohmann P, Tobey RA, Gurley LR (1976) Phosphorylation of distinct regions of f1 histone, relationship to cell cycle. J Biol Chem 251: 3685–3692

Hokin MR, Hokin LE (1953) Enzyme secretion and the incorporation of P^{32} into phospholipids of pancrease slices. J Biol Chem 203: 967–977

Holub BJ, Kuksis A, Thompson W (1970) Molecular species of mono-, di-, and tri-phosphoinositides of bovine brain. J Lipid Res 11: 558–564

Huang LC, Huang C (1975) Rabbit skeletal muscle protein kinase. Conversion from cAMP dependent to independent form by chemical perturbations. Biochemistry 14: 18–24

Inoue M, Kishimoto A, Takai Y, Nishizuka Y (1976) Guanosine 3':5'-monophosphate-dependent protein kinase from silkworm, properties of a catalytic fragment obtained by limited proteolysis. J Biol Chem 251: 4476–4478

Inoue M, Kishimoto A, Takai Y, Nishizuka Y (1977) Studies on a cyclic nucleotide-independent protein kinase and its proenzyme in mammalian tissues. II. Proenzyme and its activation by calcium-dependent protease from rat brain. J Biol Chem 252: 7610–7616

Irvine RF, Hemington N, Dawson RMC (1978) The hydrolysis of phosphatidylinositol by lysosomal enzymes of rat liver and brain. Biochem J 176: 475–484

Jonge De HR, Rosen OM (1977) Self-phosphorylation of cyclic guanosine 3':5'-monophosphate-dependent protein kinase from bovine lung. J Biol Chem 252: 2780–2783

Kemp P, Hübscher G, Hawthorne JN (1961) Phosphoinositides 3. Enzymatic hydrolysis of inositol-containing phospholipids. Biochem J 79: 193–200

Kishimoto A, Takai Y, Nishizuka Y (1977) Activation of glycogen phosphorylase kinase by a calcium-activated, cyclic nucleotide-independent protein kinase system. J Biol Chem 252: 7449–7452

Kishimoto A, Mori T, Takai Y, Nishizuka Y (1978) Comparison of calcium-activated, cyclic nucleotide-independent protein kinase and adenosine 3':5'-monophosphate-dependent protein kinase as regards the ability to stimulate glycogen brakdown in vitro. J Biochem (Tokyo) 84: 47–53

Krebs EG, Beavo JA (1979) Phosphorylation-dephosphorylation of enzymes. Annu Rev. Biochem 48: 923–959

Kumon A, Yamamura H, Nishizuka Y (1970) Mode of action of adenosine 3':5'-cyclic phosphate on protein kinase from rat liver. Biochem Biophys Res Commun 41: 1290–1297

Kumon A, Nishiyama K, Yamamura H, Nishizuka Y (1972) Multiplicity of adenosine 3':5'-monophosphate-dependent protein kinases from rat liver and mode of action of nucleoside 3':5'-monophosphate. J Biol Chem 247: 3726–3735

Kuo JF, Greengard P (1970a) Cyclic nucleotide-dependent protein kinases. VI. Isolation and partial purification of a protein kinase activated by guanosine 3',5'-monophosphate. J Biol Chem 245: 2493–2498

Kuo JF, Greengard P (1970b) Cyclic nucleotide-dependent protein kinases. VII. Comparison of various histones as substrates for adenosine 3',5'-monophosphate-dependent and guanosine 3',5'-monophosphate-dependent protein kinases. Biochim Biophys Acta 212: 79–91

Kuroda Y, Hashimoto E, Nishizuka Y, Hamana K, Iwai K (1976) Phosphorylated sites of calf thymus H2B histone by adenosine 3':5'-monophosphate-dependent protein kinase from bovine cerebellum. Biochem Biophys Res Commun 71: 629–635

Langan TA (1968) Histone phosphorylation: Stimulation by adenosine 3',5'-monophosphate. Science 162: 579–580

Langan TA (1971) Cyclic AMP and histone phosphorylation. Ann NY Acad Sci 185: 166–180

Lincoln TM, Flockhart DA, Corbin JD (1978) Studies on the structure and mechanism of activation of the guanosine 3':5'-monophosphate-dependent protein kinase. J Biol Chem 253: 6002–6009

Lipmann F (1933) Über die Bindung der Phosphorsäure in Phosphorproteinen. Biochem Z 262: 3–8

Lipmann F. Levene PA (1932) Serine phosphoric acid obtained on hydrolysis of vitellic acid. J Biol Chem 98: 109–114

Michell RH (1975) Inositol phospholipids and cell surface receptor function. Biochim Biophys Acta 415: 81–147

Michell RH (1979) Inositol phospholipids in membrane function. Trends Biochem Sci 4: 128–131

Nimmo HG, Cohen P (1977) Hormonal control of protein phosphorylation. Adv Cyclic Nucleotide Res 8: 145–266

Nishizuka Y, Takai Y, Hashimoto E, Kishimoto A, Inoue M, Takeda M (1976) Protein phosphokinases and mode of action of guanosine 3',5'-monophosphate. In: Criss WE, Ono T, Sabine JR (eds) Control mechanism in cancer, pp 139–152. Raven Press, New York

Nishizuka Y, Takai Y, Kishimoto A, Hashimoto E, Inoue M, Yamamoto M, Criss WE, Kuroda Y (1978) (1978) A role of calcium in the activation of a new protein kinase system. Adv Cyclic Nucleotide Res 9: 209–220

Nishizuka Y, Takai Y, Hashimoto E, Kishimoto A, Kuroda Y, Sakai K, Yamamura H (1979) Regulatory and functional compartment of three multifunctional protein kinase systems. Mol Cell Biochem 23: 153–165

Rabinowitz M, Lipmann F ((1960) Reversible phosphate transfer between yolk phosphoprotein and adenosine triphosphate. J Biol Chem 235: 1043–1050

Ray TK (1970) A modified method for the isolation of the plasma membrane from rat liver. Biochim Biophys Acta 196: 1–9

Reimann EM, Walsh DA, Krebs EG (1971) Purification and properties of rabbit skeletal muscle 3':5'-monophosphate-dependent protein kinases. J Biol Chem 246: 1986–1995

Rubin CS, Rosen OM (1975) Protein phosphorylation. Annu Rev Biochem 44: 831–887

Seeman P (1972) The membrane actions of anesthetics and tranquilizers. Pharmacol Rev 24: 583–655

Sherod D, Johnson G, Chalkley R (1974) Studies on the heterogeneity of lysine-rich histone in dividing cells. J Biol Chem 249: 3923–3931

Sutherland EW, Rall TW (1960) The relation of adenosine-3',5'-phosphate and phosphorylase to the actions of catecholamines and other hormones. Pharmacol Rev 12: 265–299

Takai Y, Nishiyama K, Yamamura H, Nishizuka Y (1975) Guanosine 3':5'-monophosphate-dependent protein kinase from bovine cerebellum, purification and characterization. J Biol Chem 250: 4690–4695

Takai Y, Nakaya S, Inoue M, Kishimoto A, Nishiyama K, Yamamura H, Nishizuka Y (1976) Comparison of mode of activation of guanosine 3':5'-monophosphate-dependent and adenosine 3':5'-monophosphate-dependent protein kinases from silkworm. J Biol Chem 251: 1481–1487

Takai Y, Kishimoto A, Inoue A, Nishizuka Y (1977a) Studies on a cyclic nucleotide-independent protein kinase and its proenzyme in mammalian tissues. I. Purification and characterization of an active enzyme from bovine cerebellum. J Biol Chem 252: 7603–7609

Takai Y, Yamamoto M, Inoue M, Kishimoto A, Nishizuka Y (1977b) A proenzyme of cyclic nucleotide-independent protein kinase and its activation by calcium-dependent protease from rat liver. Biochem Biophys Res Commun 77: 542–550

Takai Y, Kishimoto A, Iwasa Y, Kawahara Y, Mori T, Nishizuka Y (1979a) Calcium-dependent activation of a multifunctional protein kinase by membrane phospholipds. J Biol Chem 254: 3692–3695

Takai Y, Kishimoto A, Iwasa Y, Kawahara Y, Mori T, Nishizuka Y, Tamura A, Fujii T (1979b) A role of membranes in the activation of a new multifunctional protein kinase system. J Biochem (Tokyo) 86: 575–578

Tao M, Salas ML, Lipmann F (1970) Mechanism of activation by adenosine 3':5'-monophosphate of a protein phosphokinase from rabbit reticulocytes. Proc Natl Acad Sci USA 67: 408–414

Yamamoto M, Takai Y, Inoue M, Kishimoto A, Nishizuka Y (1978) Characterization of cyclic nucleotide-independent protein kinase produced enzymatically from its proenzyme by calcium-dependent neutral protease from rat liver. J Biochem (Tokyo) 83: 207–212

Yamamura H, Takeda M, Kumon A, Nishizuka Y (1970) Adenosine 3':5'-cyclic phosphate-dependent and independent histone kinases from rat liver. Biochem Biophys Res Commun 40: 675–682

Yamamura H, Kumon A, Nishizuka Y (1971) Cross reactions of adenosine 3':5'-monophosphate dependent protein kinase systems from rat liver and rabbit skeletal muscle. J Biol Chem 246: 1544–1547

Yamamura H, Nishiyama K, Shimomura R, Nishizuka Y (1973) Comparison of catalytic units of muscle and liver adenosine 3',5'-monophosphate-dependent protein kinases. Biochemistry 12: 856–862

Walsh DA, Perkins JP, Krebs EG (1968) An adenosine 3',5'-monophosphate-dependent protein kinase from rabbit skeletal muscle. J Biol Chem 243: 3763–3765

2. Effect of Catabolite Repression on Chemotaxis in Salmonella Typhimurium

D.E. KOSHLAND, Jr. and M.J. ANDERSON

A. Abstract

The chemotaxis of *Salmonella typhimurium* LT2-ST1, grown with different sole carbon sources toward ribose, allose, serine, aspartate, and glucose, was examined. Maximum chemotaxis toward serine, aspartate, and glucose were unchanged with either glucose or citrate as the carbon source. The height of the peak in the ribose response curve was 25% of a standard serine response when the bacteria were grown on citrate. However, when the bacteria were grown on glucose, the ribose response was 1% of the serine standard, and when on glycerol, 5%. Chemotaxis toward allose, a competitive inhibitor of ribose chemotaxis, was similarly affected. The amount of ribose binding protein, the receptor for ribose chemotaxis, was determined in bacteria grown on citrate, glucose, or glycerol. The amount of ribose-binding protein released from cells grown on glucose or glycerol was appreciably less than the amount released from cells grown on citrate. These data suggest that the observed effect on ribose chemotaxis is due to catabolite repression of ribose-binding protein and another component of the chemotactic system.

B. Introduction

To participate in a symposium dedicated to Fritz Lipmann is a particular pleasure for me. I first met him during my postdoctoral fellowship at Harvard when I was making a transition from organic chemistry to biochemistry. Fritz Lipmann welcomed a very naive young chemist into his laboratory and trained him in some of the intricacies of protein purification. More importantly, he spent many hours discussing his concepts of biochemistry which were invariably insightful and integrative. There followed a friendship of many years including periods as colleagues at Rockfeller University and as co-members of a study section. On all occasions, both formal and informal, Fritz Lipmann is an inspiration to those around him both in his warm personal friendships and his extraordinary ability and enthusiasm for solving the puzzles of nature.

(D.E. Koshland, Jr.)

Department of Biochemistry, University of California, Berkeley, California 94720, USA

The chemotactic response of bacteria to chemical signals is mediated through receptor proteins (Adler 1969). Moreover, it is apparent that the strength of response varies considerably from one chemical to another (Adler 1975; Koshland 1977). Since we are now aware that these receptors feed into a signalling system of multiple components (Ordal and Adler 1974; Parkinson 1974; Strange and Koshland 1976), several potential mechanisms are available to explain these different quantitative results. One of these is that the number of receptor molecules or other components of the signaling system varies from one chemoeffector to another. A second is that the number of receptor molecules remains the same but their interaction with the next component in the signaling system varies. Other mechanisms could be suggested as well. To investigate these alternatives, it would be of interest to determine the effect of variation in the quantity of a particular receptor molecule on the response to its substrate. Growth of bacteria on glucose or any other compound which can serve as a source of intermediary metabolites can reduce the rate of formation of glucose-sensitive proteins relative to the rate of synthesis of other proteins. We therefore examined the effects of growth on glucose and glycerol on chemotaxis toward several attractants, and found that these response systems react in different ways to catabolite repression.

C. Materials and Methods

I. Bacterial Strains

The bacteria were derived from *Salmonella typhimurium* LT2. ST1 was selected for motility on tryptone semisolid agar plates [1% tryptone (Difco), 0.5% NaCl, 0.35% agar] by allowing the bacteria to swim out from a drop of a culture of LT2 placed in the center of the plate, and picking those bacteria which swam the greatest distance (Armstrong et al. 1967). The wet weight of bacteria is estimated from the fact that an absorbance at 650 nm of 0.1 is equal to 235 μg wet weight bacteria/ml.

II. Media

Bacteria were grown at 30°C with shaking on Vogel-Bonner citrate minimal medium (VBC) (Vogel and Bonner 1956) or on media with single replacements as, for example, VB + 0.2% glucose, meaning the same composition as VBC but with citrate replaced by 0.2% glucose.

III. Assay for Ribose-Binding Protein

The micro-complement fixation technique utilized has been described (Wasserman and LeVine 1961) and the protocol followed here was described in detail by Champion et al. (1973). Antiserum was prepared in New Zealand white rabbits by injection intradermally and in the footpads in the presence of Freund's complete adjuvant. Before use in the micro-complement fixation assay, it was diluted 1:1500 in isotris buffer. The amount of cellular protein in the first dilution of the assay was adjusted in each case to be the amount obtained from 1.87 mg wet weight bacteria, using the same A_{650} for all the cultures. Subsequent twofold dilutions were made to determine the point of maximum complement fixation.

IV. Capillary Assay for Chemotaxis

Chemotaxis was assayed according to the capillary method of Adler (1973) except that the bacterial suspension was contained in short lengths of glass tubing (3.7 × 0.7 cm) plugged at one end with a stopper through which the capillary which contains attractant could be put into the solution. The bacteria were allowed to accumulate for 30 min. The contents of the capillary were then diluted, plated and counted to quantitate the response. The suspension of bacteria for the assay was obtained by first allowing the bacteria to grow to A_{650} = 0.4, and then diluting the culture to approximately 5×10^6 bacteria/ml. The bacteria were diluted for the assay, and the attractant solutions prepared, in the medium in which the bacteria were grown. All assays were performed at 30°C.

V. Protein Determination

Protein concentration was determined by the method of Lowry et al. (1951).

D. Results

I. Ribose and Serine Chemotactic Responses of ST1 Grown on Citrate, Glucose, and Glycerol

The chemotactic responses to ribose of Salmonella strain ST1 grown on citrate, glucose, and glycerol as sole carbon source were tested to examine the effects of catabolite repression on the response. Since the motility of the bacteria varied, a standard chemotactic response which is not affected by carbon source was used in order to allow valid comparisons between experiments carried out on different days. It was found that the responses to serine are equal (i.e., show the same number of bacteria accumulate) in bacteria grown in VB + 0.2% glycerol, VB + 0.2% glucose, and VBC. Therefore, this behavior was used as the standard.

The ribose and serine responses of ST1 grown on citrate, glycerol, and glucose as sole carbon source are shown in Fig. 1. At the peak in the ribose response curve of bacteria grown in citrate, about 25% as many bacteria are accumulated in a capillary tube as are accumulated in the standard, i.e., when the attractant is 10^{-1} M serine. When the bacteria are grown on glycerol (VB + 0.2% glycerol), 5% as many bacteria accumulate. When the bacteria are grown on glucose (VB + 0.2% glucose), the response is about 1% of that to serine. Thus, a ratio of 25:5:1 exists for the ribose response of bacteria grown on citrate, glycerol, and glucose. The results are consistent with catabolite repression of ribose chemotaxis.

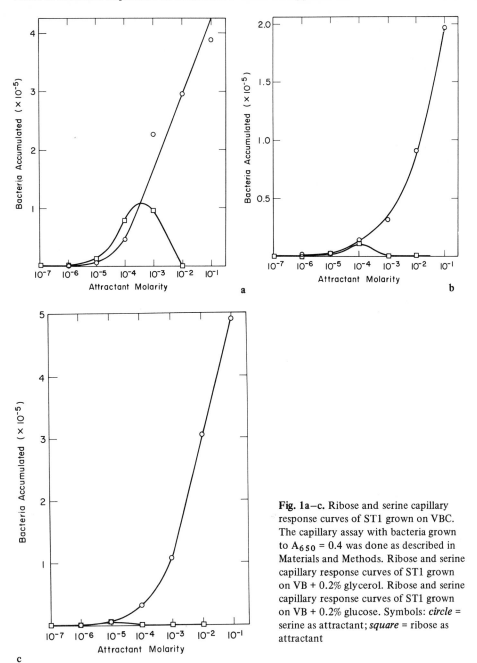

Fig. 1a–c. Ribose and serine capillary response curves of ST1 grown on VBC. The capillary assay with bacteria grown to A_{650} = 0.4 was done as described in Materials and Methods. Ribose and serine capillary response curves of ST1 grown on VB + 0.2% glycerol. Ribose and serine capillary response curves of ST1 grown on VB + 0.2% glucose. Symbols: *circle* = serine as attractant; *square* = ribose as attractant

The fact that the number of bacteria responding in the capillary assay falls off at high attractant concentrations has been previously explained by Mesibov et al. (1973). When the concentration of attractant is much higher than the dissociation constant for the binding of attractant to the chemoreceptor, the receptor affinity becomes saturated. The bacteria swim into the neighborhood of the mouth of the capillary but essentially see no gradient, and therefore do not swim into the capillary. The response to serine is unusual in that saturation is not reached even at 10^{-1} M serine.

II. Allose Response of Cells Grown on Citrate and Glucose

Specificity studies on ribose binding protein and ribose chemotaxis identified allose as the only compound known to bind competitively with ribose and to compete with ribose in a capillary assay for chemotaxis (Aksamit and Koshland 1974). Since it also acts as an attractant, it was interesting to see whether the results would be similar to those with ribose chemotaxis. The capillary responses toward serine and allose of ST1 grown in citrate and glucose were compared. The results are qualitatively similar to those observed in the case of ribose, i.e., the response drops dramatically when the bacteria assayed have been grown with glucose in place of citrate as the sole carbon source (from 22% of the serine maximum to about 1%). Although the maximum in the capillary response curve shifts to higher attracting concentrations because allose binds more weakly to the ribose receptor by a factor of 10^3, the peak heights and the strength of the response are about the same. Apparently, the amount of signal conveyed to the rest of the chemotactic system is the same whether ribose or allose is bound to the protein.

III. Other Chemotactic Responses of ST1 Grown on Citrate and Glucose or Glycerol

The effect of growth on glucose or glycerol on chemotaxis toward other attractants was examined to determine if the ribose case is unique or if some simple pattern exists among those responses subject to the glucose effect and those which are not, e.g., which are sugar chemoreception systems being repressed and amino acid systems not? Cultures of ST1 were grown on VBC and VB + 0.2% glucose and their chemotactic responses to aspartate (neutralized with NaOH) were assayed. Adler and co-workers have determined that the aspartate and serine responses are independent in *Escherichia coli* (Adler et al. 1973; Mesibov and Adler 1972). Both cultures gave a very strong response to aspartate (Fig. 2), so it appears that this chemoreception system is not susceptible to the glucose effect. It is similar to the serine response in this respect.

The glucose chemotactic response was also measured in comparison with the serine response. Bacteria grown on VBC and on VB + 0.2% glycerol were assayed. Both responses are weak, about 2% of the serine maximum, but the peak heights in comparison with the serine response are not significantly affected by the change in sole carbon source. It is reasonable that the glucose chemoreception system is not subject to a glucose effect on synthesis.

Chemotactic responses to maltose, N-acetyl glucosamine, and galactose were also assayed. The responses by bacteria grown on VBC were all extremely weak (near the limits of reliable detection with the capillary assay for chemotaxis). It is assumed that these chemoreceptor systems need to be induced, as was found in the case of *Escherichia coli*. Therefore, they are inappropriate for the study of chemoreception systems

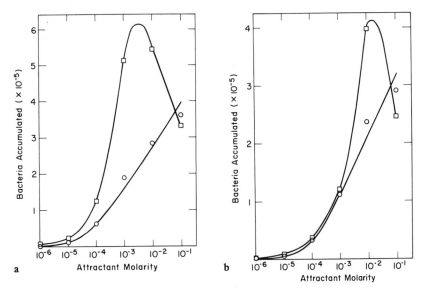

Fig. 2a, b. Serine and aspartate (neutralized with NaOH) capillary response curves of ST1 grown on VBC. Serine and aspartate (neutralized with NaOH) capillary response curves of ST1 grown on VB + 0.2% glucose. Symbols: *circle* = serine as attractant; *square* = aspartate as attractant

affected by growth on glucose because induction of the systems in this particular organism requires growth in the presence of the attractant. This could not be done without potentially obscuring the results of the glucose repression.

IV. Amount of Ribose-Binding Protein in ST1 Grown on Citrate, Glucose, and Glycerol

Duplicate cultures of ST1 were grown at 30°C on several carbon sources to the density which is used in chemotaxis assays. They were then subjected to osmotic shock followed by sonication and centrifugation. The osmotic shock fluid and the sonicate supernatant were assayed for ribose-binding protein by micro-complement fixation to determine the effect of carbon source on levels of ribose binding protein, and to determine if there is a correlation between the amount of ribose-binding protein in the cells and the strength of the chemotactic response to ribose. The results are presented in Table 1. Growth on

Table 1. Amounts of ribose-binding protein in ST1 grown on citrate, glycerol, and glucose

Growth medium	Ribose-binding protein (μg/mg cellular protein)
VBC	7.4
VB + 0.2% glycerol	4.1
VB + 0.2% glucose	2.7

Determined by micro-complement fixation assay on osmotically shocked and sonicated bacteria (grown to $A_{650} = 0.4$)

citrate yields bacteria with higher levels of ribose-binding protein than does growth on glucose or glycerol. The results are consistent with catabolite repression of ribose-binding protein synthesis. The system is not totally repressible, however, because there is always a substantial amount of ribose-binding protein remaining.

E. Discussion

It has been shown that the bacterial chemotactic responses to ribose and allose are affected differently by growth on catabolite repressing media than are the serine and aspartate responses. Chemotaxis toward ribose or allose is subject to catabolite repression while chemotaxis toward serine or aspartate is not. There is a rough correlation between the amount of ribose-binding protein present in the bacteria and the strength of the chemotactic response to ribose. A reduction in the number of receptors accompanies the reduced sensitivity of the bacteria to the chemical gradient. The difference in the amounts of ribose binding protein is similar to that observed for catabolite repression of other proteins (de Crombrugghe et al. 1969).

However, there is always a substantial amount of ribose binding protein present even when the bacteria are grown on glucose, the best carbon source for catabolite repression. The presence of this amount of the protein is not reflected in the presence of a substantial chemotactic response to ribose: the response drops to 5% of that for bacteria grown on VBC. A likely explanation is, therefore, that another component in the chemotactic response is more severely repressed, and that its control also contributes to the effects seen on the strength of the chemotactic response to ribose. This interpretation is supported by results obtained by Fahnestock and Koshland (1978). They found that growth of *Salmonella typhimurium* in the presence of ribose substantially increased the amount of ribose binding protein in the cells, as demonstrated by SDS polyacrylamide gel electrophoresis of extracts of whole bacteria, but the chemotactic re response was not precisely a linear function of amount of ribose binding protein. Thus, it appears in this case, also, that the limiting factor in the response is not simply the number of ribose receptors but includes also another component in the response mechanism.

Another possible explanation for the data, that might be considered, is that the protein plays more than one role in the bacteria and the roles are not interchangeable, e.g., some of the protein is "utilized" in transport and can not readily be converted to use in chemotaxis. Also, it may be that the minimum amount of ribose binding protein necessary for chemotaxis to occur is a fairly large one.

It has been shown that the production of flagella and hence chemotactic mobility are subject to catabolite repression in *Escherichia coli* (Adler and Templeton 1967; Dobrogosz and Hamilton 1971). This phenomenon is not involved in these experiments, however, since the chemotactic response of the bacteria to serine does not vary.

The smaller amounts of ribose binding protein in cells grown on glucose and glycerol can explain the shift of the peak in the chemotactic response curves to lower ribose concentrations. The smaller amounts of ribose binding protein might allow less uptake and metabolism of the attractant during the course of the assay. The bacteria in a capillary assay are responding to a gradient formed by two effects: (1) the diffusion of attractant from the mouth of the capillary, and (2) metabolism of the attractant by the bacteria which are responding. Thus, if there is a smaller amount of the attractant consumed, the gradient sensed by the bacteria will be different, and the peak in the response curve will appear at a lower attactant concentration.

Acknowledgment. This work was supported by U.S. Public Health Service Grant AMO9765 and National Science Foundation Grant PCM75-16410.

References

Adler J (1969) Chemoreceptors in bacteria. Science 166: 1588–1597

Adler J (1973) A method for measuring chemotaxis and use of the method to determine optimum conditions for chemotaxis by *Escherichia coli.* J gen Microbiol 74: 77–91

Adler J (1975) Chemotaxis in bacteria. Ann Rev Biochem 44: 341–356

Adler J, Templeton B (1967) The effect of environmental conditions on the motility of *Escherichia coli.* J Gen Microbiol 46: 175–184

Adler J, Hazelbauer GL, Dahl MM (1973) Chemotaxis toward sugars in *Escherichia coli.* J Bact 115: 824–847

Aksamit R, Koshland DE Jr (1974) Identification of the ribose binding protein as the receptor for ribose chemotaxis in *Salmonella typhimurium.* Biochemistry 13: 4473–4478

Armstrong JB, Adler J, Dahl MM (1967) Nonchemotactic mutants of *Escherichia coli.* Genetics 61: 61–66

Champion AB, Prager EM, Wachter D, Wilson AC (1973) In: Wright CA (ed) Biochemical and immunochemical taxonomy of animals. Academic Press, London

deCrombrugghe B, Perlman RL, Varmus EH, Pastan I (1969) Regulation of inducible enzyme synthesis in *Escherichia coli* by cyclic adenosine 3',5'-monophosphate. J Biol Chem 244: 5828–5835

Dobrogosz WJ, Hamilton PB (9171) The role of cyclic AMP in chemotaxis in *Escherichia coli.* Biochem Biophys Res Comm 42: 202–207

Fahnestock M, Koshland DE Jr (1979) Control of the receptor for galactose taxis in *Salmonella typhimurium.* J Bacteriol 137: 758–763

Koshland DE Jr (1977) Sensory response in bacteria. Adv in Nuerochem 2: 277–341

Lowry OH, Rosebrough NJ, Farr AL, Randal RJ (1951) Protein measurement with the folin phenol reagent. J Biol Chem 193: 265–275

Mesibov R, Adler J (1972) Chemotaxis toward amino acids in *Escherichia coli.* J Bact 112: 315–326

Mesibov R, Ordal GW, Adler J (1973) The range of attractant concentrations for bacterial chemotaxis and the threshold and size of response over this range. J gen Physiol 62: 203–223

Ordal GW, Adler J (1974) Properties of mutants in galactose taxis and transport. J Bact 117: 509–516

Parkinson JS (1974) Data processing by the chemotaxis machinery of *Escherichia coli.* Nature 252: 317–319

Strange PG, Koshland DE Jr (1976) Receptor interactions in a signalling system: competition between ribose receptor and galactose receptor in the chemotaxis response. Proc Natl Acad Sci USA 73: 762–766

Vogel HJ, Bonner DM (1956) Acetylornithinase of *Escherichia coli:* partial purification and some properties. J Biol Chem 218: 97–102

Wasserman E, Levine L (1961) Quantitative micro-complement fixation and its use in the study of antigenic structure by specific antigen-antibody inhibition. J Immunol 87: 290–295

3. Subunit Interaction of Adenylylated Glutamine Synthetase

E. R. STADTMAN, R. J. HOHMAN, J. N. DAVIS, M. WITTENBERGER, P. B. CHOCK, and S. G. RHEE

A. Introduction

The activity of *Escherichia coli* glutamine synthetase (GS) is regulated by the cyclic adenylylation and deadenylylation of a unique tyrosyl group in each subunit (for review, see Stadtman and Ginsburg 1974). As illustrated in Fig. 1, the adenylylation reaction is catalyzed by an adenylyltransferase (AT) and involves transfer of adenylyl groups from ATP to the enzyme with the concomitant formation of PPi. Since GS is composed of twelve identical subunits, and adenylylated subunits are catalytically inactive under most physiological conditions, the catalytic potential of the enzyme is nearly inversely proportional to the average number, $\bar{n}$, of covalently bound adenylyl groups per enzyme molecule (Kingdon et al. 1967). Although the adenylylation reaction is reversible in the presence of PPi (Mantel and Holzer 1970), removal of the adenylyl group from GS under physiological conditions is achieved by phosphorolysis of the adenylyl-O-tyrosyl bond to yield ADP and unmodified GS (Anderson and Stadtman 1970). Because the adenylylation and deadenylylation reactions are catalyzed at separate, noninteract-

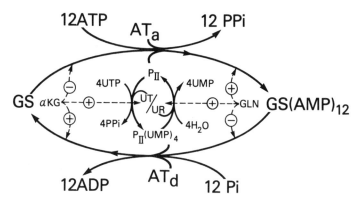

Fig. 1. Interrelationship between the uridylylation and deuridylylation of the P_{II} protein and the adenylylation and deadenylylation of glutamine synthetase (GS), and the reciprocal controls of these interconversions by glutamine (Gln) and α-ketoglutarate (α KG). (+) indicates stimulation, (−) indicates inhibition

Laboratory of Biochemistry, National Heart, Lung, and Blood Institute, National Institutes of Health, Building 3, Room 222, Bethesda, Maryland 20205, USA

ing catalytic sites on a single adenylyltransferase (Anderson et al. 1970; Rhee et al. 1978; Hennig and Ginsburg 1971), it is evident that in the absence of appropriate regulation the two reactions will be tightly coupled resulting simply in senseless phosphorolysis of ATP to ADP and PPi. As shown in Fig. 1, indiscriminate coupling of the adenylylation and deadenylylation reaction is prevented by the action of Shapiro's regulatory protein, P_{II}, (Shapiro 1969) which also exists in an unmodified form, P_{IIA}, and a modified (uridylylated) form, P_{IID} (Brown et al. 1971; Mangum et al. 1973; Adler et al. 1975). The capacity of ATase to catalyze the adenylylation of GS is dependent on the concentration of P_{IIA}, whereas the capacity to catalyze the deadenylylation reaction is determined by the concentration of P_{IID} (Brown et al. 1971; Adler et al. 1975). The steady-state distribution between P_{IIA} and P_{IID} is determined by the relative contributions of uridylyltransferase (UT) and a uridylyl-removing enzyme (UR). These activities are contained in a single protein or protein complex (Mangum et al. 1973; Brancroft et al. 1978). In the final analysis, however, the adenylylation and deadenylylation of GS is regulated by the combined effects of over 40 different metabolites that affect the catalytic activities of one or more of the regulatory enzymes AT, UT, and UR (Stadtman and Chock 1978a; Engleman and Francis 1978). By far the most important of these metabolites are α-ketoglutarate and glutamine which exhibit opposite and reciprocal effects on the adenylylation-deadenylylation and on the uridylylation-deuridylylation pairs of reactions (Stadtman and Ginsburg 1974; Stadtman and Chock 1978a; Engleman and Francis 1978).

B. The Bicyclic Cascade System

I. Dynamic Concept

The interaction between the adenylylation of GS, and the uridylylation cycle of P_{II} can be visualized as a bicyclic cascade system (Fig. 2). With the assumption that the modification and demodification steps are unintermittent, it follows that for any given metabolic condition a steady state will be attained in which the rate of the forward step in each cycle is equal to the rate of the regeneration step in that cycle. A theoretical analysis (Stadtman and Chock 1978a; Stadtman et al. 1979b) of a model based on such a cascade shows that the fraction of GS subunits that are adenylylated in the steady state is determined by the values of 16 different parameters; i.e., specific rate constants of the various steps and the stability constants that govern dissociation of the various protein-protein and effector-protein complexes. Since each of these parameters is susceptible to modulation by allosteric and substrate interactions of metabolites with one or more of the cascade enzymes, the GS cascade is endowed with remarkable flexibility with respect to metabolite control. Moreover, because the state of adenylylation is a multiplicative function of eight different parameter ratios, even small changes in several of these parameters can lead to enormous changes in the response of adenylylation to primary allosteric stimuli (Stadtman and Chock 1978a). It is not possible to describe in detail here either the theoretical analyses of the GS cascade model, or the experimental evidence that has been obtained to support the predictions derived from it. The dynamic concept upon which the theoretical analysis is based is supported (1) by in vivo studies (Senior 1975) showing that the steady-state level of adenylylation of GS varies from 0

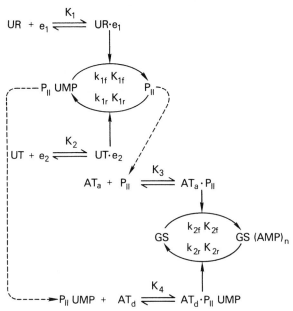

Fig. 2. The bicyclic cascade model of glutamine synthetase (GS) regulation. See Stadtman and Chock (1978a) for definition of terms

to 12, depending upon the nutritional state of the organism, and (2) by in vitro experiments (Segal et al. 1974) showing that GS assumes a steady-state level of adenylylation when it is incubated in a reaction mixture containing all of the cascade enzymes and the effectors ATP, UTP, Pi, α-ketoglutarate, glutamine, Mg^{2+} and Mn^{2+}; moreover, a change in the concentration of any one of the effectors results in a shift in the steady-state level of adenylylation.

II. In vitro Analysis of the Cascade

Inasmuch as the sole function of the UR/UT complex is to regulate the distribution of P_{IIA} and P_{IID}, in vitro studies of the adenylylation cascade have been greatly simplified by substituting mixtures of pure P_{IIA} and P_{IID} for the UR/UT system (Rhee et al. 1978). The net effect is to convert the bicyclic cascade into the experimentally more manageable monocyclic system illustrated in Fig. 3. A theoretical analysis of such a system shows that the steady-state level of adenylylation is a function of the mol fraction ($[P_{IIA}]_{mf}$) of P_{IIA} and the parameter ratio α_f/α_r, where α_f is the product of several constants that govern the forward (adenylylation) step, and α_r is the product of the corresponding constants that govern the regeneration (deadenylylation) step. Experimentally, $(P_{IIA})_{mf}$ can be varied at will by altering the proportions of pure P_{IIA} and P_{IID}, and the α_f/α_r ratio can be varied by varying the relative concentrations of α-ketoglutarate and and glutamine, which exhibit opposite and reciprocal effects on the adenylylation and deadenylylation reactions. A few rather remarkable regulatory characteristics of the GS cascade have been disclosed by a detailed experimental analysis of this model system.

As can be seen from the data in Fig. 4A, the steady-state level of adenylylation can vary enormously depending upon both the $(P_{IIA})_{mf}$ and the ratio of α-ketoglutarate to

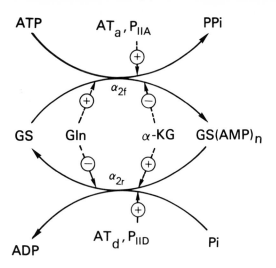

Fig. 3. Modified monocyclic cascade illustrating the regulatory roles of P_{IIA}, P_{IID}, α-ketoglutarate, and glutamine on the adenylylation of GS. From Rhee et al. (1978)

glutamine. Bearing in mind the fact that, in vivo, the $(P_{IIA})_{mf}$ is determined by the relative activities of UR and UT, the data illustrate the important contribution of the uridylylation cycle to the overall control of GS activity. Moreover, the fact that at any given $(P_{IIA})_{mf}$ value the state of adenylylation can vary from almost 0 to 12 illustrates the extraordinary flexibility of the system to allosteric control. This is further illustrated by the data in Fig. 4B. The *solid lines* show how the value of $\bar{n}$ varies as a function of

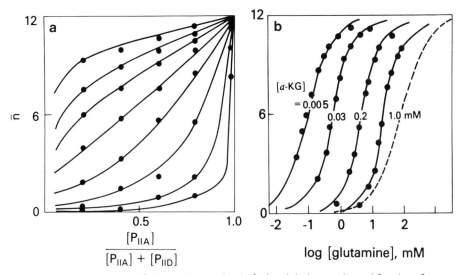

Fig. 4a, b. The dependence of the steady-state level of adenylylation on the mol fraction of P_{IIA} and the ratio of α-ketoglutarate and glutamine. **A** Experimental data from studies in which $(P_{IIA})_{mf}$ was varied by varying the proportions of P_{IIA} and P_{IID}, and the ratio α_{2f}/α_{2r} was varied by changing the relative concentration of α-ketoglutarate and glutamine as described by Rhee et al. (1978a). **B** Steady-state levels of $\bar{n}$ as a function of glutamine concentration. $(P_{IIA})_{mf}$ is 0.6 and the concentration of α-ketoglutarate as indicated on the *curves*. The *broken line* indicates the calculated fractional saturation of AT with glutamine (Kd = 80 mM)

glutamine concentration at each of several different α-ketoglutarate levels, when $(P_{IIA})_{mf}$ is held constant at 0.6. Since, in all instances, the effect of glutamine results from its binding to the adenylyltransferase (AT), the *broken line* in Fig. 4B is presented to show how the fractional saturation of AT varies with increasing glutamine concentration. A comparison of the broken line with the experimental curves discloses two important features of the cascade system: (1) The concentration of a primary allosteric effector required to sustain a steady state in which 50% of the GS is adenylylated can be orders of magnitude lower than the concentration that is required to achieve 50% saturation of the converter enzyme (AT) with which it reacts directly (Chock and Stadtman 1978; Stadtman and Chock 1978b). In other words, the cascade is endowed with an enormous signal amplification potential (Stadtman and Chock 1978a). (2) The linear portions of the experimental curves are considerably steeper than the linear portion of the broken line, which was calculated on the assumption that the binding of glutamine to AT follows a normal Michaelis-Menton hyperbolic function. This shows that the kinetic order of the response of $\bar{n}$ to glutamine concentration is greater than 1.0. In other words, the cascade elicits a "cooperative" response to glutamine concentration. This apparent cooperativity arises from the fact that glutamine not only stimulates the adenylylation step, but it also inhibits the deadenylylation reaction (Chock and Stadtman 1978).

In sum, the experimental data confirm predictions of the theoretical bicyclic cascade model in showing that (1) the cascade is extremely flexible with respect to allosteric control; (2) it is endowed with enormous signal amplification capacity; and (3) it is capable of eliciting a cooperative type of response to increasing allosteric effector concentration.

C. Multimolecular Forms of Glutamine Synthetase

The above studies show that, depending upon metabolic conditions, the state of adenylylation of GS will vary from 0 to 12. It follows that if the adenylylation and deadenylylation reactions are random processes, multimolecular forms of GS will be generated which differ from one another, not only by the number of adenylylated subunits, but also by the distribution of adenylylated and unadenylylated subunits within single molecules. Several years ago, M.S. Raff and W.C. Blackwelder (pers. comm.) showed that 382 uniquely different forms of GS are possible. In the meantime, several lines of experimental evidence have demonstrated that naturally occurring GS preparations, as well as those generated in vitro by dissociation and reassociation of mixtures of $E_{\bar{0}}$ and $E_{\overline{12}}$, consist of enzyme molecules containing both adenylylated and unadenylylated subunits (Ciardi et al. 1973), and further that heterologous interactions between adenylylated and unadenylylated subunits within these hybrid molecules affect certain catalytic parameters (Stadtman et al. 1975). Nevertheless, until recently, efforts to resolve these mixtures into more or less uniquely defined molecular entities by means of conventional chromatographic, electrophoretic, or electrofocusing techniques have failed. A partial solution to this problem has now been achieved by the use of anti-AMP antibodies, which were obtained by im-

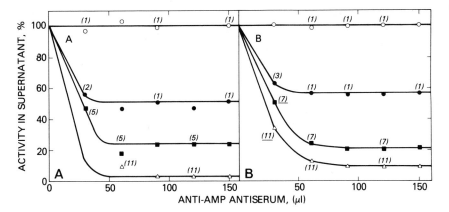

Fig. 5A, B. Immunoprecipitability of glutamine synthetase with anti-AMP antiserum. **A** Titration of native enzyme preparations $GS_{\bar{1}}$, *open circles;* $GS_{\overline{11}}$ *open triangles;* equimolar mixture of $GS_{\bar{1}}$ and $GS_{\overline{11}}$, *closed circles;* $GS_{\bar{6}}$, *closed squares.* **B** Titration of enzyme preparations following dissociation in 7 M urea and reassociation upon dialysis. $GS_{\bar{1}}$, *open circles;* $GS_{\overline{11}}$, *open triangles;* equal amounts of $GS_{\bar{1}}$ and $GS_{\overline{11}}$ after each separately had been subjected to dissociation and reassociation conditions, *closed circles;* equal amounts of $GS_{\bar{1}}$ and $GS_{\overline{11}}$ were mixed and then subjected to dissociation and reassociation conditions ($\bar{n}$ value = 8.0), *closed squares.* The *numbers in parenthesis* indicate the state of adenylylation of the nonprecipitable fraction at the indicated times

munizing sheep with adenylylated bovine serum albumin prepared by reacting AMP with albumin in the presence of carbodiimide (Hohman and Stadtman 1978).

I. Immunoprecipitation Studies

The data in Fig. 5 show that the anti-AMP antibodies will precipitate over 90% of highly adenylylated enzyme ($GS_{\overline{11}}$), but are unable to precipate unadenylylated enzyme ($GS_{\bar{1}}$); moreover, when a $GS_{\bar{6}}$ preparation obtained by mixing equal amounts of $GS_{\bar{1}}$ and $GS_{\overline{11}}$ is titrated with partially purified antiserum, only the $GS_{\overline{11}}$ component is precipitated. Direct analysis of the nonprecipitable fraction showed that it contained only $GS_{\bar{1}}$. In contrast, when a native $GS_{\bar{6}}$ preparation was titrated with antiserum, 75% of the enzyme was precipitated. Of particular significance, however, was the finding that the fraction (25%) which did not precipitate contained on the average five adenylyl groups per molecule. These data therefore show that the antibodies can clearly distinguish between more or less fully adenylylated and fully unadenylylated enzyme in mixtures, but can precipitate only a fraction of native preparations, which, as will be shown later, are complex mixtures of enzyme molecules containing 2.0 to 9.0 adenylylated subunits.

1. Precipitation of Artificial Hybrid Forms. In earlier studies, Ciardi et al. (1973) showed that when a mixture of $GS_{\bar{0}}$ and $GS_{\overline{12}}$ is incubated with 7 M urea, complete dissociation of the subunits occurs, and that subsequent removal of the urea by dialysis leads to reassociation of the subunits to form dodecameric aggregates that possess stability characterisics similar to those exhibited by partially adenylylated native enzyme preparations. These results were interpreted to indicate that dissociation and reassociation of GS$_{\bar{0}}$ and GS$_{\overline{12}}$ mixtures leads to the production of hybrid enzyme molecules containing both adenylylated and unadenylylated subunits. Since immunoprecipitation tests with anti-AMP antibodies can clearly distinguish between partially adenylylated native GS preparations and mixtures of GS$_{\bar{1}}$ and GS$_{\overline{11}}$ (Fig. 5A), it appeared possible that such tests might be used to verify the conclusion that dissociation and reassociation of mixtures of GS$_{\bar{1}}$ and GS$_{\overline{11}}$ yield hybrid molecules. This is indeed so. Immunotitrations of either native GS$_{\bar{6}}$ (Fig. 5A) or a GS$_{\bar{8}}$ preparation obtained by dissociation and reassociation of a mixture of GS$_{\bar{1}}$ and GS$_{\overline{11}}$ (Fig. 5B) led to precipitation of 75–80% of the enzyme, and in both cases, the nonprecipitable fraction (20–25%) contained about the same number of adenylylated subunits per molecule as did the starting material. In contrast, only the GS$_{\overline{11}}$ component (= 50%) of the GS$_{\bar{1}}$ plus GS$_{\overline{11}}$ mixtures was precipitated by the antibodies (closed circles in Figs. 5A and B); the nonprecipitable fraction contained only unadenylylated (GS$_{\bar{1}}$) enzyme. The results with the urea-treated mixture of GS$_{\bar{1}}$ and GS$_{\overline{11}}$ (Fig. 5B, closed squares) are not attributable to urea-induced alteration of protein conformation because the GS$_{\bar{1}}$ and GS$_{\overline{11}}$ preparations used in the control experiments (Fig. 5B, open circles, open triangles, and closed circles) had also been subjected, independently, to the dissociation-reassociation conditions. The results of these studies provide substantial support for the conclusion that the fractional adenylylation of GS which occurs under physiological steady-state conditions leads to a complex mixture of hybrid molecules and that a similar mixture of hybrids is generated when mixtures of adenylylated and unadenylylated GS preparations are subjected to dissociating and reassociating conditions in vitro.

2. Fluorescence Measurement of Antibody-Adenylylated Subunit Interaction. The failure to obtain quantitative precipitation of native GS$_{\bar{6}}$ is not due to inability of the antibodies to react with all adenylylated subunits, but rather to the fact that not all of the antibody-enzyme complexes are capable of forming macromolecular lattices. This conclusion is indicated by the fact that the nonprecipitable fraction of native GS$_{\bar{6}}$ (Fig. 5A, closed squares) is precipitated by addition of rabbit anti-sheep IgG antibodies (data not shown), and also by direct measurement of the rapid (10 s) quenching of the intrinsic tryptophan fluorescence which occurs when anti-AMP antibodies bind to adenylylated GS subunits.

As shown in Fig. 6A, when equivalent amounts of native GS$_{\bar{1}}$, GS$_{\bar{6}}$, and GS$_{\overline{11}}$ are titrated with antibodies, the amplitude of the observed fluorescence change is proportional to the average number of adenylylated subunits present. This suggests that the antibodies can react with all adenylylated subunits, irrespective of the state of adenylylation. This interpretation is supported further by the data in Fig. 6B, which show that the change in intrinsic fluorescence observed when antibodies react with an artificial GS$_{\bar{6}}$ preparation (obtained by mixing equal amounts of GS$_{\bar{1}}$ and GS$_{\overline{11}}$) is the same as that observed with a native GS$_{\bar{6}}$ preparation.

These fluorescence quenching measurements indicate that the primary interaction of antibodies with adenylylated subunits is not affected by the state of adenylylation. Nevertheless, lattice formation, as measured by precipitability of the antibody-enzyme complex (Fig. 5) is obviously very much dependent upon the adenylylation state.

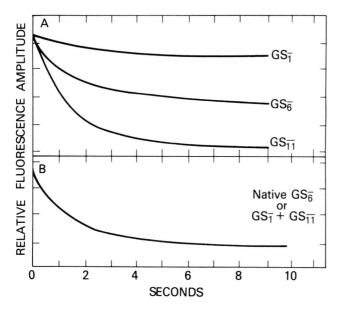

Fig. 6A, B. Tryptophan fluorescence quenching due to reaction between glutamine synthetase and anti-AMP antibodies. **A** 0.33 mg/ml of either $GS_{\bar{1}}$, $GS_{\bar{6}}$, or of $GS_{\overline{11}}$ were mixed together with 0.23 mg/ml of an affinity column purified anti-AMP antibody preparation (Hohman and Stadtman 1978) in a stopped-flow fluoremeter, and the change in fluorescence was monitored by means of oscillograph tracing over a period of 10 s. **B** Similar to (A), but a native $GS_{\bar{6}}$ preparation (0.58 mg) and an artificial $GS_{\bar{6}}$ preparation (obtained by premixing 0.29 mg each of $GS_{\bar{1}}$ and $GS_{\overline{11}}$) were each mixed with 0.23 mg/ml of purified antibodies. The changes in fluorescence obtained with the native and artificial $GS_{\bar{6}}$ preparation were identical, as shown by the *curve*

3. Measurement of Lattice Formation by Light Scattering Techniques. In order to study the kinetics of lattice formation in greater detail, the change in light scattering, which occurs when antibodies are mixed with GS, was monitored over a period of several minutes. As can be seen from the data in Fig. 7, the rate of macromolecular aggregate formation (lattice formation) increases enormously as the state of adenylylation is increased over the range of 4 to 12.

Furthermore, when GS preparations with different states of adenylylation are examined at the same protein concentration, the results obtained (Fig. 7B) are very similar to those obtained when the protein concentrations are adjusted such that the number of adenylylated subunits is the same in all cases (Fig. 7C). This shows that the increase in aggregation rate obtained by increasing the state of adenylylation is due mainly to an increase in the density of adenylylated subunits per enzyme molecule, rather than to an increase in the total number of antigenic sites present. Whether these differences in lattice formation (aggregation) reflect conformational variations related to differences in the distribution of adenylylated subunits within GS molecules, or are due to differences in the relative rates of intra- vs intermolecular reactions of enzyme with bivalent antibodies, remains to be determined. Several ways by which antibodies might react with adenylylated GS subunits are illustrated in Fig. 8A. An idealized representation of how the bivalent reaction of antibodies with adenylyl groups on two different

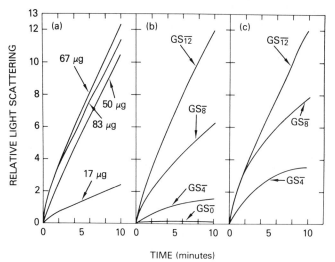

Fig. 7a-c. Effect of adenylylation state and protein concentration on macromolecular aggregation as measured by light scattering. 200 μg of anti-AMP antibodies were mixed with the various GS preparations, and the $90^\circ C$ light scattering at 400 nm was followed as a function of time. **a** 17, 50, 67, or 83 μg of $GS_{\overline{12}}$ as indicated. **b** 67 μg of either $GS_{\overline{0}}$, $GS_{\overline{4}}$, $GS_{\overline{8}}$, or $GS_{\overline{12}}$, as indicated. **c** 70 μg $GS_{\overline{12}}$, 105 μg $GS_{\overline{8}}$, or 210 μg $GS_{\overline{4}}$, as indicated. Otherwise conditions were as described by Hohman and Stadtman (1978), from which this figure was taken

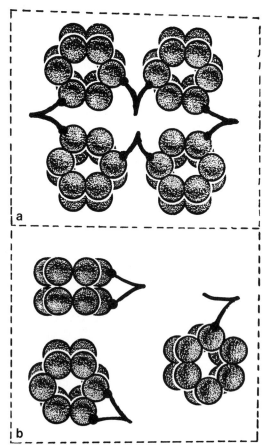

Fig. 8a, b. Idealized diagrammatic representation of the anti-AMP antibody–glutamine synthetase interactions. **a** Intermolecular reactions leading to lattice formation. **b** Intramolecular interactions that preclude lattice formation

GS molecules could lead to lattice formation is shown in the *upper part* of the figure. However, as is shown in Fig. 8B, if an antibody reacts with either two adjacent adenylylated subunits within the same hexagonal ring, or with two adenylylated subunits occupying superimposed positions in different hexagonal rings, then the capacity of that antibody molecule to link together two different GS molecules is neutralized, and the adenylylated subunits thus sequestered are no longer available for intermolecular lattice formation. The extent of lattice formation would therefore be determined by the number of adjacent adenylylated subunits within a given GS molecule, as well as the propensities of bivalent antibodies to form intra- vs intermolecular bridges. Hopefully, further studies now in progress will clarify the mechanisms involved.

II. Affinity Chromatography on Affi-Blue Sepharose

1. Separation of Adenylylated GS from Unadenylylated GS by Affinity Chromatography. As noted above, previous efforts to separate fully adenylylated GS from unadenylylated enzyme by conventional means were unsuccessful. A new experimental approach to this problem was suggested by the demonstration that Cibacron Blue binds to the nucleotide substrate site on the enzyme and can be displaced by ADP (E.R. Stadtman and M. Federici, unpublished results), and also by the fact that under certain conditions the affinity of the unadenylylated subunits for ADP is several orders of magnitude greater than is the affinity of adenylylated subunits for this ligand. Accordingly, it appeared feasible to separate adenylylated GS from unadenylylated GS by adsorption on sepharose columns to which Cibacron Blue was covalently attached, followed by selective elution with buffer containing an appropriate concentration of ADP. To test this pssibility, a mixture of $GS_{\overline{0.7}}$ and (^{14}C-adenylyl)-$GS_{\overline{12}}$ was adsorbed on a column of Affi-Blue Sepharose. The column was subsequently washed with pH 7.0 buffer containing 10 mM imidazole, 100 mM KCl, 1.0 mM $MnCl_2$, and 10 mM glutamate, and then the GS was eluted with the same buffer containing ADP whose concentration was increased stepwise from 10^{-5} M to 10^{-3} M. As shown in Fig. 9A, two major and one minor widely separated peaks of GS activity were covered. The first peak was eluted with 10^{-5} M ADP, and a minor peak, which was eluted with 10^{-4} M ADP, contained only unadenylylated GS as determined by direct enzyme assay (Stadtman et al. 1979a) and by the absence of radioactivity, whereas the third major peak, which was eluted with 10^{-3} M ADP, contained only ^{14}C-adenylylated GS.

2. Resolution of Native $GS_{\overline{6}}$ into Multimolecular Species. When a native $GS_{\overline{6}}$ preparation was adsorbed on an Affi-Blue Sepharose column and then eluted with an ADP gradient ranging from 10^{-5} M to 10^{-3} M, a broad peak of enzyme activity was observed, and the state of adenylylation across this peak increased progressively from $GS_{\overline{2}}$ to $GS_{\overline{9}}$ (Fig. 9B). Rechromatography of pooled fractions 48–63 which encompassed enzyme in the $GS_{\overline{6}}$ to $GS_{\overline{7.5}}$ range, using a more shallow ADP gradient, led to the separation of two more or less homogeneous preparations with average states of adenylylation ranging from 4–5 and from 6–7.0 (data not shown). These results clearly demonstrate that partially adenylylated native enzyme preparations are mixtures of hybrid molecules containing from 2 to 9 adenylylated subunits and that such preparations can be resolved

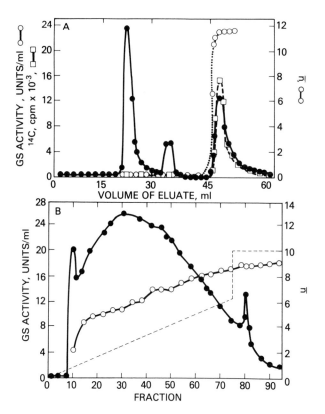

Fig. 9A, B. Affinity chromatography of glutamine synthetase on Affi-Blue-Sepharose columns.
A 0.19 ml containing 1.0 mg each of $GS_{\overline{1}}$ and (^{14}C-adenylyl)-$GS_{\overline{12}}$ was placed on an Affi-Blue
Sepharose column (0.9 × 10 cm) which had been preequilibrated with buffer containing 10 mM
imidazole (pH 7.0), 2 mM $MnCl_2$, 100 mM KCl, and 10 mM glutamate. The column was eluted as
follows: first, with 15 ml of the same buffer; second, with 15 ml of the same buffer containing 10^{-5}
M ADP, third, with 15 ml of the buffer containing 10^{-4} M ADP; fourth, with 15 ml of the buffer
containing 10^{-3} M ADP. 1.5 ml fractions were collected and assayed for GS activity, $\overline{n}$ and ^{14}C.
B 2.0 ml containing 24 mg of native $GS_{\overline{6}}$ were placed on an Affi-Blue column as above, except the
buffer contained 10 mM Hepes (pH 7.5) in place of imidazole. The sample was washed through
with the same buffer. The GS was eluted with a linear gradient of ADP in the same buffer ranging
from 10^{-5} to 6.3 × 10^{-4} M ADP (fractions to 0–75), then with buffer containing 10^{-3} M ADP (fractions 76–95), as indicated by the *broken line*. Each 1.5 ml fraction was assayed for GS activity
(*closed circles*) and for the state of adenylylation, $\overline{n}$, (*open circles*)

into more or less uniquely adenylylated molecular entities by means of affinity chromatography.

In other experiments, which will be described elsewhere, it has been shown that the
apparent affinity of adenylylated subunits in hybrid glutamine synthetase species for
the substrate glutamate increased from 1 to 13 mM as the proportion of adenylylated
subunits increases. These results are therefore consistent with those of earlier studies
(Ginsburg et al. 1970) with partially adenylylated native enzyme preparations, and
demonstrate that heterologous interactions between adenylylated and unadenylylated
subunits molecules affect the affinity of the enzyme for glutamate.

D. Summary

The glutamine synthetase (GS) activity in *Escherichia coli* is regulated by a bicyclic inter-convertible enzyme cascade which involves the cyclic adenylylation (inactivation) and deadenylylation (activation) of GS on the one hand, and the modulation of these processes by the uridylylation and deuridylylation of Shapiro's regulatory protein on the other. The specific activity of GS in a given metabolic state is determined by the fraction of its subunits that are adenylylated, and this fraction is determined by the concentration of over 40 metabolites. Through allosteric and substrate interactions with one or more of the cascade enzymes, these metabolites alter the rates of the covalent modification and demodification reactions. By means of immunoprecipitation studies with anti-AMP specific antibodies, it has been established that the partially adenylylated glutamine synthetase, which is present in a given steady state, is a mixture of hybrid molecules containing different numbers and possibly distributions of adenylylated subunits. Partial separation of these hybrid mixtures has been achieved by affinity chromatography on Affi-Blue Sepharose columns. From immunochemical studies it is evident that anti-AMP antibodies can react with adenylylated subunits of all molecular species of GS, but that the capacities of these primary antigen-antibody reactions to yield precipitable aggregates is very dependent on the number of adenylylated subunits per molecule, and much less so upon the total concentration of adenylylated subunits present. The studies suggest that precipitability is a function either of the distribution of adenylylated subunits within hybrid species, or of the kinetics of intra- vs intermolecular bivalent interactions.

References

Adler SP, Purich D, Stadtman ER (1975) Cascade control of *Escherichia coli* glutamine synthetase. Properties of the P_{II} regulatory protein and the uridylyltransferase-uridylyl-removing enzyme. J Biol Chem 250: 6264–6272

Anderson WB, Stadtman ER (1970) Glutamine synthetase deadenylylation: A phosphorolytic reaction yielding ADP as nucleotide product. Biochem Biophys Res Commun 41: 704–709

Anderson WB, Hennig SB, Ginsburg A, Stadtman ER (1970) Association of ATP: Glutamine synthetase adenylyltransferase activity with the P_I component of the glutamine synthetase deadenylylation system Proc Natl Acad Sci USA 67: 1417–1424

Bancroft S, Rhee SG, Neumann C, Kuster S (1978) Mutations that alter the covalent modification of glutamine synthetase in *Salmonella typhimurium*. J Bacterial 134: 1046–1055

Brown MS, Segal A, Stadtman ER (1971) Modulation of glutamine synthetase adenylylation and deadenylylation is mediated by metabolic transformation of the P_{II}-regulatory protein. Proc Natl Acad Sci USA 68: 2949–2953

Chock PB, Stadtman ER (1978) Superiority of interconvertible enzyme cascades in metabolic regulation: Analyses of multicyclic systems. Proc Natl Acad Sci USA 74: 2766–2770

Ciardi JE, Cimino F, Stadtman ER (1973) Multiple forms of glutamine synthetase. Hybrid formation by association of adenylylated and unadenylylated subunits. Biochemistry 12: 4321–4330

Engleman EG, Francis SH (1978) Cascade control of *E. coli* glutamine synthetase. II. Metabolite regulation of the enzymes in the cascade. Arch Biochem Biophys 191: 602–612

Ginsburg A, Yeh J, Hennig SB, Denton MD (1970) Some effects of adenylylation on the biosynthetic properties of the glutamine synthetase from *Escherichia coli*. Biochemistry 9: 633–649

Hennig SB, Ginsburg A (1971) ATP: Glutamine synthetase adenylyltransferase from *Escherichia coli:* Purification and properties of a low molecular weight form. Arch Biochem Biophys 144: 611–627

Hohman RJ, Stadtman ER (1978) Use of AMP specific antibodies to differentiate between adenylylated and unadenylylated *E. coli* glutamine synthetase. Biochem Biophys Res Commun 82: 865–870

Kingdon HS, Shapiro BM, Stadtman ER (1967) Regulation of glutamine synthetase, VIII. ATP: Glutamine synthetase adenylyltransferase, an enzyme that catalyzes alterations in the regulatory properties of glutamine synthetase. Proc Natl Acad Sci USA 58: 1703–1710

Mangum JH, Magni G, Stadtman ER (1973) Regulation of glutamine synthetase adenylylation and deadenylylation of enzymic uridylylation and deuridylylation of the P_{II} regulatory protein. Arch Biochem Biophys 158: 514–525

Mantel M, Holzer H (1970) Reversibility of the ATP: Glutamic synthetase adenylyltransferase reaction. Proc Natl Acad Sci USA 65: 660–667

Rhee SG, Park R, Chock PB, Stadtman ER (1978) Allosteric regulation of monocyclic interconvertible enzyme cascade systems: Use of *E. coli* glutamine synthetase as an experimental model. Proc Natl Acad Sci USA 75: 3138–3142

Segal A, Brown MS, Stadtman ER (1974) Metabolite regulation of the state of adenylylation of glutamine synthetase. Arch Biochem Biophys 161: 319–327

Senior P (1975) Regulation of nitrogen metabolism in *Escherichia coli* and *Klebsiella aerogenes:* Studies with the continuous culture technique. J Bacteriol 123: 407–418

Shapiro BM (1969) The glutamine synthetase deadenylylation system from *Escherichia coli.* Resolution into two components, specific nucleotide stimulation, and cofactor requirements. Biochemistry 8: 659–670

Stadtman ER, Chock PB (1978a) Interconvertible enzyme cascades in metabolis regulation. In: Horecker BL, Stadtman ER (eds) Current topics in cellular regulation, vol 13, pp 53–95. Academic Press, New York

Stadtman ER, Chock PB (1978b) Superiority of interconvertible enzyme cascades in metabolic regulation: Analysis of monocyclic systems. Proc Natl Acad Sci USA 74: 2761–2765

Stadtman ER, Ginsburg A (1974) The glutamine synthetase of *Escherichia coli:* Structure and control. In: Boyer PD (ed) The enzymes, vol X, 3rd edn, pp 755–807. Academic Press, New York

Stadtman ER, Ciardi JE, Smyrniotis PZ, Segal A, Ginsburg A, Adler SP (1975) Role of adenylylated glutamine synthetase enzymes and uridylylated regulatory protein enzymes in the regulation of glutamine synthetase activity in *Escherichia coli.* In: Markert CL (ed) Isoenzymes, vol II, pp 715–732. Academic Press, New York

Stadtman ER, Smyrniotis PZ, Davis JN, Wittenberger M (1979a) Enzymic procedures for determining the average state of adenylylation of *Escherichia coli* glutamine synthetase. Arch Biochem 95: 275–285

Stadtman ER, Chock PB, Rhee SG (1979b) Allosteric control of *E. coli* glutamine synthetase is mediated by a bicyclic nucleotidylation cascade system. In: Russell TR, Brew K, Sultz J, Faber H (eds) XI Miami Winter Symposium. Academic Press, New York

4. Dynamic Compartmentation

B. HESS, A. BOITEUX, and E. M. CHANCE

A. Introduction

Our classical picture of compartments has a static structure. Cellular compartments are surrounded by stable membranes, which represent a controlled diffusion barrier for chemicals and establish a well-defined reaction space for turnover. Today, also dynamic, that is time-dependent, compartmentation must be considered. These phenomena are described as dissipative spatial structures.

Dynamic compartments evolve from critical coupling between chemical reactions and diffusion. In far from equilibrium conditions of chemical processes, spatially homogeneous distribution of chemical reactants by free diffusion becomes rather improbable, if the rate laws of chemical processes are nonlinear and have self-activating properties. Indeed, instability with respect to diffusion has been described by Turing in case of morphogenesis (Turing 1952) and by Prigogine and his colleagues for general physical and chemical systems (Nicolis and Prigogine 1977). It has been shown theoretically that autocatalytic reactions generating instability or explosion will inevitably initiate also heterogeneous spatial distribution of chemical components. Such reactions are common in biochemical systems. They are examplified in excitation phenomena and oscillation of chemical reactions and ion fluxes (Boiteux and Hess 1974; Hess et al. 1978). Recently, the development of suitable analytical techniques allowed to demonstrate experimentally the time-dependent evolution and maintenance of oscillating structures in the spatial distribution of the concentration of pyridine nucleotides in an oscillating glycolysing yeast extract. These studies are complemented by theoretical analysis of a diffusion-coupled model of oscillating glycolysis indicating boundary conditions for the occurrence of various types of dynamic compartments. Here, we summarise some of our findings.

B. Biochemical Analysis

The development of spatial gradients of pyridine nucleotide concentrations in initially homogeneous, oscillating yeast extracts has been observed some time ago with a photometric scanning method (Boiteux and Hess 1971). However, a direct visualisation of the

1 Max-Planck-Institut für Ernährungsphysiologie, Rheinlanddamm 201, 4600 Dortmund, FRG
2 University College London, London, WC1E 6BT, Great Britain

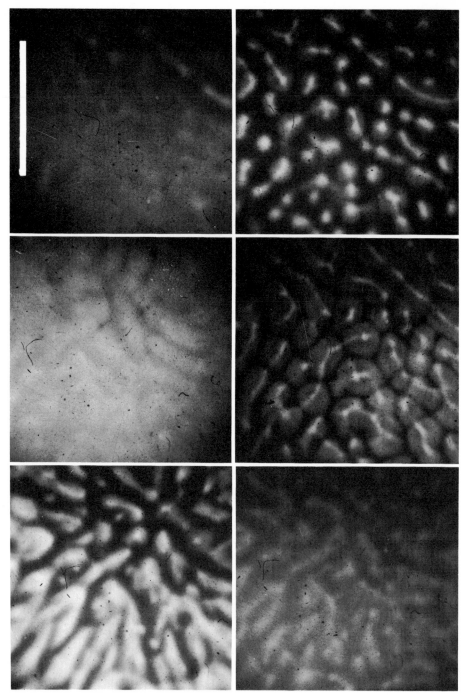

Fig. 1a–f. Initial phase of pattern formation in glycolysing yeast extract. The photographs were taken at min 2, 3, 4, 6, 7 and 9 after mixing an oscillating yeast extract to homogeneity. The sequence of pictures starts in the *upper left corner* proceeding clockwise. Vertical dimension: d = 0.25 cm, horizontal dimension: *calibration bar* $\triangleq$ 1 cm. *Dark areas* represent high concentrations of reduced pyridine nucleotides

formation of concentrational pattern in homogeneous reaction space was achieved some years later by means of high-resolving UV photographic techniques (Boiteux and Hess 1978).

The experimental set-up is as follows: A thin layer (d = 0.1–0.2 cm) of glycolysing yeast extract in a quartz dish is illuminated by a superactinic light source and the changes of optical density (λ 340–390 nm) in the system are photographed via UV-selective filters and optics on high contrast document film. All pictures on the following pages have been obtained by this method. The photographs were taken in sequence at the indicated times after mixing a glycolysing extract to homogeneity. Each set of pictures represents one experiment.

The series of pictures in Fig. 1 demonstrates evolution, development, and rearrangement of structural elements in the system after the initial mixing of glycolysing yeast extract. The upper left picture taken at min 2 after mixing shows the bulk extract in the reduced state, while locally the first traces of oxidised pyridine nucleotide shine up. As can be seen in the next two photographs, the inital structure develops into a rather regular pattern, increasing contrast as well as regularity. Then a rather dramatic rearrangement takes place during min 6 to 7, changing the inital pattern to a more complex structure which finally disappears when the bulk solution reaches the oxidized state (min 9).

It is interesting to note that the phase of rapid structural rearrangement has been observed so far only during the first minutes after mixing the extract. In consecutive oscillatory periods of the bulk extract the observed pattern and its dynamics are usually similar to the following series of pictures.

In Fig. 2 the first photograph was taken 6.5 min after mixing the glycolysing extract. As can be seen from the relatively high density of the picture area, at that time the bulk solution is in the reduced state except a network of feeble thin lines. In the next two pictures the pattern of oxidised pyridine nucleotides continues to develop and to broaden while the relative contrast is increased. At min 15.5 the bulk extract is in a highly oxidised state. Here the contrast has decreased to a point where the original pattern is barely discernible. However, as shown on the last two pictures at min 16.5 and 17.5, contrast and pattern reappear in phase with the oscillation of the bulk solution when the oscillating system is on its way back to the reduced state. The sequence of Fig. 2 is an example for stable temporal and spatial oscillation: while the oscillation of the bulk solution carries on, its structure appears and vanishes periodically in the indicated manner.

Spatial structures of NADH concentrations in yeast extracts have been observed so far only in the feedback-controlled glycolysing state, in other words under conditions where the system also exhibits temporal oscillations. This state needs a controlled input of glycolysable substrates to the system and can be maintained experimentally using either trehalose or amylose as substrates. On the other hand, lack of substrates as well as saturation or inhibition of glycolytic enzymes leads to rapid fading and disappearing of all spatial structures.

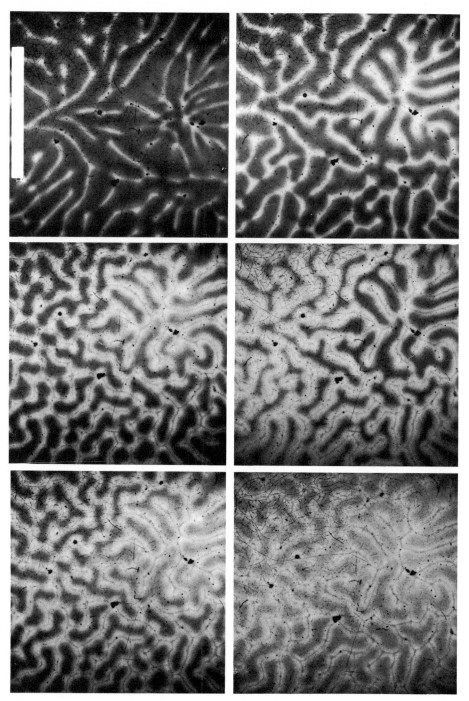

Fig. 2a–f. Transformation of pyridine nucleotide pattern in oscillating yeast extract during one cycle. The photographs were taken at min 6.5, 11, 14, 15.5, 16.5 and 17.5 after mixing an oscillating yeast extract to homogeneity. The sequence of pictures starts in the *upper left corner* proceeding clockwise. Vertical dimension: d = 0.12 cm, horizontal dimension: *calibration bar* ≙ 1 cm. *Dark areas* represent high concentrations of reduced pyridine nucleotides

Table 1. Oscillatory glycolytic diffusion in two dimensions

$$D\alpha\left(\frac{\delta^2\alpha}{\delta x^2}\right)_{t,y} + D\alpha\left(\frac{\delta^2\alpha}{\delta y^2}\right)_{t,x} - \left(\frac{\delta\alpha}{\delta t}\right)_{x,y} - v\alpha + A \cdot \sigma_1 = 0 \qquad (1)$$

$$D\gamma\left(\frac{\delta^2\gamma}{\delta x^2}\right)_{t,y} + D\gamma\left(\frac{\delta^2\gamma}{\delta y^2}\right)_{t,x} - \left(\frac{\delta\gamma}{\delta t}\right)_{x,y} + v\alpha - A \cdot \sigma_2 \cdot \gamma = 0 \qquad (2)$$

$$v = V_m \frac{\alpha\left(1 + \frac{\alpha}{E}\right) \quad (1 + \gamma)^2}{L(1 + c\alpha)^2 + (1 + \gamma)^2 + (1 + \frac{\alpha}{E})^2} \qquad (3)$$

C. Theoretical Studies

Earlier studies have shown that the dynamic behaviour of oscillating glycolysis can be reduced to the kinetic properties of phosphofructokinase, the allosteric master enzyme of glycolysis (Boiteux and Hess 1974). Indeed, when in oscillating yeast extracts all reactants are experimentally kept in homogeneous distribution, their complex kinetics can be described quantitatively on the basis of a feedback-controlled allosteric enzyme operating in the concerted mode of Monod and colleagues (Boiteux et al. 1975). However, the complexity of the system increases dramatically when diffusion processes come into play and are allowed to modify and control enzymic reactions, leaving little chance for quantitative treatment of the system.

Here the analysis of simplified model systems proves to be helpful. When in the most simple model one-dimensional diffusion was coupled to the enzymic reaction, the concentration of reactants turned out to be distance-dependent (Goldbeter 1973). For more realistic conditions we selected a two-dimensional approach (Chance et al. 1979). The two-dimensional partial differential Eqs. (1, 2) given in Table 1 denote the necessary flux terms, where α, γ represent the concentrations of substrate and product, and σ_1, σ_2 the rates of source and sink, respectively. Eq. (3) describes the saturation function of the enzyme with the usual notation.

Discretisation of the system in both dimensions to a 26×26 square mesh reduced the problem to the simultaneous solution of 1352 linear first order differential equations. The substrate is injected at each mesh point. The equations are solved by means of the computer programme FACSIMILE in computing time equal to experimental time. Boundary values for concentrations of α (ATP), γ (ADP), and the injection rate (σ_1) were fixed within the range of experimental observations. This system produces spatio-temporal oscillation of substrate concentrations with a periodicity of 3 min.

Our computer solutions allow to simulate a variety of different boundary conditions. Their effect on the wave form of the oscillation are summarised in Table 2. Depending on the conditions, the computation yields travelling waves, standing waves, or temporal oscillations of the entire area, which is in fair agreement with experimental observations in yeast extracts (Boiteux and Hess, recent data). Standing waves and dynamic compartmentation similar to the pictures in Fig. 2 are generated with no glycolysis but diffusion in the boundary domain. The pictures of Fig. 1, where the experimental system

Table 2. Effect of boundary conditions on wave form

Boundary condition	Area of mesh domain mm^2	Area of boundary domain mm^2	Glycolysis in boundary domain	Diffusion in boundary domain	Wave form
Steady state	3.14	4.030	Yes	No	Travelling[a]
Variable	3.14	0.805	Yes	No	Travelling[a]
Variable	3.14	0.161	Yes	No	Travelling[b]
Variable	3.14	0.834	No	Yes	Standing[c]
No transport across boundary	3.14	0.0805	Yes	Yes	Temporal[d]

[a] Travelling wave from boundary to centre
[b] Travelling wave from centre to boundary
[c] Standing wave
[d] Without diffusion in mesh domain: temporal oscillation of bulk solution

has not yet reached a stable state, correspond to computer solutions obtained before the model system merges into the limit cycle.

The spatial concentration gradients in the model are as steep as in the experimental system. This is demonstrated in the isometric drawings of Fig. 3 which gives a typical example of the wave forms of a travelling wave obtained from the model after reaching the limit cycle. The isometric drawings show a quarter of the total solution by symmetry from corner forward position and corner right position (90° rotation) to give a comprehensible presentation of the wave form in these dimensional perspectives. The waves of ATP (*top*) and ADP (*bottom*) proceed from right to left in each period and are represented at 5 s intervals. The figure shows a snapshot of the wave progress roughly halfway through the given territory, with ADP forming a steep ridge with gradients in front and rear, whereas ATP travels like a reversed flood with a steep front gradient followed by a trough such as are seen in weather maps. The different wave forms of the two components result from the fact that the removal of α from the system is proportional to γ^2 so that when γ rises sufficiently, α decreases rapidly, promoting the wave propagation. Such experiments are only the first step to an understanding of a rather complex dynamic biochemical situation. However, they stress the importance of looking more closely at the ratio between boundary volumes and reactivity space in biological organisations.

D. Outlook

The organisation of biochemical processes in time and in compartments is based on a hierarchical principle of reaction space and boundary systems. On isolation each of the various levels of organisation retains in general its own autonomous functions and structures, preserving its own reaction space and boundary. Such levels are the microreaction space at the active sites of enzymes or of membrane-bound reactive proteins. Then the

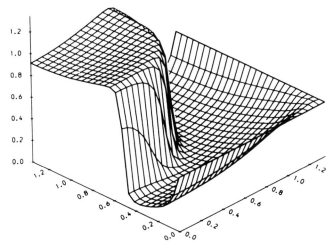

ALPHA AT TIME   1430
PERCENT OF MAXIMUM 64

Fig. 3a, b. Isometric perspective of an intermediary state of oscillatory glycolytic diffusion in two dimensions. The *coordinates* of the *x,y-plane* are given in mm. The *z-axis* represents the normalized concentrations of α and γ, scaled to obtain isometric perspective. Time is computer time in seconds. For further explanation see text

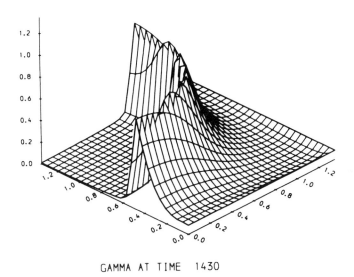

GAMMA AT TIME   1430
PERCENT OF MAXIMUM 16

subcellular macrospace of mitochondria, lysosomes, peroxisomes, vacuols, and the so-called cytoplasmic space is recognized. Finally, the whole cell forms the basic spatial unit of biology. At higher levels, networks of cells, organised at random, in specific nets and higher organs, are defined as reaction territories. Although each of these compartments is quasi autonomous, preserving most of its functions, only the coupling of the various functions of all levels of this organisation through transport and diffusion yields the very physiological properties, which after all are more than the sum of its subparts. Thus chemicals from each reaction space are transferred and distributed at transport rates fitted to the rates of their generation and consumption. Without a knowledge of the relationship between the rates of transport and chemical turnover understanding is distorted.

Indeed, in addition to this static organisation of reaction spaces in biological systems a dynamic compartmentation must be recognised, which is dependent on the dynamic state of metabolic processes as well as on transport phenomena. These dynamic compartments are lost on isolation and standard biochemical analyses yield only average concentrations in fairly large volumes. Our experiments demonstrate that dynamic heterogeneity can be observed even in cell-free extracts, such as the glycolysing yeast system. It is expected that such phenomena occur in general.

References

Boiteux A, Hess B (1971) Unpublished experiments
Boiteux A, Hess B (1974) Oscillations in glycolysis, cellular respiration and communication. Faraday symposia of the chemical Society 9: 202–214
Boiteux A, Hess B (1978) Visualization of dynamic spatial structures in glycolysing cell-free extracts of yeast. In: Dutton PL, Leigh J, Scarpa A (eds) Frontiers of biological energetics, pp 789–798. Academic Press, New York
Boiteux A, Goldbeter A, Hess B (1975) Control of oscillating glycolysis of yeast by stochastic, periodic, and steady source of substrate: A model and experimental study. Proc Natl Acad Sci USA 72: 3829–3833
Chance EM, Boiteux A, Hess B (1979) Unpublished experiments
Goldbeter A (1973) Patterns of spatiotemporal organization in an allosteric enzyme model. Proc Natl Acad Sci USA 70: 3255–3259
Hess B, Goldbeter A, Lefever R (1978) Temporal, spatial and functional order in regulated biochemical and cellular systems. Adv Chem Physics 38: 363–413
Nicolis G, Prigogine I (1977) Self-organization in nonequilibrium systems. Wiley, New York
Turing AM (1952) The chemical basis of morphogenesis. Phil Trans R Soc London Ser B 237: 37–72

5. The Genes for and Regulation of the Enzyme Activities of two Multifunctional Proteins Required for the De Novo Pathway for UMP Biosynthesis in Mammals

M.E.JONES

A. Introduction

The biosynthesis of the first pyrimidine nucleotide, UMP, universally requires six enzymes (Reichard 1959): a glutamine-utilizing carbamyl-P synthetase (Levenberg 1962); aspartate transcarbamylase; dihydroorotase; a dihydroorotate dehydrogenase (Taylor et al. 1966) that couples with the respiratory chain (and not with NAD); orotate phosphoribosyltransferase and orotidylate decarboxylase (Fig. 1).

B. Genes for the Six Mammalian Enzymes Required for de novo Biosynthesis of UMP

In most bacteria these six enzyme centers are associated with six structural genes which map at distinct sites on the bacterial chromosome and code for six distinct and separable proteins (O'Donovan and Neuhard 1970; Taylor and Trotter 1972; Makoff and Radford 1978). In fungi, however, the first two enzyme centers are part of a multi-enzymic (or multifunctional) protein that is coded in a single gene, i.e., *pyr-3* in *Neurospora* (Davis 1967; Denis-Duphil and Lacroute 1971; Jones 1972; Makoff and Radford 1978). This single *Neurospora* gene has two distinct areas, *pyr-3a* and *pyr-3d,* and, if point mutants are obtained at one or the other area of this gene, the resulting mutants can lack only one of the two enzyme activities coded for by the wild-type gene. The studies show that the *pyr-3d* area of the *pyr-3* gene codes for aspartate transcarbamylase activity and the *pyr-3a* area codes for carbamyl-P synthetase activity. However, many point mutations occurring anywhere across the entire gene can also cause a loss or diminution of both enzyme activities (see Jones 1972; Makoff and Radford 1978). This behavior is typical of a gene which codes for a single polypeptide that has more than one domain, each of which is capable of catalyzing a different enzyme reaction, such as the *pyr-3* gene of *Neurospora* (Fink 1971).

Jones hypothesized in 1971 (Jones 1971; Shoaf and Jones 1971) that the initial three enzyme activities and the final two enzyme activities for UMP biosynthesis might

Department of Biochemistry and Nutrition, School of Medicine, University of North Carolina, Chapel Hill, NC 27514, USA

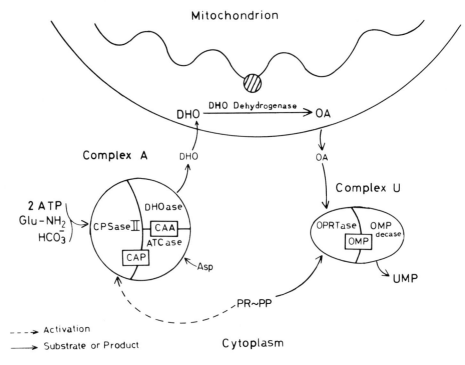

Fig. 1. Intracellular location of the six mammalian enzymes for UMP biosynthesis and the two potential sites where phosphoribosyl pyrophosphate (PR∿PP) concentrations could regulate the pathway. Complex A has the active centers for carbamyl phosphate synthetase (CPSase II, EC 2.7.2.9); aspartate transcarbamylase (ATCase, EC 2.1.3.2), and dihydroorotase (DHOase, EC 3.5.2.3). Complex U has the active centers for orotate phosphoribosyl transferase (OPRTase, EC 2.4.2.10) and orotidylate decarboxylase (OMPdecase, EC 4.1.1.23). Dihydroorotate dehydrogenase (DHO dehydrogenase, EC 1.3.3.1) is on the outer surface of the inner membrane of the mitochondrion. The substrates are ATP, L-glutamine (Glu-NH₂), HCO_3^-; L-aspartate (asp); 5-phosphoribosyl-l-pyrophosphate (PR∿PP). The intermediates of the pathway are carbamyl phosphate, CAP; carbamyl L-aspartate, CAA; dihydroorotate, DHO; orotate, OA; orotidylate, OMP and uridylate, UMP

be in two proteins similar to the carbamyl∿P synthetase: aspartate transcarbamylase complex of yeast and *Neurospora,* and that these two proteins might be determined by two genes. At that time the terms multifunctional or multienzymic were unknown, although it was appreciated by our and other laboratories that some special clustering of these and other enzyme centers had to exist (Davis 1967; Jones 1971, 1972; Fink 1971; Kirschner and Bisswanger 1976).

These two proteins, which have been called Complex A and U (Fig. 1), do have the first three biosynthetic enzyme activities and the last two enzyme activities, respectively (Jones 1971; Shoaf and Jones 1971, 1972; Hoogenraad et al. 1971; Mori and Tatibana 1975, 1978; Coleman et al. 1977, 1978; Hatfield and Wangaarden 1964; Kasbekar et al. 1964; Appel 1968; Reyes and Guganig 1975; Kavipurapu and Jones 1976; Traut and Jones 1978, 1979). The fourth enzyme of the pathway, dihydroorotate dehydrogenase, is located on the outer surface of the inner membrane of the mitochondrion (Matsuura and Jones 1974; Miller 1975; Chen and Jones 1976). Therefore, only three genes are necessary for the six enzyme centers required for UMP biosynthesis.

The correlation of the anatomy of a gene with the anatomy of a polypeptide is most easily made in prokaryotes and lower eukaryotes where the relationship of the site of a given genetic mutation is easily related to an amino acid change in a polypeptide. Nontheless, a considerable body of evidence has accumulated since 1971 that, in my view, makes it reasonably certain that Complex A is a multienzymic polypeptide having three enzyme activities which are encoded on a single gene (Jones 1971, 1972; Shoaf and Jones 1971, 1972; Coleman et al. 1977, 1978; Norby 1970; Rawls and Fristrom 1975), while complex U probably is also a single multienzymic polypeptide which possesses the last two enzyme centers for UMP biosynthesis encoded by a single gene (Smith et al. 1966; Pinsky and Krooth 1967a, b; Jones 1971, 1972; Shoaf and Jones 1972; Traut and Jones 1977a, b, 1978, 1979; Levinson et al. 1979; Suttle and Stark 1979).

For Complex A, the unit polypeptide has a molecular weight of 200,000 (Mori and Tatibana 1975; Coleman et al. 1977), while for Complex U the unit size may be 45,000 (Traut and Jones 1979). Both polypeptides form proteins that are ellipsoidal and therefore polymers of the monomeric polypeptide. Complex A (Coleman et al. 1977) and U (Traut and Jones 1979) have unusual behavior in chromatography that uses molecular sieves (Brown et al. 1975; Traut and Jones 1979).

Complex A monomer can aggregate into trimers, hexamers, nanomers and dodecamers of the 200,000 dalton subunit (Coleman et al. 1977) and Complex U monomer (S_W^{20} = 3.6) can aggregate to a dimer (S_W^{20} = 5.1) which has a changed f/fo coefficient from the monomer, as well as two other species having S values of 4.7 and 5.6 which are probably related to the monomer and dimer (Traut and Jones 1979). These four rapidly equilibrating species are formed in response to the anions in the solvent: the anions that are effective are competitive inhibitors versus OMP for the decarboxylase active center of Complex U (Traut and Jones 1979 and unpublished experiments). In addition, Complex U forms some stable aggregates of the monomer in the absence of thiols (Brown et al. 1975; Traut and Jones 1979). This complex picture helps explain the various molecular sizes reported earlier for Complex A (Shoaf and Jones 1972; Mori and Tatibana 1975, 1978a, b) and for Complex U (Becker et al. 1974; Brown et al. 1975; Grobner and Kelley 1975; Reyes and Guganig 1975; Kavipurapu and Jones 1976; Brown and O'Sullivan 1977; Camplbell et al. 1977; Traut and Jones 1977, 1979; Jones et al. 1978; Reyes and Intress 1978).

The characteristics which support the concept that Complex A is a multienzymic protein containing the first three enzyme centers for UMP biosynthesis include the fact that these three enzyme activities cosediment in a centrifugal field (Shoaf and Jones 1972; Mori and Tatibana 1975), copurify to homogeneity (Mori and Tatibana 1975; Coleman et al. 1977), and are coinduced in response to growth of cells in the presence of N-(phosphonacetyl)-L-aspartate (PALA), a bisubstrate analog of aspartate transcarbamylase (Kempe et al. 1976), or in rats fed orotic acid (Bresnick et al. 1968). The induction of Complex A by orotate, observed by Bresnick et al., could have been due to increased cellular pools of carbamyl aspartate (Hager and Jones 1965; Chen and Jones 1979), which form when a cellular pool of orotate inhibits dihydroorotate dehydrogenase (Chen and Jones 1979; Christopherson and Jones 1979).

The two enzyme activities of Complex U are usually both absent in the human hereditary disease, orotic aciduria (Smith et al. 1966). When cells are grown in the presence of azauridine, the two enzyme activities of Complex U increase in fibroblasts from such patients, and from the heterozygous parents as well as in cells from normal subjects (Pinsky and Krooth 1967a, b), and in a Simian virus 40 transformed baby hamster kidney cell line (Suttle and Stark 1979). Azauridine is converted by these cells to aza-uridine monophosphate (Aza-UMP), which is a potent inhibitor of the orotidylate decarboxylase of Complex U (Pasternak and Handschumacher 1959; Traut and Jones 1977a), and a weak inhibitor of the phosphoribosyl-transferase activity of Complex U (Traut and Jones 1977a). In addition, these two enzyme activities copurify (Hatfield and Wangaarden 1964; Kasbekar et al. 1964; Appel 1968; Reyes and Guganig 1975; Kavipurapu and Jones 1976), cosediment in a centrifugal field (Grobner and Kelley 1975; Kavipurapu and Jones 1976; Traut and Jones 1979a), pass through a molecular sieve together (Grobner and Kelley 1975; Brown et al. 1975; Traut and Jones 1979) and comigrate in an electrical field (Kavipurapu and Jones 1976).

As yet there is no conclusive structural evidence that either Complex U or Complex A is a multifunctional protein. The genetic evidence cited above is most compelling; the structural studies of these proteins are incomplete and the limited data available can be interpreted in several ways (Coleman et al. 1977; Mori and Tatibana 1978a; Brown and O'Sullivan 1977; Traut and Jones 1979a). The least definitive studies involve situations where protein denaturation and/or proteolysis (Brown and O'Sullivan 1977; Mori and Tatibana 1978a) could occur to produce derived proteins with only a fraction of the original activity consisting of a smaller polypeptide that has only a single enzyme activity. The active centers of Complex A and U can be separated from one another by limited proteolysis with elastase (Mori and Tatibana 1973; Reyes and Guganig 1975).

It is particularly significant that if the mitotic rate of a tissue increases or decreases there is a corresponding change in the activities of these enzyme centers (Smith and Baker 1960; Kim and Cohen 1965; Smith et al. 1966; Young et al. 1967; Hager and Jones 1967b; Tatibana and Ito 1969; Galofre and Kretchmer 1970; Jones 1970; Yip and Knox 1970; Sweeney et al. 1971; Weber et al. 1971; Ito and Uchino 1971) and, in addition, mutant cells can be derived by growth of cells in the presence of an antimetabolite. Cells grown on an inhibitor of the aspartate transcarbamylase center of Complex A have only increased levels of the enzyme activities of Complex A (Kempe et al. 1976); those grown on an inhibitor of Complex U can have increased levels of the enzyme activities of Complex U (Pinsky and Krooth 1967a, b; Levinson et al. 1979; Suttle and Stark 1979) or Complex A (Ennis and Lubin 1963). Mutant cells selected by growth on 5-fluorouracil, an unnatural substrate for the transferase of Complex U (Reyes and Guganig 1975), have reduced levels of Complex U (Levinson et al. 1979).

C. Kinetic Importance of the Multienzymic Proteins for UMP Biosynthesis

Although the regulation of the structural genes for Complex A and U may prove to be the more important way to regulate the synthesis of UMP, there are important effects at the metabolite level also.

I. Channeling of Intermediates

Perhaps the most important of these is "channeling" of intermediates of the pathway; indeed it was this phenomenon which led to the discovery of Complex A and U (Hager and Jones 1965; Jones 1971; Shoaf and Jones 1971, 1972).

Channeling is the effective use of a metabolite for a specific biosynthetic pathway such that this metabolite does not enter the solvent (either in vitro or in vivo), rather it is efficiently converted to a single end product. Channeling can explain the lack of a pool of a metabolite when one enzyme center of a multienzymic protein synthesizes an intermediate that is utilized by a second enzyme center of the same protein. It can also explain why an intermediate, such as carbamyl∿P that is normally formed by two different enzymes in the same cell, is not available to both biosynthetic pathways, i.e., the arginine and pyrimidine pathways, when one of the two enzymes synthesizing carbamyl∿P is inactive, as in mutants (Davis 1967; Williams et al. 1970; Jones 1972).

When Hager and Jones (1965) searched for and found a second carbamyl∿P synthetase in mammals that would have provided carbamyl∿P for UMP biosynthesis, we made an observation that intrigued me. When intact rat Ehrlich ascites cells were incubated with [^{14}C]-bicarbonate (to label C-2 of the uridine ring), glutamine to serve as the source of the N-atom of carbamyl∿P and glucose to serve as an energy source and possibly as a source of the carbon atoms of aspartic acid, radioactivity was found in UMP but not in carbamyl aspartate (CA-asp), dihydroorotate (DHO), orotate (OA) or orotidylic acid (OMP), the chemically stable intermediates between the substrate [^{14}C]-bicarbonate and the end product [^{14}C]-UMP (or its nucleotide derivatives). It was as if the bicarbonate had entered a black box until it appeared as UMP. Shoaf and Jones (1971, 1972) thought this effect might be due to substrate channeling.

We examined the kinetic data then available to see if one might expect one of these intermediates to accumulate. The apparent Km values for the intermediates of the UMP biosynthetic pathway between bicarbonate and UMP are shown in Fig. 2. The values listed are those for the compound when it served as substrate in the biosynthetic direction, i.e., for CA-asp as a substrate for dihydroorotase. Two sets of apparent Km values are given; those available in 1969, when Dr. Shoaf and I decided to investigate this "black box" phenomena in more detail, and for comparison these same values are listed as they have been refined in 1979. The relative activity values listed are apparent V_{max} values for the various enzymes in crude homogenates; these values were measured at or near pH 7.4. The data indicate that except for the high rate of the aspartate transcarbamylase reaction (enzyme ②), the other enzymes have similar V_{max} values. Therefore, beyond enzyme ②, pooling of the intermediates would be expected principally where the Km value for an intermediate is high. One markedly high Km value exists, namely the value for CA-asp as a substrate for dihydroorotase. Figure 2 also illustrates whether a reaction is essentially irreversible (——→) or reversible (←——→), and whether the equilibrium for a reversible reaction favors the substrates (←⊢—→) or the products (←—⊢→). The formation of a pool of CA-asp is favored by the equilibria of reactions ② and ③ (Reichard 1956; Christopherson et al. 1978; Christopherson and Jones 1979) as well as by the Km values for CA-asp when it is a substrate for the dihydroorotase (Bresnick and Blatchford 1964; Christopherson and Jones 1979). Despite these considerations, no pool of [^{14}C]-CA-asp is observed in intact ascites cells utilizing [^{14}C]-bicarbonate

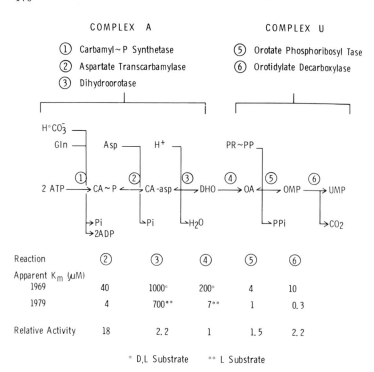

Reaction	②	③	④	⑤	⑥
Apparent K_m (µM)				
1969	40	1000°	200°	4	10
1979	4	700°°	7°°	1	0.3
Relative Activity	18	2.2	1	1.5	2.2

° D,L Substrate   °° L Substrate

Fig. 2. Six enzyme reactions of de novo UMP biosynthesis indicating the substrates, the reversibility of the reaction, and whether the equilibrium favors the biosynthetic substrate or product (see text for description of this notation). The Km values are given for the biosynthetic substrate and are µM. The relative activity values are nmols of product formed/min/mg protein at pH 7.4 with saturating substrates. These values for enzymes 2 and 3 are from unpublished data of Dr. Richard I. Christopherson; for enzyme 4 from Matsuura and Jones (1974), Chen and Jones (1976), and for enzymes 5 and 6 from Jones et al. (1978), Traut and Jones (1979b). The 1969 Km values are from Smith and Baker (1959) for enzymes 2, 3, 5, and 6, or from Bresnick and Blatchford (1964). These values were at the optimal pH for the enzyme center

in the normal production of UMP (Fig. 3A; Hager and Jones 1965; Chen and Jones 1978, 1979a). If orotate is added to the incubation medium (Hager and Jones 1965) or its generated within cells by the inhibition of orotidylate decarboxylase by azaUMP (Chen and Jones 1979b), a pool of CA-asp is obserrved, because OA inhibits both dihydroorotase (K_i value of 200 µM; Kennedy 1974; Christopherson and Jones, unpublished data) and dihydroorotate dehydrogenase (K_i value of 8 µM; Chen and Jones 1976). It is the latter inhibition which causes the accumulation of CA-asp (Chen and Jones 1979a). The amount of the CA-asp pool formed in the experiment of Hager and Jones (1965) was equal to three-fourths of the amount of UMP that would have accumulated if OA had not been added; two-thirds of this CA-asp remained in the cells while one-third was in the medium. These results indicate that a CA-asp pool could have been observed if it occurred when OA was not added to the medium. The results also indicate that CA-asp is not readily degraded in this cell by the reversal of the aspartate transcarbamylase reaction or by other reactions, such as a hydrolysis to carbamate and aspartate.

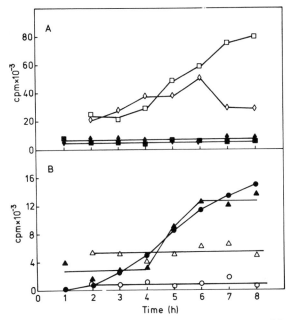

Fig. 3A, B. The accumulation of radioactivity from $NaH[^{14}C]CO_3$ into nucleic acids, uridine mononucleotide **A**, and precursors of UMP **B** in Ehrlich ascites cells in the presence or absence of 6-azaUMP. Separations were done by PEI-cellulose thin layer chromatography (Chen and Jones 1979b). In the absence of 6-azauridine: nucleic acids (*white diamonds*) and UMP (*white squares*) for (**A**); CAA-h and DHO(*white triangles*) and OA and orotidine (*white rings*) for (**B**). In the presence of 6-azauridine: nucleic acids (*black diamonds*) and UMP (*black squares*) for (**A**); CAA-h and DHO (*black triangles*) and OA and orotidine (*black balls*) for (**B**)

Dr. Richard I. Christopherson and I are now carrying out experiments to see if CA∿P and CA-asp are channeled by Complex A. In the cell DHO is normally removed so efficiently by DHO dehydrogenase that, if a pool of CA-asp and/or DHO exists, these pools are normally very small and were below the level of detection in the experiments of Hager and Jones (1965) and of Chen and Jones (1978, 1979a). CA-asp accumulates in cells when the OA concentration is sufficiently high to inhibit DHO dehydrogenase or when an OA pool is created in the cells by the presence of azauridine in the medium (Chen and Jones 1979b; Figs. 3 and 4).

In experiments that have been carried out in vitro with partially purified Complex A in the absence of the last three enzymes of the pathway, a pool of both CA-asp and DHO accumulates (Christopherson, unpublished results). After 60 min the CA-asp was 14 μM; the DHO was about 2 μM. The equilibrium constant for dihydroorotase (Christopherson et al. 1978) at pH 7.4 would predict a ratio of CA-asp/DHO of 16.6; the ratio of CA-asp/DHO observed was 7, a value that is predictably lower since CA-asp and DHO were constantly being formed from bicarbonate, ATP, glutamine, and aspartate. Chen and Jones (1979b) have observed a similar ratio of CA-asp/DHO in cells made permeable to orotate by treatment with dextran sulfate. However, when an acceptor enzyme system was added to convert DHO, the final product of the Complex A reaction to OA (to

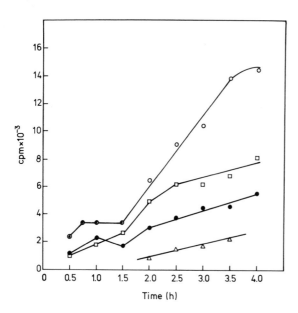

Fig. 4. Effect of the addition of 6-azauridine to the culture medium of cultured Ehrlich ascites cells on the accumulation of metabolites in the acid-soluble fraction of cells. These results were obtained by using a Bio-Gel P-2 column (Chen and Jones 1979b) DHO, *white triangles;* OA, *black balls;* orotidine, *white squares;* CAA-h, *white rings.* ⊕, samples which we do not believe are CAA-h, although the radioactivity migrated as CAA-h (see text of Chen and Jones 1979b)

mimic the removal of DHO as it occurs in the cell), then the CA-asp pool was smaller (5 μM). The rate at which CA-asp accumulated during the first 5 min was the same whether or not the acceptor system was added. After 5 min, however, the two experiments differed. In the absence of the acceptor system CA-asp accumulation increased in a nonlinear fashion with time and reached a concentration of 14 μM after 60 min. In the presence of the acceptor system, the CA-asp concentration remained at about 5 μM from 10 through 60 min while the OA concentration increased linearly for 20 min and stopped increasing after 40 min at a concentration of about 25 μM (presumably because a substrate had been exhausted). The DHO pool was 2 μM without the acceptor system, but was less than 0.1 μM with it. It seems highly significant to Dr. Christopherson and myself that the CA-asp pools seen in both of these experiments are markedly below the apparent Km of L-CA-asp for the dihydroorotase of Complex A, i.e., the CA-asp pool was 1/20th the Km value in the absence of an acceptor system, and 1/140th the value when the acceptor was present. We believe these preliminary results indicate that Complex A channels CA-asp and does not readily release this intermediate into the solution. In addition, no significant amount of CA∿P was released into the solution at any time in either experiment; its concentration remained at 0.4 μM or less. It would seem therefore that CA∿P may also be channeled by Complex A. We would also predict that with a sufficiently sensitive assay system one might find a pool of both CA∿P and CA-asp in intact cells but the concentrations would be less than 1 μM for CA∿P and about 1 μM for CA-asp.

Channeling of OMP by Complex U has been demonstrated by Traut and Jones (1977b). In this case the data show that [^{14}C]-OMP, synthesized from [^{14}C]-OA and phosphoribosyl-pyrophosphate (PR∿PP) by the transferase center of Complex U, is not readily released from the protein into the solvent. Rather, the [^{14}C]-OMP is transferred rather directly from the transferase center of Complex U to the decarboxylase center of the same molecule of Complex U where it is converted to UMP. Both in vitro (Traut and Jones 1977b) and in vivo (Hitchings 1973), the solvent levels of OMP never exceed a concentration of 0.05 to 0.1 μM, even when very high amounts (50 μM) of OA and excess amounts of PR$\smallsmile$PP are used as substrates in vitro. This observation is particularly dramatic since inorganic pyrophosphatase is present in these crude extracts to convert the second product of the transferase reaction, inorganic pyrophosphate, to orthophosphate. The nonaccumulation of OMP can not, therefore, be due to a reversal of the transferase reaction (Traut and Jones 1977b). Solvent CA$\smallsmile$P, CA-asp, and OMP are utilized readily by Complex A and U so that the aspartate transcarbamylase, dihydroorotase, and OMP decarboxylase activities of Complex A and U, respectively, can easily be measured and can utilize these substrates if they are available. Therefore, it is the slow release of OMP by Complex U (and probably a similar slow release of CA$\smallsmile$P and CA-asp from Complex A) that does not allow the accumulation of these intermediates in the intact cell.

II. Possible Enhancement of the Rate of the Enzyme Centers of Complex A and U

What has been accomplished by the channeling of these three intermediates other than a directing of the intermediate to the intended products? It is possible that a rate advantage has been achieved because the enzyme centers may have been broght sufficiently close to one another in an alignment that might permit an extremely effective transfer of the product of one reaction, such as OMP formed by the transferase of Complex U, to the substrate site of the next active center, i.e., the active site of the decarboxylase of Complex U. Such an increase in rate could result from a marked increase in the local concentration of the substrate because it remains in the very small volume of solvent present between the active sites. Levitsky and Koshland (1971) have calculated a rate advantage for cytidine triphosphate synthetase of approximately 300 with regard to the production of nascent ammonia from glutamine versus the ability of the enzyme to utilize NH_4^+ from the solvent. This example may be directly relevant to the carbamyl phosphate synthetase of Complex A, which is also a glutamine-utilizing synthetase. A similar effect may result from the sequestering of CA$\smallsmile$P, CA-asp, and OMP from solvent by Complex A and U, respectively. At the moment we cannot measure this potential rate advantage. To do so one would need pure Complex A and U. One might also wish to cleave the five active centers from these two intact proteins in order to compare these centers in isolation from one another; with the native protein, for Complex A, one could make various combinations of the three enzyme centers. There is evidence for both Complex A and U that mild treatment with a protease will release the active centers (Mori and Tatibana 1973; Reyes and Guganig 1975). In the case of Complex U, OMP activates the complex (Traut and Jones 1977) so that it is possible that the rate that we currently measure is higher than the one the component centers would have when they are separated from one another. Eventually, studies on the three-dimensional structure of multienzymic proteins should be most informative, as it has been for immunoglobulins.

III. Substrate and Energy Conservation Resulting from Channeling

The sequestering of intermediates can protect a chemically unstable product, like CA∿P; this might conserve the amount of ATP required for its synthesis to that imposed by the stoichiometry of the reaction. In addition, since one of the products of the chemical (but not the enzymatic) hydrolysis of CA∿P is cyanate (Allen and Jones 1964), the cell is protected from this toxic chemical if channeling does occur.

In contrast to CA∿P, OMP is chemically stable; however, in mammalian extracts it is rapidly degraded to orotidine by a phosphatase or a nucleotidase (Pasternak and Hand-schumacher 1959; Cardoso et al. 1961; Fallon et al. 1962; Buttoo et al. 1965; Fox et al. 1970; Kelley and Beardmore 1970; Janeway and Cha 1977; Chen and Jones 1979b; Traut 1980). Therefore, channeling of OMP by Complex U in mammalian cells preserves this intermediate and conserves the amount of ATP required for the synthesis of the PR∿PP and for CA∿P, both of which are required to produce OMP. A comparison of the metabolism of OA and OMP in yeast and Ehrlich ascites cells (Table 1) clearly shows that the enzymes concerned with the metabolism of these intermediates are very differ-ent (Traut 1980). In yeast the two enzyme centers of Complex U are present in two distinct proteins (Jund and Lacroute 1972). Therefore, OMP has to be released by orotate phosphoribosyltransferase to the solvent and then removed from the solvent by oroti-dylate decarboxylase, to be converted to UMP. Neither of these nucleotides is dephos-phorylated by yeast extracts (Natalini et al. 1979; Traut 1980); presumably levels of

Table 1. Comparison of UMP synthesis from orotate in cell extracts prepared from Baker's yeast or from Ehrlich ascites cells

| | Percent of [6-^{14}C]orotate converted to products | | | |
Time (min)	OMP	Orotidine	UMP	Uridine
A. *Baker's yeast*				
10	24.4	0	5.0	0
20	32.8	0	10.0	0
40	38.4	0	16.2	0.2
100	42.2	0.6	32.4	1.0
B. *Ehrlich ascites cells*				
10	1.0	0	51.2	8.4
20	1.0	0	58.6	19.6
40	0	0	55.2	41.8
100	0	0	21.4	76.4

Yeast cells were freeze-thawed, and then further disrupted with glass beads in a Vir-Tis homogenizer (medium speed, 2 min). Ehrlich ascites cells were homogenized as described previously (Jones et al. 1978). Both cell homogenates were centrifuged at 700 g for 20 min; the supernatant fractions were then dialyzed twice versus 100 vol of the standard buffer (pH 7.4 at 37°), containing Tris HCl; 20 mM; and dithiothreitol, 2 mM. Aliquots of these preparations, containing 5.7 μg protein (*A*) or 615 μg protein (*B*), were incubated in a final volume of 50 μl, containing standard buffer, 5 mM MgCl$_2$, 300 μM P-Rib-PP, and 100 μM (6-^{14}C) orotate (10 m Ci/mol). The reactions were stopped and the products analyzed by thin layer chromatography on polyethyleneimine cellulose plates as described previously (Jones et al. 1978)

Table 2. Degradation of OMP by a nucleotidase activity in Ehrlich ascites cell extracts

| | | Percent of substrates converted to products | | |
Time (min)	OMP	Orotidine	UMP	Uridine
A. *With [7-^{14}C]OMP as substrate:*				
2	–	8.9	25.3	–
4	–	18.2	50.2	–
7	–	29.8	62.0	–
10	–	32.8	62.6	–
B. *With [6-^{14}C]orotate as substrate, and azaUMP to inhibit UMP synthesis:*				
10	48.9	10.2	1.3	0
20	50.2	25.8	2.7	0.9
40	27.6	47.1	5.3	2.6
100	3.6	68.9	1.8	6.2

In *A*, ascites cell extract, as described in the legend to Table 1, containing 750 μg protein was incubated in a final volume of 50 μl, containing standard buffer, 5 mM MgCl$_2$, and 100 μM [7-^{14}C] OMP (2 Ci/mol). Orotidine was measured by thin layer chromatography. Since UMP was measured as the amount of ^{14}CO$_2$ released in a combined assay that has been described elsewhere (Traut and Jones 1977b), it was not possible to measure uridine. In *B*, ascites cell extract, containing 500 μg protein, was incubated in a reaction mixture, as described in the legend to Table 1, that included 100 μM 6-azaUMP, an inhibitor of orotidylate decarboxylase

nucleotidase, or a phosphatase cleaving OMP and UMP to their respective nucleosides, are low or lacking in yeast cells. In contrast, in the experiment with extracts of Ehrlich ascites cell (Table 1) OMP is channeled by Complex U so it barely appeared in the solvent; rather UMP accumulated at the early time periods of this experiment when the limiting substrate, OA, was still present. When OA was completely converted to UMP, the UMP concentration decreased because there is a phosphatase or nucleotidase present that degrades UMP to uridine. The action of the degradative enzyme might not have been as marked if ATP had been present to convert UMP to other nucleotides, such as UDP, UTP, CTP, etc. When OMP is the substrate for Ehrlich ascites extract (Table 2), it is rapidly converted to either UMP or orotidine. About 60% of the OMP was converted to the biosynthetic product, UMP, but 33% was converted to orotidine. Orotidine cannot be utilized unless it is reconverted to OA by loss of the ribose group. The data of Table 2 suggest that this does not happen; orotidine is known to be excreted by patients given azauridine (Cardoso et al. 1961; Fallon et al. 1962; Buttoo et al. 1965) and patients with orotic aciduria (Smith et al. 1966). OMP also accumulates if aza-UMP, which inhibits orotidylate decarboxylase, is added besides the substrates OA and PR∿PP to the incubation medium containing Complex U (Table 2 B). In this situation very little OMP can be or is converted to UMP and uridine, rather OMP accumulates, but it is rather rapidly converted to orotidine. In mammalian cells therefore the normal channeling of OMP by Complex U, which keeps the cellular concentration of OMP low, does not allow this nucleotidase, which we presume plays an important role in base recovery or in the excretion of the products of nucleic acid catabolism, to degrade OMP.

The channeling of CA-asp may have more subtle advantages, since CA-asp appears not to be degraded readily in mammalian cells. It does pass out of the intact cell, for Hager and Jones (1965) found that in 10 min a third of the CA-asp pool formed by the Ehrlich ascites cells (kept in a medium with 5 mM orotate) was in the medium surrounding the cells. If formed in sufficiently large amounts, CA-asp could affect the acid-base balance. It could also act as an acetylaspartate analog if it passes the blood-brain barrier. Acetyl aspartate is present in avian and mammalian brains during and after myelinization (Tallan et al. 1956); the role of acetyl aspartate in the mammalian brain is not known. It is possible therefore that the channeling of CA-asp could be physiologically important, since it appears not to be further metabolized and its release in sufficient quantities might burden normal metabolism.

IV. Regulatory Control of the Multienzymic Proteins for Mammalian UMP Biosynthesis

Allosteric multienzymic proteins may make feedback and feedforward regulation more complete and effective than it is when only the first committed enzyme of a series of independent enzymes is regulated by ligands affecting protein conformation. There are as yet no firm data concerning this suggestion. Complex A contains the first committed enzyme for UMP biosynthesis. ATP is an allosteric substrate effector for the carbamyl∿P synthetase of Complex A and a sigmoid velocity curve results when the ATP concentration is increased (Hager and Jones 1967a; Mori and Tatibana 1975). The apparent Km for ATP is 5 mM in the absence of other effectors (Mori and Tatibana 1975). UTP is an effective feedback inhibitor for the synthetase of Complex A. When 2 mM UTP is present to bind to an effector site, it increases the apparent Km for ATP to 20 mM. PR∿PP is a feedforward activator that decreases the apparent Km for ATP to 0.7 mM when the PR∿PP concentration is 50 μM. The suggestion made above is that these effectors may change the conformation of the three enzyme centers of Complex A, i.e., not only that of the carbamyl∿P synthetase, so that all centers become inactive simultaneously.

For Complex U the aggregation state is changed from monomer to dimer by substrates, substrate analogs, and inhibitors that compete with the substrates (Traut and Jones 1978, 1979). In addition, the dimer state is more active than the monomer state (Traut and Jones, unpublished data). Dimer formation is aided by the product of the first enzyme, OMP; OMP also activates the enzyme centers of Complex U (Traut and Jones 1977b). It is possible that Complex U has very little activity in the monomeric state; for technical reasons this hypothesis is difficult to test. Agents that promote dimerization include PR∿PP particularly in the presence of Mg^{2+}, OMP, AMP, UMP, Aza-UMP, and anions such as P_i and Cl, which act as competitive inhibitors of these two active centers (Jones et al. 1978). In addition, the data suggest that an unknown protein may bind to the dimer and it is possible that this protein has a regulator effect on the reaction catalyzed by Complex U (Traut and Jones 1979).

D. Regulation of UMP Biosynthesis in the Cell

The aim of our studies has been to understand this sequence of six enzymes well enough to anable us to explore their regulation in the intact cell. Such studies are essential for the development of sophisticated drugs which will produce minimal side effects. In whole-cell studies we have found that 6-azauridine produces a "ripple" effect which leads to the accumulation of intermediates in the pathway. Aza-uridine is converted in cells to aza-UMP, which is a potent inhibitor of orotidylate decarboxylase (Pasternak and Handschumacher 1959; Traut and Jones 1977a) and a weak inhibitor of orotate phosphoribosyl transferase (Traut and Jones 1977a). When cells are incubated with aza-uridine (Chen and Jones 1979b), a complete inhibition of UMP synthesis and of UMP incorporation into nucleic acids resulted (Fig. 3A); rather the intermediates CA-asp, DHO, OA, and orotidine accumulated (Fig. 3B). A major product was orotidine, which could only have been formed from OMP released into solution where it was dephosphorylated. The second major product was CA-asp, which accumulated because dihydro-orotate dehydrogenase was inhibited (Chen and Jones 1979b) by orotate which also accumulated but in lesser amounts than CA-asp or orotidine (Fig. 4). In orotic aciduria one can observe the excretion of orotate, orotidine, and CA-asp (Smith et al. 1966; Smith and Gilmour 1975). The experiments with the cultured cells, therefore, mimic to some degree the events occurring in the metabolism of these patients and possibly in patients receiving allopurinol for gout (Fox et al. 1970; Kelley and Beardmore 1970) who also excrete orotic acid and orotidine.

Studies by Chen and Jones (1978, 1979a) have shown that, when one lowers the PR∿PP levels in cells by adding adenine to the medium surrounding the cultured cells, the first enzyme of the pathway that is sensitive to the lowered intracellular PR∿PP concentration is carbamyl∿P synthetase (Chen and Jones 1978). The other enzyme that could have been affected was orotate phosphoribosyltransferase which utilizes PR∿PP as substrate. A second series of experiments showed that this second enzyme center is also sensitive to the lowering of the PR∿PP concentration by adenine (Chen and Jones 1979a). One cannot easily tell which site is more sensitive to the intracellular PR∿PP levels, since the sensitivity of the synthetase should be dependent on the cellular ATP and UTP concentration (Mori and Tatibana 1975; Chen and Jones 1979a). In addition, Complex U can only function when a supply of OA is available; therefore, if both sites are equally sensitive to the PR∿PP concentration, it would appear that only Complex A was affected. More experiments are necessary in which the levels of the other effectors, such as ATP, AMP, UTP, and UMP, and the substrates, are measured as well as the PR∿PP level. We believe, however, that sophisticated drug design requires such studies.

E. Summary

UMP biosynthesis requires six enzyme activities. Five of these enzyme centers are clustered into two multienzymic proteins which are known to, or appear to, sequester the intermediates carbamyl∿P, carbamyl aspartate and orotidylic acid. The advantages of sequestering these intermediates appear to be a conservation of energy, since two intermediates, carbamyl∿P and orotidylate, might otherwise be rapidly degraded in mammalian cells. Carbamyl-aspartate appears not to be degraded rapidly in mammalian cells but it can pass into the blood and could possibly disrupt brain metabolism by acting as an acetylaspartate analog, if it passes the blood-brain barrier. For this, and possibly for other reasons, there may be advantages to the fact that these intermediates are not readily released from Complex A and U. In addition, these multienzymic proteins may have other kinetic advantages, some of which have been discussed above.

Studies with intact cells illustrate that azauridine, a chemical designed originally as an antineoplastic drug, produces a "ripple" effect when it inhibits the last enzyme of this pathway which leads to a sequential accumulation of pools of the various intermediates or their metabolites. This same agent increases the amount of some of the enzymes of this biosynthetic pathway in cells exposed to this drug. Both of these effects can negate the effectiveness of this potential antineoplastic drug. Sophisticated drug design may depend on whole-cell studies, such as those discussed here, in addition to the classic studies on the inhibition of a single enzyme center to select drugs that may be without significant side effects when they are finally tested in animals.

Acknowledgments. I am grateful for the grant support my laboratory has continuously received from the National Science Foundation and the National Institutes of Health. I appreciate the suggestions and criticism of this manuscript offered by Michael Black, Richard I. Christopherson, Ronald W. McClard, and Thomas W. Traut.

References

Allen CM Jr, Jones ME (1964) Decomposition of carbamyl phosphate in aqueous solutions. Biochemistry 3: 1238–1247

Appel SH (1968) Purification and kinetic properties of brain orotidine 5'-phosphate decarboxylase. J Biol Chem 243: 3924–3929

Becker MA, Argubright KF, Fox RM, Seegmiller JE (1974) Oxipurinol-associated inhibition of pyrimidine synthesis in human lymphoblasts. Mol Pharmacol 10: 657–668

Bresnick E, Blatchford K (1964) Studies on dihydroorotase activity in preparations from Novikoff ascites hepatoma cells. Arch Biochem Biophys 104: 381–386

Bresnick E, Mayfield ED Jr, Mosse H (1968) Increased activity of enzymes for *de novo* pyrimidine biosynthesis after orotic acid administration. Mol Pharmacol 4: 173–180

Brown GK, O'Sullivan WJ (1977) Subunit structure of orotate phosphoribosyltransferase – orotidylated decarboxylase complex from human erythrocytes. Biochemistry 16: 3235–3242

Brown GK, Fox RM, O'Sullivan WJ (1975) Interconversion of different molecular weight forms of human erythrocyte orotidylate decarboxylase. J Biol Chem 250: 7352–7358

Buttoo AS, Israel MCG, Wilkinson JF (1965) Hypocholesterolaemia and orotic aciduria during treatment with 6-azauridine. Brit Med J 1: 552–554

Campbell MT, Gallagher ND, O'Sullivan WJ (1977) Multiple molecular forms of orotidylate decarboxylase from human liver. Biochem Med 17: 128–140

Cardoso SS, Calabresi P, Handschumacher RE (1961) Alterations in human pyrimidine metabolism as a result of therapy with 6-azauridine. Cancer Res 21: 1551–1556

Chen J-J, Jones ME (1976) The cellular location of dihydroorotase dehydrogenase: Relation to *de novo* biosynthesis of pyrimidine. Arch Biochem Biophys 176: 82–90

Chen J-J, Jones ME (1978) The regulation of *de novo* pyrimidine biosynthesis by cellular phosphoribosyl pyrophosphate levels in cultured Ehrlich ascites cells. In: Srere PA, Estabrook RW (eds)

Microenvironments and metabolic compartmentation, pp 211–226. Academic Press, New York San Francisco London

Chen J-J, Jones ME (1979a) Effect of 5-phosphoribosyl-1-pyrophosphate on *de novo* pyrimidine biosynthesis in cultured Ehrlich ascites cells made permeable with dextran sulfate 500. J Biol Chem 254: 2697–2704

Chen J-J, Jones ME (1979b) Effect of 6-azauridine on *de novo* pyrimidine biosynthesis in cultured Ehrlich ascites cells. J Biol Chem 254; 4908–4914

Christopherson RI, Jones ME (1979) Interconversion of carbamyl-L-aspartate and L-dihydroorotate by dihydroorotase from mouse Ehrlich ascites carcinoma. J Biol Chem 254: 12506–12512

Christopherson RI, Matsuura T, Jones ME (1978) Radioassay of dihydroorotase utilizing ion-exchange chromatography. Anal Biochem 89: 225–234

Coleman PF, Suttle DP, Stark GR (1977) Purification from hamster cells of the multifunctional protein that initiates *de novo* synthesis of pyrimidine nucleotides. J Biol Chem 252: 6379–6385

Coleman PF, Suttle DP, Stark GR (1978) Purification of a multifunctional protein bearing carbamyl-phosphate synthetase, aspartate tran carbamylase, and dihydroorotase enzyme activities from mutant hamster cells. In: Hoffee PA, Jones ME (eds) Methods in enzymology: Purine and pyrimidine nucleotide metabolism, vol LI, pp 121:134. Academic Press, New York

Davis RH (1967) Channeling of *Neurospora* metabolism. In: Vogel HJ, Lampen LO, Bryson V (eds) Organizational biosynthesis, pp 303–322. Academic Press, New York

Denis-Duphil M, Lacroute F (1971) Fine structure of the ura_2 locus in Saccharomyces cerevisiae. I In vivo complementation studies. Mol Gen Genet 112: 354–364

Ennis HL, Lubin M (1963) Capacity for synthesis of a pyrimidine biosynthetic enzyme in mammalian cells. Biochim Biophys Acta 68: 78–83

Fallon HJ, Lotz M, Smith LH Jr (1962) Congenital orotic aciduria: demonstration of an enzyme defect in leukocytes and comparison with drug induced orotic aciduria. Blood 20: 700–709

Fink GR (1971) Gene clusters and the regulation of biosynthetic pathways in Fungi. In: Vogel HJ (ed) Metabolic pathways, vol V, Metabolic regulation, pp 200–223. Academic Press, New York

Fox RM, Royse-Smith D, O'Sullivan WJ (1970) Orotidinuria induced by allopurinol. Science 168: 861–862

Galofre A, Kretchmer N (1970) Biosynthesis of pyrimidines by various organs of the chick during embryogenesis. Pediat Res 4: 55–62

Grobner W, Kelley WN (1975) Effect of allopurinol and its metabolic derivatives on the configuration of human orotate phosphoribosyltransferase and orotidine 5-phosphate decarboxylase. Biochem Pharmacol 24: 379–384

Hager SE, Jones ME (1965) Initial step in pyrimidine synthesis in Ehrlich ascites carcinoma in vitro. I. Factors affecting incorporation of ^{14}C-bicarbonate into carbon 2 of the uracil ring of the acid-soluble nucleotides of intact cells. J Biol Chem 240: 4556–4569

Hager SE, Jones ME (1967a) Initial steps in pyrimidine synthesis in Ehrlich ascites carcinoma in vitro. II. The synthesis of carbamyl phosphate by a soluble glutamine dependent carbamyl phosphate synthetase. J Biol Chem 242: 5667–5673

Hager SE, Jones ME (1967b) A glutamine-dependent enzyme for the synthesis of carbamyl phosphate for pyrimidine biosynthesis in fetal rat liver. J Biol Chem 242: 5674–5680

Hatfield D, Wyngaarden JB (1964) 3-Ribosylpurines. I. Synthesis of (3-ribosyluric acid) 5'-phosphate and (3-ribosylxanthine) 5'-phosphate by a pyrimidine ribonucleotide pyrophosphorylase of beef erythrocytes. J Biol Chem 239: 2580–2586

Hitchings GH (1973) Indications for control mechanisms in purine and pyrimidine biosynthesis as revealed by studies with inhibitors. In: Weber G (ed) Advances in enzyme regulation, vol XII, pp 121–129. Pergamon Press, Oxford New York

Hoogenraad NJ, Levine RL, Kretchmer N (1971) Copurification of carbamoyl phosphate synthetase and aspartate transcarbamylase from mouse spleen. Biochem Biophys Res Commun 44: 981–988

Ito K, Uchino H (1971) Control of pyrimidine biosynthesis in human lymphocytes. Induction of glutamine-utilizing carbamyl phosphate synthetase and operation of orotic acid pathway during blastogenesis. J Biol Chem 246: 4060–4065

Janeway CM, Cha S (1977) Effects of 6-azauridine on nucleotides, orotic acid, and orotidine in
    L5178Y mouse lymphoma cells in vitro. Cancer Res 37: 4382–4388
Jones ME (1970) Vertebrate carbamyl phosphate synthetase I and II. Separation of the arginine-
    urea and pyrimidine pathways. In: Schmidt-Nelson B (ed) Urea and the kidney, pp 35–47. Ex-
    cerpta Medica Foundation, International Congress Series No. 195, Amsterdam
Jones ME (1971) Regulation of pyrimidine and arginine biosynthesis in mammals. In: Weber G (ed)
    Advances in enzyme regulation, vol IX, pp 19–49. Pergamon Press, Oxford New York
Jones ME (1972) Regulation of uridylic acid biosynthesis in eukaryotic cells. In: Horecker B, Stadt-
    man ER (eds) Current topics on cellular regulation, vol VI, pp 227–265. Academic Press, New
    York
Jones ME, Kavipurapu PR, Traut TW (1978) Orotate phosphoribosyltransferase: orotidylate de-
    carboxylase (Ehrlich ascites cell). In: Hoffee PA, Jones ME (eds) Methods in enzymology: Purine
    and pyrimidine nucleotide metabolism, vol LI, pp 155–167. Academic Press, New York
Jund R, Lacroute F (1972) Regulation of orotidylic acid pyrophosphorylase in *Saccharomyces
    cerevisiae.* J Bacteriol 109: 196–202
Kasebekar DK, Nagabhushanam A, Greenberg DM (1964) Purification and properties of orotic acid-
    decarboxylating enzyme from calf thymus. J Biol Chem 239: 4245–4249
Kavipurapu PR, Jones ME (1976) Purification, size and properties of the complex of orotate phos-
    phoribosyltransferase: orotidylate decarboxylase from mouse Ehrlich ascites carcinoma. J Biol
    Chem 251: 5589–5599
Kelley WN, Beardmore TJ (1970) Allopurinol: alteration in pyrimidine metabolism in man. Science
    169: 388–390
Kempe TD, Swyryd EA, Bruist M, Stark GR (1976) Stable mutants of mammalian cells that over-
    produce the first three enzymes of pyrimidine nucleotide biosynthesis. Cell 9: 541–550
Kennedy J (1974) Dihydroorotase from rat liver: Purification properties and regulatory role in
    pyrimidine biosynthesis. Arch Biochem Biophys 160: 358–365
Kim S, Cohen PP (1965) Transcarbamylase activity in fetal liver and in liver of partially hepatectomiz-
    ed parabiotic rats. Arch Biochem Biophys 109: 421–428
Kirschner K, Bisswanger H (1976) Multifunctional proteins. Annu Rev Biochem 45: 143–166
Levenberg B (1962) Role of L-glutamine as donor of carbamyl nitrogen for the enzymatic synthesis
    of citrulline in *Agaricus bisporus.* J Biol Chem 237: 2590–2598
Levinson BB, Ullman B, Martin DW Jr (1979) Pyrimidine pathway variants of cultured mouse
    lymphoma cells with altered levels of both orotate phosphoribosyltransferase and orotidylate
    decarboxylase. J Biol Chem 254: 4396–4401
Levitsky A, Koshland DE Jr (1971) Cytidine triphosphate synthetase. Covalent intermediates and
    mechanisms of action. Biochemistry 10: 3365–3371
Makoff AJ, Radford A (1978) Genetics and biochemistry of carbamoyl phosphate biosynthesis
    and its utilization in the pyrimidine biosynthetic pathway. Microbiol Rev 42: 307–328
Matsuura T, Jones ME (1974) Subcellular localization in the Ehrlich ascites cells of the enzyme which
    oxidizes dihydroorotate to orotate. In: Richter D (ed) Lipmann Symposium: Energy, biosyn-
    thesis and regulation in molecular biology, pp 422–434. de Gruyter, Berlin New York
Miller RW (1975) A high molecular weight dehydroorotate dehydrogenase of *Neurospora crassa.*
    Purification and properties of the enzyme. Can J Biochem 53: 1288–1300
Mori M, Tatibana M (1973) Dissociation by elastase digestion of the enzymes complex catalyzing
    the initial steps of pyrimidine biosynthesis in rat liver. Biochem Biophys Res Commun 54:
    1525–1531
Mori M, Tatibana M (1975) Purification of homogeneous glutamine-dependent carbamyl phosphate
    synthetase from ascites hepatoma cells as a complex with aspartate transcarbamylase and di-
    hydroorotase. J Biochem 78: 239–242
Mori M, Tatibana M (1978a) Multi-enzyme complex of glutamine-dependent carbamoyl-phosphate
    synthetase with aspartate carbamoyltransferase and dihydroorotase from rat ascites-hepatoma
    cells. Eur J Biochem 86: 381–388
Mori M, Tatibana M (1978b) A multienzyme complex of carbamoyl-phosphate synthetase (glutamine):
    aspartate carbamoyltransferase: dihydroorotase (rat ascites hepatoma cells and rat liver). In:
    Hoffee PA, Jones ME (eds) Methods in enzymology: Purine and pyrimidine nucleotide metabolism,
    vol LI, pp 111–121. Academic Press, New York

Natalini P, Ruggieri S, Santarelli I, Vita A, Magni G (1979) Baker's yeast UMP: pyrophosphate phosphoribosyltransferase. Purification enzymatic and kinetic properties. J Biol Chem 254: 1558–1563

Nørby S (1970) A specific nutritional requirement for pyrimidine in rudimental mutants of Drosophila melanogaster. Hereditas 66: 205–214

O'Donovan GA, Neuhard J (1970) Pyrimidine metabolism in microorganisms. Bacteriol Rev 34: 278–343

Pasternak CA, Handschumacher RE (1959) The biochemical activity of 6-azauridine: interference with pyrimidine metabolism in transplantable mouse tumors. J Biol Chem 234: 2992–2997

Pinsky L, Krooth RS (1967a) Studies on the control of pyrimidine biosynthesis in human diploid cell strains. I. Effect of 6-azauridine on cellular phenotype. Proc Natl Acad Sci USA 57: 925–932

Pinsky L, Krooth RS (1967b) Studies on the control of pyrimidine biosynthesis in human diploid cell strains. II. Effects of 5-azaorotic acid, barbituric acid and pyrimidine precursors on cellular phenotype. Proc Natl Acad Sci USA 57: 1267–1274

Rawls JM, Fristrom JW (1975) A complex genetic locus that controls the first three steps of pyrimidine biosynthesis in Drosophila. Nature 255: 738–740

Reichard P (1959) The enzymatic synthesis of pyrimidines. In: Nord FF (ed) Advances in enzymology, vol XXI, pp 263–294. Interscience, New York

Reichard P, Hanshoff G (1956) Aspartate carbamyl transferase from *Escherichia coli*. Acta Chem Scand 10: 548–566

Reyes P, Guganig ME (1975) Studies on a pyrimidine-phosphoribosyl-transferase for murine leukemia P1534J. Partial purification, substrate specificity, and evidence for its existence as a bifunctional complex with orotidine 5'-phosphate decarboxylase. J Biol Chem 250: 5097–5108

Reyes P, Intress C (1978) Coordinate behavior of orotate phosphoribosyl-transferase and orotidylate decarboxylase in mouse liver and brain. Life Sci 22: 577–582

Shoaf WT, Jones ME (1971) Initial steps in pyrimidine synthesis in Ehrlich ascites carcinoma. Biochem Biophys Res Commun 45: 796–802

Shoaf WT, Jones ME (1972) Uridylic acid synthesis in Ehrlich ascites carcinoma. Properties, subcellular distribution, and nature of enzyme complexes of the six biosynthetic enzymes. Biochemistry 12: 4039–4051

Smith LH Jr, Baker FA (1959) Pyrimidine metabolism in man. I. The biosynthesis of orotic acid. J Clin Invest 38: 798–809

Smith LH Jr, Baker FA (1960) Pyrimidine metabolism in man. III. Studies on the leukocytes and erythrocytes in pernicious anemia. J Clin Invest 39: 15–20

Smith LH Jr, Gilmour L (1975) Determination of urinary carbamyl-aspartate and dihydroorotate in normal subjects and in patients with hereditary orotic aciduria. J Lab Clin Med 86: 1047–1051

Smith LH Jr, Huguley CM Jr, Bain JA (1966) Hereditary orotic aciduria. In: Stanbury JB, Wyngaarden JB, Fredrickson DS (eds) The metabolic basis of inherited disease, 2nd ed, pp 739–758. McGraw-Hill, New York

Suttle DP, Stark GR (1979) Coordinate overproduction of orotate phosphoribosyltransferase and orotidine-5'-phosphate decarboxylase in hamster cells resistant to pyrazofurin and 6-azauridine. J Biol Chem 254: 4602–4607

Sweeney MJ, Hoffman DH, Poore GA (1971) Enzymes in pyrimidine biosynthesis. Adv Enzyme Regul 9: 51–61

Tallan HH, Moore S, Stein WH (1956) N-Acetyl-L-aspartic acid in brain. J Biol Chem 219: 257–264

Tatibana M, Ito K (1969) Control of pyrimidine biosynthesis in mammalian tissues. I. Partial purification and characterization of glutamine-utilizing carbamyl phosphate synthetase of mouse spleen and its tissue distribution. J Biol Chem 244: 5403–5413

Tatibana M, Shigesada K (1972) Activation by 5-phosphoribosyl l-pyrophosphate of glutamine-dependent carbamyl phosphate synthetase from mouse spleen. Biochem Biophys Res Commun 46: 491–497

Taylor AL, Trotter CD (1972) Linkage map of *Escherichia coli* strain K12. Bacteriol Rev 36: 504–524

Taylor WH, Taylor ML, Eames DF (1966) Two functionally different dihydroorotic dehydrogenases in bacteria. J Bacteriol 91: 2251–2256

Traut TW (1980) Significance of the enzyme complex that synthesizes UMP in Ehrlich ascites cells. Arch Biochem Biophys 200: 590–594

Traut TW, Jones ME (1977a) Inhibitors of orotate phosphoribosyltransferase and orotidine-5'-phosphate decarboxylase from mouse Ehrlich ascites cells: a procedure for analyzing the inhibition of a multi-enzyme complex. Biochem Pharmacol 26: 2291–2296

Traut TW, Jones ME (1977b) Kinetic and conformational studies of the orotate phosphoribosyltransferase: orotidine-5'-phosphate decarboxylase enzyme complex from mouse Ehrlich ascites cells. J Biol Chem 252: 8374–8381

Traut TW, Jones ME (1978) Mammalian synthesis of UMP from orotate: the regulation of and conformers of Complex U. In: Weber G (ed) Advances in enzyme regulation, vol XVI, pp 21–41. Pergamon Press, Oxford New York

Traut TW, Jones ME (1979) Interconversion of different molecular weight forms of the orotate phosphoribosyltransferase orotidine-5'-phosphate decarboxylase enzyme complex from mouse Ehrlich ascites cells. J Biol Chem 254: 1143–1150

Weber G, Queener SF, Ferdinandus JA (1971) Control of gene expression in carbohydrate, pyrimidine and DNA metabolism. In: Weber G (ed) Advances in enzyme regulation, Vol IX, pp 65–95. Pergamon Press, Oxford New York

Williams LG, Bernhardt PH, Davis RH (1970) Copurification of pyrimidine-specific carbamyl phosphate synthetase and aspartate transcarbamylase of *Neurospora crassa*. Biochemistry 9: 4329–4335

Yip MCM, Knox WE (1970) Glutamine-dependent carbamyl phosphate synthetase. Properties and distribution in normal and neoplastic rat tissue. J Biol Chem 245: 2199–2204

Young JE, Prager MD, Atkins IC (1967) Comparative activities of aspartate transcarbamylase in various tissues of the rat. Proc Soc Exp Biol Med 125: 860–862

6. Regulation of Muscle Contraction by Ca Ion

S. EBASHI, Y. NONOMURA, K. KOHAMA, T. KITAZAWA, and T. MIKAWA

A. Introduction

There is no doubt that the contractile processes of every kind of muscle are solely regulated by Ca ion. The fundamental evidence for this concept was furnished by the laboratory of Professor Fritz Lipmann in 1959.

In 1948 Kielley and Meyerhof presented a new ATPase preparation of skeletal
muscle, called "a new magnesium-activated adenosine phosphatase", clearly different
from actomyosin ATPase. Although they could not connect this ATPase with a specified
function, they emphasized that it has much higher specific activity than actomyosin
ATPase; even its total activity was fairly comparable to that of actomyosin ATPase
under certain conditions, suggesting its important physiological role. This was undoubtedly the most significant work of Otto Meyerhof after his coming to the United States.
Indeed, this ATPase fraction was identified by Kumagai, Ebashi, and Takeda-Ebashi
(1955) with the relaxing factor, which was originally discovered by Marsh (1951) as a
factor inducing the swelling of muscle homogenete, and then recognized to be the system to induce the relaxation in living muscle (cf. Weber 1966; Ebashi and Endo 1968).

When this conclusion was reached, I (S.E.) felt that the final goal would be very
close. However, the subsequent 3 years passed without success. I inquired once into
the possibility that Ca^{2+} should play some role, but no experimental result obtained by
me lived up to such an expectation. On the other hand, most researchers at that time
assumed that a "true" relaxing factor, i.e., a soluble factor of small molecular weight,
would be produced by the ATPase, and therefore their efforts were concentrated on
isolating such a true relaxing factor.

When I asked Professor Lipmann in 1958 to accept me in his laboratory, I was very
pessimistic about the future of the studies on the relaxing factor, viz., Kielley-Meyerhof's
Mg-activated ATPase. So I wrote him a letter that I had no particular intention but only
wished to learn from him, with a tacit expectation that he would give me a thesis concerning enzymology or protein synthesis. Therefore, when I received his letter indicating
that I should just continue my work in Japan, I was rather disappointed, fearing that I
would have to struggle another year with the relaxing factor in vain.

One day early in 1959, however, owing to the inspiring guidance of Professor Lipmann, I suddenly became convinced that the relaxing factor should combine Ca in the

Department of Pharmacology, Faculty of Medicine, University of Tokyo, Hongo, Bunkyo-ku,
Tokyo 113, Japan

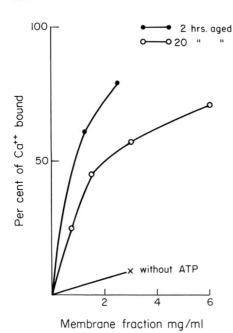

Fig. 1. ATP-dependent Ca^{2+} binding of "relaxing factor" (quoted from Ebashi and Lipmann 1962). The legend in the original article was: ATP-dependent calcium ion binding with various concentrations of normal and aged membrane fraction. The "membrane fraction" was spun down in the presence of 10 mM Mg, 2 mM ATP, 144 mM KCl Tris-maleate (pH 6.8) and 10 μM $CaCl_2$ including ^{45}Ca, and the radioactivity in the pellet was expressed relative to the total radioactivity; the amounts of membrane fraction were expressed by wet weight, assumed to be five times the dry weight

presence of ATP (Ebashi and Lipmann 1962; cf. Ebashi 1960, 1961). Figure 1 shows the data of my first experiment along this line, which was so far the only experiment that fulfilled my expectations at once.

B. Regulatory Mechanisms

I. General Aspects

The establishment of the role of Ca^{2+} on the biochemical basis has soon led to the discovery of the "third component", i.e., the troponin-tropomyosin system (Ebashi 1963; cf. Ebashi and Endo 1968) which is located in the actin filament of vertebrate skeletal muscle (Ebashi et al. 1969). This work seemed to have established the actin-linked nature of regulatory mechanism in general; this concept appeared to coincide with the rather auxiliary role of actin in muscle contraction, as compared with the role of myosin.

However, the regulatory system of scallop striated muscle (Kendrick-Jones et al. 1970), in which the Ca-binding site resides in the myosin molecule itself, has strongly indicated the diverse nature of the regulatory mechanisms in general (Table 1).

In fact, the regulatory system of vertebrate smooth muscle has been shown to be entirely different from the above two systems (Table 1) (Ebashi et al. 1976), although its actual mechanism has been the subject of heated controversy.

Table 1. Comparison of three kinds of muscle from a regulatory point of view

	Basic state of actin-myosin-ATP interaction	Role of Ca	Biphasic response to MgATP[a]	Requirement of tropomyosin for regulation	Remarks
Troponin-dependent	Contracting	De-repressor	+++	Yes	Only system in vertebrates; per-haps so in deu-terostomias
Myosin-linked	Contracting	De-repressor	++++	No	In protostomias; often coexists with troponin system
Leiotonin-dependent	Relaxed	Activator	+	Yes	In vertebrate smooth muscle

[a] Almost synonymous with "substrate-inhibition type response"

II. Regulation in Smooth Muscle

Several groups of research workers have claimed that smooth muscle myosin becomes active only when its light chain is phosphorylated by a Ca-dependent light chain kinase and the myosin thus phosphorylated is inactivated by dephosphorylation with a specific phosphatase. Thus, the regulatory system consists of the Ca-dependent light chain kinase and the phosphatase specific for this dephosphorylation (Bremel et al. 1977; Aksoy et al. 1976; Chacko et al. 1977; Small and Sobieszek 1977; Ikebe et al. 1978; Sherry et al. 1978; cf. Casteel et al. 1977).

The view concerning the regulatory mechanism in smooth muscle based on the studies carried out by T. Toyo-oka, M. Hirata, K. Saida and three of us (S.E., Y.N., and T.M.) (Ebashi et al. 1976; Mikawa et al. 1977; Mikawa et al. 1978; cf. review articles below) is entirely different from the above. Since we have recently published several review articles (Ebashi et al. 1977; Ebashi 1979; Ebashi et al. 1979), only a brief summary will be given below so as to avoid overlapping with these articles.

Ca regulation in smooth muscle is of actin-linked nature like that in striated muscles (Mikawa 1979). The regulatory system is composed of tropomyosin and a new type of regulatory protein, called leiotonin; the latter consists of leiotonin A, a fairly neutral protein with a molecular weight of about 80,000 dalton, and leiotonin C, an acidic Ca-binding protein of 18,000 dalton. In Table 2 the properties of leiotonin are compared with troponin.

It should be emphasized that the myosin-actin-ATP interaction of smooth muscle is quiescent unless provided with regulatory proteins. Thus the relaxation is the ground state of the interaction (Table 1). In sharp contrast with this, the principal response of the myosin-actin-ATP interaction of striated muscle is contraction, which does not need the aid of regulatory proteins. On the contrary, the crucial role of regulatory proteins, viz., troponin and tropomyosin, is to suppress the interaction, giving rise to relaxation.

Table 2. Properties of regulatory factors

	Molecular weights	Subunits	Affinity for	Ratio to actin
Troponin	80,000	3 (T, I, C)	Tropomyosin (Actin)	1 : 7
Leiotonin	100,000	2 (A, C)	Actin	1 : 70

The situation in scallop striated muscle is essentially the same as in vertebrate striated muscle, in spite of a sharp difference in their regulatory mechanisms.

It is also worthy of note that leiotonin A has a strong affinity for actin but not for tropomyosin, and that leiotonin can fully regulate the actomyosin system at a molar ratio to actin of only 1/70th or less.

III. Ca-Binding Proteins

The inquiry into the regulatory system of smooth muscle has brought about a new Ca-binding protein, i.e., leiotonin C (Mikawa et al. 1978). Now we have three kinds of Ca-binding proteins with a definite physiological function, viz., troponin C, leiotonin C, and modulator protein (Table 3).

Modulator protein (CDR or calmodulin) has been found as the activator of phosphodiesterase (Kakiuchi et al. 1970; Cheung et al. 1970). It is now recognized as the general mediator of Ca-related intracellular reactions. However, in muscle regulation, though it can replace the role of physiological Ca-binding proteins in in vitro experiments (Amphlett et al. 1976; Mikawa et al. 1978), modulator protein is not utilized in physiological processes, but specialized proteins such as leiotonin C and troponin C have been developed for a physiological purpose. The reason for this is not yet clear, but it has been speculated (Ebashi et al. 1979; Ebashi 1980) that the nature of modulator protein easily detachable from its parent proteins in the absence of Ca ion, although it may be convenient for regulating a number of Ca-dependent cytoplasmic processes

Table 3. Specificities of Ca-binding proteins

	Phospho-diesterase activation	Troponin C-like action	Leiotonin C-like action	In absence of Ca^{2+}
Modulator protein	(1)	Yes[a]	Yes	Easily dissociable
Troponin C	0.01	. . . .	No	Not dissociable
Leiotonin C	0.01	?	. . . .	Hardly dissociable

[a] Troponin T is not required

carried out by soluble proteins, is not suitable for the contractile processes which require preciseness, rapidity and reporducibility.

Anyway, it is an interesting hypothesis that modulator protein is the ancestral protein of muscle Ca-binding proteins as proposed by Kakiuchi (cf. Mikawa et al. 1978). Further investigation along this line will provide a perspective view for the physiological and evolutional significance of Ca-regulated processes, not only of muscle but also in general.

C. Some Facets of the Troponin-Tropomyosin System

I. Physiological Significance of High-Affinity Site(s) for Ca^{2+}

Ebashi et al. (1968) have shown that troponin of fast skeletal muscle has four Ca-binding sites, two of them having higher affinities for Ca ion, and the other two lower affinities. They simply believed that the four Ca-binding sites would be directly involved in the regulatory processes in a qualitatively similar manner. The relationship between the tension development of skinned fibers, or superprecipitation of myosin B and the Ca^{2+} concentrations, shifted to the direction of higher Ca^{2+} concentrations with increase in the $MgCl_2$ concentration from 1 mM to 8 mM (cf. Figs. 2 and 3 in Ebashi and Endo 1968). They put these data in their review with the tacit understanding that this might be due to the competitive effect of Mg^{2+} on Ca-binding sites of troponin.

Based on their elegant studies on Ca and Mg binding to troponin, Potter and Gergely (1975) have concluded, however, that the "high affinity sites", which have also a high affinity for Mg^{2+}, are not directly involved in the regulatory processes, and the regulation is carried out by Ca^{2+}-binding to the sites termed as the "low-affinity sites", or "Ca^{2+}-specific sites".

Their view seems now widely accepted by muscle scientists. However, since the change in Mg^{2+} concentration has no effect on the superprecipitation of desensitized myosin B and reconstituted actomyosin (unpublished), the Mg^{2+} effect mentioned above should be concerned with some Ca^{2+}-related process(es).

Kohama (1980) has studied Ca and Mg binding to troponin conjugated with Sepharose 4B (Kohama 1979). Although his results are largely in accord with theirs, rather minor discrepancies have come out which lead to an interpretation of the roles of Ca-binding sites, which is different from the above.

First, his results favor the idea that two classes of Ca-binding sites exist even in the presence of Mg^{2+} concentrations higher than 1 mM, in accordance with the results of Ebashi et al. (1968); the binding constant of the high-affinity sites is at least ten times higher than that of low-affinity sites. Second, extraordinary strong affinity for Mg^{2+} of the high-affinity site(s) (Potter and Gergely 1975) can be confirmed, but it disappears in the presence of high Mg^{2+} and Ca^{2+}. From the two *curves* in Fig. 2, showing Ca^{2+} binding at 1.3 mM and 13 mM $MgCl_2$, apparent Mg-binding constants of low- and high-affinity sites are calculated to be about $7 \times 10\ M^{-1}$ and $5 \times 10^2\ M^{-1}$, respectively. These binding constants may well explain the results of Ebashi and Endo (1968) referred to above.

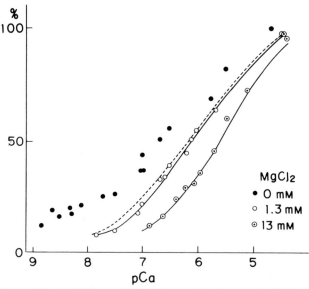

Fig. 2. Effect of Mg ion on Ca binding to skeletal troponin. Chicken skeletal troponin was immobilized by conjugation with sepharose 4 B resin. Specified amount of the immobilized troponin was taken into a small tube which contained 0.1 M KCl, 20 mM Tris-maleate buffer and various concentrations of $MgCl_2$, GEDTA (glycoletherdiaminetetraacetic acid, EGTA), and $CaCl_2$ containing ^{45}Ca. After incubation at $23°C$ for 30 min, the immobilized troponin in the incubation mixture was spun down. The amount of bound Ca was then determined by atomic absorption spectrophotometry and scintillation counting. *Dotted line* represents theoretical relationship between pCa and Ca binding in the absence of Mg^{2+} calculated from the relationships in the presence of 1.3 and 13 mM $MgCl_2$ on the assumption that Ca^{2+} and Mg^{2+} compete for the same binding sites

Kohama (1979) has already shown that cardiac troponin has two Ca-binding sites, of which the binding constants are quite different, the one being about 500 times higher than the other. Its high-affinity site has properties similar to those of fast skeletal troponin, exhibiting a high affinity for Mg^{2+}. Based on the fact that Sr^{2+} induced cardiac contractile processes at its low concentrations, which affected only the high-affinity site, he suggested that the high-affinity site would play a definite role also in the physiological contraction of cardiac muscle. This view is substantiated by the result shown in Fig. 3, i.e., if the total Mg concentration is low enough, viz., 0.1 mM, Ca^{2+} in the concentration ranging from 10^{-9} M to 10^{-7} M — a range in which only the high-affinity sites are capable of binding Ca^{2+} — can modify the contractile processes. Essentially the same result is shown with fast skeletal myosin B (Kohama 1980).

As a whole, we are now of the opinion that the four Ca-binding sites have qualitatively the same properties and can actively participate in the regulatory processes in accordance with the previous assumption by Ebashi et al. (1968).

Whether the high-affinity sites are actually involved or not in the regulation of fast skeletal muscle under ordinary physiological conditions, is another problem to be inquired into. This question can be more simply expressed: Are the sites already saturated with Ca at the threshold Ca^{2+} concentration for contraction? This is certainly the case of superprecipitation and contraction (Fig. 4). However, if we use Ca^{2+}-regulated ATPase activity as the criterion of contraction, the high-affinity sites are not saturated by Ca

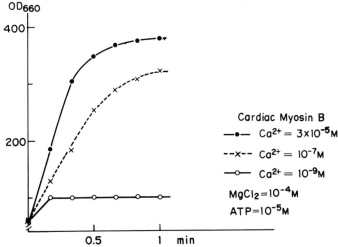

Fig. 3. Superprecipitation at low Ca^{2+}. Superprecipitation of chicken cardiac myosin B was induced in the presence of 35 mM KCl, 20 mM Tris-maleate buffer (pH 6.8), 0.1 mM $MgCl_2$, 50 μM ATP with ATP regenerating system of pyruvate kinase and phosphoenolpyruvic acid. *Open rings:* 10^{-9} M Ca^{2+}; *dotted line:* 10^{-7} M Ca^{2+}; *black balls:* 3×10^{-5} M Ca^{2+}

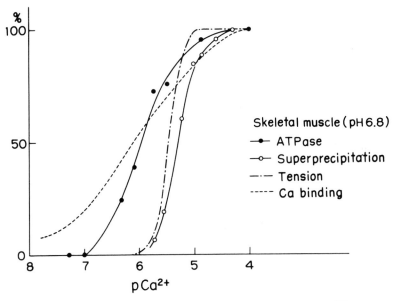

Fig. 4. ATPase, superprecipitation, tension and Ca binding versus Ca^{2+} concentration. Relative rates of ATPase activity and superprecipitation of fast skeletal myosin B were plotted against pCa together with tension development of glycerinated fast skeletal muscle (Kitazawa 1976) and Ca binding of fast skeletal troponin (Kohama 1979). Notice the resemblance in the profile of curves between superprecipitation and tension and also that between Ca binding and ATPase

ion and virtually no Ca binding to the low-affinity sites takes place at the threshold Ca^{2+} concentration (Fig. 4). This discrepancy between the ATPase activity and the tension development or superprecipitation is by itself an interesting phenomenon, as will be discussed in the next section, but no matter how the ATPase activity is related to contraction, it may not be absurd to assume that the preliminary step of contraction has already started when a definite increase in the ATPase activity is observable.

In conclusion, the high-affinity sites of fast skeletal troponin have essentially the same quality as that of the low-affinity sites and play active roles in Ca^{2+} regulation of physiological contractile processes.

II. ATPase, Superprecipitation and Contraction

There is no doubt that actin-activated myosin ATPase underlies muscle contraction; without the ATPase no contraction can take place. However, it is a moot point whether or not the ATPase is the contraction itself, i.e., whether or not the actin-activated myosin ATPase in in vitro system corresponds perfectly to the contraction of fiber models such as glycerinated and skinned fibers.

The extracted actomyosin system exhibits a high ATPase activity even in the state which corresponds to relaxation (an example is incidentally shown in Fig. 5; at low Sr^{2+} concentrations, the ATPase activity reaches a low plateau value which exceeds by one-fifth the maximum activity; the same is true for Ca^{2+}). The fraction of this ATPase, let us say "residual" ATPase, increases as myosin is degraded to H-meromyosin and H-meromyosin-SF-1. This residual ATPase, not regulated by Ca^{2+}, is very low in fiber models and virtually zero in living fibers, so it is generally considered as a kind of artifact, though it is a serious problem how to interpret actin-activated H-meromyosin ATPase. In the following, this residual ATPase will be disregarded for the sake of simplicity and discussions will be confined to Ca^{2+}-regulated ATPase.

In the previous section we have already seen a discrepancy between the ATPase activity and contractile processes, i.e., the threshold Ca^{2+} concentrations for the increment of the ATPase is much lower than those for the superprecipitation or tension development. Solaro et al. (1974) and Ronald et al. (1976) have made similar observations, using cardiac glycerinated fibers and skeletal skinned fibers, respectively. Therefore, the question raised above certainly deserves further investigation.

It is well known that the contractile system of cardiac muscle is more sensitive to Sr^{2+} than skeletal muscle so far as the contraction of fiber models and superprecipitation are concerned (Ebashi et al. 1968; Kitazawa 1976). However, such a marked difference could not be seen in the case of actin-activated ATPase and Sr^{2+} binding to troponin between these two muscles (Fig. 5), as pointed out by Berson (1974). Such a discrepancy between the behaviors of the in vitro system and the fiber model, as seen with the Sr^{2+}-induced contraction, is also observed in the cases of the effects of pH and Mg^{2+} on contractility, as listed in Table 4.

Although the qualities of the items examined are very much different from one another, the results show similar tendencies. In all cases the fiber models are most sensitive to the change in the milieu. One might argue that the fiber models, more native than the extracted protein system, should contain some factor(s) which would make them sensitive to the environment, but this can be refuted by the fact that the results with superprecipitation are more similar to the results with fiber models than to those derived from ATPase determinations. It is worthy of note that the ATPase activity appears to reflect the mode of Ca^{2+} (or Sr^{2+}) binding to troponin.

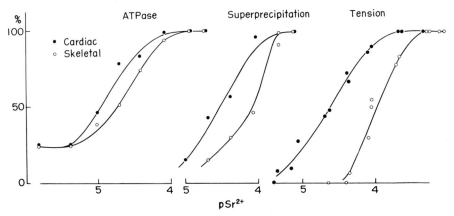

Fig. 5. Sr^{2+} sensitivity of cardiac and skeletal myosin B and glycerinated fibers. Relative rates of ATPase activity and superprecipitation of myosin B and relative tension of glycerinated fibers were plotted against pSr. Notice that the difference in Sr^{2+} sensitivities between cardiac and skeletal systems is largest in tension and smallest in ATPase activity

Table 4. Various factors influencing Ca^{2+} sensitivities of actomyosin ATPase, superprecipitation and contraction and Ca binding to troponin

	Mg^{2+} (10 mM/1 mM)[a]	pH (6.8/7.4)[b]	Sr^{2+} (skeletal/cardiac)[c]
Skinned fibers	—	—	4.9[d]
Glycerinated fibers	~4	4.3	4.5
Superprecipitation of myosin B	4.0	4.0	2.9
ATPase of myosin B	3.9	2.3	1.7
Binding to troponin	2.5	2.6	1.0

[a] The ratio of the Ca^{2+} concentration to give half maximum activity at 10 mM Mg^{2+} to those at 1 mM Mg^{2+}; if the Mg^{2+} concentration used is not 1 mM or 10 mM, correction was made the assumption that Mg^{2+} would be simply competitive
[b] The ratio of that at pH 6.8 to that at 7.4
[c] The ratio of Sr^{2+} concentration to give half maximum activity of fast skeletal myosin B to that of cardiac myosin B
[d] The value of cardiac skinned fibers was compared with that of fast skeletal glycerinated fibers

Since superprecipitation and ATPase activity are dealing with the same material, the fact that the results with superprecipitation are more akin to those with the fiber models implies that superprecipitation represents more crucial aspects of contraction than does the ATPase. In other words, the ATPase activity includes a fraction which cannot be directly converted to contraction. At present we cannot identify the processes which distinguish between the actin-activated ATPase and superprecipitation.

The uncoupling may provide a partial explanation for Solandt's effect (1936), i.e., a small increase in K^+ in the medium of living muscle fibers gives rise to an enormous increase in oxygen consumption without contraction, undoubtedly due to the increase in ATP breakdown. Depolarization due to K^+ increase may elevate Ca^{2+} concentration so as to stimulate the actomyosin ATPase but not the contractility.

Summarizing, a part of actin-activated myosin ATPase under Ca^{2+} control cannot be coupled with superprecipitation and contraction. This uncoupling, modified by various conditions, indicates the presence of controllable step(s) in converting the energy of ATP into contraction.

D. Concluding Remarks

The regulatory role of Ca^{2+} in intracellular phenomena was first established in muscle contraction (cf. Ebashi and Endo 1968), but Ca^{2+} is no longer confined to muscle contraction and is widely recognized as the most fundamental regulatory factor in various kinds of cellular functions.

Ca^{2+} has unique characteristics compared with cyclic nucleotides: (1) Ca^{2+} is an indispensable consituent of the surrounding medium of cells, modulating the activity of excitability of the cell membrane (in some kinds of muscle it acts as the charge carrier itself of the action current; consequently, Ca^{2+} outside the cell can directly affect intracellular processes), (2) no chemical change is required for the function of Ca^{2+} and, therefore, the ion can join it in the very fast biological reactions; (3) Ca^{2+} cannot be chemically converted to inactive form; accordingly, it is usually segregated from its targets by membraneous systems, e.g., the plasma membrane or sarcoplasmic reticulum. Therefore, the rate-limiting step of Ca^{2+} action is usually the change in the membrane permeability; (4) when life was first created, Ca^{2+} was undoubtedly present in the outer medium, viz., the sea. Since even primitive organisms somehow utilize Ca^{2+} for their cellular movement, Ca^{2+} must have started its regulatory role in a very early stage of evolution.

It is certain that the number of Ca^{2+}-dependent processes will increase with the advance of research. The mechanism how Ca^{2+} regulates the biological processes may not be simple but diverse and complicated, as revealed by the studies on muscle contraction. Modulator protein (calmodulin), perhaps the ancestral protein of most Ca^{2+}-binding proteins, is often considered to be common to most Ca^{2+}-dependent reactions, but as exemplified by the presence of specialized Ca^{2+}-binding proteins in muscle, there must be some other Ca^{2+}-dependent processes independent of modulator protein.

We believe that the mode of action of Ca^{2+} in muscle contraction will give useful suggestions to those who also look for Ca^{2+}-dependent biological processes in other fields.

Acknowledgments. We heartily congratulate Prof. Fritz Lipmann on his birthday. Without the days in his laboratory in 1959, I (S.E.) could not of think of a career as a scientist.

The work presented in this paper is supported in part by grants from the Muscular Dystrophy Association; the Ministry of Education, Science and Culture, Japan; the Ministry of Health and Welfare, Japan; and the Iatrochemical Foundation.

References

Aksoy MO, Williams D, Sharkey EM, Hartshorne DJ (1976) Relationship between Ca^{2+} sensitivity and phosphorylation of gizzard actomyosin. Biochem Biophys Res Commun 69: 35–41

Amphlett GW, Vanaman TC, Perry SV (1976) Effect of the troponin C-like protein from bovine brain (Brain Modulator Protein) on the Mg^{2+}-stimulated ATPase of skeletal muscle actomyosin. FEBS Lett 72: 163–168

Berson G (1974) Ca^{2+}, Sr^{2+} and Ba^{2+} sensitivity of tropomyosin-troponin complex from cardiac and fast skeletal muscles. In: Drabikowski W, Strzelecka-Golaszewska H, Carafoli E (eds) Calcium binding proteins, pp 197–201. PWN-Polish Scientific Publishers, Warszawa

Bremel RD, Sobieszek A, Small JV (1977) Regulation of actin-myosin interaction in vertebrate smooth muscle. In: Stephens NL (ed) Biochemistry of smooth muscle, pp 533–549. Univ. Park Press, Baltimore

Casteels R, Godfraind T, Rüegg JD (1977) Excitation contraction coupling in smooth muscle. Elsevier/North-Holland, Amsterdam

Chacko S, Conti MA, Adelstein RS (1977) Effect of phosphorylation of smooth muscle myosin on actin activation and Ca^{2+} regulation. Proc Natl Acad Sci USA 74: 129–133

Cheung WY (1970) Cyclic 3', 5'-nucleotide phosphodiesterase: Demonstration of an activator. Biochem Biophys Res Commun 38: 533–538

Ebashi S (1960) Calcium binding and relaxation in the actomyosin system. J Biochem 48: 150–151

Ebashi S (1961) Calcium binding activity of vesicular relaxing factor. J Biochem 50: 236–244

Ebashi S (1963) Third component participating in the superprecipitation of 'Natural actomyosin'. Nature 22: 1010–1012

Ebashi S (1979) Ca^{2+} ion and muscle contraction. In: Stoclet JC (ed) Advances in pharmacology and therapeutics, vol III, pp 81–98. Pergamon Press, Oxford

Ebashi S (1980) Regulation of muscle contraction. Proc R Soc London Ser B 207: 259–286

Ebashi S, Endo M (1968) Calcium ion and muscle contraction. Prog Biophys Mol Biol 18: 123–183

Ebashi S, Lipmann F (1962) Adenosine triphosphate-linked concentration of calcium ions in a particulate fraction of rabbit muscle. J Cell Biol 14: 389–400

Ebashi S, Kodama A, Ebashi F (1968) Troponin, I. Preparation and physiological function. J Biochem 64: 465–477

Ebashi S, Endo M, Ohtsuki I (1969) Control of muscle contraction. Q Rev Biophys 2: 351–384

Ebashi S, Nonomura Y, Toyo-oka T, Katayama E (1976) Regulation of muscle contraction by the calcium-troponin-tropomyosin system. In: Duncan CJ (ed) Calcium in biological systems; Symp. of the Soc. for Exp. Biol, vol XXX, pp 349–360. Cambridge Univ. Press, London

Ebashi S, Mikawa T, Hirata M, Toyo-oka T, Nonomura Y (1977) Regulatory proteins in smooth muscle. In: Casteels R, Godfraind T, Rüegg JD (eds) Excitation contraction coupling in smooth muscle, pp 325–334. Elsevier/North-Holland, Amsterdam

Ebashi S, Nonomura Y, Mikawa T, Hirata M, Saida K (1979) Regulatory mechanisms of muscle contraction. In: Hatano T, Sato H, Ishikawa H (eds) Cell motility, molecule and organization, pp 225–237. Univ. of Tokyo Press, Tokyo

Ikebe M, Aiba T, Onishi H, Watanabe S (1978) Calcium sensitivity of contractile protein from chicken gizzard muscle. J Biochem 83: 1643–1656

Kakiuchi S, Yamazaki T, Nakajima H (1970) Properties of a heat stable phosphodiesterase activating factor isolated from brain extract. Proc Japan Acad 46: 587–592

Kendrick-Jones J, Lehman W, Szent-Györgyi AG (1970) Regulation in molluscan muscles. J Mol Biol 54: 313–326

Kielley WW, Meyerhof O (1948) Studies on adenosinetriphosphatase of muscle. II. A new magnesium activated adenosinetriphosphatase. J Biol Chem 176: 591–601

Kitazawa T (1976) Physiological significance of Ca uptake by mitochondria in the heart in comparison with that by cardiac sarcoplasmic reticulum. J Biochem 80: 1129–1147

Kohama K (1979) Divalent cation binding properties of slow skeletal muscle troponin in comparison with those of cardiac and fast skeletal muscle troponins. J Biochem 86: 811–820

Kohama K (1980) Role of the high affinity Ca binding site of troponin. J Biochem 88: 591–599

Kumagai H, Ebashi S, Takeda F (1955) Essential relaxing factor in muscle other than myokinase and creatine phosphokinase. Nature 176: 166–168

Marsh BB (1951) A factor modifying muscle fibre syneresis. Nature 167: 1065–1066

Mikawa T (1979) 'Freezing' of the calcium-regulated structures of gizzard thin filaments by glutaraldehyde. J Biochem 85: 879–811

Mikawa T, Toyo-oka T, Nonomura Y, Ebashi S (1977) Essential factor of gizzard 'Troponin' fraction. J Biochem 81: 273–275

Mikawa T, Nonomura Y, Hirata M, Ebashi S, Kakiuchi S (1978) Involvement of an acidic protein in the regulation of smooth muscle contraction by the leiotonin-tropomyosin system. J Biochem 84, 1633–1636

Potter JD, Gergely J (1975) The calcium and magnesium binding sites on troponin and their role in the regulation myofibrillar adenosine triphosphatase. J Biol Chem 250: 4628–4633

Ronald ML, Umazume Y, Kushmeric MJ (1976) Ca^{2+} dependence of tension and ADP production in segments of chemically skinned muscle fibers. Biochim Biophys Acta 430: 352–365

Sherry JMF, Gorecka A, Aksoy MO, Dabrowska R, Hartshorne DJ (1978) Roles of calcium and phosphorylation in the regulation of the activity of gizzard myosin. Biochemistry 17: 4411–4418

Small JV, Sobieszeck A (1977) Ca-regulation of mammalian smooth muscle actomyosin *via* a kinase-phosphatase-dependent phosphorylation and dephosphorylation of the 20,000-Mr light chain of myosin. Eur J Biochem 76: 521–530

Solandt DY (1936) The effect of potassium on the excitability and resting metabolism of frog's muscle. J Physiol 86: 162–179

Solaro RJ, Weise RM, Shiner JS, Briggs NL (1974) Calcium requirements for cardiac myofibrillar activation. Circ Res 34: 525–530

Weber A (1966) Energized calcium transport and relaxing factors. In: Sanadi DR (ed) Current topics in bioenergetics, pp 203–254. Academic Press, New York London

7. Why is Phosphate so Useful?

M.S.BRETSCHER

In getting life going, Nature had to solve a lot of problems. The central one – how a piece of nucleic acid could assist the production of a polypeptide which itself would favour the multiplication of that particular piece of nucleic acid in some way – is still as elusive as ever. Once that had been achieved, everything else could (at least, in principle) be left to Natural Selection.

But at a less sophisticated level there are lots of other problems, and here I should like to think aloud about some of these from a chemist's point of view. Everyone knows that, if you want to make ethyl acetate in the lab, you mix ethanol and acetic anhydride. But if a cell wished to make ethyl acetate, it would not do it that way. The principal reason for this is that acetic anyhydride is far too reactive – the activation energy for acetylating any amino groups, hydroxyl groups, or other nucleophiles existing in the cell is too low. Any pool of acetic anhydride in the cell would acetylate everything. This means that Nature is restricted to using reagents whose spontaneous reaction with other things inside the cell is minimal and which do not spontaneously react with water (which would be wasteful). Nature's reagents must, therefore, be designed so that any reaction in which they participate has a high activation energy; in that way no reaction can occur unless catalyzed by an enzyme (which simply lowers the activation energy and hence increases the rate of the desired reaction). The point here is, then, that Nature must operate within a chemistry of high activation energies but also, if a reaction is to be driven far, of a large loss of free energy. These two requirements – a high free energy of reaction yet low spontaneous (uncatalyzed) rate – usually work against one another. An interesting case exists in the activation of amino acids during protein biosynthesis. The key intermediates – aminoacyl adenylates – are highly reactive compounds and wonderful aminoacylating agents. Nature could not afford to have these intermediates at any appreciable concentration inside the cell. It has overcome this dilemma by designing the activating enzymes so that not only do they make the aminoacyl adenylates, but they also hold on to them and utilise them themselves by transferring the amino acid to tRNA. In other words, the free aminoacyl adenylate concentration inside the cell is kept very low indeed. Nevertheless, this reactive intermediate will sometimes escape from the activating enzyme and thereby cause damage elsewhere. Nature can not be perfect but, to aim at this, it is likely that she has evolved a battery of enzymes to take care of the products arising from spontaneous chemical reactions. The true function of such enzymes would be hard to determine, but I have no doubt that they exist, as cellular house-cleaners.

MRC Laboratory of Molecular Biology, Hills Road, Cambridge, CB2 2QH, Great Britain

But to return to ethyl acetate. A cell would never wish to make such a molecule, nor could it use acetic anhydride to do so: both molecules would diffuse out of the cell through its plasma membrane in no time at all. Nature is then faced with yet another general problem. The small molecules upon which she operates must be designed so that they do not diffuse out of the cell (unless they are waste products or perfumes), which generally means they must have strongly charged groups at pH 7 or several hydrophobic-excluding groups on them. It would be rather useless for a cell to make acetyl p-nitro-phenyl ester as a general intermediate for acetylation reactions – rather, some charge on the molecule, such as acetyl-phosphate or acetly-coA have, would serve to keep the reagent inside the cell.

And this is why phosphates are so terribly useful. Phosphoric acid has three hydroxyl groups with pKs around 2, 7 and 13. The linkage in methyl-phosphate (or AMP) has the character of an ether linkage, leaving two hydroxyl groups with pKs around 2 and 7. In dimethyl phosphate only one hydroxyl group is left (with a pK of around 2) and the linkages are both very stable (as in DNA). Acetyl methyl phosphate behaves as a good acetylating agent (rather like acetyl-N-hydroxy succinimide) and also retains its hydroxyl group with a pK of about 2. Thus, phosphoric acid's two hydroxyl groups at pH 6 can be used to produce compounds of widely varying reactivity and stability, and also to crosslink molecules (as in nucleic acids, lipids, and some coenzymes). But most important is the fact that this still leaves one strong negative charge on the phosphate – and that keeps the molecule soluble and inside the cell. The operation of this principle is nicely shown by the way glycerol is utilized. Glycerol gets across a lipid bilayer quite easily by passive diffusion – so as soon as it is inside the cell it is phosphorylated to glycerol phosphate, and this anchor seems to keep it, and its derivatives (dihydroxy acetone phosphate, glyceraldehyde phosphate, etc.), inside the cell.

Finally, the ability of phosphates to link with each other to form pyrophosphates (like ATP), which have relatively high free energies of hydrolysis but are quite stable at neutral pH, makes them useful as a standard currency of energy – the famous squiggle-pee.

We see, therefore, that Nature has exploited phosphates in a variety of contexts. This, I believe, has had a special effect on how other cell components have been fashioned. Nucleic acids are polyanions. In order to prevent nonspecific adsorption of other molecules, especially proteins, to them, Nature has had to make these other molecules negatively charged. This is reflected by the enzymologist's wide use of DEAE at neutral pH – almost all proteins bind to it at low salt, whereas few (and these are often proteins which act on nucleic acids) bind to anionic matrices at neutral pH. In a similar way, membranes have to be negatively charged and hence we have such lipids as phosphatidyl-serine and phosphatidyl-glycerol (but no common cationic lipids), presenting the membrane's cytoplasmic face with a negatively charged surface. In a sense, once Nature had used phosphates to link nucleosides to form nucleic acids, there has been a natural se-lection for negatively charged molecules, leading to the further use of phosphate.

It is a great tribute to Fritz Lipmann that he was one of the first to recognize this versatility of P.

8. ppGpp, a Signal Molecule

J. SY

A. Introduction

Faced with a shortage of nutrient source, bacteria rapidly adjust their metabolism and adapt to a more frugal life style. Of the doubtless many overlapping and hierarchical controls required for such an adjustment, a simple mechanism mediated by the peculiar nucleotide guanosine 5'-diphosphate,3'-diphosphate (ppGpp) will be reviewed and its metabolism discussed.

ppGpp and pppGpp were discovered in *Escherichia coli* by Cashel and Gallant (1969) and have now been found in all bacterial species so far tested but are absent in eukaryotic organisms, with the possible exception of yeast and *Chlamydomonas reinhardtii* (Silverman and Atherly 1979). These nucleotides are rapidly accumulated during stringent response where, following starvation for a required amino acid, the bacteria restrict the transcription of rRNA and tRNA, as well as the synthesis of ribosomal proteins, phospholipids, lipids, carbohydrates, polyamines and nucleotides, the phosphorylation of glycolytic intermediates, and the transport of nucleobases and glycosides. The intracellular concentration of ppGpp in *E. coli* rises from basal μM level to a maximum of up to 4 mM, following 3–5 min of amino acid starvation. The level then gradually falls to a new basal concentration characteristic of the starving condition and the strain used (Cashel 1969).

There is much in vivo evidence to indicate that ppGpp may be the pleotropic effector that coordinately controls the diverse phenomena found during stringent response. (1) Mutants that fail to exhibit stringent response, the relaxed mutants, also fail to accumulate ppGpp. Second-site revertants of some relaxed mutants regain the ability to accumulate ppGpp (Fiil et al. 1977). (2) Accumulation of ppGpp occurs within seconds of amino acid removal and immediately before the slowdown of rRNA synthesis (Cashel 1969). (3) As noted above, ppGpp is accumulated at a rapid rate; it also has a high turnover rate (Cashel 1969). This property is in accord with a signal molecule whose concentration should swiftly respond to outside stimulus. (4) Starvation for a carbon source leads to accumulation of ppGpp and a reduction of rRNA synthesis (Lazzarini et al. 1971). (5) There is an inverse correlation between ppGpp concentration and rate of growth, except at extremely slow or fast rates (Nierlich 1978). In vitro experiments provide further support for the contention that ppGpp is a pleotropic effector. The in

The Rockefeller University, New York, New York 10021, USA

vitro effect can be divided into that which affects the transcriptional process and that which affects the enzymatic process.

B. Transcriptional Effects

ppGpp has a complex effect on in vitro transcription, inhibiting some genes while stimulating others (Table 1). Most of the inhibited genes code for parts of the protein synthesizing machinery. There is currently no consensus as to whether ppGpp directly or indirectly inhibits rRNA gene transcription. There are reports that with cloned rRNA genes and RNA polymerase a direct and specific inhibition of rRNA transcription can be shown by the addition of ppGpp (Gilbert et al. 1979). Other workers (Glaser and Cashel 1979) using similar systems have not found this effect. There is also a dispute with respect to in vitro tRNA transcription (Yang et al. 1974; Debenham and Travers 1977). Using a coupled transcription and translation system, there is general agreement on the inhibitory effect of ppGpp on the transcription of genes for ribosomal proteins, elongation factors, and the α-subunit of RNA polymerase (Lindahl et al. 1976). The in vivo synthesis of these proteins, as with the synthesis of rRNA and tRNA, has been found to be under stringent control.

Table 1. Effect of ppGpp on transcriptional processes

Gene or operon	Stringent response	In vitro ppGpp effect	Reference
rRNA	Inhibition	Inhibition (0.2 mM) No inhibition	Gilbert et al. (1979) Glaser and Gashel (1979)
tRNA	Inhibition	Inhibition (10 μM) No inhibition	Debenham and Travers (1977) Yang et al. (1974)
Ribosomal proteins	Inhibition	Inhibition (0.2–0.3 mM)	Lindahl et al. (1976)
EF-T	Inhibition	Inhibition (0.2–0.3 mM)	Lindahl et al. (1976)
EF-G	Inhibition	Inhibition (0.2–0.3 mM)	Lindahl et al. (1976)
α-subunit RNA pol.	?	Inhibition (0.2–0.3 mM)	Lindahl et al. (1976)
arg	?	Inhibition (0.1 mM)	Yang et al. (1974)
his	Stimulation	Stimulation (0.1–0.2 mM)	Stephens et al. (1975)
ilv	?	Stimulation (0.1–0.3 mM)	Smolin and Umbarger (1975)
trp	?	Stimulation (0.1 mM)	Yang et al. (1974)
CM-acetyl-transferase	?	Stimulation ().1 mM)	De Crombrugghe et al. (1973)
lac	Stimulation	Stimulation (0.1 mM)	Yang et al. (1974)
ara	Stimulation	Stimulation (0.1 mM)	Yang et al. (1974)

A rather unexpected finding in the in vitro transcriptional studies was the stimulation by ppGpp of some gene transcriptions. This indicated that ppGpp is not only a negative effector but it can act as a positive one as well. In general, stimulation occurs only if the operon or gene is already de-repressed. For example, transcription of the *lac* operon, as measured by the synthesis of β-galactosidase, or the *ara* operon is stimulated by ppGpp only in the presence of both inducer and cAMP (Yang et al. 1974). ppGpp produces its maximal transcriptional effect on either inhibition or stimulation in the concentration range of 0.1–0.2 mM. These findings indicate that ppGpp as a signal of amino acid or energy source depletion will slow down the production of under-utilized macromolecules, the ribosome and its ancillary proteins, and at the same time prime the transcriptional system so that it will be highly efficient for the synthesis of needed molecules when additional signals such as cAMP or inducers are received.

C. Enzymatic Effects

A variety of enzymes are directly affected by ppGpp. Most of these are regulated enzymes in the various metabolic pathways. A partial list of these enzymes is shown in Table 2. ppGpp may act by competition with known GTP or GDP binding sites, such as ornithine decarboxylase, uracil phosphoribosyl transferase, and adenylosuccinate syn-

Table 2. Effect of ppGpp on enzymatic processes

System	Target enzyme	Stringent response	In vitro ppGpp effect	Reference
Fatty acid synthesis	Acetyl-CoA carboxylase	Inhibition	Inhibition (1–2 mM)	Polakis et al. (1973)
Polyamine synthesis	Ornithine decarboxylase	Inhibition	Inhibition (1 mM, compete with GTP)	Holtta et al. (1974)
Histidine synthesis	ATP-phosphoribosyl-transferase	?	Inhibition depends on *his*	Morton and Parsons (1977)
TCA cycle intermediate	PEP-carboxylase	?	Stimulate (3–4 mM)	Taguchi et al. (1977)
Phospholipid synthesis	sn-Glycerol-3-PO_4 acetyltransferase	Inhibition	Inhibition (2–4 mM)	Lueking and Goldfine (1975)
Uracil uptake	Uracil phosphoribosyl-transferase	Inhibition	Inhibition (Compete with GTP)	Fast and Skold (1977)
Glycogen synthesis	ADP-glucose synthetase	?	Inhibition (1–2 mM)	Dietzler and Leckie (1977)
AMP synthesis	Adenylosuccinate synthetase	?	Inhibition (compete with GTP)	Stayton and Fromm (1979)

thetase; or it can act as an allosteric effector. Acetyl-CoA carboxylase, the first enzyme in fatty acid synthesis, is specifically inhibited at the carboxyltransferase reaction but not the biotin-dependent reaction (Polakis et al. 1973). The first key enzyme in histidine biosynthesis, ATP-phosphoribosyl transferase, is inhibited by ppGpp only in the presence of histidine, which means that *his* biosynthesis will be suppressed when another amino acid is missing (Morton and Parsons 1977). The maximum effect on the enzyme reaction usually occurs at 1–4 mM ppGpp, while 0.1–0.4 mM is effective in influencing the transcriptional process. This difference in effective concentration indicates that the transcriptional processes are the primary targets of ppGpp control. The influence of ppGpp on enzymatic reactions will be more transient since, in vivo, ppGpp concentration drops after reaching a maximum of 1–4 mM.

Although the results accumulated to date point to a central role of ppGpp as a signal during nutritional stress, it is by no means certain that ppGpp is the direct pleotropic effector. There is no genetic evidence showing ppGpp to interact directly with the transcriptional or the enzymatic systems. For example, there is no known mutation in the rRNA promoter that will lead to insensitivity to ppGpp control.

D. Biosynthesis of pppGpp and ppGpp

The ribosome-mRNA translational complex serves as a sensory device for the detection of amino acid deficiency and the production of the signal molecule, ppGpp. Starvation for any one of the 20 amino acids will readily lead to the accumulation of specific uncharged tRNA species. These molecules can then bind to the amino acid site of the ribosome-mRNA complex when the proper codon is reached. The uncharged tRNA in the ribosomal complex immediately activates the stringent factor, the product of the *relA* gene (Table 3), and catalyzes the formation of pppGpp from GTP and ATP via a pyrophosphoryl transfer. pppGpp is assumed to be the primary product of the stringent factor reaction, since in vivo GTP is present in a 20-fold higher concentration than is GDP. The classical relaxed mutants (*relA⁻*) are those that have lost the ability to synthesize pppGpp by the ribosome-dependent mechanism.

The proper interaction of stringent factor with the ribosomal complex is lost in *relC* mutants, which contain stringent factor but cannot activate it during amino acid starvation (Friesen et al. 1974). Therefore, they have a relaxed phenotype. The *relC* gene is equivalent to the *rplK* gene, and codes for the large ribosomal subunit, protein L 11. The exact mechanism by which L 11 participates in the ribosome-dependent synthesis of pppGpp is currently unknown. A detailed review of *relA*-dependent synthesis of ppGpp has recently been made by Richter and Isono (1977).

The classical *relA⁻* mutants, although incapable of accumulating ppGpp during amino acid starvation, are able to do so at a normal rate when starved for a carbon source. Recently, nonsense and deletion mutants at the *relA* gene locus have been isolated (Atherly 1979) which, having no functional stringent factor, still accumulate ppGpp during carbon source shift. These results indicate that *E. coli* contain another ppGpp-synthesizing pathway. However, since the discovery of the *relA*, ribosome-dependent ppGpp synthesis system, and despite the efforts of various laboratories, no demonstrable

Table 3. Genes involved in ppGpp metabolism

Gene	Location (min)	Gene product	Function
relA	59	ATP:GTP(GDP) pyrophosphotransferase, stringent factor	Ribosome-dependent synthesis of (p)ppGpp
relB	34.5	?	Delayed relaxed
relC	88	Ribosomal protein L 11	Ribosome-stringent factor interaction
relS	?	? (ppGpp synthetase)?	? (Ribosome-independent (p)ppGpp synthesis)?
relX	59.4	?	Control basal synthesis of ppGpp
gpp	83	pppGpp 5'-phosphohydrolase	pppGpp $\rightarrow$ ppGpp + P_i
spoT	81	ppGpp 3'-pyrophosphohydrolase	ppGpp $\rightarrow$ ppG + P_i
shf	2	?	Control ppGpp synthesis during carbon source shift

ppGpp synthesis has been shown in cell-free extracts of *relA*⁻ cells. More recently, a mutant was found called *relS* which, in the presence of *relA*⁻, contained a nondetectable ppGpp basal level (Engel et al. 1979). This double mutant also failed to accumulate ppGpp when subjected to carbon starvation. It is presumed, therefore, that *relS* codes for the second ppGpp-synthetase. The absence of ppGpp in *relA*⁻, *relS*⁻ cells indicates that ppGpp is not essential for the viability of the organism.

The failure to find an in vitro ppGpp-synthesizing system in *E. coli* that is *relA*-independent has led us to search for this biosynthetic mechanism in other organisms. We have now demonstrated that *Bacillus brevis,* an organism extensively used in Dr. Lipmann's laboratory, contains two distinct ppGpp synthetic pathways: a ribosome-dependent, *relA* type, and a ribosome-independent, presumably *relS* system (Sy 1979). Both enzymes catalyze the transfer of the β,γ-pyrophosphate group of ATP to the 3'-OH of GTP and GDP. The *relA* type enzyme is bound to the ribosome, has a molecular weight of 76,000, and requires the ribosome-mRNA-uncharged tRNA complex for activity but, in contrast to the *E. coli* stringent factor, it is not activated by methanol. The ribosome-independent enzyme is a supernatant enzyme, has a molecular weight of 55,000, and is inhibited by methanol. The *relA* type enzyme activity is greatly reduced when sporulating cells are assayed, whereas the ribosome-independent enzyme activity remains relatively constant throughout the growth phase. It is currently unknown whether the ribosome-independent enzyme is under the control of any metabolic process.

E. The Conversion of pppGpp to ppGpp

The elongation factors T and G (EF-T and EF-G) can convert pppGpp to ppGpp very rapidly in the presence of the ribosome. However, it is not known whether they do so in vivo. Recently Somerville and Ahmed (1979) have reported a new class of mutants (*gpp*) that have a much higher pppGpp to ppGpp ratio than normal and, at the same time, exhibit a reduced rate of (p)ppGpp turnover. They have found that this mutant is deficient in a pppGpp 5'-phosphohydrolase which hydrolyzes pppGpp to ppGpp. The enzyme is specific for pppGpp since it does not hydrolyze GTP, ATP, or UTP. It is a supernatant enzyme and does not require the ribosome for activity.

F. The Degradation of ppGpp

The degradation of ppGpp occurs at a high rate, with a half-life of approximately 20 s following readdition of the missing amino acid. Laffler and Gallant (1974) have identified a mutant, called *spoT⁻*, which has a 20-fold reduction in the rate of ppGpp degradation. The cellular mechanism of ppGpp degradation has been a puzzling and controversial subject for several years due, primarily, to the absence of a cell-free system that can degrade ppGpp. Recently, our laboratory (Sy 1977) as well as Richter's (Heinemeyer and Richter 1977) have developed an in vitro system that has the ability specifically to hydrolyze ppGpp. The hydrolytic enzyme is associated with the ribosomal fraction but does not need the ribosome for activity. Degradation proceeds by cleavage of the 3'-pyrophosphate group as a unit, yielding inorganic pyrophosphate and 5'-GDP. The reaction specifically requires Mn^{2+}; neither Mg^{2+}, Zn^{2+}, Co^{2+}, Fe^{2+}, nor Ca^{2+} can substitute. Possible the Mn^{2+}. ppGpp complex is the true substrate for this enzyme. The enzyme will also hydrolyze both pppGpp and pGpp to GTP and GMP, respectively, and it has a slower activity toward ppGp, the recently described MS-3 nucleotide (Pao and Gallant 1979). That ppGpp 3'-pyrophosphohydrolase is the product of the *spoT* gene is demonstrated by the absence of this enzyme in *spoT⁻* ribosomal fractions. Recently, the *spoT* gene has been cloned and expressed in minicells. The gene product has a molecular weight of 80,000 (An et al. 1979).

Carbon source starvation results in a great reduction in the in vivo rate of ppGpp degradation. At present, there is no known mechanism for coupling carbon starvation with ppGpp 3'-pyrophosphohydrolase activity. The control mechanism is probably highly complex since it involves at least two other genes: *relX*, which governs ppGpp basal level (Pao and Gallant 1978), and *shf*, which affects the ability of *E. coli* to accumulate both cAMP and ppGpp during carbon source shift (Gallant 1976).

G. The ppGpp Cycle

The biosynthetic and degradative pathways of ppGpp may be summarized:

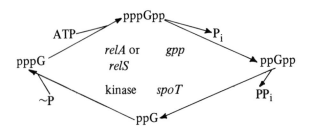

There may be a minor degradative pathway by which P_i is sequentially removed from the 3'-end of ppGpp to form first ppGp (Pao and Gallant 1979) and then ppG. However, the *gpp* and *spoT* gene products should dominate the ppGpp metabolic cycle, since the double mutant *gpp⁻, spoT⁻* is nonviable (Somerville and Ahmed 1979) presumably because of excessively high pppGpp concentration.

Bacteria use two separate mechanisms, depending on the kind of nutritional deficiency, to achieve rapid accumulation of the signal molecule, ppGpp. During amino acid deprivation, the biosynthetic rate is greatly increased by stimulating ribosome-dependent ppGpp synthetase. During carbon source deprivation, the degradation rate is greatly reduced by inhibiting ppGpp 3'-pyrophosphohydrolase, while synthesis occurs via a ribosome-independent pathway. The increased concentration of ppGpp, signaling metabolic hard times, slows down production of underutilized macromolecules as well as the activity of some metabolic pathways. By readjusting their cellular metabolism, bacteria insure that they can survive until there is a better nutritional environment.

Acknowledgments. These studies were supported by a grant from the National Science Foundation (PCM77-17683) to the author, and from the National Institutes of Health (GM-13972) to Fritz Lipmann.

References

An G, Justesen J, Watson RJ, Friesen JD (1979) Cloning the *spoT* gene of *E. coli.* Identification of the *spoT* gene product. J Bacteriol 137: 1100–1110

Atherly AG (1979) *E. coli* mutant containing a large deltion from *relA* to *argA.* J Bacteriol 138: 530–534

Cashel M (1969) The control of RNA synthesis in *E. coli.* J Biol Chem 244: 3133–3141

Cashel M, Gallant J (1969) Two compounds implicated in the function of the RC gene of *E. coli.* Nature (London) 221: 838–841

Debenham P, Travers A (1977) Selective inhibition of tRNA^{Tyr} transcription by ppGpp. Eur J Biochem 72: 515–523

De Crombrugghe B, Pastan I, Shaw WV, Rosner JL (1973) Stimulation by cAMP and ppGpp of chloramphenicol acetyl transferase synthesis. Nature New Biol (London) 241: 237–239

Dietzler DN, Leckie MP (1977) Regulation of ADP-glucose synthetase, the rate limiting enzyme of bacterial glycogen synthesis, by the pleotropic nucleotides ppGpp and pppGpp. Biochem Biophys Res Commun 77: 1459–1467

Engel JA, Sylvester J, Cashel M (1979) ppGgg is a dispensible metabolite. In: Koch G, Richter D (eds) Regulation of macromolecular synthesis by low molecular weight mediators, p 25. Academic Press, New York

Fast R, Skold D (1977) Biochemical mechanism of uracil uptake regulation in *E. coli* B. J Biol Chem 252: 7620–7624

Fiil NP, Willumsen BM, Friesen JD, von Meyenburg K (1977) Interaction of alleles of the *relA, relC* and *spoT* genes in *E. coli:* Analysis of the interconversion of GTP, ppGpp and pppGpp. Mol Gen Genet 150: 87–101

Friesen JD, Fiil NP, Parker JM, Haseltine WA (1974) A new relaxed mutant of *E. coli* with an altered 50S ribosomal subunit. Proc Natl Acad Sci USA 71: 3465–3469

Gallant J (1976) Elements of the downshift servomechanism. In: Control of ribosome synthesis (Alfred Benzon Synposium IX), pp 385–396. Copenhagen: Munksgaard 1976

Gilbert SF, de Boer HA, Nomura M (1979) Identification of initiation sites for the in vitro transcription of rRNA operons *rrnE* and *rrnA* in *E. coli.* Cell 17: 211–224

Glaser G, Cashel M (1979) In vitro transcripts from the *rrnB* ribosomal RNA cistron originate from two tandem promoters. Cell 16: 111–121

Heinemeyer E, Richter D (1977) In vitro degradation of ppGpp by an enzyme associated with the ribosomal fraction from *E. coli.* FEBS Lett 84: 357–361

Holtta E, Janne J, Pispa J (1974) The regulation of polyamine synthesis during the stringent control in *E. coli.* Biochem Biophys Res Commun 59: 1104–1111

Laffler T, Gallant J (1974) *spoT,* a new genetic locus involved in the stringent response in *E. coli.* Cell 1: 27–30

Lazzarini RA, Cashel M, Gallant J (1971) On the regulation of ppGpp pools in stringent and relaxed strains of *E. coli.* J Biol Chem 246: 4381–4385

Lindahl L, Post L, Nomura M (1976) DNA-dependent in vitro synthesis of proteins, protein elongation factors, and RNA polymerase subunit α: Inhibition by ppGpp. Cell 9: 439–448

Lueking DR, Goldfine H (1975) The involvement of ppGpp in the regulation of phospholipid biosynthesis in *E. coli.* J Biol Chem 250: 4911–4917

Morton DP, Parsons SM (1977) Synergistic inhibition of ATP phosphoribosyltransferase by ppGpp and histidine. Biochem Biophys Res Commun 74: 172–177

Nierlich DP (1978) Regulation of bacterial growth, RNA and protein synthesis. Annu Rev Microbiol 32: 393–432

Pao CC, Gallant J (1978) A gene involved in the metabolic control of ppGpp synthesis. Mol Gen Genet 158: 271–277

Pao CC, Gallant J (1979) A new nucleotide involved in the stringent response in *E. coli:* guanosine 5 5'-diphosphate 3'-monophosphate. J Biol Chem 254: 688–692

Polakis SF, Guchhait RB, Lane MD (1973) Stringent control of fatty acid synthesis in *E. coli.* J Biol Chem 248: 7957–7966

Richter D, Isono K (1977) The mechanism of protein synthesis-initiation, elongation and termination in translation of genetic messages. In: Current topics in microbiology and immunology, vol 76, pp 83–125. Springer, Berlin Heidelberg New York

Silverman RH, Atherly AG (1979) The search for ppGpp and other unusual nucleotides in eucaryotes. Microbiol Rev 43: 27–41

Smolin DE, Umbarger HE (1975) Specificity of the stimulation of in vitro RNA synthesis by ppGpp. Mol Gen Genet 141: 277–284

Somerville CR, Ahmed A (1979) Mutants of *E. coli* defective in the degradation of pppGpp. Mol Gen Genet 169: 315–323

Stayton MM, Fromm HJ (1979) ppGpp inhibition of adenylosuccinate synthetase. J Biol Chem 254: 2579–2581

Stephens JC, Artz SW, Ames BN (1975) ppGpp: positive effector for histidine operon transcription and general signal for amino acid deficiency. Proc Natl Acad Sci USA 72: 4389–4393

Sy J (1977) In vitro degradation of ppGpp. Proc Natl Acad Sci USA 74: 5529–5533

Sy J (1979) Biosynthesis of guanosine tetraphosphate in *Bacillus brevis.* In: Koch G, Richter D (eds) Regulation of macromolecular synthesis by low molecular weight mediators, p 95. Academic Press, New York

Taguchi M, Izui K, Katsuki H (1979) Activation of *E. coli* phosphoenolpyruvate carboxylase by ppGpp. FEBS Lett 77: 270–272

Yang HL, Zubay G, Urm E, Reiness G, Cashel M (1974) Effects of ppGpp, pppGpp, and β-γ-methylenyl-guanosine pentaphosphate on gene expression of *E. coli* in vitro. Proc Natl Acad Sci USA 71: 63–67

9. Gramicidin S-Synthetase: On the Structure of a Polyenzyme Template in Polypeptide Synthesis

H. KLEINKAUF and H. KOISCHWITZ

A. The "Lipmann Model" of Sequential Polymerization of Amino Acids

In the pre-RNA history of protein biosynthesis two models of protein formation were proposed (Campbell and Work 1953): Enzymic transpeptidation and the "template theory".

The stepwise addition or transpeptidation catalyzed by specific enzymes had been termed "polyenzyme model" by Spiegelman (Spiegelman 1956). It had been speculated that protein synthesis occurred by stepwise elongation of polypeptides mediated by a series of specific enzymes. This model suffered from the dilemma of proteins catalyzing the synthesis of proteins. It was finally rejected by Crick in his statement of the "Central Dogma" (Crick 1958) that "once 'information' has passed into protein *it cannot get out again.*"

The Lipmann model (Lipmann 1954, Fig. 1a) somehow was a link connecting a linear template (nucleic acid) and enzymic carboxyl group activation as a prerequisite of polymerization. Lipmann himself considered it to be a template model (Lipmann 1956), although it did not relate the amino acid sequence to DNA, as had been done by Dounce (Dounce 1952) in his activated nucleic acid template (Fig. 1b). He supposed that a nucleic acid template is first phosphorylated to react with either amino acid or nucleoside. Linear polymerization then gives a peptide chain or a replica of the template. While the Dounce model of protein formation has been forgotten, the Lipmann model has been revived in the late sixties in the nucleic acid-free biosynthesis of peptides (Lipmann 1971).

The model does not yet show a direction of polymerization. Thus, initiation may occur at the carboxyl end of the peptide, as indicated by the numbering in Fig. 1a, leaving a free or blocked carboxyl throughout the elongation process. The second possibility of a free or blocked amino terminal was proposed by Lipmann in 1956 (Lipmann 1956) on the basis of fatty acid chain elongation. This model of polymerization has been termed headward chain growth (Lipmann 1968) and has indeed been found in protein biosynthesis and gramicidin S formation. The control of sequence of reactions, however, has not yet been solved. If we do have activated amino acids lined up on a template, is there a transport system like tRNA, and a single reaction center like peptidyltransferase, or do we have multiple reaction centers with a peptidyl transport system?

Max-Volmer-Institut für Physikalische Chemie und Molekularbiologie, Technische Universität Berlin, 1000 Berlin 10, FRG

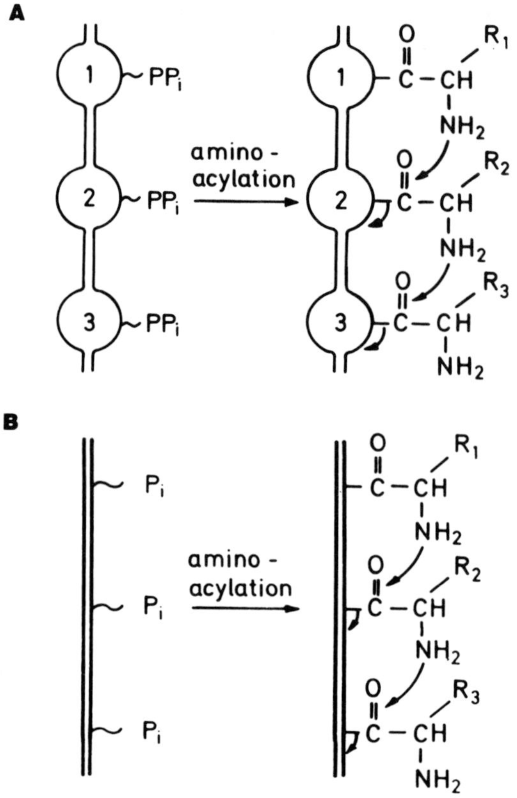

A

amino-
acylation

B

amino-
acylation

Fig. 1a, b. Early template models of protein biosynthesis: The activated polyenzyme (**a**), and the activated nucleic acid (**b**). The polyenzyme model (Lipmann 1954), however, has not been related to informational nucleic acid structure. More recently, polyenzymes have been found in nucleic acid-free peptide biosynthesis

The second possibility has been proposed for gramicidin S formation (Gilhuus-Moe et al. 1970; Kleinkauf et al. 1971), while more recently a peptidyltransferase scheme has been discussed in tyrocidine biosynthesis (Lee and Lipmann 1977). In both models peptidyl transfer is mediated by the covalently attached cofactor 4'-phosphopantetheine by successive transthiolation and transpeptidation reactions (Fig. 2).

Thus the movement of a transport function along a template may select the sequence of reactions. A further extension of the protein template assumes a random movement of pantetheine. Control of sequence of reactions is thought to be directed by specific intermediate binding; each "activation spot" now resembles a specific peptide synthetase (Koischwitz 1975; Kleinkauf 1979). This system very much resembles the old concept of "enzymic transpeptidation".

B. Gramicidin S-Synthetase

The two enzymes catalyzing formation of gramicidin S (Fig. 3) have been isolated from late log phase cells of *Bacillus brevis* ATCC 9999. Their properties have been reviewed recently (Kleinkauf and Koischwitz 1978, and references therein; Kleinkauf and Koischwitz 1979a). In this section we give a short summary including some unpublished results.

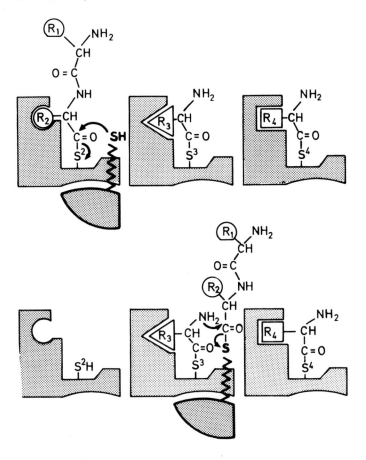

Fig. 2. Extension of the polyenzyme model by introduction of a transport function, the coenzyme A-fragment 4'-phosphopantetheine. *Top,* Transthiolation reaction; *Bottom,* Transpeptidation = chain elongation

2 L Phe		D - Phe ← Pro ← Val ← Orn ← Leu
+ 2 L - Pro		↓ ↑
+ 2 L - Val	Gramicidin S -	Leu → Orn → Val → Pro → D - Phe
+ 2 L - Leu	Synthetase	
+ 2 L - Orn		+ 10 AMP
+ 10 M^{2+} ATP		+ 10 M^{2+} PP$_i$

Fig. 3. Reactions catalyzed by gramicidin S-Synthetase (M = Mg, Mn or Ca)

I. Catalytic Properties

1. Amino Acid Activation

The constituent amino acids are activated as aminoacyl adenylates according to the scheme in Fig. 4. These are then accepted by enzyme thiol groups to form active esters. Each active site, once thiolaminoacylated, may catalyze the formation of a second aminoacyl adenylate.

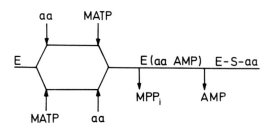

Fig. 4. Amino acid activation by gramicidin S-Synthetase enzymes. Binding of amino acid (*aa*) and ATP is random, and leads to the noncovalently bound aminoacyl adenylate (*aa AMP*) followed by the aminoacylation of an enzyme thiol (*E-S-aa*)

2. Amino Acid Polymerization

The reaction sequence of gramicidin S formation is shown in Fig. 5 and will be discussed in the text.

a) Initiation. The multienzyme GS 1 (E_1) racemizes phenylalanine at the thiolester step to give a 7:3 ratio of D-form to L-form. If the multienzyme GS 2 (E_2) is charged with proline, a loose initiation complex is formed, and D-Phe is transferred giving the enzyme-bound dipeptide D-Phe-Pro. This dipeptide may be lost by piperazinedione formation if the third amino acid valine is not present. This reaction can be used as an assay of the initiation reaction.

b) Elongation. Once a dipeptide is formed, it is thought to be transported by phosphopantetheine via transthiolation to the following elongation site (Fig. 2). This concept originates from the isolation of dipeptide pantetheine fragments (Kleinkauf et al. 1971). An "exclusion principle" has been postulated; i.e., only peptides but not amino acids may be accepted by the pantetheine thiol (Kleinkauf and Koischwitz 1978).

c) Termination. The "head-to tail" cyclization of two enzyme-bound nascent pentapeptides could proceed intermolecularly by complex formation of two GS 2 multienzymes, or intramolecularly if two pentapeptides are formed by a single enzyme. This question has not been settled (Kleinkauf 1979).

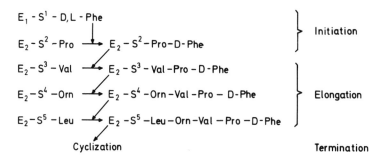

Fig. 5. Reaction sequence leading to gramicidin S. The two multienzymes comprising gramicidin S-synthetase are E_1 or GS 1 (100,000 daltons, also called light enzyme or phenylalanine racemase, E.C.5.1.11) and E_2 or GS 2 (280,000 daltons, also called heavy enzyme or fraction I)

II. Structure

Isolation and characterization of the two multienzymes have shown that they do not contain subunits demonstrable by protein dissociating agents (Kleinkauf and Koischwitz 1974, 1979b). These results suggest that we do have multifunctional polypeptide chains catalyzing four reactions (activation, aminoacylation, racemization, and transfer of Phe) in the case of GS 1, and at least 16 reactions (activation, aminoacylation, peptide transfer, elongation and termination) in the case of GS 2. Activation and aminoacylation are considered as separate reactions, since it has been possible selectively to damage the aminoacylation site of GS 1 by mutation (Kanda et al. 1978).

1. A Multifunctional Polypeptide Chain Without Free Subunits

Evidence for covalent linkage of catalytic subunits of GS 2 has been obtained by the observation of a single protein species under denaturing conditions (dodecylsulfate, urea, guanidine, reduction with dithiothreitol, modification with sulfite or citraconic acid anhydride) with an unchanged size. Since other types of protein linkage are not known to be present, the conclusion that there is a single polypeptide is most likely. Chemical evidence from structural studies of the protein is needed, however, to confirm this assumption. Analysis of the arrangement of subunits has been carried out by limited proteolysis, and by study of defective enzymes obtained from mutants.

2. Enzyme Fragment Analysis

Limited proteolysis of multifunctional chains may produce partially active enzyme fragments or "domains". A possible interdomain-directed cleavage would suggest that interdomain structures are highly susceptible to proteolysis. Studies of protease action on the multienzyme GS 2 have revealed, however, that cleavage quite often occurs within a "functional domain" (Table 1). Generally the activation of at least one of the four amino acids is lost rapidly (Fig. 6). Upon fragmentation, dissociation of fragments occurs, and fragments carrying partial activities can be isolated by conventional methods (Kleinkauf and Koischwitz 1979b). A preliminary characterization has been carried out by combining DEAE-cellulose chromatography and SDS-polyacrylamide gel electrophoresis, to correlate protein bands with amino acid activation reactions. The order of elution of fragments from DEAE-cellulose corresponds to the gramicidin S sequence. Proline activating units are eluted first, while leucine activating units are most tightly bound. Some of these results are shown in Fig. 7.

Another approach to the structure of multienzymes is based on structural alterations obtained by mutagenesis. So far two mutants producing GS 2-multienzymes of reduced size have been isolated (Kambe et al. 1974). One no longer activates leucine, the last amino acid, and has an apparent molecular mass of 2.6×10^5 daltons (Kleinkauf and Koischwitz 1979b), while the other activates only the second amino acid, proline, and catalyzes gramicidin S initiation. If these two polypeptides originate from early termination, then leucine activation should be located at the C-terminal part of the polypeptide,

Table 1. Limited proteolysis of the multienzyme GS 2: Effect on amino acid activation reactions

Proteinase	$T^{\circ}C^{b}$	Observed decrease of amino acid activation reaction[a] as measured by ATP-PPi-exchange dependent on:			
		Pro	Val	Orn	Leu
1 Chymotrypsin	20	+	++	0	+
2 Trypsin	20	++	+	++	+
3 Papain	30	+	+	++	+:
4 Crude pancreatic protease	27	(+)	+	++	+
5 Subtilisin Carlsberg	27	(+)	(+)	++	+
6 Protease from *Tritirachium album*	27	+	+	++	++
7 Bromelain	30	+	++	+	++
8 Protease from *Bacillus brevis*	4	++	+	0	+

[a] Explanation: Loss of activation reaction:
$$0 : <10\%, (+) : <20\%, + : <70\%, ++ : >70\%$$
[b] Conditions were 30 min incubation at indicated temperature with 1% proteinase, except for 8 where proteolysis was observed after 15 h at $4^{\circ}C$

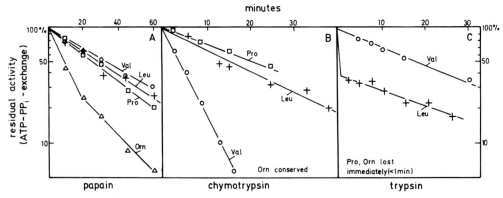

Fig. 6. Limited proteolysis of GS 2. Loss of catalysis of activation reactions as measured by amino acid-dependent ATP-PPi-exchange reaction. Under the reaction conditions employed, we observe, respectively, conservation of function (effect of chymotrypsin on Orn activation), slow loss rapid loss (effect of chymotrypsin on Val activation), a two-phase rapid loss (effect of papain on Orn activation), an initial very rapid, then slow loss (effect of trypsin on Leu activation), and very rapid inactivation (effect of trypsin on Pro and Orn activation). Conditions were: *A,* 200 μg GS 2, 2 μg papain, in 800 μl of 20 mM phosphate, pH 7.2, 30°C: the reaction was slowed down by addition of 2.5 mg/ml casein before the activation assay. *B,* 200 μg GS 2, 10 μg chymotrypsin in 1.2 ml of 50 mM TES-buffer, pH 7.0, at 20.5°C: reaction stopped with phenylmethyl sulfonylfluoride (PMSF). *C,* 500 μg GS 2, 8 μg trypsin in 1 ml 50 mM phosphate, pH 7.2, at 19°C: reaction stopped with PMSF

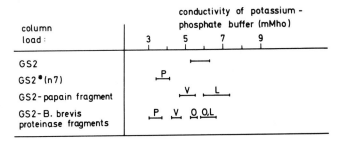

Fig. 7. Order of elution of fragments of the multienzyme GS 2 from DEAE-cellulose. Explanation: The elution range indicated corresponds to peak width at 50% of the activity maximum. Elutions have been performed with a linear potassium-phosphate gradient. Activation was estimated by amino acid-dependent ATP-PPi-exchange reaction (P = proline, V = valine, O = ornithine, L = leucine)

Table 2. Some properties of GS 2-multienzymes derived from mutants of *Bacillus brevis* (Kleinkauf and Koischwitz 1978, 1979b and references therein)

Mutant	Partial reactions					App. mol mass		
	Activation of				Initiation	Native		Denatured
	Pro	Val	Orn	Leu	reaction	S_0	Kd	Kd
Wild type	+	+	+	+	+	12.2[a,b]	280[c]	280[d]
BII-3	–	+	+	+	–	12.2[a]		
BI-3	+	–	+	+	+	12.2[a]		
BI-6	+	–	+	+	+			
BII-1	+	–	+	+	+			
E-4	+	–	+	+	+			
E-5	+	–	+	+	+			
BI-9	+	+	+	–	+	12.2[a]		
BI-4	+	+	+	–	–			
C-3	+	+	+	–	–			
E-2	+	+	+	–	–			
BI-2	+	+	+	+	+			
n7	+	–	–	–	++	6.1[a]		
hh	+	+	+	–	(+)	11.3[a]		260[d]

[a] sucrose gradient centrifugation; [b] analytical ultracentrifuge; [c] gel chromatography; [d] SDS-polyacrylamide gel electrophoresis

while proline activation should be the first catalytic function to be translated, and thus be found at the N-terminal part. Some data on mutant-derived enzymes are shown in Table 2.

To summarize the results obtained so far from fragment analysis: the sequence of activating units on the GS 2-polypeptide corresponds to the amino acid sequence of gramicidin S. The proposed structure (Fig. 8) thus resembles the Lipmann polyenzyme model. The position of the unit carrying the cofactor remains to be established.

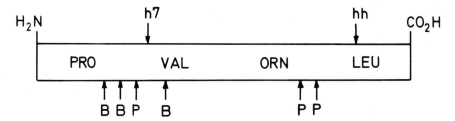

Fig. 8. Proposed structure of the multienzyme GS 2 of gramicidin S-synthetase. Positions of early termination mutants n7 and hh together with fragmentation patterns of papain (*P*) and a *Bacillus brevis* proteinase (*B*) are indicated

III. Control of Sequence and Specificity of Reactions

The control of a sequence of reactions has been described by Lipmann (Lipmann 1956) in his essay *Currents in biochemical research:*

"... it seems important to realize that the synthesis of a fixed sequence can be effected methodologically either by a predominant space pattern, a template, or by a timing device analogous to the assembly line. In reality it most likely will be effected by a mixture of both."

The two elements, template and direction, could be compared to the linear poly-enzyme template and a directed transport system, as indicated in Fig. 2. The difficulty, however, of realizing a directed transport or timing device comparable to mRNA trans-location, in a globular protein molecule, leads us to the proposal of a more complicated space pattern: a template carrying details of the peptide structure beyond the line-up of activation spots.

1. Activation Reactions

Studies on the activation of amino acids by gramicidin S-synthetase have been carried out by measuring the ATP-PPi exchange reaction, or the amino acid-dependent pyro-phosphate formation from ATP ("active site titration"), and by isolation of enzymes charged with adenylates or thiol esters. Compared to aminoacyl-tRNA ligases, the affinity of substrates (K_m), and the reaction rates (V_{max}), are reduced by at least one order of magnitude. Specifity of binding of amino acid is less restricted than in tRNA-aminoacyla-tion.

Activation of phenylalanine by GS 1 is not stereospecific. It has not been established, however, whether one or two binding sites are involved. Inhibitors like phenylalanine-amide or N-acetyl-phenylalanine, whether used in the L- or D-form, affect the activation reactions of L-Phe and D-Phe equally. Substituents on the aromatic ring, like fluoro or hydroxyl groups, are tolerated, and even p-amino-phenylalanine is activated. Amino acids with other R groups, such as β-thienyl-alanine or tryptophan, are activated at high rates. The adenylate transferred to a thiol is epimerized to a final 2:1 ratio of D:L.

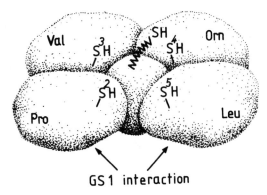

Fig. 9. Model of the GS 2 multienzyme of gramicidin S-synthetase. The amino acid activating "subunits" are arranged around the pantetheine-carrying protein. The linear arrangement predicts that the subunits for proline and leucine activation may be in contact. This prediction is supported by properties of some mutant derived enzymes (Table 2) damaged in leucine activation. These mutants show no (B1-4, C-3, E-2) or reduced (hh) initiation reaction. Interaction of the enzymes GS 1 and GS 2 may involve sites on both of the activating units

In studying the activation of proline, valine, ornithine, and leucine by GS 2, we assume that there is no interaction between these activation centers. This goes back to the observation that the sum of the individual rates of the amino acid-dependent ATP-PPi exchange reactions does not deviate significantly from the observed rate of all activation reactions. Complications arise, however, in investigations of substrate analogs, especially when one compares the sites of valine and leucine binding. Limited proteolysis may be necessary for the study of individual reaction sites. So far, isolation of valine and leucine activating fragments from papain-treated GS 2 has been achieved (Altmann et al. 1979). The proline activation site of GS 2 is available in a fragment from mutant n7 of *B. brevis.*

Studies have hitherto been restricted to the multienzyme. The activation of Val and Leu is not stereospecific, but the D-form reacts much more slowly. Many related structures like isoleucine, norleucine, alloisoleucine, and norvaline can be activated at both sites with high rates. Several compounds are activated at low rates but are incorporated at high rates into the peptide. This has been shown in the case of the substitution of ornithine by lysine and arginine, and in the case of the substitution of valine by leucine. The rate of peptide synthesis is generally about 2.5% of the ATP-PPi-exchange reaction.

To summarize available data, we observe random addition of substrates. The ATP-metal ion complex in apparently accepted in the *anti* conformation, since 2-chloroadenosinetriphosphate is a substrate, while the 8-substituted analogs (Br, N_3, SH) are not recognized. Specificity of substrate binding is low, so that extensive substitutions are permitted. Aminoacylation of enzyme thiols may be followed by formation of a second adenylate, while thiol blocking by iodoacetamide or N-ethyl-maleimide also destroys the catalytic ATP-PPi-exchange reaction.

2. Initiation Reaction

It has been found that the charging of GS 2 with proline is a requirement for the initiation reaction. This has been interpreted to mean that a conformational change is needed for interaction with GS 1. The exclusive transfer of D-Phe leads to the formation of enzyme-bound D-Phe-Pro. Since a small Phe-transfer has also been observed with GS 2 adsorbed to a prolyl-column (Pass et al. 1974) or in the presence of pyrrolidine and N-acetyl-proline, the existence of a primary acceptor has been proposed. The nature of this transferred Phe has not been established, however.

3. Elongation Reaction

The control of elongation could take place during transport or during peptide bond formation. Though a specific amino acid is activated as a thiolester, it may not accept the peptide to be elongated. This has been observed for D-valine, D-leucine, and D-iso-

Table 3. Rates of incorporation of amino acid analogs into gramicidin S-like peptides

Amino acids					Rate (relative)
F	P	V	O*	L	100
mFF	P	V	O*	L	95
F	AZ	V	O*	L	86
F	TP	V	O*	L	31
F	P	V	R	L*	23
F	P	V	K	L*	69
mFF	AZ	V	O	L*	47
mFF	TP	V	O	L*	14
mFF	P	V	R	L*	16
mFF	P	V	K	L*	24
F	AZ	V	R	L*	16
F	AZ	V	K	L*	56
F	TP	V	R	L*	7
F	TP	V	K	L*	19
mFF	AZ	V	R	L*	15
mFF	AZ	V	K	L*	7
mFF	TP	V	R	L*	3
mFF	TP	V	K	L*	7

F: Phe, P: Pro, V: Val, O: Orn, L: Leu, mFF: m-fluoro-Phe, AZ: azetidine-2-carboxylic acid, TP: thioproline, R: Arg, K: Lys

Rates have been estimated by the millipore filter assay using partially purified enzymes. The concentration of the labeled amino acid (O*, L*) has been rate limiting; analogs have been used with 1,5 mM final concentrations. Linear rates have been observed for 30 min, with about 25% incorporation of the label in the control experiment (set = 100)

leucine. The exclusion of these elongation reactions is obviously related to the D-form of these amino acids. A control mechanism related to the transport of activated peptide intermediates has so far not been established. It is expected that each peptide is attached by transthiolation of the thiol of 4'-phosphopantetheine, and then, by diffusion, reaches its elongation site, where it is slowed down or bound so that it can be accepted by the activated amino acid. Experiments investigating possible elongations of activated peptides or binding studies of peptides should clarify this concept. Evidence for a control beyond activation and exclusion of peptidation has been obtained from studies with amino acid analogs. Replacements of Phe by D, L-m-fluoro phenylalanine, Pro by azetidine-2-carboxylic acid and thioproline, and Orn by lysine and arginine lead to the rates of peptide formation shown in Table 3.

If we observe a decreased rate of peptide formation with one amino acid analog, we would expect an even slower rate with two or three analogs. Quantitatively, if the rates were controlled only at the activation level, a rate R_1 with the substitution 1, and a rate R_2 with the substitution 2, should reduce to a rate $R_1 R_2$ with both substitutions. As is clearly shown in Table 2, this prediction does not apply to the observed rates. More data on partial reactions are obviously needed.

Acknowledgments. The authors thank M. Altmann and R. Kittelberger for helpful discussions. Some experiments have been carried out with the excellent technical assistance of G. Podjaski, A. El-Samaraie and D. Schmidt. The work was supported by grants Kl 148/16 from the Deutsche Forschungsgemeinschaft and PTB 8013 from the Bundesministerium für Forschung und Technologie.

References

Altmann M, Koischwitz H, Salnikow S, Kleinkauf H (1978) (3,3'-Leu)-Gramicidin S formation by gramicidin S-synthetase. FEBS Lett 93: 247–250

Altmann M, Koischwitz H, Kittelberger R (1979) Isolation of partially active structures of the multienzyme gramicidin S-synthetase. 11th Inter Congr Biochem Toronto 1979. Abstract 04-5-S 107. Nat Res Council Canada, Can Biochem Soc, Int Union Biochem, Ottawa

Campbell PN, Work TS (1953) Biosynthesis of proteins. Nature 171: 997–1001

Crick FHC (1958) On protein synthesis. In: The biological replication of macromolecules, pp 138–163. Symp Soc Exp Biol No 12. University Press, Cambridge

Dounce AL (1952) Duplicating mechanism for peptide chain and nucleic acid synthesis. Enzymologia 15: 251–258

Gilhuus-Mode CC, Kristensen T, Bredesen JE, Zimmer T-L, Laland SG (1970) The presence and possible role of phosphopantothenic acid in gramicidin S-synthetase. FEBS Lett 7: 287–290

Kambe M, Imae Y, Kurahashi K (1974) Biochemical studies on gramicidin S nonproducing mutants of *Bacillus brevis* ATC 9999. J Biochem 75: 481–493

Kanda M, Hori K, Kurotsu T, Miura S, Nozoe A, Saito Y (1978) Studies on gramicidin S-synthetase. Purification and properties of the light enzyme obtained from some mutants of *Bacillus brevis*. J Biochem 84: 435–441

Kleinkauf H (1979) Antibiotic polypeptides-biosynthesis on multifunctional protein templates. Planta Med 35: 1–18

Kleinkauf H, Koischwitz H (1974) Gramicidin S-synthetase: Active form on the multienzyme complex is undissociable by sodium dodecylsulfate. In: Richter D (ed) Lipmann Symposium: Energy, biosynthesis and regulation in molecular biology. de Gruyter, Berlin New York

Kleinkauf H, Koischwitz H (1978) Peptide bond formation in nonribosomal systems. In: Hahn FE (ed) Progress in molecular and subcellular biology, vol VI. Springer, Berlin Heidelberg New York

Kleinkauf H, Koischwitz H (1980a) Gramicidin S-synthetase. In: Bisswanger H, Schmincke-Ott E, (eds) Multifunctional proteins, chap 8. Academic Press, New York

Kleinkauf H, Koischwitz H (1980b) Gramicidin S-synthetase 1979. In: Proceedings leopoldina symposium. Nover L, Lynen F, Mothes K (eds) Cell compartmentation and metabolic channeling, pp 147–158. Gustav Fischer, Jena, Elsevier, Amsterdam New York Oxford

Kleinkauf H, Roskoski R Jr, Lipmann F (1971) Pantetheine-linked peptide intermediates in gramicidin S and tyrocidine biosynthesis. Proc Natl Acad Sci USA 68: 2069–2072

Koischwitz H (1975) Non-ribosomal polypeptide formation: A model for peptide sequence fixation on protein templates. 10th FEBS Meet Paris 1975, abstract 860. Societe de Chimie Biologique, Paris

Koischwitz H (1979) Zur Struktur der Proteinmatrize von Gramicidin S. Hoppe-Seyler's Z Physiol Chem 360: 307

Lee SG, Lipmann F (1977) Isolation of amino acid activating subunit-pantetheine protein complexes: Their role in chain elongation in tyrocidine synthesis. Proc Natl Acad Sci USA 74: 2343–2347

Lipmann F (1954) On the mechanism of some ATP-linked reactions and certain aspects of protein synthesis. In: McElroy WD, Glass B (eds) The mechanism of enzyme action, pp 599–604. Hopkins, Baltimore

Lipmann F (1956) Attempts at the formulation of some basic biochemical questions. In: Green DE (ed) Currents in biochemical research, pp 241–250. Interscience, New York

Lipmann F (1968) The relation between the direction and the mechanism of polymerization. Essays Biochem 4: 1–23

Lipmann F (1971) Attempts to map a process evolution of peptide biosynthesis. Science 173: 875–884

Pass L, Zimmer T-L, Laland SG (1974) On the use of affinity chromatography in demonstrating the transfer of thioester-bound D-phenylalanine from the light enzyme of gramicidin S-synthetase to an acceptor site for this amino acid on the heavy enzyme. Eur J Biochem 47: 607–611

Spiegelmann S (1954) Nucleic acids and the synthesis of proteins. In: McElroy WD, Glass B (eds) The chemical basis of heredity, pp 232:249. Hopkins, Baltimore

10. A Molecular Approach to Immunity and Pathogenicity in an Insect-Bacterial System

H.G.BOMAN

A. Introduction

In the Lipmann Symposium 5 years ago I called my contribution *Why is insect immunity interesting?* (Boman et al. 1974a). At that time we had only discovered the basic phenomena: if insects are given an injection of a nonpathogenic bacteria, they will respond by producing an antibacterial activity highly effective against a number of different bacteria both pathogenic and nonpathogenic. By the use of actinomycin D and cycloheximide we had indirectly shown that RNA and protein synthesis was required, but we had not succeeded in a biochemical characterization of any of the material synthesized as the result of an infection.

We started our work with Drosophila (Boman et al. 1972) and we have later tried seven different insects. However, the giant silk moth, *Hyalophora cecropia,* has turned out to be a most convenient model system: it is large, it has a diapause which lasts 6–8 months and it is fairly well studied in many other respects. The diapause, that is the winter hibernation of the insects, has provided us with a unique experimental situation. The reason for it being that a diapausing insect can be immunized despite the fact that its general metabolism is turned down to a few percent. Thus, by injecting a pupa with viable bacteria and labeled uridine or amino acids we can selectively label the RNA and the proteins which are made from the immunity genes without much background from the rest of the biosynthetic machinery. So far, the immunization of diapausing Cecropia has enabled us to isolate both a complex RNA fraction and to identify ten different immune proteins, six of which we have now purified.

B. Basic Properties of the Humoral Immune System of Cecropia

We could early show that the induction of immunity occurred simultaneously with the synthesis of eight proteins, designated P1–P8. These eight proteins were produced as a response to an infection with either a Gram-positive or a Gram-negative bacterium (Fig. 1). More recently, we discovered two additional low molecular weight proteins, P9A and P9B, which were overlooked before because they moved out of the gels ahead of the tracer dye (Hultmark et al. 1980).

Dept. of Microbiology, University of Stockholm, 106 91 Stockholm, Sweden

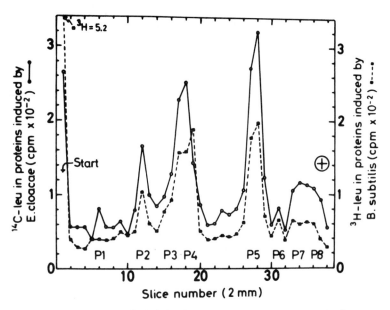

Fig. 1. Co-electrophoresis of labeled polypeptides in hemolymph samples from two pupae of *Hyalophora cecropia*, one injected with ^{3}H-leucine and viable *Bacillus subtilis* (*black squares*), the other with ^{14}C-leucine and viable *Enterobacter cloacae* (*white balls*). The polyacrylamide gel concentration was 7.5% and the buffer contained 0.1% sodium dodecyl sulfate (SDS). The different proteins were designated *P1–P8* as indicated from their respective mobilities (Faye et al. 1975)

We soon found that the antibacterial activity induced showed a rather broad specificity. A more quantitative estimation of the killing activities against ten different bacterial species are given in Table 1. These data show no obvious relationship between bacterial taxonomy and susceptibility to the immune system. Some Gram-positive and Gram-negative species like *Bacillus megaterium* and *Escherichia coli* K12 are killed very fast while other species like *Bacillus thuringiensis* and *Serratia marcescens* are very resistant to the immune system. Since the two latter organisms are pathogenic for many insect species, we recognized that resistance to the immune system could be a cardinal property for an insect pathogen. As we will show later, such resistance can be either passive or active.

The immune system of Cecropia has certain superficial similarities with the mamalian complement system (Faye et al. 1975). From the beginning we considered therefore the possibility that the immune proteins in a cooperative way jointly killed all susceptible bacteria. However, we first found that lipopolysaccharide (LPS) from *E. coli* selectively inhibited the killing of *E. coli* without affecting the activity against *B. subtilis*. When we investigated simultaneously the virulence of different strains of *B. thuringiensis*, we discovered that this insect pathogen produces two different inhibitors of the Cecropia immune system (designated InA and InB), each with a distinct specificity. Altogether we found four different inhibitors (Table 2) which affected one bacterium but not two others. These data strongly argued against our initial idea of a cooperative

Table 1. Susceptibility of different bacteria to killing by immune hemolymph from *H. cecropia*

Organism and strain	Bacterial concentration	Hemolymph concentration	Number of exp.	Killing time (min)	
				Range	Average
Bacillus megaterium, Bm11	4.9×10^5	1	2	0.3– 0.4	0. 35
Escherichia coli K12, D21	8.2×10^5	1	8	0.7– 2.0	1.2
Bacillus subtilis, Bs11	7.5×10^5	1	4	2.0– 4.0	2.8
Pseudomonas aeruginosa, OT97	5.5×10^5	1	2	2.8– 3.3	3.1
Enterobacter cloacae, β11	6.3×10^5	1	2	16 – 18	17
Salmonella typhimurium, LT2	3.8×10^5	5	2	5.7– 6.3	6.0
Bacillus cereus, Bc11	1.9×10^4	90	2	75– 85	80
Bacillus thuringiensis, Bt7	3.0×10^3	95	2	80–120	100
Staphylococcus aureus, Cowan 1	7.5×10^4	95	2	> 120	>120
Serratia marcescens, Db11	1.0×10^4	95	4	> 120	>120

Killing time is the time required for the killing of 90% of the bacteria. Bacterial concentration is given as the average number of viable cells in 0.1 ml. Hemolymph concentration is given as percent (v/v). The variations in killing is believed to be mainly variations between samples of hemolymph from different pupae. Most of the data are from Rasmuson and Boman (1977). *S. marcescens*, Db11 is a streptomycin resistant mutant of a parental strain which was isolated from a culture of sick Drosophila (Flyg, Kenne and Boman, 1980)

Table 2. Selective effects of four microbial inhibitors on the in vitro killing of three bacteria by Cecropia immune hemolymph

Test organism	Inhibitor			
	LPS	Zymosan	InA	InB
E. coli, D31	+	–	+	–
B. subtilis, Bs11	–	–	–	–
B. cereus, Bc11	–	+	–	+

The inhibitors were tested in our standard assay for antibacterial activity (Boman et al. 1974b). + indicates more than 90% inhibition; – less than 10% inhibition in time curves with at least 4 points. LPS was prepared from *E. coli* K12, strain D21, Zymosan was from Sigma Chem. Co. and InA and InB were prepared from *B. thuringiensis*, Bt75

effect between all the ten proteins and instead left us with some less exciting and more tedious questions: Which of the ten proteins found to date are responsible for the killing of the different bacteria listed in Table 1? How many species does each protein kill or do we have a combination of 2–3 proteins which kill certain groups of bacteria?

Table 3. Properties of immune proteins purified from *Cecropia* hemolymph

Protein	"Peak"	Mol. weight	pI	Antisera	Other properties
P5	L	4 × 24,000	6.6	Yes	Enhancing killing
P4	L	48,000	8.2	Yes	Preexisting
P7	S	~ 16,000	~11	Yes	Lysozyme
P8	S	< 14,000	8.6	No	Bacteriostatic?
P9A	S	~ 7,000	> 8.6	No	Lytic for *E. coli*
P9B	S	~ 7,000	> 8.6	No	Lytic for *E. coli*

Data are taken from Pye and Boman (1977) for P5, from Rasmuson and Boman (1979) for P4, from Hultmark et al. (1980) for P7, P9A and P9B. Data for P8 are unpublished observations of Hultmark. L and S refer to large and small peaks in polyacrylamide gel electrophoresis like Fig. 1

C. Efforts to Obtain Reconstitution and in vitro Synthesis

The answer to the above questions would have to come from experiments with purified proteins. Since an early aim was to reconstitute the system from purified components, we started several years ago to purify the two main components, P5 and P4 (cf. Fig. 1). In addition, we have purified four other immune proteins and a summary of their properties is given in Table 3. As indicated, P5 is composed of four equal-sized subunits and it enhances the killing of *E. coli* without having an activity of its own (Pye and Boman 1977). P4 was found to be present both in normal hemolymph and in some tissue but so far we have failed to demonstrate any function for this purified protein (Rasmuson and Boman 1979). We were more successful in finding functions for three of the minor components, P7, P9A, and P9B. Protein P7 turned out to be a conventional lysozyme: it is strongly basic, it lyses *Micrococcus lysodeikticus* but not *E. coli*, and in amino acid composition and pH-profile it resembles the insect lysozymes previously purified (Powning and Davidson 1973). However, proteins P9A and P9B may be unique because as far as we know they differ in properties from other eukaryotic proteins which kill and lyse *E. coli*. Both forms of P9 are strongly basic, their molecular weights may be as low as 7000, they both lack cysteine and they may differ only in their contents of glutamic acid and methionine, They are highly active in killing *E. coli* but they also work against *Enterobacter cloacae* and *Pseudomonas aeruginosa*. It is not yet clear whether they differ in function, but out of four individual pupae tested all were found to have both forms of P9. It is therefore likely that they originate from a gene duplication.

Insect pupae are essentially a closed system and every injection will create an injury. Injury effects have long been known in *Cecropia* and we therefore compared earlier the response to an infection and the response to an injection of a salt solution (Faye et al. 1975). It was found that the protein pattern produced as a response to such a sham injection was partly the same as the immune response but both qualitative and quantitative differences were recorded (Fig. 2). This result brought us our first questions in the area of cell and molecular biology: Do we have different types of cells which produce the respective responses or do we have control systems on the transcriptional or the translational level?

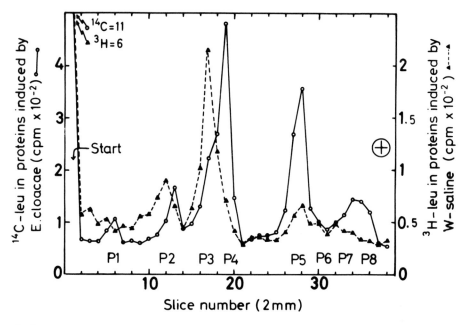

Fig. 2. Co-electrophoresis of labeled polypeptides in hemolymph samples from two *Hyalophora cecropia* pupae, one injected with ^{14}C-leucine and viable *Enterobacter cloacae* (*white balls*), the other injected with ^{3}H-leucine and sterile, physiological salt solution (*black triangles*). Experimental details were as described for Fig. 1

During my stay in the Lipmann Laboratory I was introduced to the RNA field and therefore it was natural for me to start and prepare a total RNA fraction from Cecropia pupae. It turned out to be rather difficult because the pupae were full of uric acid and in a 5 g insect there was less than 1 mg of total RNA. Assuming the majority of this RNA to be ribosomal, the messenger could be expected to be only 10–100 μg/pupa. Moreover, the uric acid must come from larval RNA and the pupae could still contain significant amounts of RNase. To minimize RNA degradation, I started the preparation by emptying the pupae of hemolymph and intestinal content and then by grinding the rest of the animal in liquid nitrogen. The dry frozen powder was extracted twice in a conventional way with phenol-chloroform buffer in the presence of urea and different RNase inhibitors. However, a clear phase separation was obtained only when male pupae were used; with females the aqueous phase was always cloudy. The combined aqueous phases contained large quantities of uric acid and were a very dilute solution of RNA. With only a 10% increase in volume we could from this solution precipitate all RNA by the addition of 0.1 M lanthanum nitrate, a very convenient method to precipitate nucleic acids, discovered by Hammarsten 50 years ago. Lanthanum and a number of impurities could then be extracted from the precipitate with a mixture of 67% of ethanol and 33% of 0.1 M EDTA. The remaining material was dissolved in one-tenth of the original volume and at this stage the RNA could be precipitated with ethanol in a conventional way. The total RNA was finally fractionated by affinity chromatography to obtain polyA containing messenger RNA. Such fractions isolated from immunized and in-

222   H.G. Boman

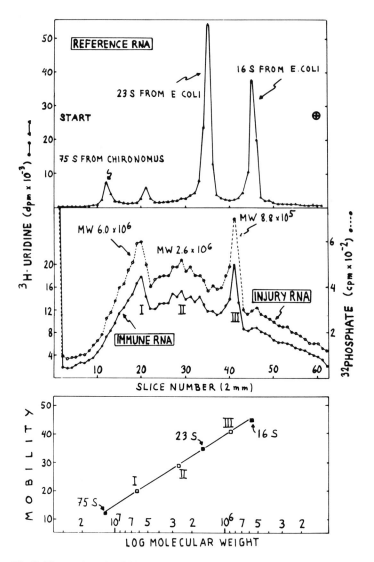

Fig. 3. Electrophoresis of RNA in gels composed of agarose-polyacrylamide (0.5% and 1.9%, respectively) in buffer containing 40 mM tris, 0.1% of SDS, 0.01 mM EDTA, pH 8.0. The time of the electrophoresis was 30 min longer than needed for brom phenol blue to leave the gel. *Upper part:* reference samples of ribosomal RNA from *E. coli* and 75S RNA from *Chironomus tentans* (the latter kindly provided by Bertil Daneholt). *Middle part:* the second peak from poly(U)-Sepharose chromatography of an RNA-preparation containing ^{3}H-uridine labeled immune RNA (*black balls*) and ^{32}P-labeled injury RNA (*white balls*). *Lower part:* plot of electrophoretic mobility versus log of molecular weight, assuming a straight line relationship

jured pupae were compared by polyacrylamide gel electrophoresis. Figure 3 shows that the preparation was rather heterogenious with respect to molecular weights, with some of the RNA molecules as large as 6×10^{6} daltons. No significant differences were found between immune and injury RNA. Thus, we had no evidence for any control on the

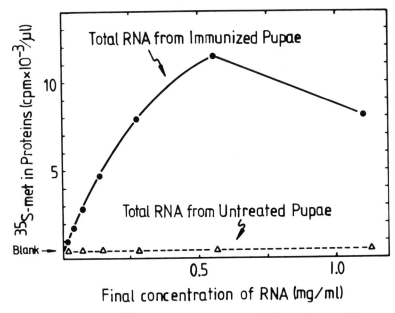

Fig. 4. Incorporation of ^{35}S-methionine into TCA-insoluble proteins for different amounts of Cecropia RNA added. The pupae were immunized with *E. cloacae* according to the standard procedure, and total RNA was prepared 3 days later. The control pupae received no treatment but were incubated during the same temperature and light conditions as the immunized pupae. The rabbit reticulocyte system was used for translation

transcriptional level and we were left with two alternatives: (1) Our method of investigation was too insensitive to detect differences in RNA synthesis, or (2) the control machinery was operating on either translational level or at a post-translational processing of the proteins.

In order to discriminate between the above alternatives, we have recently together with Andrew Pigon begun a study of the translation of the RNA. We first found that RNA fractionated on poly(U)-Sepharose with yeast carrier RNA was inactive. However, the total RNA preparation from immunized pupae gave a good translation, while the corresponding RNA fraction obtained from untreated pupae was inactive (Fig. 4). Further experiments revealed that also injury RNA gives a good translation. However, the respective products formed have not yet been characterized in a satisfactory way. It is therefore too early to conclude at which stage expression of immunity is controlled. *Note added in proof:* A full account of the RNA work is given in a recent paper by Boman, Boman and Pigon, Insect Biochem. (1980), in press.

D. Ways Bacteria Can Escape Immunity

The most common explanation for pathogenicity has always been the production of toxins. In consonance the most studied insect pathogen, *Bacillus thuringienses,* is known to produce both an endo- and an exotoxin which are well characterized (Faust 1975).

We began our work on *B. thuringiensis* by collecting strains deficient in each toxin as well as sporulation negative mutants because the endotoxin was known to be present also in the spore coat. Although the virulence decreased, toxinless mutants were still found to grow well in insects and ultimately to kill them. This gradually led us to the discovery of the two different immune inhibitors already presented in Table 2 (Edlund et al. 1976). The purification of these inhibitors turned out to be hard but for different reasons. For InA we first had to isolate a mutant (Bt75) with a reduced level of proteolytic enzymes, only from then on could we detect protein bands in polyacrylamide gels. Purified InA was found to be a single polypeptide chain with a molecular weight of 78,000 daltons (Sidén et al. 1979). This protein blocks selectively the killing of *E. coli* but does not affect the activity against *B. subtilis* or *B. cereus* (cf. Table 2) and two of its targets in Cecropia are most likely the immune proteins P9A and P9B. InB is an acid polysaccharide which has been purified by a series of chemical extractions followed by gel filtrations and chromatography on DEAE-Sepharose (Lee et al., unpublished). The molecular weight is around 46,000 and the substance blocks selectively the killing of *B. cereus* and *B. thuringiensis* but its mechanism of action is not yet understood.

B. thuringiensis is only slowly killed by the immune system of Cecropia (killing time around 100 min), and we believe this resistance contributes to the virulence (Edlund et al. 1976). When investigating another insect pathogen, *Serratia marcescens,* we found it to be totally resistant to 2 h of incubation with 95% immune hemolymph (Table 1). From this strain of Serratia we isolated a handful of spontaneous mutants resistant to two bacteriophages, ØJ and ØK. Figure 5 shows that two of these mutants DB1109 and Db1101, had become sensitive to the immune hemolymph. There was also a clear difference in susceptibility between the two mutants. Since Serratia is a pathogen, we could investigate if the resistance to the immune system was a virulence factor. Rather than testing this possibility in the large Cecropia, we turned to the small *Drosophila.*

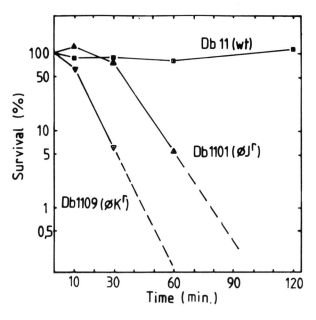

Fig. 5. Susceptibility of different strains of *Serratia marcescens* to incubation with 90% of immune hemolymph from *Cecropia.* Db11 is a streptomycin-resistant mutant which is pathogenic for *Drosophila* (cf. Fig. 6). Db1101 and Db1109 are spontaneous mutants of Db11 selected as resistant to two closely related bacteriophages, ØJ and ØK

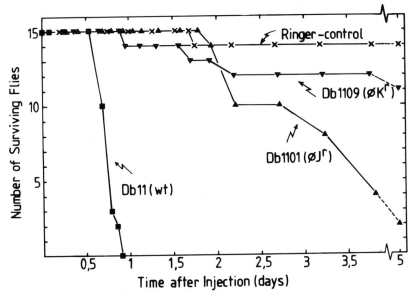

Fig. 6. Survival of *Drosophila* injected with a dose of about eight viable cells of wild-type and two phage-resistant mutants of *Serratia marcescens*. The strains are the same as used in Fig. 5

Figure 6 shows that about eight injected bacteria of the parental strain Db11 killed the flies within 24 h. The most susceptible mutant, Db1109, was found to be the least virulent strain with 73% of the flies surviving 5 days. Db1101 was something interme- diate which killed most flies within 5 days (Flyg, Kenne and Boman 1980). Thus, there was a clear correlation between the resistance to the immune hemolymph of Cecro- pia and the virulence to Drosophila. To complete the picture, it must also be stated that both the wild-type and the two mutants of *S. marcescens* produce equal amounts of an immune inhibitor which seems to be of the A type, although it is clearly different from the InA produced by *B. thuringiensis*.

From this section we can conclude that resistance to the immune system of the host can be an important property of insect pathogens. It is also clear that such resistance can be actively manifested by the production of immune inhibitors or simply passively expressed as having a target unaccessible to the immune system.

E. Discussion

Most scientists who take up a fresh problem have an ambition to find something new and exciting. In the case of problems which are obviously fundamental there will rapidly be a cluster of scientists. For insect immunity we have a different situation: at present very few people work on the problem, probably because it is still considered a gamble. It can turn out to be mainly a repetitive problem; on the other hand there is a chance of discover something where nobody looks. After 5 years it is reasonable to ask if we have found something different or if we have variations on known themes. My answer is to point to the following:

1. The Model System. The humoral immune system of diapausing Cecropia is composed of only ten different proteins which are selectively synthesized in cells which have a capacity for making thousands of proteins. Since this induced immunity is composed of several different bacteriocidal mechanisms (cf. Table 2), we have an increasing number of functions we can assay the proteins for. When compared to the heat-shock proteins of Drosophila (Tissières et al. 1974; Keinikoff and Meselson 1977; Spradling et al. 1977; Schedi et al. 1978), or the chorion proteins in the egg shells of *Antheraea polyphemus* (Gelinas and Kafatos 1977), we have a less well-defined RNA but a better understanding of the functions of the proteins. We therefore feel that Cecropia immunity could be an interesting model system for the study of induction, synthesis, and control of eukaryotic RNA and proteins.

2. The Bacteriocidal Action. The fact that protein P7 turned out to be a conventional lysozyme was of course an example of a repetitive finding and therefore also a disappointment. However, proteins P9A and P9B, which kill and lyse *E. coli,* may represent something new and interesting. One could argue that the substrates used, *Micrococcus* and *E. coli,* are artificial because these bacteria do not represent a threat to the host. However, against this reasoning argues the fact that *E. coli* becomes a pathogen if the induction of immunity is blocked by actinomycin D (Boman et al. 1974a). Moreover, there may be bacteriocidal mechanisms in vivo and in whole hemolymph which we cannot account for in terms of purified proteins. One example is the killing of *B. cereus* and *B. thuringiensis* which we cannot yet reconstitute despite numerous efforts. The clarification of this mechanism still represents a challenge.

3. The Understanding of Pathogenicity. At the beginning of this work I did not expect that there should be a connection between immunity and pathogenicity. We recognized then that passive resistance to the immune system could be an important property for an insect pathogen. Still it came as a surprise that *B. thuringiensis* produced immune inhibitors. We learned later that also parasites can cause immuno suppressions in mammals although the mechanisms are unknown (see review by Bloom 1979). Since we have purified both InA and two of its targets, P9A and P9B, we hope to be able to understand the molecular mechanism of InA and in this way contribute to the overall view on virulence factors.

4. Phylogenetic Aspects of Immunity. More than 10^6 species of insects are known and the number of individuals has been estimated to be as large as 10^{18} (see Wigglesworth 1968). These high numbers show that during evolution insects have been highly successful in competing with other forms of life, including microorganisms. Since insects also represent extreme adaptation to different ecological niches, one can predict that comparative studies of their immunity will illustrate their evolutionary development. One special fact is worth emphasizing: many insect species have bacterial symbionts which are needed for their growth and reproduction. Some of these insects are also vectors of human pathogens and an infected tsetse fly is an example of a living 4-component system (fly, two different symbionts and Trypanosomes). One can therefore predict that the immune system in these species must be constructed in such a way that invading microorganisms are attacked without the destruction of symbionts or parasites.

Acknowledgments. This paper is dedicated to Dr. Lipmann with gratitude for a 2 years stay at the Rockefeller Institute — the single most important part of my scientific education. The work described was supported by grants from the The Swedish Natural Science Research Council and from the Knut and Alice Wallenberg Foundation.

References

Bloom BR (1979) Games parasites play: how parasites evade immune surveillance. Nature 279: 21–26

Boman HG, Nilsson I, Rasmuson B (1972) Inducible antibacterial defence system in Drosophila. Nature 237: 232–235

Boman HG, Nilsson-Faye I, Rasmuson T (1974a) In: Lipmann Symposium: Energy, biosynthesis and regulation in molecular biology, pp 103–114. de Gruyter, Berlin New York

Boman HG, Nilsson-Faye I, Paul K, Rasmuson T (1974b) Insect immunity. I. Characteristics of an inducible cell-free antibacterial reaction in hemolymph of samia cynthia pupae. Infect Immun 10: 136–145

Edlund T, Sidén I, Boman HG (1976) Evidence for two immune inhibitors from Bacillus thuringiensis interfering with the humoral defense system of saturniid pupae. Infect Immun 14: 934–941

Faust RM (1975) In: Briggs JD (ed) Biological regulation of vectors, pp 31–48. A conference report. U.S. Department of Health, Education, and Welfare

Faye I, Pye A, Rasmuson T, Boman HB, Boman IA (1975) Insect immunity. II. Simultaneous induction of antibacterial activity and selective synthesis of some hemolymph proteins in diapausing pupae of Hyalophora cecropia and Samia cynthia. Infect Immun 12: 1426–1438

Flyg C, Kenne K, Boman HG (1980) Insect pathogenic properties of Serratia marcescens: Phage resistant mutants with a decreased resistance to Cecropia immunity and a decreased virulence to Drosophila. J Gen Microbiol, in press

Gelinas RE, Kafatos FC (1977) The control of chorion protein synthesis in silkmoths: mRNA production parallels protein synthesis. Dev Biol 55: 179–190

Henikoff S, Meselson M (1977) Transcription at two heat shock loci in Drosophila. Cell 12: 441–451

Hultmark D, Steiner H, Rasmuson T, Boman HG (1980) Insect immunity. VI. Purification and properties of three inducible bactericidal proteins from hemolymph of immunized pupae of Hyalophora cecropia. Eur J Biochem 106: 7–16

Powning RF, Davidson WJ (1973) Studies on insect bacteriolytic enzymes – I. Lysozyme in hemolymph of Galleria mellonella and Bombyx mori. Biochem Physiol 45B: 669–686

Pye AE, Boman HG (1977) Insect immunity. III. Purification and partial characterization of immune protein P5 from hemolymph of Hyalophora cecropia pupae. Infect Immun 17: 408–414

Rasmuson T, Boman HG (1977) In: Solomon JB, Horton JD (eds) Developmental immunobiology, pp 83–90. Elsevier/North-Holland, Amsterdam

Rasmuson T, Boman HG (1979) Insect immunity – V. Purification and some properties of immune protein P4 from haemolymph of Hyalophora cecropia pupae. Insect Biochem 9: 259–264

Schedl P, Artavanis-Tsakonas S, Steward R, Gehring WJ (1978) Two hybrid plasmids with D. melanogaster DNA sequences complementary to mRNA coding for the major heat shock protein. Cell 14: 921–929

Sidén I, Dalhammar G, Telander B, Boman HG (1979) Virulence factors in Bacillus thuringiensis: Purification and properties of a protein inhibitor of immunity in insects. J Gen Microbiol 114: 45–54

Spradling A, Pardue ML, Penman S (1977) Messenger RNA in heat-shocked Drosophila cells. J Mol
    Biol 109: 559–587
Tissières A, Mitchell HK, Tracy UN (1974) Protein synthesis in cellular salivary glands of Drosophila
    melanogaster: Relation to chromosomes puffs. J Mol Biol 84: 389–398
Wigglesworth VB (1968) The life of insects. The new American Library, New York

C. Nucleic Acid – Protein Interactions; Mutagenesis

1. Structure of the Gene 5 DNA Binding Protein from Bacteriophage fd and its DNA Binding Cleft

A. McPHERSON[1], A. WANG[2], F. JURNAK[1], I. MOLINEUX[3], and A. RICH[2]

A. Introduction

The gene 5 product of bacteriophage fd is a small DNA binding protein of 9,800 daltons containing 87 amino acids (Alberts et al. 1972; Oey and Knippers 1972). Its primary physiological role is the stabilization and protection of the single-strand DNA daughter virions from duplex formation following replication in the host (Salstrom and Pratt 1971). Because of the highly cooperative nature of its binding to DNA, it has the capacity to unwind or destabilize native DNA. Under low ionic strength conditions in vitro, it will melt double-stranded homopolymers and will reduce the melting temperature of native double-strand calf thymus DNA by 40°C (Salstrom and Pratt 1971).

The protein is made in about 10,000 copies per infected *E. coli* cell and can be isolated in substantial amounts by DNA cellulose chromatography. Its sequence (Fig. 1) has been determined (Nakashima et al. 1974; Cuypers et al. 1974) and extensive biochemical and biophysical characterization has been carried out (Coleman et al. 1976; Anderson et al. 1975; Day 1973; Pretorius et al. 1975). Evidence from these studies indicates that both the electrostatic interactions between basic residues of the protein and the phosphate groups of the polynucleotide backbone are involved in binding, and also that aromatic residues are likely to intercalate or stack upon the purine and pyrimidine bases during complex formation (Coleman et al. 1976; Pretorius et al. 1975).

Electron microscope studies on the gene 5 protein (Alberts et al. 1972; Pratt et al. 1974), which exists predominantly as a dimeric species in solution (Cavalieri et al. 1976), suggest that the protein binds to DNA strands running in opposite directions so that it would cross-link opposing strands of a duplex or opposite sides of circular single strands of DNA. The mechanism for DNA duplex unwinding is simply a linear aggregation along the two opposite strands deriving from the highly cooperative nature of the lateral binding interaction (Dunker and Anderson 1975). This high degree of cooperativity presumably is a product of strong protein—protein forces between adjacent molecules of the gene 5 protein along the DNA strands.

1 Department of Biochemistry, University of California at Riverside, Riverside, California 92502, USA
2 Department of Biology, Massachusetts Institute of Technology, Cambridge, Massachusetts 02139, USA
3 Department of Microbiology, The University of Texas, Austin, Texas 78712, USA

MET–ILE–LYS–VAL–GLU–ILE–LYS–PRO–SER–GLN–ALA–GLN–PHE–THR–THR–ARG–SER–GLY–VAL–SER–ARG–GLN–
10   20

GLY–LYS–PRO–TYR–SER–LEU–ASN–GLY–GLN–LEU–CYS–TYR–VAL–ASP–LEU–GLY–ASN–GLU–TYR–PRO–VAL–LEU–
30   40

VAL–LYS–ILE–THR–LEU–ASP–GLU–GLY–GLN–PRO–ALA–TYR–ALA–PRO–GLY–LEU–TYR–THR–VAL–HIS–LEU–SER–
50   60

SER–PHE–LYS–VAL–GLY–GLN–PHE–GLY–SER–LEU–MET–ILE–ASP–ARG–LEU–ARG–LEU–VAL–PRO–ALA–LYS
70   80

Fig. 1. Amino acid sequence of the gene 5 protein

B. Crystals and the Molecular Structure

Approximately two years ago we were able to obtain this protein in a form suitable for single crystal X-ray diffraction analysis (McPherson et al. 1976). The crystals are in the monoclinic space group C2 with one gene 5 monomer per asymmetric unit. This immediately suggested that the gene 5 dimer might contain a molecular dyad axis relating its two halves. The unit cell dimensions are $a = 76.5$, $b = 28.0$, $c = 42.5$ Å and $\beta = 108°$. The resolution of the diffraction data extends to at least 1.2 Å Bragg spacings.

The structure has been solved and the technical details are presented elsewhere (McPherson et al. 1979). The gene 5 monomer is roughly 45 Å long, 25 Å wide and 30 Å high. It is essentially globular, with an appendage of density closely approaching the molecular dyad and tightly interlocking with an identical symmetry-related appendage on the second molecule within the dimer. The major portion of the molecular density creates an overhanging ledge of density that serves in part to create an extended shallow groove banding the outside waist of the monomer. In the dimer the two symmetry-related grooves, each about 30 Å in length, run antiparallel courses and are separated by roughly 25 Å.

The course of the polypeptide chain in the gene 5 monomer is shown in Fig. 2, as deduced from the 2.3 Å electron density map. The protein is composed entirely of antiparallel β structure with no α helix whatsoever. This is as expected from spectroscopic measurements (Day 1973) and sequence-structure rules (Anderson et al. 1975). There are three basic elements of secondary structure in the molecule, as shown in Fig. 3; a three-stranded antiparallel β sheet arising from residues 12–49, a two-stranded antiparallel β ribbon formed by residues 50–70, and a second two-stranded antiparallel β ribbon derived from residues 71–82. It is the first of the two β loops (50–70) that creates the appendage of density near the molecular dyad and maintains the dimer

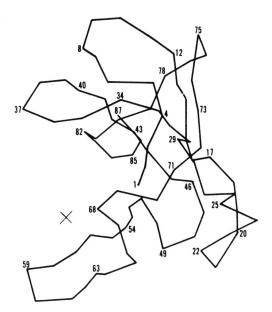

Fig. 2. Polypeptide tracing of the gene 5 DNA unwinding protein based on α-carbon coordinates measured from a Kendrew model. The view is roughly along the crystallographic a axis

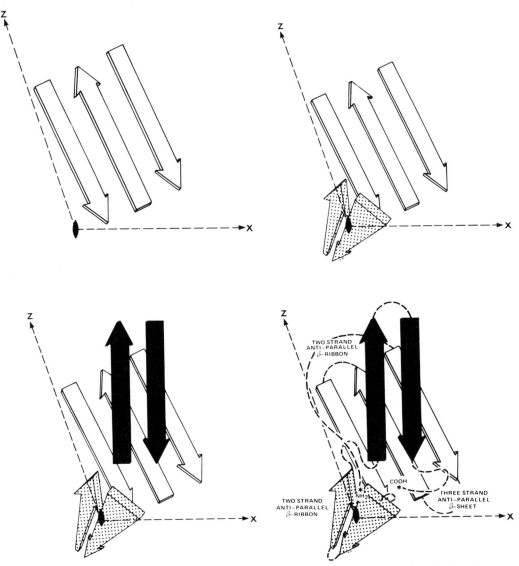

Fig. 3. Secondary structural elements of the gene 5 molecule. Beginning with the N terminal portion of the sequence, it consists of a three-stranded β sheet which forms a major part of the DNA binding region, a two-stranded β ribbon principally responsible for maintaining the molecule as a dimer in solution, and a second β ribbon that we believe is the primary participant in the lateral interactions from which the cooperativity of protein binding arises. These are illustrated in a series of steps

specie in solution. The second β loop (71–82) forms the top surface of the molecule and, we believe, is most involved in producing the neighbor–neighbor interactions responsible for the cooperative protein binding. The central density of the molecule is created by the severely twisted three-stranded β sheet made up of residues 12–49. As a result of the distortion from planarity of these three strands, a distinct concavity is produced on the underside of this sheet. Enhanced in part by density from the β ribbon (50–70) near the dyad, this concavity is extended and deepened to provide the long 30 Å groove.

The long groove beneath the three-stranded sheet by its shape and extent suggests it to be the single-strand DNA binding interface. There is no other passage through the density that would be consistent with a long polynucleotide binding region. Given this to be the site, then the mode of cross-strand attachment of the gene 5 protein would be that shown in Fig. 4. The two monomers within the dimer bind to strands of opposite polarity across the duplex DNA, with the molecular dyad roughly perpendicular to the plane of the two bound strands which are separated in the complex by about 25 Å.

C. The Presumptive DNA Binding Site

Since refinement of the structure is not yet complete, the exact placement of amino acid side chains is still tentative. Nonetheless, there remains a number of interesting features that can be described and which are not likely to be seriously revised. The tetra-nucleotide binding trough in the gene 5 protein is composed primarily of the amino acid side chains arising from residues 12–49 of the antiparallel β sheet. These strands

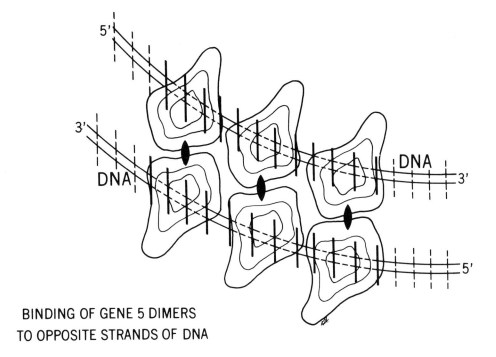

BINDING OF GENE 5 DIMERS
TO OPPOSITE STRANDS OF DNA

Fig. 4. Cross-chain binding of the gene 5 dimers to opposing strands of a DNA duplex or opposite sides of a circular single-stranded DNA molecule. The distance between opposing DNA single strands would be about 25 Å

run more or less parallel with the direction of the DNA strand as it would bind in the trough. The surface of the trough is also comprised in part of residues 50–56 and 66–69, from the interior portions of the two strands forming the β loop near the molecular dyad. A stereo drawing of the gene 5 monomer showing the binding region is shown in Figs. 5 and 6.

Aromatic amino acid side chains have been implicated in the binding of DNA to the gene 5 molecule by chemical modification and NMR studies. These show that tyrosines 26, 41, and 56 lie near the surface of the protein and are readily nitrated by tetranitromethane which prevents DNA binding (Anderson et al. 1975). Conversely, binding of

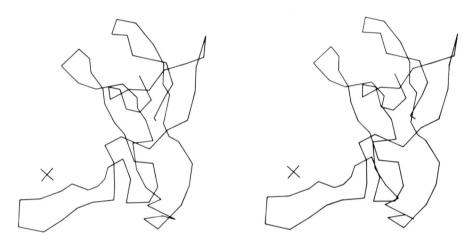

Fig. 5. Stereo photograph of the polypeptide backbone of the gene 5 protein rotated so that the view is roughly along the course of the DNA binding groove. This groove is approximately 25 Å in length and runs more or less parallel with the strands of the β sheet

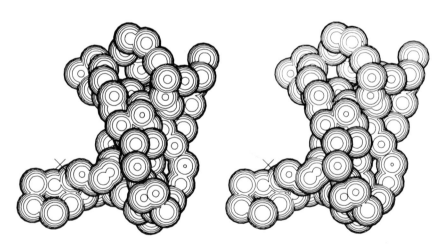

Fig. 6. Stereo photograph of the polypeptide backbone of the gene 5 protein as in Fig. 5, but each α-carbon is represented by a *sphere* of 3.0 Å radius to give a space-filling effect. The DNA binding region is the pronounced *groove* running roughly perpendicular to the plane of the drawing

oligonucleotides or DNA prior to reaction prevents nitration of these residues. ^{19}F NMR of the fluorotyrosyl containing protein confirms these results and further suggests that these tyrosyls intercalate or stack with the bases of the DNA (Coleman et al. 1976). Similar kinds of results have been obtained with deuterated protein that implicates at least one phenylalanine residue in a similar fashion (Coleman and Armitage 1979). Spectral data lend further support to the idea that the aromatic residues of the protein stack upon or intercalate between the bases (Pretorius et al. 1975).

A number of aromatic residues are arrayed along the binding surface, and these include tyrosines 26, 41, 34, and 56 and phenylalanines 13 and 68. The distribution is not uniform, one end of the trough appearing considerably richer than the other and bearing both phenylalanines as well as tyrosines 34, 41, and 56. The opposite end of the trough, that closest to the viewer in Fig. 6, contains only tyrosine 26. The aromatic side chains, with the exception of tyrosine 56 and phenylalanine 68, do not protrude into the binding cleft, but are turned away. Each can, however, be brought down into the binding groove by an appropriate rotation about the β carbon. Of particular interest are the side groups of tyrosine 41, tyrosine 34, and phenylalanine 13 which form a triple stack with phe 13 innermost, tyr 41 fully on the outside, and tyr 34 sandwiched in between. The stacking is also not precisely one atop the other, but the rings are fanned out like three playing cards. These rings are on the upper edge of the trough, below them on the lower edge, and actually positioned in the mouth of the groove is tyrosine 56. Coleman et al. (1976) note from their NMR data that in the uncomplexed protein a number of tyrosyl proton resonances show upfield shifts suggesting some ring current effects due to stacking. They hypothesized that the tyrosyl residues involved mith be in some organized array, such as we observe. These resonances are lost on oligonucleotide binding, suggesting a disruption of the pattern as the residues begin interacting with the bases of the DNA.

Tyrosine 26 is near the β bend between strands 1 and 2 of the antiparallel β sheet. This bend appears to be a very flexible elbow of density extending out away from the central mass of the molecule and making up one end of the binding region. Even in the crystal, it projects into a large solvent area and seems to be rather mobile and free to move. It is the only tyrosine that we were able to iodinate in the crystal.

We noted that three of the tyrosines in the molecule, 26, 41, and 56 fall adjacent or one removed from a proline residue. The backbone structure of the protein is engaged in β structure and one would expect that this hydrogen bonding network might restrict the freedom of many bulky side groups. By virtue of their proximity to a natural structure disrupting amino acid, proline, these three tyrosines are endowed, however, with more freedom than they might otherwise enjoy. Because of the proline residues, the tyrosine side chains can rotate from one side of the sheet to the other through the trap door created by the neighbor.

Cysteine 33 is on the inside surface of the binding groove and could certainly interact with the DNA strand. In the conformation that we observe, however, the -SH group is turned up into the interior of the molecule away from the solvent. It is not in contact with the neighboring tyrosine 34. Although inaccessible to the bulkier Ellmans' reagent, the single cysteine can be reacted with mercuric chloride. Mercuration of cysteine 33 prevents nucleotide or DNA binding to the protein and conversely, complexation with oligonucleotides prevents reaction with mercury (Anderson et al. 1975). This is consistent with its placement in the binding groove, as is the finding that this -SH group can be photo cross-linked to thymidine residues of bound nucleic acids.

Acetylation of the ϵ-amino groups of the seven lysyl residues destroys the binding of gene 5 protein to oligonucleotides and DNA, but these groups are not protected by the presence of DNA from reaction (Anderson et al. 1975). In addition, NMR spectra show that the ϵ-CH$_2$ groups do not undergo chemical shift or line broadening upon complexation and seem to remain highly mobile. This was interpreted as indicating that the ϵ-amino groups provide a neutralizing charge cloud for the negative phosphate backbone of the nucleotide, but do not form highly rigid salt bridges or hydrogen bonds (Coleman et al. 1976). Resonances from the δ-CH$_2$ groups of the arginyl residues do undergo chemical shifts and line broadening on DNA complexation, and this could represent direct interaction of the guanidino groups with the phosphate backbone.

The DNA binding trough has over its interior surface a fairly large number of basic amino acid side chains which, because of the length and flexibility of these residues, reach into the groove though originating at disparate locations within the molecule. The basic residues most clearly apparent in the cleft are arginines 21, 80, and 82, and lysines 24 and 46. These are all found on the interior surface of the trough, so that the cleft is also something of a positively charged pocket in the protein. It should be noted that other basic amino acids could conceivably approach the binding region but in the conformation we observe in the crystal they are elsewhere. In particular, arginine 16 and lysine 46 are certainly in close proximity to the interface, but we see them turned away from the groove rather than toward it.

The DNA binding cleft of about 30 Å length and formed principally by the underside of an antiparallel β sheet has been tentatively identified in the gene 5 protein (Fig. 7). This assignment is based on the general shape and size of the groove and the distribution of amino acid residues on its surface that have been implicated in DNA binding by NMR, optical spectra, and chemical modification studies in solution.

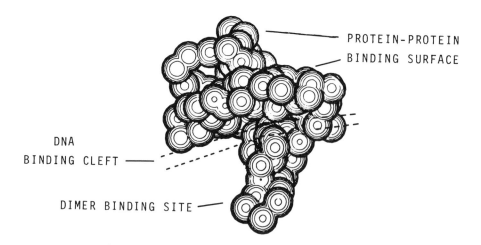

Fig. 7. Proposed interaction regions of gene 5 DNA binding protein. The postulated binding sites of the gene 5 protein are indicated schematically

The binding cleft is very interesting in that the positively charged residues of lysine and arginine are distributed predominantly over the innermost surface, while the aromatic residues are arrayed primarily along the exterior edges. Thus it appears that the negatively charged polyphosphate backbone of the single-stranded DNA is first recognized by the protein and that it is drawn and fixed to the interior of the groove by the charge interactions. This is followed by rotation of the aromatic groups down and into position to stack upon the bases of the DNA which are now splayed out toward the exterior of the protein. This is consistent with the finding of Day (1973) from micrograph and spectral data that the DNA in the gene 5 complex is completely unstacked and stretched along the filament axis. The cleft then acts as an elongated pair of jaws that draws the DNA between them by charge interactions involving the phosphates with the interior lysines and arginines. The jaws then close around the DNA strand through small conformation changes and the rotation or aromatic side chains into position to stack upon the purines and pyrimidines. That small, but not gross, conformation changes occur in the protein upon DNA binding is in agreement with the NMR studies of Coleman et al. (1976) on the α-CH and oliphatic methyl groups, which suggest that gene 5 must contain a large percentage of fixed structure without large regions of flexible polypeptide chain. Day's (1973) spectral evidence also indicates that only small changes occur in overall protein structure on binding.

That the interaction between gene 5 protein and DNA is to a great extent electrostatic is clear from the finding that moderate divalent and monovalent cation concentrations cause the complex to dissociate and that binding capacity is lost when the arginines and lysines are chemically modified (Anderson et al. 1975). The involvement of the aromatic groups, however, is also quite clear from the NMR and spectral data. The minor conformation changes in the gene 5 protein involving other residues and possibly even main chain atoms, is consistent with the physical and chemical studies. Therefore, although our binding mechanism is somewhat speculative, it is to our knowledge entirely consistent with the structure as we visualize it, and the evidence at hand from noncrystallographic analyses.

The binding of the gene 5 DNA unwinding protein to deoxy oligonucleotides is nonspecific in that complexation will occur with oligomers of any sequence. It is found, however, that the gene 5 protein binds oligomers of different sequence with differing affinities and these may vary over two orders of magnitude (Coleman et al. 1976). Thus, the protein does distinguish between different binding possibilities, and this could be the basis for recognition of specific nucleation sites on the fd DNA. Until the structure of a gene 5 oligomer complex has been directly visualized, we will not be able to confidently establish the interactions which confer the differential binding affinities. The possibility exists, however, that when the interactions of the gene 5 protein with specific sequences of DNA are completely defined, they will suggest how, by only minor structural alterations, a high degree of recognition specificity might be achieved.

Acknowledgments. This research was supported by grants from the NIH, NSF, NASA, and the American Cancer Society. A.H.-J.W. is supported by a grant from the M.I.T. Cancer Center. F.A.J. is an NIH postdoctoral fellow.

References

Alberts B, Frey L, Delius H (1972) Isolation and characterization of gene 5 protein of filamentous bacterial viruses. J Mol Biol 68: 139–152

Anderson RA, Nakashima Y, Coleman JE (1975) Chemical modifications of functional residues of fd gene 5 DNA-binding protein. Biochemistry 14: 907–917

Cavalieri S, Goldthwait DA, Neet K (1976) The isolation of a dimer of gene 9 protein of bacteriophage fd. J Mol Biol 102: 713–722

Coleman JE, Armitage IM (1980) Biochemistry (in press)

Coleman JE, Anderson RA, Ratcliffe RG, Armitage IM (1976) Structure of gene 5 protein-oligo-deoxynucleotide complexes as determined by ^{1}H, ^{19}F and ^{31}P nuclear magnetic resonance. Biochemistry 15: 5419–5430

Cuypers T, van der Oudera FJ, DeJong WW (1974) The amino acid sequence of gene 5 protein of bacteriophage M 13. Biochem Biophys Res Commun 59: 557–563

Day LA (1973) Circular dichroism and ultra-violet absorption of a deoyribonucleic acid binding protein of filamentous bacteriophage. Biochemistry 12: 5329–5339

Dunker AK, Anderson EA (1975) The binding of the fd gene-5 protein to single-stranded nucleic acid. Biochim Biophys Acta 402: 31–34

McPherson A, Molineux I, Rich A (1976) Crystalization of a DNA unwinding protein: Preliminary X-ray analysis of fd bacteriophage gene-5 product. J Mol Biol 106: 1077–1081

McPherson A, Jurnak FA, Wang AH-J, Molineux I, Rich A (1979) Structure at 2.3 Å resolution of the gene 5 product of bacteriophage fd: A DNA unwinding protein. J Mol Biol 134: 379–400

Nakashima Y, Dunker AK, Marion DA, Konigsberg W (1974) The amino acid sequence of a DNA binding protein, the gene 5 product of fd filamentous bacteriophage. FEBS Lett 40: 290–292

Oey JL, Knippers R (1972) Properties of the isolated gene-5 protein of bacteriophage fd. J Mol Biol 68: 125–128

Pratt D, Laws P, Griffity J (1974) Complex of bacteriophage M 13 single-stranded DNA and gene 5 protein. J Mol Biol 82: 425–439

Pretorius HT, Klein M, Day LA (1975) Gene V protein of fd bacteriophage. Dinner formation and the role of tyrosyl groups in DNA binding. J Biol Chem 250: 9262–9269

Salstrom JS, Pratt D (1971) Role of coliphage M 13 gene 5 in single-stranded DNA production. J Mol Biol 61: 489–501

2. Recognition of Nucleic Acids and Chemically-Damaged DNA by Peptides and Protein

C. HELENE

A. Introduction

Recognition of nucleic acid base sequences or nucleic acid structures is a fundamental process at every step of genetic expression. Very specific base sequences are recognized by operators, RNA polymerases, restriction endonucleases . . . Specific recognition of single stranded nucleic acids is achieved for example by helix destabilizing proteins which are involved in DNA replication, repair, and recombination. The secondary and tertiary structure of ribonucleic acids certainly plays a key role in the recognition of tRNA's by aminoacyl—tRNA synthetases, of ribosomal RNA's by ribosomal proteins . . .

Protein—nucleic acid complexes may be divided into two general classes depending on whether or not complex formation gives rise to a chemical (enzymatic) reaction. If no enzymatic reaction takes place (as, e.g., in the association of repressor with operator) the questions which have to be answered deal with the *selective recognition* of a nucleic acid base sequence or that of the secondary or tertiary structures of the nucleic acid. On the contrary, if the association between a protein and a nucleic acid leads to a chemical reaction in the latter molecule, then the specificity of the reaction — which is the biologically important event — might rest upon *kinetic parameters* of steps following the association of the two molecules. For example, binding of tRNA's to aminoacyl—tRNA synthetases does not appear to be very highly specific but the aminoacylation reaction is much more selective. It has been postulated that a conformational change occurs in the complex formed by a synthetase with its cognate tRNA leading to the correct positioning of the terminal adenosine which is required for efficient amino acid transfer. In the complex formed by a noncognate tRNA this conformational change does not take place, this resulting in a strong reduction of the aminoacylation rate (Krauss et al. 1976; Rigler et al. 1976). This does not necessarily mean that conformational changes are unimportant in selective recognition processes. For example, *lac* repressor binding to DNA induces a conformational change in the latter as shown by circular dichroism studies (Maurizot et al. 1974). A mutant repressor (amino acid exchange in position 16) whose binding to operator is reduced by several orders of magnitude does not induce such a conformational change (J.C. Maurizot, unpublished results). It

Centre de Biophysique Moleculaire, CNRS, 1A, avenue de la Recherche Scientifique, 45045 Orleans Cedex, France and
Laboratoire de Biophysique, Museum National d'Histoire Naturelle, 61, rue Buffon, 75005 Paris, France

is not yet known whether the inability of this mutant protein to change the conformation of DNA is the cause or the consequence of weaker binding, as compared to the wild-type protein. But one has to keep in mind that association between a protein and a nucleic acid might involve several steps including equilibria between different conformers after the initial bimolecular step. Conformational changes might also contribute to the large entropy increase which accompanies protein–nucleic acid associations (Helene 1976).

The recognition of a nucleic acid may also involve a cooperative binding of the protein. If this is the case then the interpretation of specificity must take into account the cooperativity of protein interactions. On a thermodynamic basis, what should be compared are the association constants for the binding of a single protein molecule to an empty site not surrounded by occupied sites. For example helix destabilizing proteins bind to single stranded nucleic acids in a cooperative way. The cooperativity factor can reach high values: $\sim 10^3$ in the case of the protein coded for by gene 32 of phage T4 (Kelly et al. 1976).

B. Interactions Between Functional Groups

We have been interested for several years in an investigation of the molecular interactions which are involved in the recognition of nucleic acid structures and base sequences by proteins. Electrostatic, hydrogen bonding, stacking, and hydrophobic interactions provide the basis for direct "recognition" between functional groups belonging to the two macromolecules. Indirect interactions mediated through divalent metal cations might also play an important role allowing the interacting regions of the two partners to come into register inside the complex.

I. Electrostatic Interactions

Electrostatic interactions are involved in all protein–nucleic acid associations investigated so far. Dissociation can usually be achieved by increasing the ionic strength and this phenomenon has been used to determine the number of electrostatic bonds which are involved in complex formation (Record et al. 1976). Ion pair formation is accompanied by the release of ions initially condensed on the polyribose-phosphate (polyelectrolyte) backbone. Also water molecules are most probably released during the process of association. Ion and water release probably make the highest contribution to the entropy increase upon complex formation (Hélène 1976).

The presence of stoichiometric amounts of Zn^{++} ions in many enzymes participating in the metabolism of nucleic acids suggests that these cations could be involved in the association of the protein with the nucleic acid as well as in the catalytic step. A recent study has demonstrated that Zn^{++} ions can mediate interactions between polynucleotides and polypeptides containing glutamic acid residues such as $(Glu-Tyr-Glu)_n$ (Hélène 1975; Béré and Hélène 1979). Due to repulsive electrostatic effects between negative charges there is no direct interaction between these two polymers. However in the presence of

Zn^{++} ions ternary complexes are formed which are characterized by a simultaneous chelation of Zn^{++} to two glutamic acid residues, one phosphate group and one nucleic acid base. There is a selectivity in Zn^{++} binding to nucleic acid bases which decreases in the order $C > A > G$ (U or T do not show any binding) (Zimmer et al. 1974). Therefore metal cations may not only be involved in the association between proteins and nucleic acids but they may also participate in the specificity of the association.

II. Hydrogen Bonding Interactions

Hydrogen bonding interactions have been postulated to play a key role in the recognition of nucleic acid base sequences by proteins (Bruskov 1975; Hélène 1976; Seeman et al. 1976; Hélène 1977a). The association of several amino acid side chains with nucleic acid bases has been investigated in organic solvents, mostly by nuclear magnetic resonance. The emphasis has been on those amino acid side chains which can form pairs of hydrogen bonds with nucleic bases. Some of these associations can be highly specific. It has been shown, for example, that carboxylate anions (Glu or Asp) can form a strong pair of hydrogen bonds with guanine only (Lancelot and Hélène 1977). The two hydrogen bonds involve the N(1)H and $NH_2(2)$ groups of guanine. None of the other bases is able to form such a complex. Moreover, the stability of this guanine–carboxylate "pair" is about 30 times higher than that of the guanine–cytosine pair. As a matter of fact carboxylate anions displace cytosine from the G- - -C complex formed in an organic solvent (dimethylsulfoxide). Adding water to the organic solvent reduces the equilibrium constants for both G- - -carboxylate and G- - -C associations but does not change markedly their ratio.

Arginine could also play an important role in the recognition process. The side chain of this amino acid can form pairs of hydrogen bonds with cytosine in single stranded structures and with guanine both in single stranded structures and in G- - -C base pairs (Hélène 1977a). However these interactions are weak as compared to the guanine–carboxylate association. Moreover, arginine possesses a positive charge which makes it a very good condidate for electrostatic binding to phosphate groups. A model was proposed to overcome this difficulty (Hélène 1977a). It was hypothesized that the positive charge of Arg could be neutralized by Glu or Asp residues inside the protein structure. The side chain of arginine in this Arg/Glu(Asp) ion pair would still be able to form two hydrogen bonds with G or C. On the basis of model-building studies the simplest model peptide which was suggested to form such an ion pair was the dipeptide Arg-Glu (Arg-Asp should not be able to form a coplanar system of hydrogen bonds in the ion pair structure due to the too short chain of Asp as compared to Glu). A recent NMR study has provided evidence to support this hypothesis (Lancelot et al. 1979). An ion pair with two coplanar hydrogen bonds is formed between the side chains of Arg and Glu in the dipeptide Arg-Glu. This dipeptide interacts with guanine which disrupts the interaction between the side chains and binds strongly to the carboxylate group of Glu. However, if the association between guanine and carboxylate is prevented by methylation of the guanine amino group, then the ion pair Arg-Glu interacts with guanine through its arginine side chain as expected from the model.

III. Stacking Interactions

Stacking interactions involving aromatic amino acids and nucleic acid bases appear to be strongly favored in single stranded polynucleotides or nucleic acids (Toulmé and Hélène 1977; Mayer et al. 1979). Oligopeptides containing tyrosine, tryptophan or phenylalanine together with lysyl residues have been shown to bind to nucleic acids or polynucleotides at low ionic strength. Proton magnetic resonance studies have provided evidence for stacking interactions between the aromatic amino acid and nucleic acids bases. However although stacking is very efficient when the nucleic acid or the poly-nucleotide is single stranded, the different aromatic amino acids have very different stacking efficiencies in double-stranded DNA. The extent of stacking decreases in the order Trp > Phe > Tyr. Most — if not all — of the oligopeptides containing tyrosine do not exhibit any stacking interaction when bound to a double helix (Mayer et al. 1979).

Fluorescence spectroscopy can be conveniently used to study the binding of oligo-peptides containing aromatic amino acids to nucleic acids. Stacking of tryptophan with bases appears to be the only interaction which accounts for the quenching of the fluo-rescence of this aromatic residue. In the case of tyrosine and phenylalanine both stacking interactions and energy transfer to the bases contribute to fluorescence quenching. The latter mechanism does not require a direct interaction between the aromatic amino acid and the purine or pyrimidine base, since energy transfer results from a long-range coupl-ing between transition dipole moments of the donor and acceptor molecules. Therefore, fluorescence spectroscopy can be used to provide direct evidence for stacking only in the case of tryptophan (Hélène 1977b). Fluorescence data have been analyzed according to a two-step model (Brun et al. 1975)

$$\text{Trp-containing peptide + nucleic acid} \xrightleftharpoons{K_1} \text{Complex I} \xrightleftharpoons{K_2} \text{Complex II.} \qquad (1)$$

In complex I the peptide is bound in such a way that tryptophan remains free of interaction with the bases ("outside" binding). Complex II is characterized by a stacking of the aromatic residue with bases. Both complexes involve electrostatic interactions and their concentrations depend on ionic strength. However this ionic strength dependence is the same for the two complexes. In other words, the value of K_1 decreases very rapidly when the ionic strength increases, whereas K_2 is not affected. It should be noted that K_2 is a unitless quantity which represents the ratio of the concentrations of the two complexes $K_2 = (II)/(I)$. The values of K_1 and K_2 can be determined from the analysis of fluorescence data at low degrees of saturation of the nucleic acid by the oligopeptide.

In agreement with proton magnetic resonance data, the value of K_2 is much higher for denatured DNA ($K_2 \cong 5$) than for the native double helix, ($K_2 \cong 0.3$) indicating a strongly favored stacking in single-stranded denatured DNA (Toulmé and Hélène 1977).

C. Recognition of Chemically Modified DNA's

Since stacking interactions appear specific for single-stranded nucleic acids, it was of interest to determine how a DNA containing unpaired regions would behave with respect to binding of tryptophan-containing oligopeptides. In order to introduce unpaired re-gions in DNA we have used three methods:

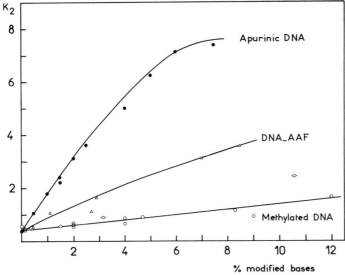

Fig. 1. Plot of K_2 values *versus* percentage of modified bases. K_2 represents the ratio at equilibrium of the concentration of complex involving stacking interactions (complex II) to that of "outside" complex (complex I) according to Eq. (1) (see text).
Native DNA was methylated by reaction with dimethylsulfate (*white circles*) and then depurinated by heating (*black circles*) (T. Alev and J.J. Toulme, unpublished results). DNA-AAF (*white triangles*) is native DNA which has been reacted with N-acetoxy-2-acetylaminofluorene (F. Toulmé, C. Hélène, R. Fuchs and M. Daune, 1980, Biochemistry 19: 870–875). The *diamonds* refer to native DNA which has been UV irradiated (see Tourmé, J.J., Charlier, M and Hélène, C., Proc. Natl. Acad. Sci. USA 71, 3185–3188, 1974)

1. UV irradiation which creates mostly pyrimidine dimers containing cyclobutane rings. Due to the local distortion of the double helix resulting from dimer formation a few base pairs are disrupted.

2. Chemical reaction of DNA with N-Acetoxy-2-Acetylamino-fluorene (AAAF) which leads to the formation of covalent adducts with guanine at the C(8) and $NH_2(2)$ positions with loss of the acetoxy group. The resulting product (DNA-AAF) possesses locally unpaired regions which lead to a destabilization of DNA (Fuchs and Daune 1972).

3. Chemical reaction of DNA with dimethylsulfate and subsequent depurination of the methylated bases (guanine is methylated at position N(7) and adenine at position N(3). Depurinated DNA has a lower stability than native DNA although the structure around "holes" is not known.

The binding of the oligopeptide Lys-Trp-Lys to these three modified DNA's was investigated by fluorescence spectroscopy. The results obtained for the K_2 values are summarized in Fig. 1. It can be seen that the value of K_2 increases with the level of base modification but that the phenomenon depends on the nature of the chemical damage of the DNA. The highest increase is observed for depurinated DNA which gives K_2 values higher than those obtained with denatured DNA when the level of depurination reaches about 4% (Alev and Toulmé, unpublished results).

One should expect that the electrostatic contribution to the association reaction changes little in the damaged regions due to the absence of separation of the two strands.

This should result in only small perturbations of the electrostatic potential of DNA as compared to the double helix. Consequently K_1 should not change markedly with the level of modification. This is what has been experimentally observed.

The K_2 value determined for a DNA containing locally unpaired regions is an average value over native and damaged regions. If K_1 is assumed to have the same value for native and damaged regions in DNA, the average K_2 value depends on the absolute K_2 value for damaged regions and on the extent of the local perturbation of the DNA structure. It should be kept in mind that the overall binding constant for peptide binding to DNA is K_1 $(1 + K_2)$. Therefore, an increase of K_2 leads to an increase of the overall binding constant resulting in a specific association of the oligopeptide with the damaged region. It is not possible to determine the exact K_2 value for damaged regions. An estimation can be made however from the corresponding DNA at the highest level of chemical modification. With UV irradiated DNA and DNA-AAF the K_2 value does not seem to be markedly different from that of denatured DNA. With depurinated DNA the K_2 value appears to be slightly higher than with denatured DNA. This might be due to the fact that apurinic sites strongly favor insertion of tryptophan whose aromatic indole ring has the same size as a purine. As compared to the situation in native DNA, no supplementary energy has to be provided to unstack bases since the "hole" left by the eliminated purine can accommodate the indole ring without further unstacking.

The results briefly described above show that aromatic residues of proteins should be able to discriminate between single-stranded and double-stranded regions in a nucleic acid. Stacking interactions of these aromatic residues with nucleic acid bases are strongly favored in single strands with tyrosine having a more pronounced discriminating power than tryptophan or phenylalanine. The binding of small oligopeptides to chemically modified DNA's shows that stacking is strongly enhanced in the damaged regions. One might therefore contemplate the possibility that proteins which recognize defects in DNA could make use of their aromatic residues to anchor along the polyribosephosphate backbone by inserting these residues between the bases of locally unpaired regions around defects. The three types of chemical modifications that have been described above affect different bases and induce different structural modifications. They all lead to a similar qualitative behavior as far as stacking interactions are concerned. However, they show large quantitative differences, with apurinic DNA exhibiting the largest increase in stacking efficiency.

Chemical damages in DNA are recognized by specific proteins during repair processes. Endonucleases specific for different damages have been described, including UV endonucleases which act on UV irradiated DNA, and apurases which react with depurinated DNA. An endonuclease specific for DNA which has been reacted with N-acetoxy-N-2-acetylamino-fluorene has also been postulated although not identified. These endonucleases recognize the damaged region in the DNA and make a single-strand break in the strand containing the defect. Then a common excision repair process which involves several enzymes (exonucleases, DNA polymerases, ligases) removes the damaged region and refills the gap using the opposite strand as a template.

Single-strand specific proteins have also been purified, which are required for DNA replication and recombination. Such proteins may also play an important role during repair processes by protecting from endonuclease attack the transiently formed single-stranded regions in a DNA. We have shown that the protein coded for by gene 32 of phage T4 does bind to UV-irradiated DNA and to DNA-AAF (Toulmé, unpublished results). Upon UV irradiation of the complex formed by T4 gene 32 protein and UV-irradiated DNA a photosensitized splitting of thymine dimers is observed (Hélène et al. 1976). This photochemical reaction, which was also observed with Lys-Trp-Lys, has been taken as evidence that at least one tryptophan residue of the protein was stacked with nucleic acid bases in the regions containing thymine dimers.

In conclusion, there is good evidence to suggest that many proteins or enzymes could utilize their aromatic residues to recognize selectively either single-stranded nucleic acids or single-stranded regions in a nucleic acid. This might also apply to the recognition of tRNA's by aminoacyl–tRNA synthetases or of ribosomal RNA's by some of the ribosomal proteins. However, the methods which have been applied in the investigation of oligopeptide binding to nucleic acids might not be applicable to biological systems. Nuclear magnetic resonance spectroscopy has been recently used to show that tyrosyl and phenylalanyl residues of the protein coded for by gene 5 of phage fd are involved in stacking interactions in gene 5 protein–oligonucleotides complexes (Coleman and Armitage 1978). This technique might be difficult to apply to more complex systems. Fluorescence quenching is always observed as a consequence of stacking interactions of aromatic amino acids with nucleic acid bases (Montenay-Garestier and Hélène 1971). However, this quenching could also result from a conformational change in the protein or from energy transfer of noninteracting tyrosines or phenylalanines to nucleic acid bases (tryptophan fluorescence cannot be quenched by such a mechanism). It is noteworthy that in most – if not all – protein–nucleic acid complexes investigated so far a quenching of tryptophan or tyrosine fluorescence has been observed (Hélène 1977b). When proteins bind to ligands other than nucleic acids, both quenching and enhancement (or no effect) have been observed.

We have recently developed a new spectroscopic probe of stacking interactions which rests upon the strong effect of heavy atoms on the spectroscopic properties of the indole ring (Hélène 1979). Using 5-mercuripyrimidines, we have shown that stacking with indole derivatives induces three very characteristic effects:

1. a quenching of indole fluorescence,
2. an enhancement of intersystem crossing which is accompanied by an increase in phosphorescence intensity,
3. a drastic shortening of the phosphorescence lifetime which is reduced by more than three orders of magnitude.

The three effects are also observed when Lys-Trp-Lys binds to poly-5-mercuriuridylic acid. Extension of this new methodology to real protein–nucleic acid complexes is under way. It should provide a rapid and convenient method to probe the existence of stacking interactions in a way similar to that developed to probe for the presence of tryptophyl residues in the binding site of sugar-binding proteins (Monsigny et al. 1978).

Acknowledgments. I wish to thank all present and former collaborators who have contributed to the work described in this review: E. Alev, A. Béré, F. Brun, M. Charlier, J.L. Dimicoli, M. Durand, G. Lancelot, J.C. Maurizot, R. Mayer, T. Montenay-Garestier, J.J. Toulmé.

References

Béré A, Hélène C (1979) Interactions between nucleic acids and polypeptides mediated by metal cations. Biopolymers 18: 2659–2672

Brun F, Toulmé JJ, Hélène C (1975) Interactions of aromatic residues of proteins with nucleic acids. III. Fluorescence studies of the binding of oligopeptides containing tryptophan and tyrosine residues to polynucleotides. Biochemistry 14: 588–563

Bruskov VI (1975) The possibility of recognition of nucleic acid bases by amino acids and peptides with the help of hydrogen bonds. Mol Biol (Moscow) 9: 304–309

Coleman JE, Armitage IM (1978) Tyrosyl-base-phenylalanine intercalation in gene 5 protein-DNA complexes. Proton nuclear magnetic resonance of selectively deuterated gene 5 protein. Biochemistry 17: 5038–5045

Fuchs R, Daune M (1972) Physical studies on deoxyribonucleic acid after covalent binding of a carcinogen. Biochemistry 11: 2659–2666

Hélène C (1975) Metal ion-mediated specific interactions between nucleic acid bases of polynucleotides and amino acid side chains of polypeptides. Nucl Ac Res 2: 961–969

Hélène C (1976) Specific recognition of nucleic acids by proteins. Studia Biophysica 57: 211–222

Hélène C (1977a) Specific recognition of guanine bases in protein-nucleic acid complexes. FEBS Lett 74: 10–13

Hélène C (1977b) Mechanisms of quenching of aromatic amino acid fluorescence in protein-nucleic acid complexes. In: Pullman B, Goldblum N (eds) Excited states in organic chemistry and biochemistry, pp 65–78. Reidel, Dordrecht

Hélène C (1979) Mise en évidence de l'effet d'atome lourd du 5-mercuri uracile sur les propriétés sepctroscopiques du tryptophane dans les complexes d'empilement. C R Acad Sci Paris 288D: 433–436

Hélène C, Toulmé F, Charlier M, Yaniv M (1976) Photosensitized splitting of thymine dimers in DNA by gene 32 protein from phage T 4. BBRC 71: 91–98

Hélène C, Mayer R, Lancelot G (1979) The recognition of nucleic acid structures and base sequences by proteins. Role of stacking and hydrogen bonding interactions. In: Rosenthal S, Bielka H, Coutelle Ch, Zimmer Ch (eds) Gene functions, vol 51, pp 63–69. FEBS Meeting, Dresden 1978. Pergamon, Oxford

Kelly RC, Jensen DE, von Hippel PH (1976) DNA "melting" proteins. IV. Fluorescence measurements of binding parameters for bacteriophage T 4 gene 32 protein to mono-, oligo-, and polynucleotides. J Biol Chem 251: 7240–7250

Krauss G, Riesner D, Maass G (1976) Mechanism of discrimination between cognate and non-cognate tRNAs by phenylalanyl-tRNA synthetase from yeast. Eur J Biochem 68: 81–93

Lancelot G, Hélène C (1977) Selective recognition of nucleic acids by proteins: the specificity of guanine interaction with carboxylate ions. Proc Natl Acad Sci USA 74: 4872–4875

Lancelot G, Mayer R, Hélène C (1979) Conformational study of the dipeptide arginylglutamic acid and of its complex with nucleic bases. J Amer Chem Soc 101: 1569–1576

Maurizot JC, Charlier M, Hélène C (1974) Lac-repressor binding to poly d(AT). Conformational changes. BBRC 60: 951–957

Mayer R, Toulmé F, Montenay-Garestier T, Hélène C (1979) The role of tyrosine in the association of proteins and nucleic acids. Specific recognition of single stranded nucleic acids by tyrosine-containing peptides. J Biol Chem 254: 75–82

Monsigny M, Delmotte F, Hélène C (1978) Ligands containing heavy atoms: Perturbation of phosphorescence of a tryptophan residue in the binding site of wheat germ agglutinin. Proc Natl Acad Sci USA 75: 1324–1328

Montenay-Garestier Th, Hélène C (1971) Reflectance and luminescence studies of molecular complex formation between tryptophan and nucleic acid components in frozen aqueous solutions. Biochemistry 10: 300–306

Record MT, Lohman TM, de Haseth PL (1976) Ion effects on ligand-nucleic acid interactions. J Mol Biol 107: 145–158

Rigler R, Pachmann U, Hirsch R, Zachau HG (1976) On the interaction of seryl-tRNA synthetase with tRNA^{Ser}. A contribution to the problem of synthetase-tRNA recognition. Eur J Biochem 65: 307–315

Seeman NC, Rosenberg JM, Rich A (1976) Sequence-specific recognition of double helical nucleic acids by proteins. Proc Natl Acad Sci USA 73: 804–808

Toulmé JJ, Hélène C (1977) Specific recognition of single-stranded nucleic acids. Interaction of tryptophan-containing peptides with native, denatured and ultraviolet-irradiated DNA. J Biol Chem 252: 244–249

Zimmer C, Luck G, Triebel H (1974) Conformation and reactivity of DNA. IV. Base binding ability of transition metal ions to native DNA and effect on helix conformation with special reference to DNA-Zn(II) complex. Biopolymers 13: 425–453

3. Specific Interaction of Base-Specific Nucleases with Nucleosides and Nucleotides

F. EGAMI, T. OSHIMA, and T. UCHIDA

A. Introduction

Twentytwo years and eleven years, respectively, have passed since we discovered RNases T_1 (Sato and Egami 1957) and U_2 (Arima et al. 1968). Both enzymes are now quite familiar, described in most textbooks of biochemistry together with RNase A as important tools for the structural analysis of RNA. This is because of their base specificity: RNase A is, as is well known, specific for pyrimidine nucleotides and, on the contrary, RNase U_2 is specific for purine nucleotides. RNase T_1 is the most specific, namely it is specific for guanylate.

The base specificity depends on the specific interaction between the enzyme proteins and respective nucleotides in RNA. So it offers an excellent model for the interaction of nucleic acids and proteins, one of the most important processes for chemical recognition in living systems.

We would like to describe briefly the specific interaction of base-specific nucleases with nucleosides and nucleotides, with special reference to RNases T_1 and U_2, mostly studied by our groups. Several reviews on the enzymes have been published (Egami et al. 1964; Egami and Nakamura 1969; Takahashi et al. 1970; Uchida and Egami 1971; Takahashi 1971a; Oshima and Imahori 1971).

B. Base-Specific Ribonuclease

I. Properties of RNases T_1 and U_2

Enzymatic and other basic properties of both enzymes are summarized in Tables 1 and 2, respectively.

As expected, the properties related to the function of both enzymes are fairly different from each other, in addition to the base specificity. The specific activity of RNase U_2 toward both RNA and best substrate in dinucleoside monophosphate is only about one-tenth of those of RNase T_1. However, both enzymes show very similar nature as proteins. They are very acidic, simple proteins having a similar molecular size and a similar content of the ordered structure. Their conformation is remarkably stable in comparison with other proteins and, further, they exert good reversibility after heating

Mitsubishi-Kasei Institute of Life Sciences, Machida-shi, Tokyo 194, Japan

Table 1. Enzymatic properties of RNases T_1 and U_2[a]

	T_1	U_2
pH optimum for RNA	7.5	4.5
N-2',3'-P	7.2	4.5
Base specificity	Guanine	Purine
Specific activity (units[b]/mg)	1.6×10^4	2.2×10^3
Molecular activity		
GpC	5.0×10^4	2.9×10^2
ApC	–	1.5×10^3
N-2',3'-P	ca. 10^2 (G)	ca. 10 (A)
Km (M)		
GpC	1.6×10^{-4}	5.3×10^{-4}
ApC	–	4.5×10^{-4}
N-2',3'-P	4.0×10^{-3}	
10^{-3} M Zn^{2+} Inhibition	+	–
Metal requirement	–	–

[a] Both are endonucleases producing 3'-nucleotide through 2',3'cyclic nucleotide
[b] One unit of RNase is provisionally defined as the amount of the enzyme causing an increase of 1.0 A_{260} upon measuring the absorbance increase of acid-soluble products from the commercial RNA under the standard assay conditions (Uchida and Egami 1967)

Table 2. Basic properties of RNases T_1 and U_2

	T_1[a]	U_2[b]
Sedimentation coefficient $S^o_{20,w}$	1.62S	2.02S[c]
Diffusion coefficient $D_{20,w}$	12.0×10^{-7} cm^{2}s^{-1}	$12.0_4 \times 10^{-7}$ cm^{2}s^{-1}[c]
Partial specific volume $\bar{v}$	(0.69_8^d)	0.67_2^c (calc. 0.69_7^d)
Molecular weight		
from amino acid sequence	11,085	12,280
Isoelectric point Ip		
from electrofocusing analysis	3.8[e]	3.3
Absorption spectrum		
Maximum	278 nm	279 nm
Minimum	251–252 nm	252 nm
Amax/Amin	3.6_3	3.7_4
$E^{0.1\%}_{280.1 \text{ cm}}$	1.9_1	1.5_5
$[\alpha]^{20}_D$	$-24°$	
Conformation		
α-helix	11%[f]	20%[g]
β-structure	38%[f]	35%[g]
	(pH 7.5, 25°C)	(pH 4.5, 25°C)
Sugar content	<0.5%	<0.175%
		(0.2 mole as Glc/mole protein)
Reaction with anti-RNase T_1[h]	+	–

[a] Takahashi et al. (1970) [b] Uchida, T. (unpublished) [c] Minato and Hirai (1979) [d] Calculated based on the amino acid analysis according to McMeekin et al. (1949) [e] Kanaya, S and Uchida, T. (unpublished) [f] Oshima, T.(unpublished) [g] Uchida and Machida (1978) [h] Uchida (1970)

or removal of the denaturing agents, such as urea, phenol etc. But neither enzymes have immunological relation to each other.

1. Base Specificity

The base specificity of both enzymes has been extensively investigated by the 3'-terminal analysis of digestion products of RNA and polyribonucleotides and by studies on the susceptibility of dinucleoside monophosphates or 2',3'-cyclic nucleotides to the enzyme. The results are summarized in Table 3 (Uchida and Egami 1971; Uchida 1971). The best substrate for RNases T_1 and U_2 is guanylate and adenylate, respectively. From the results in Table 3 the essential requirements for the preferred substrates of both enzymes are deduced as follows (Table 4): (1) nitrogen atom at N-7 of the purine base, (2) group at the 6 position (6-oxo in T_1 and 6-NH_2 in U_2), (3) no blocking at N-1 with large substituent.

Table 3. Base specificity of RNases T_1 and U_2

Position of substituent		Substituent	Base	T_1	U_2
6-	(Pu)	6-NH_2	A	–	+++
or 4-	(py)	6-Oxo	G	+++	++
			I	++	+
			X	+	–
		N^6-deriv.	ipA	–	
		6-S	s^6G	–	–
		4-NH_2	C	–	±
		4-Oxo	U	–	–
7-	(Pu)	7-N-deriv.	7-deaza I (or A)	–	–
			7-mG	–	–
1-	(Pu)	1-CH_3	1-mA (or G)	±	±
or 3-	(py)	1-O	1-oxyd A		
		CMC-deriv.[a]	CMC-G	–	–
			CMC-U	–	–
		1,2-linked	glyoxal-G	+	±
2-	(Pu)	N-2-deriv.	N-2-diMeG	±	–
		2-TNP[b]	TNP-G	–	–
8-	(Pu)	8-N	8-aza-G	++	
		8-Br	8-Br-G	++	
5-,6-	(Py)		ψU	–	–
			diHU	–	±
			5-O-AcU	–	+

[a] CMC: N-cyclohexyl-N^1-β-(4-methylmorpholinium)-ethylcarbodiimide p-toluene group
[b] TNP: Trinitrophenyl group

Table 4. Requirement for substrate

RNase	T_1	U_2
Best substrate	G	A
N-7	No blocking	No blocking
6-position	6-Oxo	6-NH_2
N-1	No large substituent group	

$\vdots$ Required region   R: ribosyl-P

Table 5. Michaelis constants of RNase T_1 and U_2 for purine nucleotidylnucleoside

	$T_1{}^a$	$U_2{}^b$
A-C		4.5×10^{-4}
A-G		10×10^{-4}
A-U		11×10^{-4}
A-A		2.0×10^{-4}
G-C	1.6×10^{-4}	5.3×10^{-4}
G-G	0.27×10^{-4}	10×10^{-4}
G-U	0.22×10^{-4}	18×10^{-4}
G-A	0.55×10^{-4}	2.1×10^{-4}

[a] Osterman and Walz (1978)
[b] Uchida and Machida (1978)

However, base specificity of cyclizing RNases is generally not absolute, but relative. The susceptibility of nucleotidyl linkages to RNase T_1 decreases in the order of $G \gg A > U > C$. But the cleavage of ApC with RNase T_1 occurs at a rate of about 1/150,000th of GpC cleavage (Uchida 1971). The order of the susceptibility to RNase U_2 is $A > G \gg C > U$ (Uchida et al. 1970), and the relative cleavage rates among four nucleotidyl linkages, ApC, GpC, CpC and UpC, are 1:1/5:1/3,000:1/23,000, respectively (Uchida and Machida 1978). Michaelis constants of RNases T_1 and U_2 for eight kinds of purine nucleotidylnucleosides (RpN) are shown in Table 5. Affinities of RNase T_1 for GpN are somewhat different from each other, depending on the second base moiety. Affinities of RNase U_2 for ApN are generally somewhat stronger than those for corresponding GpN. Km values of RNase U_2 rather decrease in the order of $U > G \gg C > A$ of the second base moiety in RpN independent of the first base. Assuming the existence of the secondary recognitions site for enzyme as reported about RNase T_1 by Walz et al. (Walz and Terenna 1976; Zabinski and Walz 1976), the subsite in RNase U_2 is suggested to interact more effectively with the amino bases than oxo bases.

RNase U$_2$ Cys-Asn-Ile-Pro-Glu-Ser-Thr-Asn-Cys-Gly-Gly-Asn-Val-Tyr-Ser-Asn-Asp-Asp-Ile-Asn-Thr-Ala-Ile-Gln-Gly-Ala-Leu-

RNase T$_1$ Ala-Cys-Asp-Tyr-Thr-Cys-Gly-Ser-Asn-Cys-Tyr-Ser-Ser-Ser-Asp-Val-Ser-Thr- - -Ala-Gln-Ala-Ala-Gly-Tyr-Gln-Leu-His-

Asp-Asp-Val-Ala-Arg-Pro-Asp-Gly-Asp-Asn-Tyr-Pro-His-Gln-Tyr-Tyr-Asp-Glu-Ala-Ser-Asp-Gln-Ile-Thr-Leu-Cys-Cys-Gly-Pro-Gly-

Glu-Asp-Glu-Glu-Thr-Val-Gly-Ser-Asn-Ser-Tyr-Pro-His-Lys-Tyr-Asn-Asn-Tyr-Glu-Gly-Phe-Asp-Phe-Ser-Val-Ser-Ser- - -Pro- - -

Ser-Trp-Ser-Glu-Phe-Pro-Leu-Val-Tyr-Asn-Gly-Pro-Tyr-Tyr-Ser-Ser-Arg-Asp-Asn-Tyr-Val-Ser-Pro-Gly-Pro-Asp-Arg-Val-Ile-Tyr-

- -Tyr-Tyr-Glu-Trp-Pro-Ile-Leu-Ser-Ser-Gly-Asp-Val-Tyr-Ser-Gly- - - - - - -Pro-Gly-Ser-Gly-Ala-Asp-Arg-Val-Val-Phe-

Gln-Thr-Asn-Thr-Gly-Glu-Phe-Cys-Ala-Thr-Val-Thr-His-Thr-Gly-Ala-Ala-Ser-Tyr-Asp-Gly-Phe-Thr-Gln-Cys-Ser

Asn-Glu-Asn-Asn- - -Glu-Leu-Ala-Gly-Val-Ile-Thr-His-Thr-Gly-Ala- - -Ser-Gly-Asn-Asn-Phe-Val-Glu-Cys-Thr

Disulfide bridges	RNase T$_1$	RNase U$_2$
	2-10	1 - 53
	6-103	54 - 95
		9 - 112

Fig. 1. Comparison of primary structure between RNase T$_1$ and U$_2$

2. Primary Structure

The primary structures of RNases T_1 and U_2 were determined by Takahashi (1965) and by Sato and Uchida (1975a), respectively.

Both RNases T_1 and U_2 consist of a single polypeptide chain of 104 and 113 amino acid residues cross-linked by two and three disulfide bonds, respectively. Their covalent structures show some similarity in that the carboxyl terminal is close to the amino terminal region because of the disulfide bridge. However, the conformation of RNase U_2 might be different from that of RNase T_1 because the central parts of the molecule are linked to the amino terminal by other two disulfide bridges (Sato and Uchida 1975b).

Comparison of both primary structures are shown in Fig. 1. In addition to the close similarities of the N- and C-terminal regions, they have four common regions as indicated by *boxes*. All these four *boxes* contain penta- or hexa-peptides with homologous sequence, except some mutual conversions between similar amino acids, such as conversion from Val to Ile. All of the important amino acid residues, which from the numerous chemical modification works, as described below, have been proved to play a crucial role in the enzymatic function, are included in these four *boxes*. Moreover, the mutual location in the T_1 molecule of these four important amino acid residues, His-40, Glu-58, Arg-77 and His-92 is also applicable in the case of U_2 molecule: that is, His-40 and Glu-61 in U_2 are located not only in the same position from N-terminal as His-40 and Glu-58 in T_1, but also Arg-84 and His-100 in U_2 in the same position from C-terminal as Arg-77 and His-92 in T_1. These common regions are expected to be important for revealing the enzymatic function. However, the two enzymes have different specificity from each other, though they catalyze the transfer and hydrolysis of phosphate group in a quite similar way. Then it is probable that the common regions should be concerned with the catalytic site, and that specific regions in both enzymes should take part in binding site. The characteristic region (positions 59–62 in U_2) around Glu-61 attached to the homologous region might be regarded as one of the specific regions.

3. Chemical Modification

Information on the active site of both enzymes obtained by the various chemical modifications are summarized in Table 6a, b. Based on the data in Table 6a, b, we have the following idea about the role of some functional groups participating in the function and the maintenance of active conformation.

a) Amino Groups. Two amino groups in RNase T_1, α-NH$_2$ group of N-terminal Ala and ϵ-NH$_2$ group of Lys-41 (Uchida and Egami 1971) and single amino agroup in RNase U_2, α-NH$_2$ group of N-terminal Cys (Sato and Uchida 1974; Minato and Hirai 1979) are not essential for the function of RNase. But Lys-41 seems to be near the active site (Takahashi 1977).

b) Active Glu Residue. Takahashi et al. (1967) found that the complete loss of RNase T_1 activity toward both RNA and Guo-2',3'-P resulted from the selective esterification of γ-carboxyl group of Glu-58 by iodoacetate. RNase U_2 was also completely inactivated

Table 6a. Chemical modification of RNase T_1

Modification reagent	Modified residue	Extent of modification (mol/mol)	Remaining activity (%)	Binding ability to Guo-3'-P Ki(mM)	(mol/mol)	Cause of inactivation	References
Native	–	–	100	0.0076	0.87		Takahashi (1972)
a) Trinitrobenzenesulfonate	Ala-1 Lys-41	1.0 1.0	67	0.0071		No relation to the activity	Kasai et al. (1969)
Maleic anhydride	Ala-1 Lys-41	1.0 1.0	5	20-fold(Km for RNA)		Steric hindrance or interaction with His-40	Takahashi (1977)
b) Monoiodoacetate Tosylglycolate at pH 5.5	Glu-58 Glu-58	1.0 0.81	0.2 5.7	0.018	0.79	Selective carboxymethylation Selective carboxymethylation	Takahashi et al. (1967) Oshima and Takahashi (1976)
at pH 8.0	Glu-58 His	0.86	2.6			Alkylation of Glu-58 and His in mutually exclusive way	
c) Ozone	Trp-59	0.98	0 at pH 7.5 19 at pH 4.75	0.62		Alteration of pH optimum, of the active conformation in binding site	Tamaoki et al. (1978)
d) N-Acetylimidazole	Tyr	7.2 3.7	88 66			No relation to the activity	Kasai et al. (1977)
e) Phenylglyoxal	Arg-77 Ala-1	0.86 0.96	5	>1.0	0.05>	Selective phenylglyoxalation	Takahashi (1970b)
TPCK-trypsin in 2M urea (with a prior nitro-troponylation)	Arg-77→ Val-78		0	–	0.00	Complete loss of native conformation	Tamaoki et al. (1976)

	Residue					Description	Reference
f) Rose-bengal catalyzed photooxidation	His-27 0.34 His-40 0.37 His-92 0.84 Trp-59 0.78	1.32	28	0.23	0.22		Takahashi (1971)
Iodoacetamide-Guo-3'-P	His-27 0.49 His-40 0.78 His-92 0.72 Ala-1 1.0	2.04	7	0.12	0.35	Selective alkylation of His-40 and His-92	Takahashi (1970a, 1973, 1976)
+Guo-3'-P	His-27 0.35 His-40 0.23 His-92 0.46 Ala-1 0.78	1.17	41	—	0	Interaction of His-40 with Guo-3'-P	
After CM-lation · Iodoacetamide+Guo-3'-P	His-27 0.31 His-40 0.86 His-92 047 Ala-1 0.32	1.57	—	—	—	Interaction of His-92 with Glu-58	

Table 6b. Chemical modifications of RNase U$_2$

Modification reagent	Modified residue	Extent of modification (mol/mol)	Remaining activity (%)	Binding ability to Ado-3'-P Ki(mM)	(mol/mol)	Cause of inactivation	References
			100	0.023	0.80		Sato and Uchida (1975c)
a) Trinitrobenzene sulfonate	Cys-1	1.0	95			No relation to the activity	Sato and Uchida (1974)
2-methoxy-5-nitrotropone	Cys-1	1.0	70	0.1			Minato and Hirai (1979)
b) Monoiodoacetate	Glu-61	1.0	0.06		0.10	Selective carboxymethylation	Uchida and Sato (1973)
d) N-Acetylimidazole	Tyr	6.6	1.6		0	Conformational change	Uchida, T. (unpublished)
	Tyr	0.65	57.9		0.57	10% reduction of β-structure	
Tetranitromethane	Tyr-77 Tyr-106	0.62 0.75 }	95			No relation to the activity	Sato, S., and Uchida, T. (unpublished)
e) Phenylglyoxal	Arg	0.93	4.8			Polymerization of the molecule	Sato S, Sato, M, and Uchida, T. (unpublished)
TPCK-trypsin	{ U-peptide (Asp-75 ~ Arg-84) RNase U$_2'$		0.2			Conformational change due to removal of U-peptide	Sato, S, Uchida T. (unpublished)
f) Ethoxyformic anhydride	His Tyr	1.0 0.5	14		parallel with His	Ethoxyformylation of His	Sato and Uchida (1975d)
Rose-bengal catalyzed photooxidation	His Trp	1.28 0.64 }	13.8			Photooxidation of His	Uchida and Sato (1973)
Iodoacetamide	His-40 His-100 Cys-1 Tyr	0.52 053 0.27 0.46	19.1		0.48	Selective alkylation of His-40 and His-100	Uchida, T. (unpublished)

by selective incorporation of one mol of carboxymethyl group into protein molecule and the carboxymethylation proved to occur also at γ-carboxyl group of Glu-61 (Uchida and Sato 1973). The rate of inactivation of RNase U_2 by iodoacetate was maximal at pH 4.5, whereas that of RNase T_1 was maximal at pH 5.5. It suggests that pK value of Glu-61 in RNase U_2 molecule shifts to the region more acidic than that of Glue-58 in RNase T_1 molecule.

Both RNases T_1 and U_2 were markedly protected from inactivation with iodoacetate in the presence of substrate analogs, such as Guo-2'-P and Ado-2'-P, respectively, showing that the reaction took place at the active site of both enzymes. Now, it is sure that the active Glu-58 or Glu-61 plays a crucial role at the active site of enzyme, but the problem that it participates at the binding site or the catalytic site remains to be elucidated. In the case of RNase T_1, however, the following two results seem to support the view that active Glu-58 is at the catalytic site: (1) The carboxymethylated T_1 with iodoacetate retained the considerable binding activity to Guo-3'-P; (2) In the tosylglycolation at Glu-58, the protective effects of guanosine were less than those of Ado-3'-P and Cyd-3'-P, showing that the phosphate moiety of these nucleotides is near the carboxyl group of Glu-58 and prevents the interaction with the reagent (Oshima and Takahashi 1976).

While in the case of RNase U_2 — considering the larger contribution of phosphate moiety than of base moiety to the binding of nucleotide to RNase U_2 (Sato and Uchida 1975c) — it would be too hasty to conclude that active Glu-61 may participate in binding site only because of a considerable loss of the binding activity to Ado-2'-P (about 90%). Our preliminary data on the modification with 2'- or 3'-bromoacetyladenosine, bringing the bromoacetyl group instead of the phosphate group of adenylate (Uchida, T., unpublished), suggested the possibility that glutamic acid was selectively modified just like His-12 upon the modification of RNase A with 2'-, or 3'-bromoacetyluridine, and then glutamic acid must be at catalytic site.

c) Trp Residue. Both enzymes have a single Trp residue near the active Glu residue. In early studies with 2-hydroxy 5-nitrobenzyl bromide and N-bromosuccinimide, the Trp residue was not readily accessible to these reagents in aqueous solution, but in 8 M urea the reagents reacted with Trp-59 to form irreversibly inactivated enzyme, suggesting that the Trp residue in native T_1 is embedded in the interior of the molecule (Uchida and Egami 1971). Recently, Tamaoki et al. (1978) obtained kynurenine-T_1 by ozone oxidation. It retained the conformation of native enzyme, but with a decreased activity toward RNA and markedly reduced binding ability to Guo-3'-P. They concluded from this experiment that Trp-59 functions in maintaining the active conformation of the protein molecule, particularly keeping the active environment for functionally important groups involved in the binding of substrate, although no direct participation of Trp in the catalytic function of RNase T_1 appears. This conclusion seems to be applicable also to Trp-59 in RNase U_2 (Minato and Hirai 1979).

d) Tyr Residues. It is well known that RNase U_2 is unstable in the alkaline region over pH 9, suggesting the presence of some labile Tyr residues. RNase U_2 was rapidly inactivated by N-acetylimidazole (Uchida, T., unpublished). As shown in Table 7, due to the acetylation of six free tyrosyl residues exposed on the surface of RNase U_2, a partial

Table 7. Acetylation of RNase U_2 with N-acetylimidazole

| | Disappearance of Tyr residues (mol/mol) | Activity toward RNA (%) | Binding activity of A2'-P (%) | Change of conformation[a] | |
				α-helix (%)	β-structure (%)
Native RNase U_2	0	100	100	21.1_6	33.3_0
AcU_2 (5 min)	5.6	19.4	43	15.6_4	31.3_0
AcU_2 (90 min)	6.6	1.6			
Isolated AcU_2	6.1	7.5			
Renatured AcU_2	0.6_5	57.9	71	21.1_7	30.3_4

[a] This measurement was carried out by a JASCO Spectropolarimeter, Model J20

unfolding of the native conformation is considered to take place resulting in the inactivation. This inactivation could be almost reversed by rapid deacetylation with a high concentration of hydroxylamine, but not so completely by spontaneous deacetylation upon storage in a freezer. Accordingly, about one Tyr residue remained acetylated. The decrease of the α-helix content by acetylation was completely reversed by spontaneous deacetylation, whereas the decrease of the pleated sheet content by acetylation remained after spontaneous deacetylation. Thus, it is suggested from the result of spontaneous deacetylation that at least one of the ten Tyr residues is located near or at the end of the β-structure part and prevents complete reversion of molecular conformation and then enzymatic activity.

By partial nitration with tetranitromethane (Sato, S. and Uchida, T., unpublished), both Tyr-77 and Tyr-106 are concluded to be so reactive to the reagents without loss of activity, indicating they are not essential for enzymatic function and are at the surface of the molecule.

In RNase T_1 the similar results are obtained from physicochemical experiments (Iida and Ooi 1969; Pongs 1970; Campbell et al. 1976) and chemical modifications (Kasai et al. 1977): (1) Two of the nine Tyr residues are at the surface and easily accessible to the reagent without much loss of activity; (2) Additional two or three Tyr residues are close enough to the surface; (3) In 8 M urea all the Tyr residues are acetylated with nearly complete loss of activity, but the fully acetylated enzyme regains full activity by deacetylation, although the original conformation of the molecule appears to be not completely regenerated.

e) Arg Residue. Takahashi (1970b) showed with the modification by phenylglyoxal that the single Arg-77 residue in RNase T_1 appears to be present at or near the active site of the enzyme. However, the phenylglyoxalation of RNase U_2 resulted in the polymerization of protein molecule with the loss of one of the three Arg residues and the significant inactivation up to 4% of the original activity (Sato, S., Sato, M. and Uchida, T., unpublished). Thus, the cause of inactivation could not be concluded to be due to the modification of Arg residue or due to polymerization.

RNase T_1 blocked the ϵ-NH_2 group of Lys-41 was selectively cleaved at Arg-77 by tryptic digestion in the presence of 2 M urea (Tamaoki et al. 1976; Takahashi and

Inoue 1977). The modified T_1, in which single Arg-77 — Val-78 peptide linkage was absent, lost completely the original conformation with loss of enzymatic functions. Similarly, by the limited tryptic digestion RNase U_2 was split into two parts, a small peptide from Asp-75 to Arg-84 and the remaining bulk of the protein. This cleavage also resulted in the complete destruction of the original ordered structure (Sato, S. and Uchida, T., unpublished). Reconstitution of the active form by mixing the two parts was a failure.

f) His Residues. From various modification experiments a common result was derived in that one, or more likely two, of the three His residues in RNase T_1 and one of the two His residues in RNase U_2 were implicated at the active site. Actually, analysis of the modified protein indicated the various extents of modification of every His residue, while no selective modification of the specific His residue was observed. In RNase U_2, however, it appears that the modification of either one of the two His residues prevents modification of the other, showing the involvement of both His residues in the active site (Uchida, T., unpublished) (Table 6b). The assigment in which His residue participates at the binding or catalytic site is very difficult. His-27 in T_1 is considered as probably not involved in the active site because of the lesser reactivity with various reagents and no homologous sequence around His-27 in RNase U_2. Takahashi (1976) suggested that both His-40 and His-92 are involved at the active site, and the former is at the binding site and the latter at the catalytic site, because the extent of alkylation at His-40 was markedly reduced in the presence of Guo-3'-*P* and that at His-92 was reduced with a prior carboxymethylation of Glu-58 (Table 6a). More recently, Arata et al. (1979) derived just the reverse speculation from NMR studies as described below. In the case of RNase U_2, His-40 and His-100 may be considered to participate in the role corresponding to His-40 and His-92 in T_1, respectively.

II. Spectrophotometric Measurements of the Interaction Between RNase T_1 and Substrate Analogs

Interactions between an enzyme and a nucleotide or its related compounds can be studied by several physicochemical methods, such as ultraviolet absorption, circular dichroism, IR and Raman spectra, fluorescence, proton-NMR, ^{13}C-NMR, ^{15}N-NMR, ^{31}P-NMR, X-ray diffraction, action kinetics, equilibrium dialysis, gel filtration, calorimetry, and the like. The binding reaction between RNase T_1 and its substrate analogs has been studied using many physicochemical methods and a variety of information has been accumulated about the chemical nature of the interactions. In the following, we would like to focus discussion on results obtained by spectrophotometric measurements such as UV, CD and NMR.

1. Ultraviolet Absorption

a) Difference Spectra. Changes in UV absorption upon binding of RNase T_1 with Guo-2'-*P* were reported by Sato and Egami (1965) in the mid-1960's. This line of study has

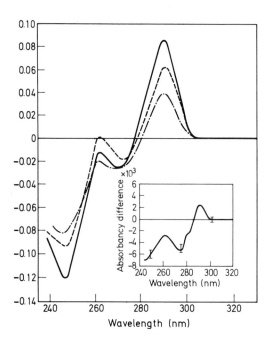

Fig. 2. Difference spectra of interaction of RNase T_1 with guanosine derivatives. The spectra were recorded at $25°C$, pH 5.5, using cells of 3 mm path length. ——: RNase T_1, 9.5×10^{-5} M + Guo-3'-P, 2.65×10^{-4} M. - - - -: RNase T_1, 7.0×10^{-5} M + Guo-2'-P, 2.2×10^{-4} M. -.-.-.-: RNase T_1, 9.2×10^{-5} M + Guo, 3.26×10^{-4} M. *Inset;* RNase T_1, 8.45×10^{-5} M + 9-methylguanine, 1.03×10^{-4} M. From Oshima and Imahori (1971a)

since been extended by several researchers (Irie 1970; Oshima and Imahori 1971a; Oshima and Imahori 1971d; Epinatjeff and Pongs 1972; Walz and Hooverman 1973; Walz, Biddlecombe and Hooverman 1975; Walz 1976; Walz and Terenna 1976; Walz 1977a, b; Sato and Uchida 1975).

Figure 2 illustrates examples of the difference spectrum recorded upon the binding of RNase T_1 with guanine derivatives. When RNase T_1 was mixed with Guo-2',3' or 5'-P, a strong peak at 291 nm and a deep trough at 247 nm were observed (Sato and Egami 1965; Oshima and Imahori 1971a; Epinatjeff and Pongs 1972; Walz and Hooverman 1973). When Guo or dGuo was used instead of the guanylates under similar conditions, the peak and the trough were greatly reduced, indicating that the affinities of these nucleosides to the enzyme are lower than those of the guanylates.

b) Assignment. It was found that the difference spectra shown in Fig. 2 resemble those for acidification of guanylates. Figure 3 shows some examples. By lowering the pH of a solution of Guo-3'-P, a peak at 292 nm and a trough around 245 nm were recorded on the difference spectrum. It is of particular interest to note that the difference spectrum of interaction between RNase T_1 and 9-methylguanine deviates somewhat from those of interactions of RNase T_1 with guanylates, but resembles the difference spectrum of 9-methylguanine upon protonation (Oshima and Imahori 1971a).

The molar extinction differences at 292 nm upon the protonation of guanosine and its nucleotides were determined to be 3.0×10^3 cm^{-1} · M^{-1}. The absorbancy difference upon binding was computed to be 3.1×10^3 cm^{-1} · (M of the complex)$^{-1}$ at 291 nm assuming that the RNase T_1 in the mixture was fully saturated when 2.65×10^{-4} M of Guo-3'-P was mixed with 9.5×10^{-5} M RNase T_1 (Kd value of Guo-3'-P is reported to be 7.7×10^{-6} M at pH 5.5 from gel filtration experiments; Takahashi 1972).

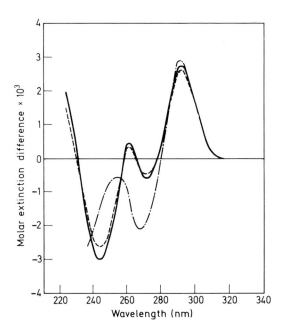

Fig. 3. Difference spectra resulting from changes of the pH of solution of the substrate analogs. *Solid line,* Guo-3'-*P* from pH 6.5 to 1.0; *broken line,* Guo from pH 6.5 to 1.35; *dots and dashes,* 9-methylguanine from pH 6.5 to 1.2. From Oshima and Imahori (1971a)

These observations indicate that the absorption change upon the binding is qualitatively and quantitatively similar to that upon protonation, and strongly support the idea that the spectral changes recorded upon the binding were primarily caused by the changes in spectrophotometric properties of the ligand. Conformational changes of Trp or Tyr residue(s) of the enzyme, if they occur at all, would have little or no detectable effect on the difference spectrum. The spectral changes could reflect the interaction between guanine base and the enzyme and be interpreted as representing the protonation of guanine at the active site.

It was reported that the protonation occurs at nitrogen seven (N-7) of the guanine base of guanosine (Miles, Howard and Frazier 1963) and of 9-methylguanine (Sobell and Tomita 1964) in acid solution. Based on these reports and the difference spectra shown in Figs. 2 and 3, Oshima and Imahori (1971a) concluded that N-7 of the guanine base is protonated in RNase T_1-substrate analog complex. This conclusion was supported by Epinatjeff and Pongs (1972). Watz and Hooverman (1973) also confirmed the results qualitatively. However, they reported quantitative difference between $\Delta\epsilon_{max}$ of the binding and that of protonation of a ligand, and concluded that the absorbancy change at around 290 nm upon the binding may reflect, in addition to N-7 protonation of guanine base, some other interaction(s) involved in the enzyme–ligand complex formation.

c) Dissociation Constants. Assuming that the molar absorbancy difference of the binding in the vicinity of 290 nm is identical to that of protonation of guanosine, that is, $3.1 \times 10^3 \text{ cm}^{-1} \cdot (\text{M of the complex formed})^{-1}$, Oshima and Imahori (1971d) estimated the amount of the complex formed in a mixture. A Scatchard plot of the interaction be-

Table 8. Dissociation constants of various RNase T_1-substrate analog complexes

	Kd (M)						$\Delta G°$ [a] (cal)
Guo-2'-P	0.9×10^{-5}	6.9×10^{-6}	7.5×10^{-5}	6.5×10^{-6}	2.4×10^{-5}	3.4×10^{-5}	6.9×10^3
Guo-3'-P	1.2×10^{-5}	1.9×10^{-5}	1.5×10^{-4}	7.7×10^{-6}	3.4×10^{-4}	6.6×10^{-5}	6.7×10^3
Guo-5'-P	3.0×10^{-5}	1.0×10^{-4}	1.3×10^{-3}	6.8×10^{-5}	1×10^{-2}	1.1×10^{-4}	6.2×10^3
Guo	3.5×10^{-4}	2.9×10^{-4}	1.8×10^{-3}	1.2×10^{-4}	2×10^{-2}	—	4.7×10^3
dGuo	1.9×10^{-3}	1.9×10^{-3}	—	7×10^{-4}	—	—	3.7×10^3
9-Me G	3×10^{-3}	—	—	—	—	—	3.4×10^3
Ado-2'-P	—	—	—	—	4.3×10^{-3}	—	—
Ado-3'-P	—	—	—	3.0×10^{-3}	—	—	—
Method	UV difference	UV difference	UV difference	Gel filtration	Gel filtration	Inhibition	
pH	5.6	5.0	5.0	5.5	5.0	7.0	5.6
Reference	Oshima and Imahori (1971a, d)	Walz and Hooverman (1973)	Epinatjeff and Pongs (1972)	Takahashi (1972)	Campbell and Ts'o (1971)	Irie (1964)	

[a] Estimated according to an equation, $\Delta G° = -RT\ln Kd$, using data reported by Oshima and Imahori

tween RNase T_1 and guanosine confirmed 1:1 complex formation. A similar conclusion was drawn by Epinatjeff and Pongs (1972), and Walz and Hooverman (1973), which is consistent with the earlier conclusion by Sato and Egami (1965).

Using difference spectra, dissociation constants can be computed. Values so far reported by several groups are summarized in Table 8. All reports gave the same order of relative affinities with RNase T_1, that is, Guo-2'-P > Guo-3'-P > Guo-5'-P > Guo > dGuo.

In Table 8, $\Delta G°$ values estimated from Kd are also listed. The binding energy contributions of the specific group in the substrate analog can be quantified by estimating selected differences in $\Delta G°$ values. For instance, comparing $\Delta G°$ values for Guo-3'-P and Guo, it can be concluded that the phosphate group at the 3' position contributes about 2.0 Kcal to the binding energy.

Epinatjeff and Pongs (1972) estimated $\Delta S°$ and $\Delta H°$ values upon the binding by measuring the effect of temperature on the difference spectra in the range of 15–60°C.

d) Interaction with CM-RNase T_1. Takahashi (1970) and Irie (1970) independently reported that the affinity of CM-RNase T_1 with Guo-3'-P was lower than that of the native enzyme. This phenomenon was confirmed by estimating the binding constant between CM-RNase T_1 and Guo-3'-P using difference spectroscopy (Oshima and Imahori 1971d). Walz (1976) extended the binding experiments with CM-RNase T_1, measuring the absorbancy difference. He estimated the binding constants for Guo and dGuo, and concluded that the 2'-OH group of Guo interacts with Glu-58 at the active site.

e) Interaction with Adenylate. Only a slight difference in UV absorption in the vicinity of 300–240 nm was observed upon mixing RNase T_1 with Ado-2'(3')-P, which is known to be a competitive inhibitor of RNase T_1 (Irie 1964). The difference spectrum failed

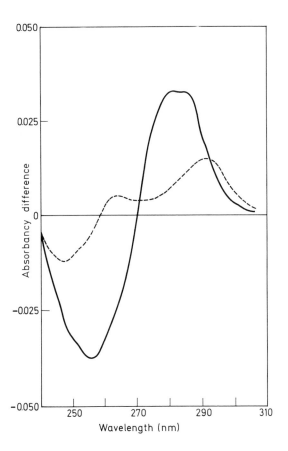

Fig. 4. Absorption difference upon mixing Ado-2'-P *(solid line)* and Guo-2'-P *(broken line)* with RNase U_2 at pH 4.5 using cells of 4.5 mm path length. Concentrations were 29 μM RNase U_2, 82 μM Ado-2'-P and Guo-2'-P. From Sato and Uchida (1975c)

to manifest the characteristic absorbancy change shown in Fig. 2. The spectrum also bears no resemblance to the difference spectrum resulting from a change of the pH of the adenylate solution (Oshima and Imahori 1971a). A similar observation was reported by Epinatjeff and Pongs (1972).

f) RNase U_2-adenylate Interaction. Sato and Uchida (1975c) studied the interaction between RNase U_2 and Guo-2'-P or Ado-2'-P using difference spectrum and gel filtration experiments. Spectral change upon binding of Guo-2'-P to RNase U_2 is similar to that of RNase T_1-guanylate in that the difference spectrum has a large positive peak at 292 nm and a deep trough at 248 nm (Fig. 4), suggesting the presence of an interaction between N-7 and the enzyme. A large spectral change was observed upon mixing RNase U_2 with Ado-2'-P. The difference spectrum observed could be mimicked by adding acid and organic solvent to a solution of Ado-2'-P. Taking into consideration that the adenosine group is protonated at either N-1 or N-7 in an acid solution (Deavin et al. 1968), and that the enzyme attacks 1-methyl-ApU but not 7-deaza-ApU (see Tables 3 and 4), it was speculated that N-7 of the adenosine group is protonated in hydrophobic environments in the enzyme-ligand complex.

2. Circular Dichroic Spectra

a) Circular Dichroic Difference. The CD spectrum of RNase T_1 and guanylate mixture was different from the graphical summation of the spectra of the components (Sander and Ts'o 1971; Oshima and Imahori 1971c and d). As an example, Fig. 5 illustrates a CD difference spectrum upon mixing RNase T_1 with Guo-3'-P. The difference spectrum can be characterized by a peak at around 250 nm and a small trough at around 280 nm. Qualitatively similar spectra were recorded when RNase T_1 was mixed with Guo-2'-P, Guo-5'-P, Guo, and dGuo.

b) An Assignment of the Dichroic Change. Taking into consideration that CD reflects the conformational and/or configurational features of a molecule, it appears that the CD difference represents the conformational change of amino acid residue(s) of RNase T_1 and/or of the ligand upon the binding. Since no amino acid side chain in the protein has absorption maximum at around 250 nm, it is most likely that the CD difference was the result of fixation of the guanosine part into a specific conformation at the binding site. Two conformations, the *syn* and *anti* forms, are known for the torsion angle of glycosidic bond of a nucleoside as illustrated in Fig. 6.

It was found that the dichroic band of Guo or Guo-3'-P in the vicinity of 250 nm is inverted upon acidification (Fig. 6). The observation can be interpreted as follows: in a solution guanosine or guanylate have freedom to rotate around the glycoside linkage, and the preferred conformation in an acid solution is opposed to one preferred in a neutral solution, therefore the CD band at 250 nm is small in magnitude and its sign is converted by changing the pH of the solution.

The CD difference spectra of RNase T_1-guanosine or guanylate complex manifested a positive cotton effect at around 250 nm just as the CD spectrum of guanosine in acid. The results suggested that the guanosine group at the binding site was fixed into the

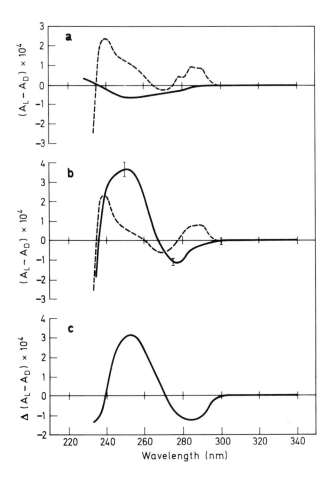

Fig. 5a−c. CD difference spectrum of the interaction of Guo-3'-*P* with RNase T$_1$ at pH 5.6 using cells of 3 mm path length. **a** Dichroic spectra of the enzyme (*broken line*) and Guo-3'-*P* (*solid line*). **b** Dichroic spectra of the mixture (*solid line*) and the algebraic sum (*broken line*). **c** Dichroic difference (the *solid line* minus the *broken line* in figure **b**). 9.5 × 10^{-5} M RNase T$_1$, 265 × 10^{-4} M Guo-3'-*P* and 0.014 M acetate buffer, pH 5.6, were used

conformation which is preferred in acid solution. Adopting many references that suggest the *anti* conformation for guanosine in neutral solutions and the *syn* in acid (Haschemeyer and Rich 1967; Guschlbauer and Courtois 1968; Danyluk and Hruska 1968; Schweizer et al. 1968; Miles et al. 1971; Son et al. 1972; Davies and Danyluk 1974; Sundaralingam 1975; Jordan and Niv 1977), it can be concluded that the guanosine residue is fixed into the *syn* conformation upon binding with RNase T$_1$. A similar conclusion was reported for the RNase U$_2$-guanylate complex (Sato and Uchida 1975).

A different interpretation was published by Sander and Ts'o (1971) indicating that the CD difference could arise from exciton formation in RNase T$_1$ interacting with the

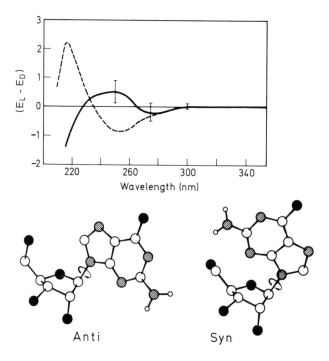

Fig. 6. The conformation of guanosine and CD spectra of guanosine at pH 6.5 (*broken line*) and at pH 0.5 (*solid line*)

analog. They observed a similar CD difference when the enzyme was mixed with 9-(2'-hydroxyethyl)-guanine 2'-phosphate or 9-(4'-hydroxybutyl)-guanine 4'-phosphate. These analogs contain no asymmetric carbon, but the extrinsic Cotton effect may be induced if these analogs are fixed at the binding site and allowed to interact closely with an asymmetric environment. So it seems possible to interpret the observed changes in circular dichroism as arising mainly from the ligands, although it is still an open question whether a part of the dichroic changes resulted from the conformational change and/or change in the electronic state of T_1 enzyme.

c) Binding of Adenylate to RNase T_1. The dichroic spectrum of the mixture of RNase T_1 and Ado-2'(3')-P was almost identical with the sum of the CD spectra of the enzyme and the ligand (Oshima and Imahori 1971c). From the Kd value in the literature, the formation of at least 5×10^{-5} M of the complex can be expected in the mixture under experimental conditions. The observation can be explained as being one of the following three cases: (1) the Ado group is not restricted, but freely rotates in the binding site, (2) the Ado group is fixed in the nearly equal amounts of the *syn* and *anti* forms, or (3) the conformation is fixed into an intermediate state between the *syn* and *anti*, that is, ϕ_{CD} is around $+60°$ or $-120°$. In any case, the Ado group does not seem to be restricted to the *syn* conformation in the binding site. In contrast, circular dichroic difference was observed upon mixing Ado-2'-P with RNase U_2 (Sato and Uchida 1975c). This observation implies a close relation between the substrate specificity and the fixation of the glycosyl bond.

3. NMR Studies

Since around 1970, NMR studies on the interactions of RNase T_1 and nucleotides have been carried out in several laboratories (Ruterjans and Pongs 1971; Fulling and Ruterjans 1978; Arata et al. 1979). Proton-NMR studies on C_2-protons of three His residues of RNase T_1 suggested that two of them interact with phosphate and guanine parts of Guo-3'-P (Ruterjans and Pongs 1971). More recently, Arata and co-workers (1979) concluded, based on ^{1}H-NMR spectra of differentially deuterated RNase T_1 and ^{31}P-NMR spectra of the enzyme-Guo-3'-P complex, that His-40 and His-92 interact with phosphate and guanine base, respectively. They also suggested that a hydrogen bond is formed between N-7 of guanine and His-92 at the binding site. Inagaki et al. (1978) carried out an elaborate ^{1}H-NMR study and reached a similar conclusion. Using CM-RNase T_1, they suggested that His-40 and His-92 interact with Glu-58 at the active site. Interaction between His-92 and Glu-58 was also suggested by Fulling and Ruterjans (1978), based on NMR spectra of photooxidized RNase T_1.

Y. Kyogoku and M. Kainosho (unpublished), in collaboration with one of us (Oshima), are carrying out ^{15}N-NMR studies on RNase T_1-^{15}N enriched Guo-3'-P complex. It is suggested that, among others, N-2 (amino group at position 2 of guanine base) strongly interacts with the protein. The signal assigned to N-2 was shifted and broadened by mixing with RNase T_1, suggesting rigid fixation of the amino group in the binding site. The result is consistent with the observation that inosine and xanthosine have lower affinity for the enzyme than guanosine, suggesting the requirement of the amino group at position 2 for tighter binding (Takahashi 1972).

Intensity of the signal of N-7 was changed, but no chemical shift was observed upon the complex formation. This finding suggests the participation of N-7 in the interaction but not protonation at this position. The signal of N-3 was also affected by the complex formation. This is the first observation which suggests the involvement of N-3 in the binding site.

4. Discussion

Studies on the binding of substrate analog to RNase T_1 suggested that an appropriate geometrical arrangement of guanine base, ribose, and phosphate is required for the highest affinity. A variety of physicochemical measurements as well as studies on competitive inhibition showed an order of binding affinity of Guo-2'-P > Guo-3'-P > Guo-5'-P > Guo > dGuo > Gua. Since Guo-5'-P showed higher affinity than Guo, it can be concluded that the enzyme binding center contains a site interacting with 5'-phosphate as well as 3'- or 2'-phosphate binding site.

From the difference between the free energy changes of binding of Guo and that of dGuo with RNase T_1 shown in Table 8, it can be seen that the 2'-OH group contributes about 1.0 Kcal to the binding energy. The participation of 2'-OH in the binding has been further investigated and an interaction of 2'-OH with Glu-58 and His-92 has been proposed, based on binding kinetics (Walz 1976; Osterman and Walz 1978; Takahashi 1976).

The participation of N-7 of guanine base in the RNase T_1 action was suggested by physicochemical measurements as well as studies on susceptibility of nucleoside 3'-phosphodiesters (Tables 3 and 4). A similar interaction of N-7 of purine base was also suggested for RNase U_2-analog complex formation (Sato and Uchida 1975).

However, the chemical nature of the interaction is controversal; UV difference spectrum studies strongly suggested the protonation of the nitrogen atom in the complex (Oshima and Imahori 1971a, d; Epinatjeff and Pongs 1972), but proton-NMR studies suggested a hydrogen bond between N-7 and the enzyme instead of protonation (Arata et al. 1979). A recent ^{15}N-NMR study did not support the protonation (Y. Kyogoku et al., unpublished data). The nature of the interaction should be elucidated in further studies.

The involvements of His-92 and His-40 in the binding were suggested from NMR studies as well as from chemical modification studies. However, speculations on the roles of these His residues in the active site based on NMR studies are contrary to those based on kinetics (Zabinski and Walz 1976; Osterman and Walz 1978) and chemical modifications (Takahashi 1976).

CD difference suggested the conformational fixation of guanosine group in the binding site, that is, the conformation of guanosine was restricted in the opposite conformation to one which is favored in a neutral solution. It can be speculated that the guanosine is fixed in the *syn* conformation upon the binding, assuming that the *anti* is favored in neutral solution (Son et al. 1972; Sundaralingam 1975), although the preferred conformation of guanosine in neutral is a contradictory proposition (Dobson et al. 1978). It seems that the difference arose mainly from the conformation of the ligand, and the change in the conformations of Tyr and Trp residues of RNase T_1 upon the binding would contribute to only a small fraction of the CD difference, since the absorbancy difference can mostly be explained by changes in spectral properties of the ligand and has no indication for detectable conformational changes in aromatic amino acid residues.

Similarly, one can speculate that Ado and Guo are fixed in the *syn* conformation in the RNase U_2-analog complex based on the reported CD difference (Sato and Uchida 1975). It is noteworthy that no remarkable CD difference was observed in the RNase T_1-adenylate complex formation. Taking the unique specificities of these enzymes and the UV and CD difference spectra upon the binding into consideration, a close correlation would be expected between the restriction of nucleoside conformation, interaction at N-7, and the enzyme recognition.

Oshima and Imahori (1971d) summarized their observations from difference spectroscopic studies (UV and CD) in the scheme shown in Fig. 7. Interactions between the enzyme and N-7 of the base, 3'- or 2'-phosphate, 5'-phosphate and 2'-OH groups have also been suggested by many other groups. In addition, there must be some other interactions between the enzyme and the substrate; for instance, studies on the base specificity strongly suggested the presence of a specific interaction between the 6-oxo group and the protein (Tables 3 and 4). In Fig. 7 the guanosine group is illustrated in the *syn* conformation according to their interpretation. However, as mentioned above, equivocal

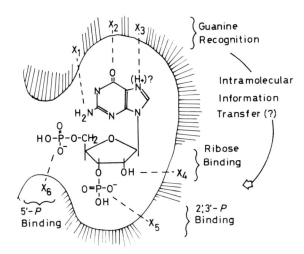

Fig. 7. Proposed interaction between RNase T_1 and the substrate analogs. Modified after Oshima and Imahori (1971b)

Guanine Recognition

Intramolecular Information Transfer (?)

Ribose Binding

2',3'-P Binding

5'-P Binding

Likely candidates for active site residues

X_2, X_3 : His-40 / His-92

X_4, X_5 : Arg-77, Glu-58, His-40/His-92

observations and different interpretations have been reported for some aspects of the binding mechanisms of substrate analogs to RNase T_1. The discrepancies should be solved in future studies.

In this context it should be added that X-ray crystallographic analysis of RNase T_1 has not yet been performed, notwithstanding the fact that the primary structure of the enzyme was elucidated 13 years ago (Takahashi 1965). This was mainly due to the difficulty in obtaining suitable single crystals. Recently, W. Saenger in Göttingen (personal communication) has succeeded in obtaining a crystal of RNase T_1 suitable for X-ray analysis. As for RNase U_2, a preliminary investigation of an orthorhombic crystal of RNase U_2 has been carried out (Matsuzaki et al. 1974). Upon the completion of the X-ray diffraction studies on RNase T_1 and the enzyme-inhibitor complex, one can discuss in an unequivocal manner the chemical nature of some of the interactions between the enzyme and the substrate analog, like RNase A-inhibitor binding. In case of RNase A the binding reaction with pyrimidine nucleotides has been interpretable, based on the combined data from chemical, physicochemical, and crystallographic studies, as briefly described below.

III. Interactions of RNase A with the Substrate Analogs

Since the elucidation of primary structure in the laboratories of Anfinsen, and of Hirs, Stein, and Moore (see Smyth et al. 1963), pancreatic RNase A has been used as an experimental material for general studies of enzyme action, and a large amount of information is available on this enzyme, including three-dimensional structure for the native enzyme and RNase S – UpcA (an analog of dinucleotide UpA, an oxygen atom in the internucleotide phosphate group being replaced by a methylene group) complex (Richards and Wyckoff 1971, 1973; Carlisle et al. 1974). Like RNase T_1, RNase A interacts in general with pyrimidine nucleotides more strongly than the corresponding nucleosides or bases, suggesting the requirement of base, ribose, and phosphate groups for tight binding.

The difference spectra upon the binding of competitive inhibitor to RNase A suggested 1:1 binding (Hummel et al. 1961), and seemed to result from changes in optical properties of the ligand because the difference spectrum observed upon binding with 4-thiouridine is shifted to the absorption region of the sulfur-containing nucleotide (Sawada and Ishii 1968). One of the interactions which affects the nucleotide chromophore (Irie 1968) could be the presence of Phe-120 near the pyrimidine ring at the binding site suggested by the X-ray diffraction study.

Based on the CD difference spectrum upon the binding of cytidine nucleotide to RNase A, Oshima and Imahori (1971b) predicted the *anti* conformation for cytidine group fixed in the enzyme. From the X-ray diffraction study for the RNase S – UpcA complex, it was shown that no tyrosine residues are in contact with the substrate analog, suggesting that the observed absorption difference and CD difference did not reflect changes in conformation of phenol residues of the enzyme protein. It was also shown that the conformation of the substrate analog is distorted from its preferred conformation. The distortions are around C4'-C5' bond and the internucleotide P-O bond torsions (Sundaralingam 1975). However, the glycosyl torsion is similar to that of unbound molecule, that is, in *anti* form. This observation, which supports the prediction based on the CD study, suggests the validity of CD difference spectroscopy for the study of nucleosides conformation bound to a protein.

As for interactions involved in the base recognition by RNase A, the X-ray crystallographic study on the enzyme-inhibitor complex suggested the presence of three hydrogen bonds between O-2 and main chain NH of Thr-45, N-3 and OH of Thr-45, and O-4 (or N-4, in case of cytidine) and OH of Ser-123. These hydrogen bondings explain the specificity of the enzyme. The study also indicated that a nitrogen atom of His-12 is close to 2'-OH of the pyrimidine nucleoside, and a nitrogen of His-119 is near 5'-OH group of the leaving nucleoside (actually an internucleotide carbon atom of Upc A in the binding site).

The implications of these two His residues as well as Lys-40 in the active site have been strongly suggested by chemical modification studies on RNase A. The X-ray study showed that Lys-40 is also close to the internucleotide phosphate.

IV. Specific DNases

So far no DNases with a simple base specificity in the sense of base specific RNases have been found. T4-endonuclease IV cleaves preferentially certain, but not all ...N-C... sequences of single-stranded DNA (Sadowski and Hurwitz 1969). Contrary to the simple base specificity of RNases, a series of restriction enzymes with much higher order of specificity are now well known. However, no investigations on the interaction of these DNases with respective nucleotide sequences have been reported.

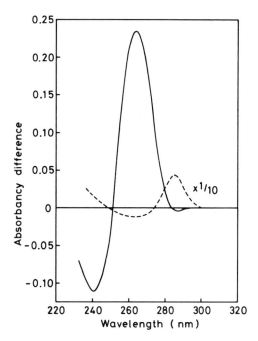

Fig. 8. Difference spectrum of interaction of DNase A and dAdo-3'-*P* or dThd-5'-*P*, recorded at 15°C. *Solid line:* 3.5 μM DNase mixed with 30 μM dAdo-3'-*P* in 0.2 M sodium acetate buffer, pH 5.0, containing 2.5% polyethylene glycol #400; *broken line:* 3.5 μM DNase mixed with 30 μM dThd-5'-*P* in 0.2 M sodium acetate buffer, pH 4.0, containing 2.5% polyethylene glycol #400

Recently, we isolated a novel DNase, designated DNase A, from a land snail, *Achatina fulica.* It is active on native and denatured DNA, poly(dA), and poly(dT), but not on poly(dG), poly(dC), RNA, and poly(A). It is an endonuclease producing oligonucleotides with an average chain length of about five nucleotides. It may be regarded as a A- or T-rich cluster-specific DNase (Yanagawa et al. 1977). We have just begun the investigation on the interaction of the enzyme with nucleotides. It was found, as expected, by the difference spectrum measurements that dAdo and dAMP interacted with DNase A much stronger than other nucleosides and nucleotides such as dGuo and dGMP. 3'- and 5'-dAMP gave a characteristic difference spectrum as shown in Fig. 8. The difference spectrum observed with the interaction of 5'-dTMP was quite different and the difference was much smaller (Fig. 8). Polyadenylates, although resistant to the enzyme, were bound strongly to the enzyme (Miyakawa, A, Yanagawa, H., and Egami, F., unpublished). So it will be excellent in opening the way to the elucidation of the interaction mechanism between adenine sequence and the enzyme protein.

References

Arata Y, Kimura S, Matsuo H, Narita K (1979) Proton and phosphorus nuclear magnetic resonance studies of ribonuclease T$_1$. Biochemistry 18: 18−24

Arima T, Uchida T, Egami F (1968) Studies on extracellular ribonucleases of *Ustilago sphaerogena;* characterization of substrate specificity with special reference to purine-specific ribonucleases. Biochem J 106: 609−613

Campbell MK, Ts'o POP (1971) Binding of purine nucleoside monophosphates by ribonuclease T$_1$. Biochim Biophys Acta 232: 427−435

Campbell MK, Shipp S, Jantzen E (1976) Location of chromophoric residues in ribonuclease T_1 by solvent perturbation difference spectroscopy. Biochem Biophys Res Commun 72: 1014–1020

Carlisle CH, Palmer RA, Mazumdar SK, Gorinsky BA, Yeates DGR (1974) The structure of ribonuclease at 2.5 Ångström resolution. J Mol Biol 85: 1–18

Danyluk SS, Hruska FE (1968) The effect of pH upon the nuclear magnetic resonance. Biochemistry 7: 1038–1043

Davies DB, Danyluk SS (1974) Nuclear magnetic resonance studies of 5'-ribo and deoxyribonucleotide structures in solution. Biochemistry 13: 4417–4434

Deavin A, Fisher R, Kemp CM, Mathias AP, Rabin BR (1968) Bovine pancreatic ribonuclease – Spectrophotometric investigations of interaction of the enzyme with nucleotides. Eur J Biochem 7: 21–26

Dobson CM, Geraldes CFGC, Ratcliffe G, Williams RJP (1978) Nuclear-magnetic-resonance studies of 5'-ribonucleotide and 5'-deoxyribonucleotide conformations in solution using the lanthanide probe method. Eur J Biochem 88: 259–266

Egami F, Nakamura K (1969) Microbial ribonucleases. (Molecular biology, biochemistry and biophysics, vol VI.) Springer, Berlin Heidelberg New York

Egami F, Takahashi K, Uchida T (1964) Ribonuclease in Takadiastase: Properties, chemical nature, and applications. In: Davidson JN, Cohn WE (eds) Progress in nucleic acid research and molecular biology, vol III, pp 59–101. Academic Press, New York London

Epinatjeff C, Pongs O (1972) Ribonuclease T_1: Spectrophotometric investigations of the interaction of the enzyme with substrate analogues. Eur J Biochem 26: 434–441

Fulling R, Ruterjans H (1978) Proton magnetic resonance studies of ribonuclease T_1. Assignment of histidine-27 C2-H and C5-H proton resonances by a photooxidation reaction. FEBS Lett 88: 279–289

Guschlbauer W, Courtois Y (1968) pH induced changes in optical activity of guanine nucleosides. FEBS Lett 1: 183–186

Haschemeyer AEV, Rich A (1967) Nucleoside conformations: An analysis of steric barriers to rotation about the glycosidic bond. J Mol Biol 27: 369–384

Hummel JP, Ver Ploeg DA, Nelson CA (1961) The interaction between ribonuclease and mononucleotides as measured spectrophotometrically. J Biol Chem 236: 3168–3172

Iida S, Ooi T (1969) Titration of ribonuclease T_1. Biochemistry 8: 3897–3902

Inagaki F, Kawano Y, Miyazawa T, Takahashi K (1978) Structure and function of active center of RNase T_1. Abstract of the 29th Symposium on protein structure, pp 65–68

Irie M (1964) Inhibition of ribonuclease T_1 by various kinds of nucleotides. J Biochem (Tokyo) 56: 495–497

Irie M (1968) A consideration on the ultraviolet spectrum of ribonuclease A-inhibitor complex. J Biochem (Tokyo) 64: 347–353

Irie M (1970) Studies on the photooxidation of ribonuclease T_1. J Biochem (Tokyo) 68: 69–79

Jordan F, Niv H (1977) C8-Amino purine nucleosides: A well-defined steric determinant of glycosyl conformational preference. Biochim Biophys Acta 476: 265–271

Kasai H, Takahashi K, Ando T (1969) Trinitorophenylation of ribonuclease T_1. J Biochem (Tokyo) 66: 591–597

Kasai H, Takahashi K, Ando T (1977) Chemical modification of tyrosine residues in ribonuclease T_1 with N-acetylimidazole and p-diazobenzenesulfonic acid. J Biochem (Tokyo) 81: 1751–1758

Matsuzaki T, Uchida T, Ambady G, Kartha G (1974) Crystallographic studies on RNase U_2: A purine specific ribonuclease. Abstract of the Summer Meeting of American Crystallographic Association

McMeekin TC, Groves ML, Hipp NJ (1949) Apparent specific volume of α-casein and β-casein and the relationship of specific volume to amino acid composition. J Am Chem Soc 71: 3298–3300

Miles DW, Townsend LB, Robins MJ, Robins RK, Inskeep WH, Eyring H (1971) Circular dichroism of nucleoside derivatives. X. Influence of solvent and substituents upon the Cotton effects of guanosine derivatives. J Am Chem Soc 93: 1600–1608

Miles HT, Howard FB, Frazier J (1963) Tautomerism and protonation of guanosine. Science 142: 1458–1463

Minato S, Hirai A (1979) Characterization of *Ustilago* ribonuclease U$_2$, Effects of chemical modification at glutamic acid-61 and cystein-1 and of organic solvents on the enzymatic activity. J Biochem (Tokyo) 85: 327–334

Oshima H, Takahashi K (1976) The structure and function of ribonuclease T$_1$. XX. Specific inactivation of ribonuclease T$_1$ by reaction with tosylglycolate. J Biochem (Tokyo) 80: 1259–1265

Oshima T, Imahori K (1971a) Difference spectral studies on the interactions between ribonuclease T$_1$ and its substrate analogues. J Biochem (Tokyo) 69: 987–990

Oshima T, Imahori K (1971b) Conformation of cytidine group in the complex of pancreatic ribonuclease A and cytidine 2'-phosphate. J Biochem (Tokyo) 70: 193–195

Oshima T, Imahori K (1971c) A change in circular dichroism due to the binding of guanosine 3'-phosphate to ribonuclease T$_1$. J Biochem (Tokyo) 70: 197–199

Oshima T, Imahori K (1971d) Ribonuclease T$_1$-substrate analog complexes. In: Funatsu M, Hiromi K, Imahori K, Murachi T, Narita K (eds) Proteins structure and function, vol I, pp 333–367. Kodansha, Tokyo

Osterman HL, Walz FG (1978) Studies and catalytic mechanism of ribonuclease T$_1$: Kinetic studies using GpA, GpC, GpG and GpU as substrates. Biochemistry 17: 4124–4130

Pongs O (1970) Influences of pH and substrate analog on ribonuclease T$_1$ fluorescence. Biochemistry 9: 2316–2322

Richards FM, Wyckoff HW (1971) Bovine pancreatic ribonuclease. In: Boyer PD (ed) Enzymes, vol IV, pp 647–806. Academic Press, New York London

Richards FM, Wyckoff HW (1973) Ribonuclease-S. Clarendon Press, Oxford

Ruterjans H, Pongs O (1971) On the mechanism of action of ribonuclease T$_1$: Nuclear magnetic resonance study on the active site. Eur J Biochem 18: 313–318

Sadowski PD, Hurwitz J (1969) Enzymatic breakage of deoxyribonucleic acid. II. Purification and properties of endonuclease IV from T4 phage-infected *Escherichia coli*. J Biol Chem 244: 6192–6198

Sander C, Ts'o POP (1971) Circular dichroism studies on the conformation and interaction of T$_1$ ribonuclease. Biochemistry 10: 1953–1966

Sato K, Egami F (1957) Studies on ribonucleases in Takadiastase I. J Biochem (Tokyo) 44: 753–767

Sato S, Egami F (1965) On the interaction of ribonuclease T$_1$ and guanosine 2'-phosphate and related compounds. Biochem Z 342: 437–448

Sato S, Uchida T (1974) Whole structure and active site of RNase U$_2$. Abstracts of the 25th Symposium on the structure of proteins, pp 21–24

Sato S, Uchida T (1975a) The amino acid sequence of ribonuclease U$_2$ from *Ustilago sphaerogena*. Biochem J 145: 353–360

Sato S, Uchida T (1975b) The disulfide bridges of ribonuclease U$_2$ from *Ustilago sphaerogena*. J Biochem (Tokyo) 77: 1171–1176

Sato S, Uchida T (1975c) On the interaction of ribonuclease U$_2$ and substrate analogues. Biochim Biophys Acta 383: 168–177

Sato S, Uchida T (1975d) Ethoxyformylation of ribonuclease U$_2$ from *Ustilago sphaerogena*. J Biochem (Tokyo) 77: 795–800

Sawada F, Ishii F (1968) Interaction between bovine pancreatic ribonuclease and 4-thiouridylic acid. J Biochem (Tokyo) 64: 161–165

Schweizer MP, Broom AD, Ts'o POP, Hollis DP (1968) Studies of inter- and intramolecular interaction in mononucleosides by proton magnetic resonance. J Am Chem Soc 90: 1042–1055

Smyth DG, Stein WH, Moore S (1963) The sequence of amino acid residues in bovine pancreatic ribonuclease: Revisions and confirmations. J Biol Chem 238: 227–234

Sobell HM, Tomita K (1964) The crystal structures of salts of methylated purines and pyrimidines. IV. 9-Methylguanine hydrobromide. Acta Cryst 17: 126–131

Son T-D, Thiery J, Guschlbauer W, Dunand J-J (1972) Nucleoside conformation. VIII. Conformation of guanosine 2'-phosphate in aqueous solution by proton magnetic resonance spectroscopy. Biochim Biophys Acta 281: 289–298

Sundaralingam M (1975) Structure and conformation of nucleosides and nucleotides and their analogs as determined by X-ray diffraction. Ann NY Acad Sci 255: 3–42

Takahashi K (1965) The amino acid sequence of ribonuclease T_1. J Biol Chem 240: PC 4117–4119

Takahashi K (1970a) The structure and function of ribonuclease T_1. X. Reactions of iodoacetate, iodoacetamide and related alkylating reagents with ribonuclease T_1. J Biochem (Tokyo) 68: 517–527

Takahashi K (1970b) The structure and function of ribonuclease T_1. XI. Modification of the single arginine residue in ribonuclease T_1 by phenylglyoxal and glyoxal. J Biochem (Tokyo) 68: 659–664

Takahashi K (1971a) Ribonuclease T_1. In: Funatsu M, Hiromi K, Imahori K, Murachi T, Narita K (eds) Proteins; structure and function, vol I, pp 285–331. Kodansha, Tokyo

Takahashi K (1971b) The structure and function of ribonuclease T_1. XII. Further studies on rose bengal-catalyzed photooxidation of ribonuclease T_1 – Identification of a critical histidine residue. J Biochem (Tokyo) 69: 331–338

Takahashi K (1972) The structure and function of ribonuclease T_1. XVIII. Gel filtration studies on the interaction of ribonuclease T_1 with substrate analogs. J Biochem (Tokyo) 72: 1469–1481

Takahashi K (1973) Evidence for the implication of histidine-40 and -92 in the active site of ribonuclease T_1. J Biochem (Tokyo) 74: 1279–1282

Takahashi K (1976) The structure and function of ribonuclease T_1. XXI. Modification of histidine residues in ribonuclease T_1 with iodoacetamide. J Biochem (Tokyo) 80: 1267–1275

Takahashi K (1977) The structure and function of ribonuclease T_1. XXIII. Inactivation of ribonuclease T_1 by reversible blocking of amino group with cis-aconitic anhydride and related dicarboxylic acid anhydrides. J Biochem (Tokyo) 81: 641–646

Takahashi K, Inoue N (1977) The structure and function of ribonuclease T_1. XXII. Tryptic cleavages of the single lysyl and arginyl bonds in ribonuclease T_1. J Biochem (Tokyo) 81: 415–421

Takahashi K, Stein WH, Moore S (1967) The identification of a glutamic acid residue as part of the active site of ribonuclease T_1. J Biol Chem 242: 4682–4690

Takahashi K, Uchida T, Egami F (1970) Ribonuclease T_1; structure and function. In: Kotani M (ed) Advances in biophysics, vol I, pp 53–98. Univ of Tokyo Press, Tokyo

Tamaoki H, Sakiyama F, Narita K (1976) Preparation and properties of trypsin-digested ribonuclease T_1 split at the single arginyl peptide bond. J Biochem (Tokyo) 79: 579–589

Tamaoki H, Sakiyama F, Narita K (1978) Chemical modification of ribonuclease T_1 with ozone. J Biochem (Tokyo) 83: 771–781

Uchida T (1970) Immunochemical properties of ribonuclease T_1 in relation to other ribonucleases. J Biochem (Tokyo) 68: 255–264

Uchida T (1971) Which part of the substrates does RNase read? Protein Nucleic Acid Enzyme 16: 1053–1059

Uchida T, Egami F (1967) Ribonuclease T_1 from Taka-Diastase. In: Colowick SP, Kaplan NO (eds) Methods in Enzymology, vol XII, pp 288-239. Academic Press, New York London

Uchida T, Egami F (1971) Microbial ribonucleases with special reference to RNases T_1, T_2, N_1 and U_2. In: Boyer PD (ed) The enzymes, 3rd ed., vol IV, pp 205–250. Academic Press, New York London

Uchida T, Machida C (1978) The specificity for dinucleoside monophosphates of RNase U_2 and urea-treated RNase U_2. Nucleic acid res, special publ, No 5, s409–412

Uchida T, Sato S (1973) Microbial cyclizing RNases. In: Zelinka J, Balan J (eds) Ribosomes and RNA metabolism, pp 453–472. Publ House Slovak Academy of Science, Bratislava

Uchida T, Arima T, Egami F (1970) Specificity of RNase U_2. J Biochem (Tokyo) 67: 91–102

Walz FG Jr (1976) Ribose recognition by ribonuclease T_1: difference spectral binding studies with guanosine and deoxyguanosine. Biochemistry 15: 4446–4450

Walz FG Jr (1977a) Spectrophotometric titration of a single carboxyl group at the active site of ribonuclease T_1. Biochemistry 16: 4568–4571

Walz FG Jr (1977b) Studies on the nature of guanine nucleotide binding with ribonuclease T_1. Biochemistry 16: 5509–5515

Walz FG Jr, Hooverman LL (1973) Interaction of guanine ligands with ribonuclease T_1. Biochemistry 12: 4846–4851

Walz FG Jr, Terenna B (1976) Subsite interactions of ribonuclease T_1: binding studies of dimeric substrate analogues. Biochemistry 15: 2837–2842

Walz FG Jr, Biddlecome S, Hooverman L (1975) The interaction of ribonuclease T_1 with DNA. Nucleic Acid Res 2: 11−20

Yanagawa H, Ogawa Y, Egami F (1977) Deoxyribonuclease A, a novel deoxyribonuclease highly active toward polydeoxyadenylic acid and polythymidylic acid from *Achatina fulica.* J Biochem 82: 519−528

Zabinski M, Walz FG Jr (1976) Subsites and catalytic mechanism of ribonuclease T_1: Kinetic studies using GpC and GpU as substrates. Arch Biochem Biophys 175: 558−564

Note added in proof: It was recently reported that X-ray diffraction studies on RNase T_1 are also under way in the United States [Martin PD, Tulinsky A, Walz FG Jr (1980) J Mol Biol 136: 95−97]

4. Structural and Dynamic Aspects of Recognition Between tRNAs and Aminoacyl-tRNA Synthetases

D. G. KNORRE and V. V. VLASSOV

A. Introduction

Interaction of transfer RNA's with cognate aminoacyl-tRNA synthetases (amino acid: tRNA ligases EC 6.1.1., abbreviated ARSases) is one of the most specific protein—nucleic acid interactions determining the fidelity of the genetic message translation (Kisselev and Favorova 1974). Chapeville, Lipmann et al. (1962) in their classical experiment have succeeded to convert cysteinyl residue enzymatically bound to tRNA^{Cys} to alanyl one by treatment with Raney Nickel. Using this Ala-tRNA^{Cys} they have clearly demonstrated that it is tRNA moiety of aminoacyl-tRNA rather than aminoacyl one which is selected by codon in the course of translation. Therefore, any erroneous aminoacylation should result almost inevitably in a mistake in the growing polypeptide chain.

In this paper we would like to summarize briefly the present state of the knowledge concerning the mechanism of ARSase—tRNA interaction, taking into account some experimental approaches and theoretical considerations used by the authors, to move a bit further toward the understanding of this extremely important as well as intriguing system.

B. Specific Problems of the ARSase-tRNA Interaction

Aminoacylation of tRNA is an enzymatic process proceeding via two distinct chemical steps (Hoagland et al. 1957). In the first step without any direct tRNA participation amino acid $NH_3^+CHR_iCOO^-$ specific for enzyme E_i is converted to aminoacyl adenylate $NH_3^+CHR_iCO\,pA$ (Hoagland et al. 1956)

$$E_i + NH_3^+CHR_iCOO^- + pppA \;\rightarrow\; E.NH_3^+CHR_iCO \sim pA + pp \qquad (1)$$

In the second step the aminoacyl residue is transferred to one of the isoacceptor tRNA's specific to this amino acid

$$E_i.NH_3^+CHR_iCO \sim pA + tRNA_j^{(i)} \;\rightarrow\; NH_3^+CHR_iCO\text{-}tRNA_j^{(i)} + E_i \qquad (2)$$

Institute of Organic Chemistry, Siberian Division of the Academy of Sciences of the USSR, 630090, Novosibirsk-90, USSR

where j is the number of tRNA$^{(i)}$ in the set of isoacceptors. There is no reason to doubt that quite similar to other enzyme-substrate systems specific complex is formed between tRNA and ARSase; that hydrophobic and electrostatic interactions as well as hydrogen bonding are the main forces stabilizing the complex; that the complex formation may be via some conformational change results in the proper orientation of the reacting OH-group of tRNA toward aminoacyl residue to be transferred and toward the catalytic groups of the enzyme. Most likely the main details of these interaction will come from the X-ray investigations, as it already happended with a number of more simple enzymes.

However, tRNA exhibits some features which make it significantly different from most others of the enourmous number of the already known substrates. Several questions arise from these features as well as from some features of the ARSases structure specific mainly for the system under consideration. These questions will be the main point of our attention in this paper.

The first group of questions is connected with the multiplicity of tRNA's. All tRNA's exhibit rather similar tertiary structure. The secondary structure of all till now sequenced tRNA's may be presented in the famous cloverleaf form proposed by Holley et al. (1965). The X-ray data for yeast tRNA^{Phe} (Kim et al. 1974; Robertus et al. 1975) clearly demonstrate that the residues most essential for tertiary interactions are those

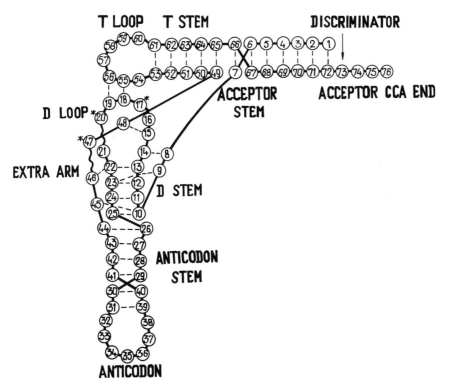

Fig. 1. L-shaped representation of tRNA structure. The most essential H-bonds supporting the secondary and tertiary structure are presented by *broken lines*. Two nucleotides (*17* and *17.1*), three nucleotides (*20, 20.1* and *20.2*) and 18 nucleotides (*47, 47.1–17*) may exist in the sites marked by asterisks

common for almost all tRNA's. In Fig. 1 the general tRNA structure is presented in a way demonstrating the L-shaped form of the molecule as well as the most essential tertiary hydrogen bondings. The numeration is given according to the last survey of the tRNA sequences (Gauss et al. 1979). The common tertiary structure of all tRNA's seems to be of extreme significance for the translation to put the aminoacyl residue of aminoacyl-tRNA in a standard manner toward the peptidyltransferase center of ribosome, irrespective of the codon to be translated.

Consequently, each ARSase meets in the living organism (as well as in the experiments with unfractionated tRNA) a rather great number of tRNA's of common tertiary structure and has to recognize among them one or a small number of species. The question arises whether ARSase completely ignores the common features of various tRNA's or, quite on the contrary, it recognizes the general appearance of tRNA's in the first unspecific step of the recognition. If so, this first step should be followed by the fine matching of tRNA to ARSase operating in native conditions only within the cognate tRNA-ARSase couple. Therefore, the recognition may be expected to be an essentially stepwise process.

The necessity of discrimination between a number of similar substrates relates ARSases to the enzymes operating in replication, transcription, and translation. However, the selection of substrate in the latter systems is performed by the template. ARSases seem to function without any template. It cannot be excluded that in the selection of proper tRNA by ARSase a similar role is played either by amino acid or aminoacyl residue of aminoacyl adenylate. Some correlation was found between the nature of the side radical of amino acid and some parts of tRNA suggested to participate essentially in the recognition, namely the middle base of anticodon and the last variable nucleotide of the tRNA polynucleotide chain (position 73) (Lestienne 1978). Thus, all tRNA's chargeable with most hydrophobic aminoacids (Val, Ile, Leu, Met, Phe) contain adenosine both in the middle of anticodon and in the position 73.

The existence of the direct specific interaction of amino acid with some parts of tRNA seems reasonable from the evolutionary point of view. According to Hoppfield (1978) the selection of amino acids by primal tRNA's was essentially based on the interaction of amino acid with the anticodon juxtaposed close to the acceptor end of the primal tRNA. This interaction could remain as a part of the overall modern recognition.

ARSase must also recognize a proper amino acid. This is mostly done in the first step of the overall aminoacylation without any visible participation of tRNA, although several Arg- and Gln-RSases were found to need the presence of tRNA at this step (for review, see Kisselev and Favorova 1974). In some cases the specificity of the enzyme at the stage of the aminoacyl adenylate formation is insufficient to discriminate between amino acids of similar structure, e.g., between valine and isoleucine or between valine and threonine. It was found that tRNA may help to amplify the selectivity inducing the hydrolysis of the unproper aminoacyl adenylate. For example tRNA[Ile] induces the hydrolysis of Val~pA erroneously formed by Ile-RSase from *E. coli* in the presence of valine (Norris and Berg 1964). Recently, the mechanism of this phenomenon was elucidated. It was demonstrated that the transfer of the noncognate aminoacyl residue is followed by rapid specific hydrolysis of aminoacyl-tRNA formed (Fersht and Kaethner 1976; von der Haar and Cramer 1976).

The next group of questions is connected with the great dimensions of tRNA as substrate. The chemical event, namely the attachment of aminoacyl residue to tRNA takes place at the 3'-end of tRNA. At the same time anticodon 80 Å away from this point is essential for recognition of some tRNA's (see below). The arising problems concerning the general mode of the mutual arrangement of tRNA and ARSase were recently discussed by Reid (1977) and we shall not deal with them in our paper. The main topic of our consideration will be the number and the position of the points of contacts between tRNA and ARSase in the course of recognition. It looks improbable that strong interaction operates along the whole tRNA molecule from anticodon to the acceptor end. The rather low association constants of tRNA-ARSase interaction, never exceeding 10^7 M^{-1} in native conditions, suggest that a rather small number of such points is scattered over the tRNA molecule.

The last group of questions arises from the stepwise mechanism of aminoacylation, complicated still more by the subunit structure of most ARSases as well as strong cooperative effects operating between active sites for different substrates and between subunits. A great number of the states of the enzyme may exist differing in the distribution of the ligands and vacant sites over the subunits. The mode of tRNA-ARSase interaction may be in some cases strongly influenced by the state of the other sites of the enzyme. The number of essentially different states of tRNA-ARSases complexes, the structural nature of the differences, the mechanism of the changes accompanying the transitions between the states, present the third group of questions inherent in the problem of the tRNA-ARSase recognition.

C. Experimental Methods of Investigation of the tRNA-ARSase Interaction

Till now no significant results concerning the structure of tRNA-ARSase complexes were obtained by two most informative physical methods — X-ray cristallography and NMR spectroscopy. No doubt, these methods will play an essential role in the future. In this section we shall discuss briefly the methods which have already led to some results in the field under consideration.

A great number of tRNA's are already sequenced (Gauss et al. 1979). The comparison of the primary structures of tRNA's which may be charged with amino acid by the same enzyme may help to elucidate the sites essential for the interaction with this enzyme. This approach looks most promising for the groups of isoacceptors which are aminoacylated by the same enzyme in natural conditions. As an example, we shall present the results of such analysis for valine-specific tRNA's from yeast and *E. coli* which may be charged by yeast Val-RSase (Axelrod et al. 1974). Considering the question we shall additionally take into account the structure of yeast $tRNA_{2b}^{Val}$ recently sequenced by the same authors (Gorbulev et al. 1977).

Figure 2 gives the generalized structure of $tRNA^{Val}$ in which only nucleotides common for all yeast and *E. coli* valine-specific tRNA's are marked. Besides the residues present in all elongator tRNA's, five points specific for the group of tRNA's under consideration are found. These points may be regarded as candidates for the sites of specific recognition. These are two letters of anticodon, secondary base pairs $C_{31}-G_{39}$ in the

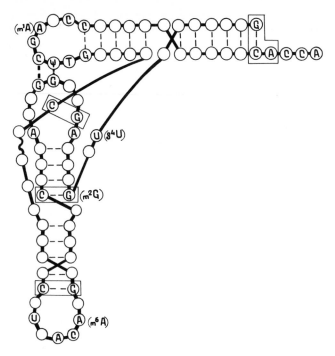

Fig. 2. The generalized structure of yeast and *E. coli* valine-specific tRNA's. Only nucleotides common to all these tRNA's are marked by *symbols*. The candidates for the sites of specific recognition are *framed*

anticodon stem and $G_{10}-C_{25}$ in the D-stem, tertiary base pair $G_{15}-C_{48}$, and base pair in the acceptor stem G_1-C_{72} with adjacent A_{73}. It is worth mentioning that among already sequenced yeast tRNA's only tRNA^{Ala} has the same combination of nucleotides in the four latter points and should be discriminated either only by the difference in the anticodon or by some additional element of structure (e.g., the structure of D-stem of yeast tRNA^{Ala} differs from those of all considered valine-specific tRNA's).

The binding to ARSase of oligonucleotides representing some parts of the tRNA molecule may serve as indication of the participation of the corresponding sites of tRNA in the interaction with ARSase. The first successful attempt of this type was described by Letendre et al. (1969). The authors have demonstrated that poly(U), which contains the anticodon sequence of tRNA^{Lys}, binds to yeast Lys-RSase and inhibits aminoacylation of tRNA^{Lys}. Recently, Schoemaker and Schimmel (1977a) have found that tritium incorporation into C-5 position of U_8 of *E. coli* tRNA^{Ile} is accelerated in the presence of *E. coli* Ile-RSase. They suggested that this may be due to the direct contact of this residue with ARSase in the tRNA-ARSase complex. This suggestion was supported by demonstration of the inhibitory action of U-A-G representing positions 8–10 of tRNA^{Ile} on the aminoacylation of tRNA (Schoemaker and Schimmel 1977b).

This approach was further improved by Vlassov and Khodyreva (1978), who studied the distribution of oligonucleotides of the pancreatic RNAase digest of *E. coli* tRNA^{Phe} between the chambers of the equilibrium dialysis cell containing and lacking *E. coli*

Table 1. Distribution of oligonucleotides of the pancreatic RNAase digest of *E. coli* tRNA[Phe] between chambers of dialysis cell after equilibration for 14 h at $0°C$. One of the chambers contained PheRSase *E. coli* (4 mg/ml) in 0.02 ml of 0.01 M cacodylate pH 6.5, 0.005 M $MgCl_2$, 2×10^{-4} M EDTA. The second chamber contained 0.15 A_{260} units of the oligonucleotide mixture in 0.02 ml of the same buffer. R is the ratio of concentrations of the oligonucleotide in the chamber with PheRSase and in the chamber lacking the enzyme

Oligonucleotide	R
A-G-Cp	1.0
A-G-Dp + G-A-Up	0.8
G-G-Dp + G-G-Tp	0.8
pG-Cp	1.0
G-A-Up + G-G-A-Up	1.1
G-G-G-Cp	0.9
A-G-A-G-Cp	1.5
G-A-A-$s^2 i^6$A-A-ψp	1.0
A-G-G-G-G-A-ψp	4.8
G-G-A-s^4Up	2.8

Phe-RSase. The ratio of the oligonucleotide amounts in both chambers is shown in the Table 1. The enhanced amount of some oligonucleotides in the Phe-RSase-containing chamber demonstrates the specific binding of these oligonucleotides to the enzyme. This technique permits to get information concerning the binding of a set of oligonucleotides in one experiment without laborious preparation of a number of oligonucleotides with defined sequences. It is a new experimental version of the general approach proposed several years ago for the elucidation of the sequences interacting with some biopolymer by extraction (e.g., by gel-filtration) of the corresponding oligonucleotides from the statistical mixture of oligonucleotides with the biopolymer (Vassilenko et al. 1972; Knorre et al. 1973).

The next approach, may be the most popular one, is based on the investigation of functional properties of tRNA's modified in some definite points. Modification may be performed either by chemical means or by mutations. It may be expected that the successes in the chemical synthesis of tRNA's (Adamiak et al. 1978; Ohtsuka et al. 1978; Werstiuk and Neilson 1976) will provide in the near future new extremely effective possibilities for obtaining modified tRNA molecules.

A highly promising version of this general approach is the dissected molecules method proposed by Bayev et al. (1967). The method is based on the observation that tRNA retains the ability to be aminoacylated by ARSase after enzymatic cleavage of a small number of internucleotide bonds. This observation permitted to obtain various combinations of tRNA fragments lacking some nucleotide residues.

Three main types of results may be got by using modified tRNAs:

1. Modification or lack of some part of tRNA molecule does not change significantly the ability of tRNA to be aminoacylated. The conclusion may be drawn that this part does not participate in the tRNA-ARSase recognition.

2. Modification of some residue strongly damages the functional activity of tRNA. This result indicates that the residue is significant for the recognition process. However, one should discriminate between two possibilities. The first is that the modified residue

directly participates in the tRNA-ARSase interaction. The second is that the tertiary structure is damaged due to modification of the residue under consideration.

3. The specificity of aminoacylation may be changed due to modification. In this case one may conclude that the modified residue takes part in the specific step of the process.

The most essential results obtained by the above mentioned methods will be taken into account in the next section in the course of discussing structural features of the tRNA-ARSase recognition.

It must be emphasized that the level of aminoacylation (acceptor activity) is an insufficient characteristic of the level of damage of tRNA due to modification. In the absence of competing native tRNA the acceptor activity may be rather high, even with kinetic parameters K_m and V_{max} changed by several orders of magnitude, thus masking the real effect of modification.

As an example of the approach, we shall present the results obtained by Mirzabekov et al. (1972) using the dissected molecules method with yeast $tRNA_1^{Val}$. It was demonstrated that $tRNA_1^{Val}$ dissected at the $I_{35}-A_{36}$ bond remains active without $\psi_{33}-U_{34}$ $-I_{35}$ fragment, but is almost completely inactive without either G_1-G_2 or $A_{36}-C_{37}$ fragment. Neither of these fragments takes part in the stabilization of the tertiary structure of tRNA and, therefore, the results may be regarded as the indication of the direct participation of anticodon and of the acceptor stem in recognition. These data are in accordance with the above mentioned results based on the primary structure comparison.

Chemical modification of biopolymer may help to identify the points of contact of this polymer with the other one in two other manners. The first is based on the elucidation of the residues protected from modification in the complex. The second is based on the cross-linking of two biopolymers in the complex with subsequent identification of the points of cross-linking.

To ARSases this approach was at first applied by Bosshard et al. (1978) who studied the protection of Tyr-RSase of *B. stearothermophilus* from acetylation with acetic anhydride by $tRNA^{Tyr}$. With tRNA's this approach was first realized using nuclease digestion as a method of modification of tRNA (Dube 1973; Hörz and Zachau 1973; Dickson and Schimmel 1975). The interpretation of the data obtained with nucleases must be done with caution due to large dimensions of the enzymes and the strong influence they exhibit in some cases on the macromolecular structure of nucleic acids (Jensen and von Hippel 1976). The first attempt to use for the same pupose chemical reagents was made in our laboratory (Ankilova et al. 1975) and by Schoemacker and Schimmel (1976) who studied the points of protection of *E. coli* $tRNA^{Ile}$ by Ile-RSase against tritium exchange at C-8 of purine bases.

Using chemical modification to study biopolymers and their complexes one meets with the danger of damage to the macromolecular structure and the binding properties of the biopolymer in the course of modification.

In order to obtain information concerning the native state of biopolymers, we have proposed to measure the initial rates of modification of various points of the biopolymer (Vlassov et al. 1972b). We have chosen the alkylating reagents R_1Cl (Grineva et al. 1970) and R_2Cl (Bogachev et al. 1970) capable to modify at N-7 the guanine residues participating in the Watson-Crick base pairing

Modification with R_1Cl with subsequent acid treatment to hydrolyze the acetal bond, as well as modification with R_2Cl followed by treatment with 2,4-dinitrofluorobenzene, result in the appearance of the UV absorbance at 350 nm and, therefore, the extent of modification of tRNA may be easily measured spectrophotometrically. The charge of modified residues remains unchanged at pH 8 and is enhanced by one at pH 3.7, as compared with guanosine. Thus any modified oligonucleotide of the pancreatic RNAase digest may be easily identified by its position at the ion-exchange chromatography profile. Using special microcolumn technique elaborated in our institute (Vlassov et al. 1972a) we could measure the extent of modification of each oligonucleotide starting from 0.3 A_{260} units of tRNA.

The alkylation with aromatic 2-chloroethylamines proceeds via intermediate formation in the rate-limiting step of the corresponding ethylene immonium cation R^+, followed by the rapid reaction of R^+ with either guanosine G_i (rate constant k_i M^{-1} s^{-1}) or with water or nucleophilic components of the buffer (specific rate a s^{-1}). The ratio k_1/a is used to characterize the reactivities of the defined guanosine residues.

The alkylation of several tRNA's was studied using mainly R_1Cl: yeast $tRNA^{Val}$ (Vlassov et al. 1972b), mammalian $tRNA^{Val}$ (Vlassov et al. 1978a), E. coli $tRNA^{Phe}$ (Vlassov and Skobeltsyna 1978). As an example, we present in Table 2 our recent data with yeast $tRNA^{Phe}$. It is seen than not only G_{18}, G_{19}, and G_{35} in the looped region but also G_{10} and G_{24} in the D-stem are readily alkylated. At the same time G_{22} and G_{57} with N-7 participating in hydrogen bonding (Rich and Rajbhandary 1976) are low reactive. The low reactivity of G_{20} and G_{45} may be connected with the participation of their N-7 in the Mg^{2+} binding (Jack et al. 1977).

Both reagents were used to modify the E. coli $tRNA^{Phe}$-PheRSase complex. The results are given in Table 3. It should be emphasized that R_2Cl is significantly more reactive that R_1Cl and, therefore, the absolute values of k_i/a are significantly greater with the former reagent. This permits to obtain the well measurable level of modifica-

Table 2. The k_i/a values for the alkylation of various guanine residues in yeast tRNA^{Phe} by R_1Cl at 37°C in the conditions of stability of tertiary structure (0.1 M NaClO$_4$, 5.8 mM Mg(ClO$_4$)$_2$, 0.03 M Tris-HClO$_4$, pH 7.8) (A). For comparison, data obtained in the conditions of destroyed tertiary structure (0.03 M Tris-HClO$_4$, pH 7.8) are presented (B)

Guanosine	k_i/a, M^{-1}		Guanosine	k_i/a, M^{-1}	
	A	B		A	B
G_1	8	24	$m_2^2G_{26}, G_{71}$[a]	8	35
G_3, G_4[a]	2	22	G_{30}	21	26
m^2G_{10}	17	35	G_{34}	39	80
G_{15}	3	22	G_{42}	2	14
G_{18}, G_{19}[a]	21	37	G_{43}, G_{45}[a]	2	15
G_{20}	1	29	G_{51}, G_{53}, G_{57}[a]	2	22
G_{22}	3	16	G_{65}	1	23
G_{24}	18	15			

[a] Average values are given

Table 3. k_i/a values for the alkylation of various points of E. coli tRNA^{Phe} in the presence and in the absence of E. coli PheRSase (37°C, 0.025 M cacodylate, pH 6.2, 0.05 M MgCl$_2$, 2×10^{-4} M EDTA)

Guanosine residue	R_1Cl		R_2Cl	
	without ARSase	with ARSase	without ARSase	with ARSase
G_{34}	20	20	210	190
$G_{18}, G_{19}, G_{52}, G_{53}$[a]	18	18	420	400
G_{10}	34	40	600	200
G_{22}	1	1	150	30
G_{24}	52	4	1500	300

[a] Average values are given

tion of G_{22} and, therefore, to detect definitely the decrease of its reactivity in the presence of ARSase. It is seen that G_{24} is protected by ARSase in the experiments with both reagents. G_{10} is protected against R_2Cl but not against R_1Cl. The reason for this difference remains unclear.

Direct cross-linking of tRNA's to ARSases induced by UV irradiation of the complexes was proposed by Schoemaker and Schimmel (1974) for investigation of the structure of complexes. This method was applied to a set of tRNA-ARSase complexes (Budzik et al. 1975; Schoemaker et al. 1975; Rosa et al. 1979). The method permits in principle fairly precise localization of the points of cross-linking. Till now only approximate localization of these points was performed by identification of oligonucleotides of the T1 RNAase digest bound to the enzyme. The main disadvantage of the method is the use of intensive irradiation at 254 nm, which inactivates the enzyme. This may give rise to various artefacts.

To avoid this disadvantage we have proposed to introduce photoreactive groups in tRNA by alkylation with R_2Cl and subsequent treatment with 2,4-dinitro-5-fluorophenyl azide.

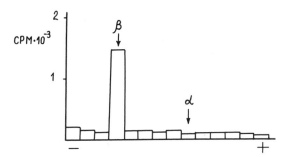

At low extent of modification (on the average two residues per tRNA molecule) *E. coli* tRNA^{Phe} still binds to *E. coli* Phe-RSase and may be cross-linked by irradiation at 350 nm without damage to either tRNA or ARSase. Using this method we could demonstrate that tRNA^{Phe} binds to β-subunit of the enzyme (Fig. 3). The result is in agreement with the data of Bartmann et al. (1974). These authors have demonstrated that β-subunit of *E. coli* Phe-RSase is alkylated in the complex with N-bromoacetyl-Phe-tRNA^{Phe}. At the same time it was recently found that Phe-tRNA^{Phe} bearing rather large chlorambucilyl residue attached to the α-amino group

$(ClCH_2CH_2)_2 N\text{-}C_6H_4\text{-}(CH_2)_3\text{-}CO\text{-}Phe\text{-}tRNA^{Phe}$

Fig. 3. The radioactivity profile of the sodium dodecyl sulfate gel electrophoresis of *E. coli* PheRSase photo-cross-linked with tRNA^{Phe} containing the randomly positioned photoreactive groups. The positions of α- and β-subunits are shown by *arrows*

alkylates in the complex the α-subunit (Lavrik and Khodyreva 1979). This strongly indicates that the acceptor end of tRNA is positioned not far from the region of contact between the subunits.

D. Structural Aspects of tRNA-ARSase Recognition

The role of the general features of the tertiary structure of tRNA's in the interaction with ARSases was recently discussed by Rich and Schimmel (1977). These authors proposed that all tRNA's are attached to ARSases along the diagonal connecting the edges of the L-shaped molecule including acceptor stem, D-stem, and anticodon arm. In different tRNA-ARSase couples either the entire area or some part of this region participates in the interaction.

The common features of the macromolecular structure of tRNA's should result in a common type of the charge distribution along the area of contact with the enzyme. One may reasonably expect that electrostatic interaction of negatively charged tRNA with some cluster of the positive charges at the enzyme surface is essential in the primary step of recognition.

The stability of tRNA-ARSase complexes decreases with the rise of the ionic strength, thus indicating the participation of the interaction between opposite charges. The number of electrostatic contacts was estimated from this dependence, using the simple electrostatic model of two interacting cylindrical polyelectrolytes, and was found to be near 6 for yeast $tRNA^{Phe}$-PheRSase and $tRNA^{Val}$-ValRSase complexes (Bonnet et al. 1975). However, it has to be realized that ARSases carry at neutral pH overall negative charge. The area of electrostatic interactions is, therefore, better represented as a positively charged area surrounded with negative charges. For this model the dependence of the association constant on the ionic strength should be less sharp and the same experimental data should correspond to a greater number of contacts. It looks likely that uneven charge distribution additionally directs tRNA molecule towards the recognition site of ARSase.

The role of one type of the carriers of positive charges, namely the lysine residues in the interaction of *E. coli* Phe-RSase with cognate tRNA, was demonstrated by Gorshkova et al. (1978). It was found that modification of the lysine residues of the enzyme with 2,4-pentanedione does not influence the aminoacyl adenylate formation, but decreases the ability of the enzyme to catalyze aminoacylation and to bind $tRNA^{Phe}$. The presence of $tRNA^{Phe}$ in the reaction mixture during the treatment of Phe-RSase with the reagent protects the enzyme against inactivation.

Further, we shall consider the data concerning the role of different parts of tRNA molecules in the interaction with ARSases.

We shall start with the acceptor stem and the adjoining nonpaired nucleotide in position 73. The latter is usually conserved within all tRNA's of the same specificity (e.g., A for all tRNA's specific to Ala,Ile,Leu, Met,Phe, Thr,Tyr,Val; G for all tRNA's specific to Asp,Asn,Glu,Gln,Ser). It is called discriminator nucleotide as it was proposed that it is used by the enzyme to discriminate between different groups of tRNA's (Crothers et al. 1972). The mutation in *E. coli* $tRNA^{Tyr}$ leading to the single base change $A_{73} \rightarrow G_{73}$ results in the conversion of tyrosine specific tRNA to glutamine specific one (Ghysen and Celis 1974).

Acceptor stem and discriminator nucleotide are the nearest neighbors of the 3'-end of tRNA accepting the amino acid and, therefore, they may be expected to play an essential role in the specific recognition step. Within each of both most extensively sequenced groups of tRNA's (E. coli tRNA's of 19 different specificities and yeast tRNA's of 15 different specifities, Gauss et al. 1979) the acceptor stems of the identical structure are met with only among isoacceptors. This may also be regarded as indirect indication of the specific role of this part of tRNA in recognition. However, the strong difference between the acceptor stems is found also within some groups of isoacceptor tRNA's, e.g., within valine-specific tRNA's, and between $tRNA_f^{Met}$ and $tRNA_m^{Met}$.

The first experimental proof of the role of the acceptor stem has come from the experiments of Schulman and Chambers (1968). The authors proposed the general chemical approach to the elucidation of the role of different parts of the tRNA molecule. The main idea of the method was to modify tRNA randomly at various points, to aminoacylate exhaustively the obtained heterogeneous population of modified tRNA's, and to separate the obtained aminoacylated molecules from the unacylated inactive tRNA's. The modified points completely absent from the aminoacylated fraction are regarded as essential for aminoacylation. Using photooxidation of the guanine residues, the authors demonstrated that the damage of G_{72} in the acceptor stem inactivates yeast $tRNA^{Ala}$.

The experiments with the mutants of $tRNA^{Tyr}$ from E. coli demonstrated that some changes in positions 1, 2, 81 of the acceptor stem result in the appearance of the glutamine acceptor activity (Celis 1979).

The data presented in the previous section concerning valine-specific tRNA's are also in favor of the specific role of the acceptor stem in the recognition.

The dinucleotide junction between the acceptor and D-stems is low variable (U-A or U-G, sometimes modified). However, this U_8 was found to be in contact with the enzyme in the above mentioned experiments of Schoemaker and Schimmel (1977a, b). E. coli $tRNA^{Phe}$ is cross-linked to Phe-RSase, being irradiated at wavelengths exceeding 300 nm, absorbed only by sU_8, thus indicating the close contact of sU_8 with the enzyme (Budker et al. 1974). G-G-A-s^4Up, representing positions 5-8 of E. coli $tRNA^{Phe}$, binds specifically with Phe-RSase (Table 1).

The structure of D-stem is not as variable as that of the acceptor stem. For example D-stem

(13)C-U-C-G(10)
 : : : :
(22)G-A-G-C(25)

is common for E. coli tRNA's specific to Ala, Ile, Lys,Met (noninitiator), Phe, Thr, Trp, Val. The D-stem

(13)C-G-C-G(m^2G)(10)
 : : : :
(22)G-C-G-C(25)

is common for yeast $tRNA^{Cys}$, *Neurospora crassa*, yeast and mammalian $tRNA_f^{Met}$, chicken cell $tRNA^{Trp}$, and yeast $tRNA^{Lys}$.

Dudock et al. (1971) have noticed that the D-stem structure of the first of the above mentioned types is common to a group of tRNA's chargeable by yeast Phe-RSase. Therefore, they proposed this region to be the recognition site for this enzyme. This proposal was later found to be inconsistent with some new data (Feldman and Zachau 1977; McCutchan et al. 1978).

However, a set of results indicate that D-stem participates in the tRNA-ARSase interaction. D-stem is photo-cross-linked with synthetases (Schoemaker and Schimmel 1974; Budzik et al. 1975). In our experiments we have demonstrated that G_{10}, G_{22}, and G_{24} of *E. coli* $tRNA^{Phe}$ are protected against alkylation by Phe-RSase (see Table 3). Oligonucleotide A-G-A-G-C, representing positions 21–25 of $tRNA^{Phe}$, binds specifically to the enzyme (Table 1).

Therefore, we may conclude that this region of tRNA participates in some step of recognition, may be the nonspecific one.

The structure of D-loops in various tRNA's are rather similar and the D-loop can not be regarded as a region of specific recognition. Most probably, D-loop does not participate at all in the tRNA-ARSase interaction. Chemical modifications and splitting off parts of the D-loop usually do not inactivate tRNA's toward aminoacylation (for review, see Rich and Schimmel 1978).

Anticodon stems of tRNA are rather variable and could be the site of specific recognition. However, any definite data concerning the role of this part of tRNA are practically absent. Photo-cross-linking results in some cases in the covalent binding of ARSase to oligonucleotides containing some part of the anticodon stem (Schoemaker et al. 1975; Rosa et al. 1979). However, these oligonucleotides overlap several other parts of cross-linked tRNA's. Therefore, it is difficult to decide whether cross-links were formed by the anticodon stem or by the adjacent part of the tRNA structure.

Anticodons always differ for tRNA's of different specificities. Therefore, they are the most attractive candidates for being the recognition sites. The role of anticodon in the tRNA-ARSase interaction was first formulated by Kisselev and Frolova (1964) and further supported by a great number of data. The most impressive are the results with *E. coli* $tRNA^{Trp}$. The conversion of anticodon C-C-A to C-U-A, either due to single base mutation (Yaniv et al. 1974) or by treatment with bisulfite (Seno 1975), results in the change of specificity from tryptophan to glutamine. Detailed consideration of the experimental evidence in favor of the role of anticodon in the recognition is presented by Kisselev and Favorova (1974).

The participation of anticodon in the recognition is probably not the common feature of all tRNA-ARSase couples. Thus, essential tyrosine acceptor activity was found in the complex of the two halves of *T. utilis* $tRNA^{Tyr}$ completely lacking anticodon nucleotides (Hashimoto et al. 1972).

E. coli PheRSase does not protect anticodon of $tRNA^{Phe}$ against alkylation with R_1Cl and R_2Cl (Table 3). However, this result demonstrates only the nonparticipation of N-7 of G_{34} in the interaction with ARSase and does not give any information concerning the other parts of the anticodon.

Large extra arm is a common feature of all till now sequenced tRNA's specific to serine and leucine, and of prokaryotic tyrosine-specific tRNA's. Two former amino acids are coded each by six codons. One might expect that the presence of a large additional part in the respective tRNA's is essential for discrimination of this group of isoacceptors by corresponding ARSases. The middle part of the sectra arm of *E. coli* tRNA^{Tyr} is photo-cross-linked with the cognate enzyme (Schoemaker and Schimmel 1974).

At the same time, the presence of the large loop does not hinder the interaction of tRNA with ARSases specific to tRNA's with small extra loops. Thus, it was already mentioned that *E. coli* tRNA^{Tyr} bearing large extra arm is converted to glutamine-specific tRNA by some single base mutations without changes in the dimension of the extra arm.

The structure of small extra arms are rather variable. Among *E. coli* and T4 tRNA's, interacting in native conditions with *E. coli* ARSases, only in few cases have two or three tRNA's of different specificity identical small extra arms (e.g., G-U-m^{7}G-U-U for tRNA^{Thr} and tRNA^{Trp}, A-G-m^{7}G-U-C for tRNA$_f^{Met}$, tRNA^{Ala}, and T4 tRNA^{Pro}). Consequently, this part of tRNA may be regarded as a candidate for the recognition site. However, the experimental data concerning the functional significance of the small extra arm are extremely poor. The single definite result was obtained by Schulman (1971) who has found that tRNA$_f^{Met}$ with modified G$_{45}$ is absent from the active fraction of tRNA$_f^{Met}$ with photooxidized guanosines. T-stem is represented by one constant and four highly variable base pairs. In some cross-linking experiments oligonucleotides, representing the part of T-stem with adjacent part of either T-loop or extra arm, bind to ARSase. G$_{50}$ of *E. coli* tRNA^{Ile} is protected by cognate ARSase against tritium exchange at C-8. Oligonucleotide A-G-G-T, representing three nucleotides of the T-stem with universal T of the T-loop, was found to be protected by beef pancreas TrpRSase against alkylation with R$_1$Cl in yeast tRNA^{Trp} (Vlassov et al. 1978b). All these data are in favor of some participation of T-stem in the interaction with ARSases.

T-loop is the least variable part of tRNA's. According to Rich and Schimmel (1977) this part does not interact with ARSases. There are no definite data which disagree with this statement. Some part of T-loop was found to be photo-cross-linked with ARSases in the yeast tRNA^{Phe}-PheRSase (Schoemaker et al. 1975) and several tRNA$_f^{Met}$-MetRSase complexes (Rosa et al. 1979). However, in all of these cases cross-linked oligonucleotides contain the adjoining part of the T-stem.

Till now none of the tRNA-ARSase couples were studied with sufficient completeness. Within the data scattered over a number of couples the greatest part is concentrated upon yeast tRNA^{Val} (see previous section), and *E. coli* tRNA$_f^{Met}$ and tRNA^{Ile}. A survey of the data concerning tRNA$_f^{Met}$ was recently presented by Schulman and Pelka (1977). This tRNA was mainly studied by the above mentioned general approach of Schulman and Chambers (1968), using several different chemical modifications. The results are given in Fig. 4 together with photo-cross-linking results of Rosa et al. (1979). The data concerning tRNA^{Ile} were obtained mainly by using the protection of purine bases against tritium exchange with the cognate enzyme. The results are presented in Fig. 5 together with photo-cross-linking data.

The main conclusion is that in all cases several separated points of tRNA participate in the interaction. That means that all tRNA's have composite recognition sites. The changes in specificity due to single-base mutation do not contradict this statement. *E. coli* tRNA^{Tyr}, tRNA^{Trp}, and tRNA^{Gln} have a number of common sites which are

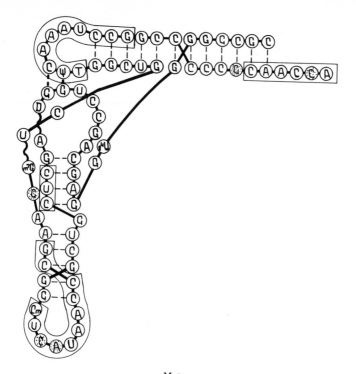

Fig. 4. Structure of *E. coli* tRNA$_f^{Met}$. The bases essential for recognition as revealed by chemical modification data are *stippeled.* Regions photo-cross-linked to *E. coli* MetRSase are *framed*

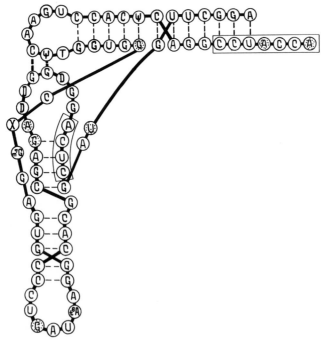

Fig. 5

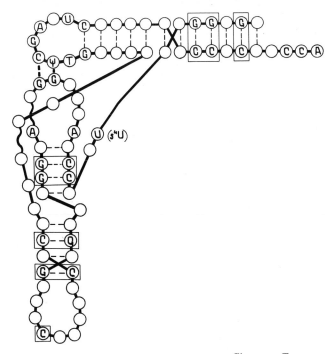

Fig. 6. The generalized structure of *E. coli* tRNA[Gln], tRNA[Trp] and tRNA[Tyr]. Only nucleotides common to all these tRNA's are marked by *symbols*. The candidates for the sites of specific recognition are *framed*

presented in Fig. 6. Some or all of them may participate in the formation of the composite recognition site.

The operation in the tRNA-ARSase recognition of the composite recognition sites seems to be generally accepted. Thus Freist and Cramer (1976) have proposed the model of recognition based on the existence of a set of binary decisions (six to eight), taking into account the differences at various parts of the tRNA structure.

Concluding this section, we would like to emphasize that not only the points of definite chemical structure but some more complicated images may be parts of the composite recognition sites. A proposal of this type was made by Ivanov (1975), who suggested that guanine residues in the minor grooves may exercise some repulsive function toward the enzyme, and that some specific types of distribution of guanines in the tRNA stems are necessary for recognitions.

Fig. 5. Structure of *E. coli* tRNA[Ile]. Purines protected from ^{3}H exchange by Ile-RSase and U_8 directly interacting with the enzyme are *stippled*. Regions photo-cross-linked to IleRSase are *framed*

E. The Dynamic Aspects of the tRNA-ARSase Recognition

The multiplicity of the recognition sites on tRNA molecules strongly suggests the multi-step character of the process. To go deeper into the problem it is essential to get information concerning the number of these steps and their kinetic characteristics.

The first attempt to follow the kinetics of the process

$$\text{tRNA} + \text{ARSase} \underset{k_{-1}}{\overset{k_1}{\rightleftharpoons}} \text{tRNA.ARSase}$$

was made by Yarus and Berg (1969). The authors have proposed to determine k_{-1} by measuring the rate of exchange between labeled aminoacyl-tRNA and unlabeled tRNA within the complex, assuming dissociation to be the rate-limiting step of the exchange. Using nitrocellulose filters to quantify the amount of *E. coli* [^{14}C] Ile-tRNA.IleRSase, they have found k_{-1} to be 7.5×10^{-3} s^{-1} at $17°$ and pH 5.5. With association constant $K_1 = k_1/k_{-1}$ measured in equilibrium conditions they have estimated k_1 as 1.3×10^6 M^{-1}. Similar estimates were obtained for *E. coli* LeuRSase (Rouget and Chapeville 1971) and ValRSase (Helene et al. 1971).

The measurements near pH optimum of ARSases have become possible by using the fast kinetics technique. Already the first data, obtained by using stopped-flow spectrophotometer adapted to fluorescence detection, have demonstrated that at pH 7.2 and $20°C$ k_1 is of the order of 10^8 M^{-1} s^{-1}, the value not differing significantly from that for the diffusion controlled process (Krauss et al. 1973). It was found that up to KCl concentration of the order of 0.1 M k_{-1} increases with the rise in the salt concentration (Pingoud et al. 1973; Rigler et al. 1976).

The temperature jump experiments have revealed that interaction of tRNA with cognate ARSase proceeds via two steps. Association is followed by some conformational transition with the equilibrium constant K_2 of the order of unity. Thus K_2 at $25°C$ and neutral pH was found to be 7 for the yeast SerRSase-tRNA^{Ser} system (Rigler et al. 1976), and 0.6 for the yeast PheRSase-tRNA^{Phe} system (Krauss et al. 1976). It was found that both forward and reverse reactions slow down when homologous Phe-RSase is changed for heterologous *E. coli* enzyme. The rate constant of the forward reaction k_2 was found to be 420 and 150 s^{-1}, and the rate constant of the reverse reaction k_{-2} to be 750 and 200 s^{-1}, respectively. The complex of yeast PheRSase with *E. coli* tRNA tRNA^{Tyr} was found to be of nearly the same stability as the cognate complex (5×10^5 M^{-1} and 8×10^8 M^{-1}, respectively). However, the formation of the complex PheRSase. tRNA^{Tyr} was not accompanied by any conformational transition. This transition was also absent in the course of interaction between yeast PheRSase with cognate tRNA lacking 3'-terminal adenosine (Krauss et al. 1977). In the formation of 2tRNA.ARSase complexes conformational transitions were demonstrated to accompany the binding of both tRNA molecules (Riesner et al. 1976).

These data clearly demonstrate that tRNA-ARSase interaction includes essentially dynamic events. The conformational change preceding the catalytic step of the tRNA aminoacylation process may be induced by some definite part of the composite recognition site. Rigler et al. (1976) have proposed that this step is connected with identification of the cognate tRNA following the unspecific scanning. However, it may also be directed by some universal part of the tRNA molecule. Thus, van der Haar and Gaertner

(1975) have proposed that 3'-terminal adenosine triggers this conformational change. This concept was further developed by van der Haar and Cramer (1978a, b). The experiments with various tRNA-ARSase couples as well as with mutant and modified tRNA's are necessary to discriminate between these alternatives.

The quantitative data obtained by the temperature jump method provide the possibility to estimate the influence of the conformational transition on the overall kinetic parameters of aminoacylation. This analysis may be of interest in connection with the data of Ebel et al. (1973) concerning the particular importance of the maximal velocity V_m in the specificity of aminoacylation.

The extremely simplified scheme of aminoacylation may be presented as

$$E + S \underset{k_{-1}}{\overset{k_1}{\rightleftharpoons}} ES \underset{k_{-2}}{\overset{k_2}{\rightleftharpoons}} (ES)^* \xrightarrow{k_3} E + P$$

where E = enzyme with preformed aminoacyl adenylate, S = tRNA, P = aminoacyl-tRNA. The Michaelis parameters of this process are given by the expressions (Knorre 1975)

$$V_m = \frac{k_3 e}{1 + \frac{k_{-2} + k_3}{k_2}}, \quad K_m = \frac{k_{-1}}{k_1} \times \frac{\frac{k_3}{k_{-1}} + \frac{k_{-2} + k_3}{k_2}}{1 + \frac{k_{-2} + k_3}{k_2}}$$

where e = the total enzyme concentration.

We shall further use the values of the rate constants for yeast PheRSase.tRNA^{Phe} measured in the presence of the analog of aminoacyladenylate and phenylalanyladenylate at 25°C (Krauss et al. 1978):

$$k_1 = 7.5 \times 10^7 \text{ M}^{-1} \text{ s}^{-1}, k_{-1} = 70 \text{ s}^{-1}, k_2 = 10 \text{ s}^{-1}, k_{-2} = 20 \text{ s}^{-1}.$$

The rate constant of the transfer of phenylalanyl residue to tRNA^{Phe} was determined by Fasiolo and Fersht (1978) as $k_3 = 11 \text{ s}^{-1}$.

As a first approximation, we shall assume that conformational transition is completely responsible for the specificity of the overall process. The k_3 values may be suggested to be the same for cognate and noncognate tRNA, for in both cases the identical universal oligonucleotide X-C-C-A interacts with the same active site of the enzyme. V_m/e and K_m are easily calculated as 2.7 s^{-1} and 7×10^7 M, respectively. It is easily seen that significant decrease of k_2 results in the nearly proportional decrease of V_m and only slight rise of K_m (to 9×10^7 M). This is in good qualitative agreement with the experimental data of Ebel et al. (1973).

The problem is still more complicated due to multiplicity of the states of tRNA-ARSase complexes. The detailed kinetic investigation of beef pancreas TrpRSase (Kiselev et al. 1978) and yeast PheRSase (Tiebe 1978) have led the authors to the conclusion that several different tRNA-containing enzyme-substrate complexes participate in the reaction mechanism. For example, the following complexes were introduced in the reaction scheme for TrpRSase: (1) tRNA.E.Trp~pA; (2) E.Trp-tRNA; (3) ATP.E. Trp-tRNA; (4) ATP.Trp.E.Trp-tRNA; (5) tRNA.Trp~pA.E.Trp-tRNA. The arrangement of tRNA in various complexes may be significantly different.

The experiments of Yarus and Berg (1969) have clearly demonstrated that the presence of low molecular weight ligands may effect the tRNA-ARSase interaction. They have found that both k_1 and k_{-1} for IleRSase.tRNA^{Ile} complex increased significantly in the presence of ATP and isoleucine. Similar data were obtained for the *E. coli* LeuRSase (Rouget and Chapeville 1971) and ValRSase systems (Helene et al. 1971). Further, the influence of the small ligands on the behavior of tRNA-ARSase complexes was demonstrated using various approaches. Thus, it was found that the presence of methionyl adenylate is essential for the binding of the second tRNA molecule to *E. coli* MetRSase (Jayat et al. 1074). It was noticed that the presence of ATP is essential to prevent the excess (over stoichiometric) binding of rat liver tRNA^{Ser} to yeast SerRSase (Maelicke et al. 1974). ATP was demonstrated to enhance the sensitivity of yeast tRNA^{Ser} in the complex with SerRSase to nuclease digestion (Packmann and Zachau 1978).

In the course of affinity labeling of *E. coli* PheRSase (Gorshkova and Lavrik 1975) and of beef pancreas TrpRSase (Akhverdyan et al. 1977) it was found that modification of the enzymes with corresponding chlorambucilyl-aminoacyl-tRNA's slows down significantly in the presence of either ATP or respective amino acid, and does not proceed at all in the presence of aminoacyl adenylate. The rate of photo-cross-linking of tRNA^{Phe}-bearing azido group attached to s^4U_8 to *E. coli* PheRSase is significantly reduced in the presence of aminoacyl adenylate (Gorshkova et al. 1976).

Strong negative cooperativity was found to exist between two tRNA molecules binding to the same enzyme molecule (Bartmann et al. 1975; Pingoud et al. 1975).

Again the question arises whether some of these interactions between active centers of ARSase bear some function essential for the specificity of recognition. Most of the above mentioned results may be explained as being due to overlapping between the active sites for the low molecular weight substrates and universal C-C-A end of tRNA (Gorshkova et al. 1976). Two tRNA binding sites may also partially overlap (Bosshard et al. 1978). The elucidation of the role of the cooperative effects in the specificity of recognition is one of the most important problems which have to be solved in the near future.

Being extremely cumbersome as compared with the enzymatic conversions of the low molecular weight substrates, tRNA-ARSase interaction represents perhaps the simplest system with the crucial role of dynamic events. The knowledge of the main features of the mechanism of recognition between tRNAs and ARSases will permit to move on to the investigation of the molecular mechanisms of more complicated systems operating in the living organisms.

References

Adamiak RW, Biala E, Grzeskowiak K, Kierzek R, Kraszewski A, Markiewicz WT, Okupniak J, Stawinski J, Wiewioroski M (1978) The chemical synthesis of the anticodon loop of an eukaryotic tRNA containing the hypermodified nucleoside N^6-(N-threonylcarbonyl)-adenosine (t^6A). Nucl Acids Res 5: 1889–1905

Akhverdyan VZ, Kisselev LL, Knorre DG, Lavrik OI, Nevinski GA (1977) Affinity labelling of tryptophanyl-transfer RNA synthetase. J Mol Biol 113: 475–501

Ankilova VN, Vlassov VV, Knorre DG, Melamed NV, Nuzhdina NA (1975) Involvment of the D-stem of tRNA^{Phe} (*E. coli*) in interaction with phenylalanyl-tRNA synthetase as shown by chemical modification. FEBS Lett 60: 168–171

Axel'rod VD, Kryukov VM, Isaenko SN, Bayev AA (1974) Nucleotide sequence in tRNA$_{2a}^{Val}$ from baker's yeast. FEBS Lett 45: 333–336

Bartmann P, Hanke T, Hammer-Raber B, Holler E (1974) Selective labelling of the β-subunit of L-phenylalanyl-tRNA synthetase from *E. coli* with N-bromoacetyl-L-phenylalanyl-tRNA^{Phe}. Biochem Biophys Res Commun 60: 743–747

Bartmann P, Hanke T, Holler E (1975) Active site stoichiometry of L-phenylalanine:tRNA ligase from *Escherichia coli* K-10. J Biol Chem 250: 7668–7674

Bayev AA, Fodor I, Mirzabekov AD, Axel'rod VD, Kazarinova LYa (1967) Functional studies of fragments of yeast valine tRNA molecule. Mol Biol (Moscow) 1: 859–871

Bogachev VS, Veniaminova AG, Grineva NI, Lomakina TS (1970) The alkylating derivatives of nucleic acids components. IX. Synthesis and properties of uridylyl- and adenylyl-(5'→N)-4-(N-2-chloroethyl-N-methylamino)-benzylamines. Izv Sib Otd Akad Nauk SSSR, ser khim nauk N°14, iss 6, 110–116

Bonnet J, Renaud M, Raffin JP, Remy P (1975) Quantitative study of the ionic interactions between yeast tRNA^{Val} and tRNA^{Phe} and their cognate amino acyl-tRNA ligases. FEBS Lett 53: 154–158

Bosshard HR, Koch GLE, Hartley BS (1978) The aminoacyl-tRNA synthetase – tRNA complex: Detection by differential labelling of lysine residues involved in complex formation. J Mol Biol 119: 377–389

Budker VG, Knorre DG, Kravchenko VV, Lavrik OI, Nevinsky GA, Teplova NM (1974) Photo-affinity reagents for modification of aminoacyl-tRNA synthetases. FEBS Lett 49: 159–162

Budzik GP, Lam SSM, Schoemaker HJP, Schimmel PR (1975) Two photo-cross-linked complexes of isoleucine specific transfer ribonucleic acid with aminoacyl transfer ribonucleic acid synthetases. J Biol Chem 250: 4433–4439

Celis JE (1979) Collection of mutant tRNA sequences. Nucl Acids Res 6: r21–r27

Chapeville F, Lipmann F, von Ehrenstein G, Weisblum B, Ray WJ, Benzer S (1962) On the role of soluble ribonucleic acid in coding for amino acids. Proc Natl Acad Sci USA 48: 1086–1092

Crothers DM, Seno T, Söll DG (1972) Is there a discriminator site in tRNA? Proc Natl Acad Sci USA 69: 3063–3067

Dickson LA, Schimmel PR (1975) Structure of transfer RNA-aminoacyl-tRNA synthetase complexes investigated by nuclease digestion. Arch Biochem Biophys 167: 638–645

Dube SK (1973) Evidence for "three-point" attachment of tRNA to methionyl-tRNA synthetase. Nature New Biol 243: 103–105

Dudock B, DiPeri C, Scileppi K, Reszelbach R (1971) The yeast phenylalanyl-transfer RNA synthetase recognition site: the region adjacent to the dihydrouridine loop. Proc Natl Acad Sci USA 68: 681–684

Ebel JP, Giege R, Bonnet J, Kern D, Bedfort N, Bollack C, Fasiolo F, Gangloff J, Dirheimer G (1973) Factors determining the specificity of the tRNA aminoacylation reaction. Non-absolute specificity of tRNA-aminoacyl-tRNA recognition and particular importance of the maximal velocity. Biochimie 55: 547–557

Fasiolo F, Fersht A (1978) The aminoacyladenylate mechanism in the aminoacylation reaction of yeast phenylalanyl-tRNA synthetase. Eur J Biochem 85: 85–88

Fayat G, Blanquet S, Waller JP (1974) Etude des sites reactionnels de la methionyl tRNA synthetase d'*Escherichia coli.* Arch Int Physiol Biochim 82: 766–767

Feldmann H, Zachau HG (1977) Charging of a yeast methionine tRNA with phenylalanine and its implication for the synthetase recognition problem. Hoppe Seyler's Z Physiol Chem 358: 891–896

Fersht AR, Kaethner MM (1976) Enzyme hyperspecificity. Rejection of threonine by the valyl-tRNA synthetase by misacylation and hydrolytic editing. Biochemistry 15: 3342–3346

Freist W, Cramer F (1976) A binary code model for substrate recignition of aminoacyl-tRNA ligases. J Theor Biol 58: 401–416

Gauss DH, Grüter F, Sprinzl M (1979) Compilation of tRNA sequences. Nucl Acids Res 6: r1–r19

Ghysen A, Celis JE (1974) Mischarging single and double mutants of *E. coli* sup 3 tyrosine transfer RNA. J Mol Biol 83: 333–351

Gorbulev VG, Axelrod VD, Bayev AA (1977) Primary structure of baker's yeast tRNA$_{21}^{Val}$. Nucl Acids Res 4: 3239–3258

Gorshkova II, Lavrik OI (1975) The influence of the ATP, amino acids and their analogs on the kinetics of the affinity labelling of phenylalanyl-tRNA synthetase. FEBS Lett 52: 135–138

Gorshkova II, Knorre DG, Lavrik OI, Nevinski GA (1976) Affinity labelling of phenylalanyl-tRNA synthetase from *E. coli* MRE-600 by *E. coli* tRNA^{Phe} containing photoreactive group. Nucl Acids Res 3: 1577–1589

Gorshkova II, Dacy II, Lavrik OI, Mamaev SV (1978) The role of the lysine residues in the phenylalanyl-tRNA synthetase substrates interaction. Mol Biol (Moscow) 12: 1096–1104

Grineva NI, Zarytova VF, Knorre DG (1970) Alkylating derivatives of nucleic acids components. VII. 2',3'-O- 4(N-2-chloroethyl-N-methylamino)-benzylidene uridine-5'-methylphosphate. Zh Obsh Khim 40: 215–222

Haar F von der, Cramer F (1976) Hydrolytic action of aminoacyl-tRNA synthetases from baker's yeast: "Chemical proofreading" preventing acylation of tRNA^{Ile} with misactivated valine. Biochemistry 15: 4131–4138

Haar F von der, Cramer F (1978a) Seryl-, Threonyl-, Valyl- and Isoleucyl-tRNA synthetases from baker's yeast: role of the 3'-terminal adenosine in the dynamic recognition of tRNA. Biochemistry 17: 3139–3145

Haar F von der, Cramer F (1978b) Valyl- and phenylalanyl-tRNA synthetase from baker's yeast: recognition of transfer RNA results from a multistep process, as indicated by inhibition of aminoacylation with modified transfer RNA. Biochemistry 17: 4509–4514

Haar F von der, Gaertner E (1975) Phenylalanyl-tRNA synthetase from baker's yeast: role of 3'-terminal adenosine of tRNA^{Phe} in enzyme-substrate interaction studied with 3'-modified tRNA^{Phe} species. Proc Natl Acad Sci USA 72: 1378–1382

Hashimoto S, Kawata M (1972) Reconstitution of an active acceptor complex which lacks the anticodon of *Torulopsis* tyrosine transfer ribonucleic acid. J Biochem 72: 1339–1349

Hélène C, Brun F, Yaniv M (1971) Fluorescence studies of interactions between *Escherichia coli* Valyl-tRNA synthetase and its substrates. J Mol Biol 58: 349–365

Hoagland MB, Keller EB, Zamecnik PC (1956) Enzymatic carboxyl activation of amino acids. J Biol Chem 218: 345–358

Hoagland MB, Zamecnik PC, Stephenson ML (1957) Intermediate reactions in protein biosynthesis. Biochem Biophys Acta 24: 215–216

Holley RW, Apgar J, Everett GA, Madison JT, Marquisee M, Merrill SH, Penswick JR, Zamir A (1965) Structure of a ribonucleic acid. Science 147: 1462–1465

Hopfield JJ (1978) Origin of the genetic code: A testable hypothesis based on tRNA structure, sequence, and kinetic proofreading. Proc Natl Acad Sci USA 75: 4334–4338

Hörz W, Zachau HG (1973) Complexes of aminoaxyl-tRNA synthetases with tRNAs as studied by partial nuclease digestion. Eur J Biochem 32: 1–14

Ivanov VI (1975) The binary code for protein-nucleic acid recognition with repulsive guanine: application to tRNA case. FEBS Lett 59: 282–286

Jack A, Ladner JE, Rhodes D, Brown RS, Klug A (1977) A crystallographic study of metal-binding to yeast phenylalanine transfer RNA. J Mol Biol 111: 315–328

Jensen DE, von Hippel PH (1976) DNA "melting" proteins. 1. Effects of bovine pancreatic RNAase binding on the conformation and stability of DNA. J Biol Chem 251: 7198–7214

Kim SH, Suddath FL, Quigley GH, McPherson A, Sussman JL, Wang AHJ, Seeman NC, Rich A (1974) Three dimensional tertiary structure of yeast phenylalanine transfer RNA. Science 185: 435–440

Kisselev LL, Favorova OO (1974) Aminoacyl-tRNA synthetases: some recent results and achievements. Adv Enzymol 40: 141–238

Kisselev LL, Frolova LYu (1964) "Recognition sites" of transfer RNAs responsible for specific interaction with aminoacyl-RNA-synthetases. Biokhimiya 29: 1177–1189

Kisselev LL, Malygin EG, Akhverdyan VZ, Zinovyev VV (1978) The mechanism of functioneering of triptophanyl-tRNA synthetase. Dokl Akad Nauk SSSR 238: 1475–1478

Knorre DG (1975) Stepwise tRNA recognition mechanism and its kinetic consequences. FEBS Lett 58: 50–52

Knorre VL, Vasilenko SV, Salganik RI (1973) Specific binding of oligoribonucleotide fractions to *E. coli* RNA polymerase. FEBS Lett 30: 229–230

Krauss G, Römer R, Riesner D, Maass G (1973) Thermodynamics and kinetics of the interaction of phenylalanine specific tRNA from yeast with its cognate synthetase as studied by the fluorescence of the Y-base. FEBS Lett 30: 6–10

Krauss G, Riesner D, Maass G (1976) Mechanism of discrimination between cognate and non-cognate tRNAs by phenylalanyl-tRNA synthetase from yeast. Eur J Biochem 68: 81–93

Krauss G, Riesner D, Maass G (1977) Mechanism of tRNA-synthetase recognition: role of terminal A. Nucl Acids Res 4: 2253–2262

Krauss G, Coutts SM, Riesner D, Maass G (1978) Mechanism of tRNA - aminoacyl-tRNA synthetase recognition: influence of aminoalkyladenylates. Biochemistry 17: 2443–2449

Lavrik OI, Khodyreva SN (1979) Modification of the α-subunit of phenylalanyl-tRNA synthetase from *E. coli* MRE-600 with N-chlorambucilyl-phenylalanyl-tRNA. Biokhimiya 44: 570–572

Lestienne P (1978) The specificity of aminoacylation: a tRNA-tRNA interaction model. J Theor Biol 73: 159–180

Letendre CH, Humphreys JM, Grunberg-Manago M (1969) Complex formation between lysyl transfer RNA synthetase and polyuridylic acid. Biochim Biophys Acta 186: 46–61

Maelicke A, Engel G, Cramer F, Staehelin M (1974) ATP-induced specificity of the binding of serine tRNAs from rat liver to seryl-tRNA synthetase from yeast. Eur J Biochem 42: 311–314

McCutchan T, Silverman S, Kohli J, Söll D (1978) Nucleotide sequence of phenylalanine transfer RNA from *Schizosaccharomyces pombe:* implications for transfer RNA recognition by yeast phenylalanyl-tRNA synthetase. Biochemistry 17: 1622–1628

Mirzabekov AD, Lastity D, Levina ES, Undritzov IM, Bayev AA (1972) The acceptor activity of dissected molecules of baker's yeast tRNA$_1^{Val}$. Mol Biol (Moscow) 6: 87–105

Norris AT, Berg P (1964) Mechanism of aminoacyl-RNA synthesis: studies with isolated aminoacyl adenylate complexes of isoleucyl RNA synthetase. Proc Natl Acad Sci USA 52: 330–337

Ohtsuka E, Nishikawa S, Markham AF, Tanaka S, Miyake T, Wakabayashi T, Ikehara M, Sigiura M (1978) Joining of 3'-modified oligonucleotides by T4 RNA ligase. Synthesis of a heptadecanucleotide corresponding to the bases 61–77 from *Escherichia coli* tRNA^{fMet}. Biochemistry 17: 4894–4899

Pachmann U, Zachau HG (1978) Yeast seryl-tRNA synthetase: interactions between the ATP binding site and the sites for tRNA^{Ser} and L-serine. Nucl Acids Res 5: 975–985

Pingoud A, Riesner D, aoehme D, Maass G (1973) Kinetic studies of the interaction of seryl-tRNA syntehtase with tRNA^{Ser} and Ser-tRNA^{Ser} from yeast. FEBS Lett 30: 1–5

Pingoud A, Boehme D, Riesner D, Kownatzki R, Maass G (1975) Anti-cooprative binding of two tRNA^{Tyr} molecules to tyrosyl-tRNA synthetase from *Escherichia coli.* Stopped-flow investigations using unmodified tRNA^{Tyr} and a fluorescent derivative of tRNA^{Tyr}. Eur J Biochem 56: 617–622

Reid BR (1977) Synthetase-tRNA recognition. In: Vogel HJ (ed) Nucleic acid-protein recognition. Academic Press, New York London

Rich A, Rajbhandary UL (1976) Transfer RNA: molecular structure, sequence and properties. Ann Rev Biochem 45: 805–860

Rich A, Schimmel PR (1977) Structural organization of complexes of transfer RNAs with aminoacyl transfer RNA synthetase. Nucl Acids Res 4: 1649–1665

Riesner D, Pingoud A, Boehme D, Peters F, Maass G (1976) Distinct steps in the specific binding of tRNA to aminoacyl-tRNA synthetase. Temperature-jump studies on the serine-specific system from yeast and the tyrosine-specific system from *Escherichia coli.* Eur J Biochem 68: 71–80

Rigler R, Pachmann V, Hirsh R, Zachau HG (1976) On the interaction of seryl-tRNA synthetase with tRNA^{Ser}. A contribution to the problem of synthetase-tRNA recognition. Eur J Biochem 65: 307–315

Robertus JD, Ladner JE, Finch JT, Rhodes D, Brown RS, Clark BFC, Klug A (1974) Structure of yeast phenylalanine tRNA at 3 Å resolution. Nature 250: 546–551

Rosa JJ, Rosa MD, Sigler PB (1979) Photocross-linking analysis of the contact surface of tRNA^{Met} in complexes with *E. coli* methionine: tRNA ligase. Biochemistry 18: 637–647

Rouget P, Chapeville F (1971) Leucyl-tRNA synthetase. Mechanism of leucyl-tRNA formation. Eur J Biochem 23: 443–451

Schoemaker HJP, Schimmel PR (1974) Photo-induced joining of a transfer RNA with its cognate aminoacyltransfer RNA synthetase. J Mol Biol 84: 503–513

Schoemaker HJP, Schimmel PR (1976) Isotope labelling of free and aminoacyl transfer RNA synthetase-bound transfer RNA. J Biol Chem 251: 6823–6830

Schoemaker HJP, Schimmel PR (1977a) Effect of aminoacyl transfer RNA synthetases on H-5 exchange of specific pyrimidines in transfer RNAs. Biochemistry 16: 5454–5460

Schoemaker HJP, Schimmel PR (1977b) Inhibition of an aminoacyl-tRNA synthetase by a specific trinucleotide derived from the sequence of its cognate transfer RNA. Biochemistry 16: 5461–5464

Schoemaker HJP, Budzik GP, Giege R, Schimmel PR (1975) Three photo-cross-linked complexes of yeast phenylalanine specific transfer ribonucleic acid with aminoacyl transfer ribonucleic acid synthetases. J Biol Chem 250: 4440–4444

Schulman LH (1971) Structure and function of *Escherichia coli* formylmethionine transfer RNA. J Mol Biol 58: 117–131

Schulman L-DH, Chambers RW (1968) Transfer RNA. II. A structural basis for alanine acceptor activity. Proc Natl Acad Sci USA 61: 308–315

Schulman LH, Pelka H (1977) Structural requirement for aminoacylation of *Escherichia coli* formylmethionine transfer RNA. Biochemistry 16: 4256–4265

Seno T (1975) Conversion of *Escherichia coli* tRNA^{Trp} to glutamine-accepting tRNA by chemical modification with sodium bisulfite. FEBS Lett 51: 325–329

Thiebe R (1978) Analysis of the steady-state mechanism of the aminoacylation of tRNA^{Phe} by phenylalanyl-tRNA synthetase from yeast. Nucl Acids Res 5: 2055–2071

Vasilenko SK, Ankilova VN, Dimitrova FF, Servo NA (1972) Formation of complexes between tRNA and trinucleotides. FEBS Lett 27: 215–218

Vlassov VV, Khodyreva SH (1978) Equilibrium screening-dialysis investigation of the nucleotide sequences in the tRNA^{Phe} recognized by phenylalanyl-tRNA synthetase (*Escherichia coli*). FEBS Lett 96: 95–98

Vlassov VV, Skobeltsyna LM (1978) Chemical modification study on the macrostructure of tRNA^{Phe}. Bioorgan Khymiya 4: 550–562

Vlassov VV, Grachev MA, Komarova NI, Kuzmin SV, Menzorova NI (1972a) Ion-excahnge chromatography and spectral analysis of oligonucleotides in a microscale. Mol Biol (Moscow) 6: 809–816

Vlassov VV, Grineva NI, Knorre DG (1972b) Exposed and buried guanosine residues in tRNA$_1^{Val}$ from yeast. FEBS Lett 20: 66–70

Vlassov VV, Pusyriov AT, Gross HJ (1978a) Evidence from chemical modification for an unusual tertiary structure of the TψC loop in rabbit liver tRNA^{Val}. FEBS Lett 94: 157–160

Vlassov VV, Tchizhikov VE, Scheinker VS, Favorova OO (1978b) Alkylation of tRNA^{Trp} in a complex with tryptophanyl-tRNA synthetase. FEBS Lett 90: 103–106

Werstiuk ES, Neilson T (1976) Oligoribonucleotide synthesis. X. An improved synthesis of the anticodon loop region of methionine transfer ribonucleic acid from *E. coli*. Can J Chem 54: 2689–2696

Yaniv M, Folk WR, Berg P, Söll D (1974) A single mutational modification of a tryptophan-specific transfer RNA permits aminoacylation by glutamine and translation of the codon UAG. J Mol Biol 86: 245–260

Yarus M, Berg P (1969) Recognition of tRNA by isoleucyl-tRNA synthetase. Effect of substrates on the dynamics of tRNA-enzyme interaction. J Mol Biol 42: 171–189

5. Recognition of Promoter Sequences by RNA Polymerases from Different Sources

Ch. LEIB, H. ERNST, and G. R. HARTMANN

A. Introduction

Some time ago Fritz Lipmann and his group have observed that a combination in vitro
of ribosomal subunits and soluble proteins isolated from taxonomically very different
microorganisms yields a hybrid system active in protein synthesis (Felicetti and Lipmann
1968; Gordon et al. 1969; Krisko et al. 1969; Lucas-Lennard and Lipmann 1966; Takeda
and Lipmann 1966). Two conclusions may be drawn from these experiments: (1) the
components of the molecular apparatus for protein biosynthesis in different microorganisms correspond to each other rather closely in function; (2) equivalent proteins of
this system from different microorganisms must be structurally very similar at least
in those regions which are required for the functional interaction of the components
of the hybrid system. Otherwise mutual substitution would be difficult to understand.
Obviously, evolution and diversification of microorganisms have not affected too heavily the kinship among the components of the machinery of protein biosynthesis.

B. Comparison of RNA Polymerase from M. Luteus and E. coli

At the celebration of Fritz Lipmann's 75th birthday in Berlin in 1974 our laboratory
reported similar observations for the enzymatic machinery of RNA synthesis in Eubacteria (Lill et al. 1974). These investigations were carried out with the subunits of the
DNA-directed RNA polymerase from *Micrococcus luteus* and *Escherichia coli,* two
Eubacteria which differ in all classical taxonomic criteria. Notwithstanding these differences, the DNA-directed RNA polymerase from both bacteria is composed in the same
stoichiometric ratio of four different subunits designated as α, σ, β and β' according
to decreasing mobility in dodecylsulfate containing gels (Lill et al. 1975; Lill et al.
1977). Similarly to the preparation of the hybrid system of protein synthesis (Felicetti
and Lipmann 1968; Gordon et al. 1969; Higo et al. 1973; Krisko et al. 1969; Lucas-
Lennard and Lipmann 1966; Takeda and Lipmann 1966), we were able to form in
vitro catalytically active hybrids by combining isolated subunits of RNA polymerase
from both organisms. Each of the four subunits could be substituted by an equivalent

Institut für Biochemie der Ludwig-Maximilians-Universität München, Karlstr. 23, 8000 München 2,
FRG

subunit from the other organism (Hartmann 1976). Since the isolated subunits are devoid of enzymatic activity and have to associate tightly for catalytic activity, a close structural similarity at least of the interacting regions (sites of contact) has to be assumed. Otherwise mutual substitution would be difficult to understand at molecular level.

C. Peptide Mapping

To test this hypothesis, we have compared the two RNA polymerases by a few simple methods. Firstly we have compared the smallest subunit α from both organisms by fingerprinting a mixture of the tryptic digest of the polypeptides (Fig. 1). Surprisingly, very few of the peptides deriving from different subunits moved with identical mobility. Clearly, the basic amino acids must be distributed rather differently in the polypeptide chains which are of almost equal length. Subunit α from *M. luteus* is only about 6% larger than subunit α from *E. coli* with a molecular weight of 36512 (Ovchinnikov et al. 1977). An analogous result was obtained when studying the distribution of methionine residues in the polypeptide chains by measuring the size of the fragments produced by cleavage with BrCN.

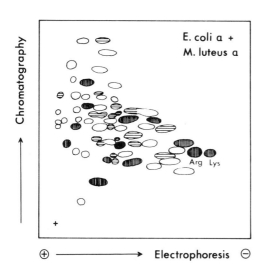

Fig. 1. Fingerprint of a mixture of the tryptic digest of subunit α from *M. luteus* and *E. coli.* The separation was carried out on thin layer plates (Sil G, Macherey, Nagel and Co, Düren) by electrophoresis at pH 1.9 (Fujiki et al. 1976) and subsequent chromatography (Bates et al. 1975). *Circle with horizontal lines* = peptides from *E. coli; blank circle* = peptides from *M. luteus; circle with vertical lines* = joint peptides with identical mobility

D. Complement Fixation

Another sensitive and simple method to detect homologies in the amino acid sequence of proteins is available by the use of specific antibodies. When applying the Ouchterlony technique we could not detect any cross-reactivity of the two RNA polymerases with the heterologous antiserum after an incubation of 24 h. However, increasing the incubation period to 48 h a very faint line of a precipitate is formed in the space between the RNA polymerase from *E. coli* and the antibody against the enzyme from *M. Luteus,* indicating a distant similarity between the two polymerases.

To quantify this relationship we have used the micro-complement fixation assay (Champion et al. 1974). The results obtained with this assay are usually expressed in units of immunological distance. This unit is defined as 100 times the logarithm of the factor by which the concentration of the antiserum has to be increased to obtain a complement binding curve with the heterologous antigen, similar to that with the specific antigen. An immunological distance of 100 corresponds to a 20% difference in amino acid sequence between polypeptide chains of equal length (Champion et al. 1974). Results obtained with the two RNA polymerases clearly demonstrate a very large immunological distance (Table 1). However, it is interesting to note that the isofunctional subunits from the two Eubacteria are distinctly more similar in antigenic properties than the complete parental enzymes. This difference suggests the notion that the regions of isofunctional subunits which are located on the outside of the holoenzyme are more dissimilar than the regions which form the sites of contact. Such regions are hidden in the interior of the holoenzyme and only exposed to antiserum in a solution of isolated subunits. With this notion it is easier to comprehend how structurally rather different but otherwise isofunctional subunits from different organisms can specifically associate to form catalytically active hybrid enzymes (Hartmann 1976).

In this context it should be mentioned that the antiserum to RNA polymerase from *E. coli* clearly inhibits the activity not only of this enzyme but also of the enzyme from *M. luteus*. This observations indicate some similarity of the antigenic groups in the region of the active center of the two enzymes.

Table 1. Immunological distance between RNA polymerase from *E. coli* and *M. luteus* and their isofunctional subunits determined with antiserum against the holoenzyme from *M. luteus* (treated with Freund's complete adjuvans)

	Immunological distance determined with antiserum against enzyme from *M. luteus*
^{e}holoenzyme / ^{m}holoenzyme	194
^{e}core enzyme / ^{m}core enzyme	176
$^{e}\beta' / {^{m}}\beta'$	180
$^{e}\beta / {^{m}}\beta$	86
$^{e}\sigma / {^{m}}\sigma$	61
$^{e}\alpha / {^{m}}\alpha$	167

E. Promoter Binding

The preceding experiments have clearly shown that distinct differences in the chemical and antigenic properties of the two RNA polymerases do exist. Although obvious differences in the general catalytic power to synthesize RNA have not been found, alterations in the structure of the polypeptide chains may influence the capacity to recognize nucleotide sequences on a template as signals for the start or the termination of RNA synthesis. Therefore the property of short double-stranded DNA's to initiate RNA synthesis with the two polymerases was compared. As source for these DNA fragments

ϕX 174 RF DNA was used, which was cleaved by restriction endonucleases such as Hae III, Hind II, or Mbo II (Sanger et al. 1977). A sensitive method to detect signals for RNA chain initiation on a DNA is the filter binding assay (Jones et al. 1977; Spiegelman and Whiteley 1978). At high ionic strength only complexes of RNA polymerase and DNA, which have started synthesis of RNA chains, are retained on a nitrocellulose filter. Upon incubation of RNA polymerase from *M. luteus* with the mixture of DNA fragments obtained by digestion of ϕX 174 RF DNA with Hae III in the presence of limited substrate (ATP, GTP, CTP), a number of fragments including fragment Hae III/7 is adsorbed to the filter. This is revealed by elution of the filter with dodecylsulfate and identification of the DNA fragments by agarose gel electrophoresis. The same experiment performed with RNA polymerase from *E. coli* leads to a similar result except that now fragment Hae III/7 is not retained on the filter. This observation hints to the possibility that RNA polymerase from *M. luteus* recognizes a signal for the start of RNA synthesis on this fragment, which is not recognized by the enzyme from *E. coli*.

A different result is obtained when a digest of ϕX 174 RF DNA with the endonuclease Hind II is used in such an experiment. Now the fragment Hind II/4, comprising the promoter of gene A, is adsorbed to the filter when RNA polymerase from *E. coli* is used, whereas only traces of this fragment are retained in an experiment with the micrococcal enzyme. These experiments indicate that the promoter of gene A is barely recognized by RNA polymerase from *M. luteus*.

F. RNA Products

To obtain more conclusive evidence for these notions, the RNA products synthesized on the isolated (Ledeboer et al. 1978) fragments were analyzed. When fragment Hae III/7 is incubated with RNA polymerase from *M. luteus* in presence of substrate, an RNA chain with a length of about 90 nucleotides is synthesized. This is revealed by electrophoretic analysis of the product in polyacrylamide gels containing 0.1% dodecylsulfate (Fig. 2). When RNA polymerase from *E. coli* is used in this experiment, no dis-

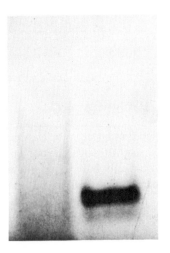

Fig. 2. Transcription of fragment Hae III/7 (Sanger et al. 1977) of ϕX 174 RF DNA by RNA polymerase from *M. luteus* (*right*) or *E. coli* (*left*) in presence of ATP, (^{14}C)UTP, CTP, and GTP. The products were separated by electrophoresis in polyacrylamid-agarose gels containing 0.1% dodecyl-sulfate (Peacock and Dingman 1968) and identified by autoradiography

tinct RNA chains are synthesized (Fig. 2). To confirm these results, another fragment obtained with the restriction endonuclease Mbo II, which comprises almost the same region of the genome, was used as template. This fragment (Mbo II/7) is 90 base pairs longer than fragment Hae III/7. Its (+)strand exceeds the (+)strand of Hae III/7 by 147 nucleotides in the 3'-direction. With this template an RNA chain of about 230 nucleotides long is synthesized by RNA polymerase from *M. luteus*. Again, no distinct product is formed with the enzyme from *E. coli*.

These findings strongly support the hypothesis of a species-specific recognition of starting signals on a DNA template by different RNA polymerases. Such results are not only obtained with fragments of the genome of ϕX 174. Similar observations have been made with Col E1 and RSF 1030 DNA (G.R. Hartmann, Report at the Deutsch-Sowjetisches Symposium on the *Structure and Function of the Genome*, Baku, USSR, 19–23 September 1977). Furthermore, RNA polymerase from *Caulobacter crescentus* recognizes other promoters on the RF I form of coliphage fd DNA than the enzyme from *E. coli* (Iida et al. 1979). Species specificity in the recognition of signals on the DNA template may have significance for experiments with recombinant DNA. Nucleotide sequences on the spliced DNA, which otherwise would not function as signals, could be recognized as such by the host polymerase. This could severely interfere with correct transcription.

On the other hand, these examples of species specificity in signal recognition do not imply that RNA polymerases from different bacteria totally differ in their capacity to recognize certain nucleotide sequences as promoters, since, e.g., the promoters A to E on the DNA of coliphage T7 are universally recognized by RNA polymerases from a number of bacteria, although with different efficiency (Wiggs et al. 1979).

G. Subunit Sigma and Species-Specific Promoter Recognition

Indeed, promoters seem to exhibit a large spectrum of rates of association with RNA polymerase (von Gabain and Bujard 1979). Hence, it is rather likely that the rate of formation of the complex between the enzyme and the nucleic acid is not only determined by the nucleotide sequence but also by the specific structure of RNA polymerase. This conjecture leads to the question, which subunit of the enzyme exerts a strong influence on the recognition of signals for the start of RNA synthesis? The most likely candidate is subunit σ, which has been affiliated with promoter recognition since its discovery (Burgess et al. 1969). To check this notion, use was made of the observation that subunit σ from *M. luteus* can replace in function the equivalent subunit in RNA polymerase from *E. coli* (Lill et al. 1977). It has been described above that the complete enzyme from *M. luteus* is able to transcribe fragment Hae III/7 of ϕX 174 RF DNA, whereas the enzyme from *E. coli*, even in the presence of its own subunit σ, cannot. However, when *E. coli* RNA polymerase lacking σ is incubated with fragment Hae III/7 and substrate in the presence of subunit σ from *M. luteus*, an RNA chain is synthesized of the same size as the RNA made by the complete enzyme from *M. luteus* (Fig. 3). Obviously, subunit σ from *M. luteus* can very efficiently direct RNA polymerase from *E. coli* to recognize different nucleotide sequences as signals of initiation. Similar ob-

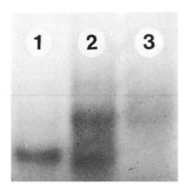

Fig. 3. Transcription of fragment Hae III/7 (Sanger et al. 1977) of ϕX 174 RF DNA by RNA polymerase from *M. luteus (1)*, by RNA polymerase (core enzyme) from *E. coli* + subunit σ from *M. luteus (2)*, and by RNA polymerase (core enzyme) from *E. coli (3)* in presence of ATP, (α-^{32}P)UTP, CTP and GTP. Separation and identification as in Fig. 2

servations have been made with other DNA's, demonstrating the universality of this species-specific effect of subunit σ from *M. luteus* (G.R. Hartmann, Report at the Deutsch-Sowjetisches Symposium on the *Structure and Function of the Genome,* Baku, USSR, 19–23 September 1977). These results strongly suggest an important contribution of subunit σ in the species-specific recognition of signals for RNA chain initiation.

Acknowledgment. These investigations have been supported by the Deutsche Forschungsgemeinschaft and the Fonds der Chemischen Industrie.

References

Bates DL, Perham RN, Coggins JR (1975) Methods for obtaining peptide maps of proteins on a subnanomole scale. Anal Biochem 68: 175–184

Burgess RR, Travers AA, Dunn JJ, Bautz EKF (1969) Factor stimulating transcription by RNA polymerase. Nature 221: 43–46

Champion AB, Prager EM, Wachter D, Wilson AC (1974) Microcomplementfixation. In: Wright CA (ed) Biochemical and immunological taxonomy of animals, pp 397–416. Academic Press, New York London

Felicetti L, Lipmann F (1968) Comparison of amino acid polymerization factors isolated from rat liver and rabbit reticulocytes. Arch Biochem Biophys 125: 548–557

Fujiki H, Palm P, Zillig W, Calendar R, Sunshine M (1976) Identification of a mutation within the structural gene for the α subunit of DNA-dependent RNA polymerase of *E. coli*. Molec Gen Genet 145: 19–22

Gabain A von, Bujard H (1979) Interaction of *Escherichia coli* RNA polymerase with promoters of several coliphage and plasmid DNAs. Proc Natl Acad Sci USA 76: 189–193

Gordon J, Schweiger M, Krisko I, Williams CA (1969) Specificity and evolutionary divergence of the antigenic structure of the polypeptide chain elongation factors. J Bacteriol 100: 1–4

Hartmann GR (1976) Austausch von Untereinheiten zwischen Enzymen aus verschiedenen Organismen in vitro: Enzymchimären. Angew Chem 88: 197–203; Angew Chem Int Ed Engl 15: 181–186

Higo K, Held W, Kahan L, Nomura M (1973) Functional correspondence between 30S ribosomal proteins of *Escherichia coli* and *Bacillus stearothermophilus.* Proc Natl Acad Sci USA 70: 944–948

Iida H, Ikehara K, Okada Y (1979) Differential transcription of fd RFI DNA by *Caulobacter crescentus* and *Escherichia coli* RNA polymerases. FEBS Lett 99: 346–350

Jones BB, Chan H, Rothstein S, Wells RD, Reznikoff WS (1977) RNA polymerase binding site in
    λ*plac*5 DNA. Proc Natl Acad Sci USA 74: 4914–4918
Krisko I, Gordon J, Lipmann F (1969) Studies on the interchangeability of one of the mammalian
    and bacterial supernatant factors in protein biosynthesis. J Biol Chem 244: 6117–6123
Lederboer AM, Hille J, Schilperoort RA (1978) An easy and efficient procedure for the isolation
    of pure DNA restriction fragments from agarose gels. Biochim Biophys Acta 520: 498–504
Lill UI, Behrendt EM, Hartmann GR (1974) Intergeneric complementation of RNA polymerase
    subunits. In: Richter D (ed) Lipmann Symposium: Energy, regulation and biosynthesis in molec-
    ular biology, pp 377–383. de Gruyter, Berlin New York
Lill UI, Behrendt EM, Hartmann GR (1975) Hybridization in vitro of subunits of the DNA-dependent
    RNA polymerase from *Escherichia coli* and *Micrococcus luteus*. Eur J Biochem 52: 411–420
Lill UI, Kniep-Behrendt EM, Bock L, Hartmann GR (1977) Purification and characterization of
    the DNA-dependent RNA polymerase and its subunit σ from *Micrococcus luteus*. Hoppe-Seyler's
    Z Physiol Chem 358: 1591–1603
Lucas-Lennard J, Lipmann F (1966) Separation of three microbial amino acid polymerization fac-
    tors. Proc Natl Acad Sci USA 55: 1562–1566
Ovchinnikov YuA, Lipkin VM, Modyanov NN, Chertov OYu, Smirnov YuV (1977) Primary struc-
    ture of α-subunit of DNA-dependent RNA polymerase from *Escherichia coli*. FEBS Lett 76:
    108–111
Peacock AC, Dingman CW (1968) Molecular weight estimation and separation of ribonucleic acid
    by electrophoresis in agarose-acrylamide composite gels. Biochemistry 7: 668–674
Sanger F, Air LM, Barrell BG, Brown NL, Coulson AR, Fiddes JC, Hutchinson CA III, Slocombe
    PM, Smith M (1977) Nucleotide sequence of bacteriophage φX 174 DNA. Nature 265: 687–695
Spiegelman GB, Whiteley HR (1978) Bacteriophage SP82 induced modifications of *Bacillus subtilis*
    RNA polymerase result in the recognition of additional RNA synthesis initiation sites on phage
    DNA. Biochem Biophys Res Commun 81: 1058–1065
Takeda M, Lipmann F (1966) Comparison of amino acid polymerization in *B. subtilis* and *E. coli*
    cell-free systems; hybridization of their ribosomes. Proc Natl Acad Sci USA 56: 1875–1882
Wiggs JL, Bush JW, Chamberlin MJ (1979) Utilization of promoter and terminator sites on bacterio-
    phage T7 DNA by RNA polymerases from a variety of bacterial orders. Cell 16: 97–109

6. DNA as a Target for a Protein Antibiotic: Molecular Basis of Action

I. H. GOLDBERG, T. HATAYAMA, L. S. KAPPEN and M. A. NAPIER

A. Protein Structure of Neocarzinostatin

The antitumor antibiotic neocarzinostatin (NSC), isolated from the culture filtrates of *Streptomyces carzinostaticus* variant F-41 (Ishida et al. 1965), is an acidic single-chain polypeptide with a molecular weight of 10,700 (Meienhofer et al. 1972a). The protein has been purified to homogeneity and its amino acid sequence (Fig. 1) (Meienhofer et al. 1972a, b; Maeda et al. 1974; Samy et al. 1977) and physical properties (Maeda et al. 1973; Samy and Meienhofer 1974) have been determined. There are high degrees of homology of some regions of NCS and the protein antibiotics actinoxanthin (Khokhlov et al. 1969, 1976) and macromomycin (Sawyer et al. 1979). NCS exists in a tight, proteolysis-resistant conformation with an antiparallel β-pleated sheet structure (Samy et al. 1974). It possesses two reduction-resistant disulfide bridges and lacks methionine and histidine. The positions of the disulfides have not yet been unambiguously assigned. NCS contains two tryptophan residues in positions 46 (buried) and 79 and one buried tyrosine residue at position 32. Oxidation of tryptophan 79 does not result in loss of biological activity (Samy et al. 1974). Similarly, acylation of the amino groups (alanine 1 and lysine 20) does not affect the activity of NCS (Maeda 1974; Samy 1977). On the other hand, modification of the carboxyl groups results in loss of activity (Samy 1977). Further, spontaneous deamidation of asparagine 83 at a weakly acidic pH generates "preneocarzinostatin" which lacks biological activity (Maeda and Kuromizu 1977). The chemically deamidated compound is thought to be the same as the material isolated from culture filtrates that antagonizes NCS activity (Kikuchi et al. 1974).

B. Presence of a Nonprotein Chromophore in Neocarzinostatin

Optically active absorption bands have been reported for NCS above 300 nm (Samy et al. 1974; Napier et al. 1979), and inactivation of NCS by irradiation with 300–400 nm light has been found (Kohno et al. 1974; Burger et al. 1978; M. Napier and I.H. Goldberg, unpublished data). Since these spectral characteristics are unusual for a purified protein, we further characterized native NCS spectroscopically and were successful in separating a previously unidentified highly fluorescent chromophore from the protein

1 Department of Pharmacology, Harvard Medical School, Boston, Massachusetts 02115, USA
2 Present address: Department of Biochemistry, Osaka City University, Mecial School, Osaka 545, Japan

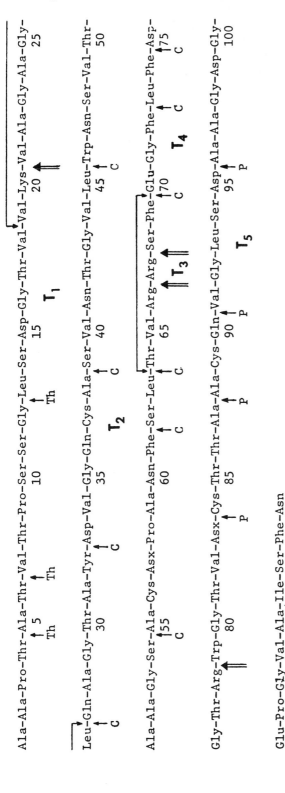

Fig. 1. Structure of neocarzinostatin. *Double arrows* indicate tryptic cleavage; *small arrows* show some of the subsequent cleavages by thermolysis (Th), chymotrypsin (C), or pepsin (P). Overlap peptides are indicated by a *bridged arrow*. (From Meienhofer et al. 1972). Copyright 1972 by the American Association for the Advancement of Science

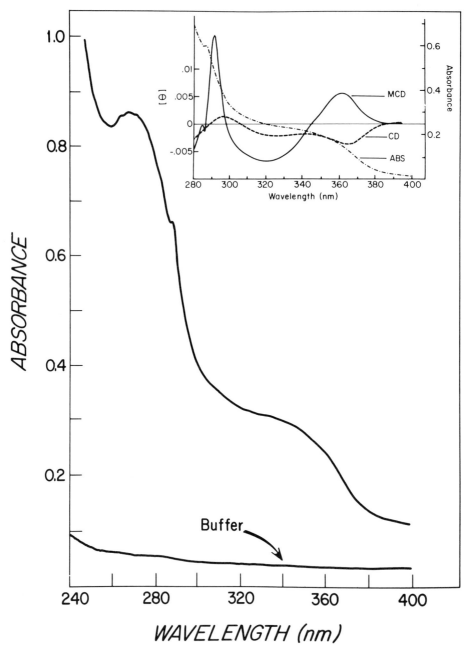

Fig. 2. Ultraviolet-visible absorption spectrum of neocarzinostatin (3.5×10^{-5} M sodium acetate, pH 4.5). *Inset:* Magnetic circular dichroism (——), circular dichroism (– – –), and absorption (–·–·–) spectra of neocarzinostatin (3.5×10^{-4} M, 0.015 M sodium acetate, pH 4.5) (from Napier et al. 1979)

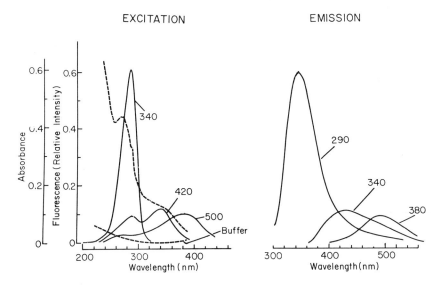

Fig. 3. Fluorescence excitation (emission wavelength = 345 nm, 420 nm, 500 nm) and emission (excitation wavelength = 290 nm, 340 nm, 380 nm) of neocarzinostatin (2×10^{-5} M, 0.05 M Tris, pH 8.0). Absorption spectrum (– – –) of neocarzinostatin (from Napier et al 1979)

(Napier et al. 1979). As shown in Fig. 2, NCS exhibits an absorption maximum at 270 nm with an abnormally high extinction coefficient (ϵ 23,000) and a broad shoulder tailing into the visible between 300 nm and 400 nm ($\epsilon_{340} \sim 8,000$). Further, strong optical activity is found above 300 nm with extremes in the magnetic circular dichroism (MCD) at 320 nm and 365 nm corresponding to the circular dichroism (CD) minima at these wavelengths (Fig. 2, inset).

The fluorescence spectrum of NCS (Fig. 3) has major excitation maxima at 285 nm, 340 nm and 380 nm with emission maxima at 345 nm, 420 nm, and 490 nm, respectively. The 345 nm emission band is attributable to tryptophan residues, but the intensity is low for the two tryptophans present, 45% of the intensity of an equivalent concentration of N-acetyltryptophanamide. The tryptophan emission is enhanced by treatment with reducing or denaturing agents. In the presence of 0.01 M 2-mercaptoethanol or 4 M guanidine-HCl, in 0.05 M Tris, pH 8.0, for 24 h, the 345 nm emission is increased 1.4-fold and 2.0-fold, respectively. S-carboxymethylated NCS, which has virtually no absorption or fluorescence excitation above 300 nm, exhibits normal tryptophan emission. Guanidine-HCl-treated NCS has increased absorption below 270 nm and a broad maximum at 380 nm. The fluorescence emission intensity is increased 1.4-fold at 420 nm and 8-fold at 490 nm. NCS treated with 2-mercaptoethanol (or sodium borohydride) exhibits no apparent change in the absorption above 300 nm. The intensity of the 420 nm and 490 nm emission bands, however, is increased 2.4-fold and 1.3-fold, respectively. The reduced emission of native NCS might be due to quenching resulting from the tightly folded structure of NCS, which is relaxed by these treatments, as measured by CD (Samy et al. 1974; M. Napier and I. Goldberg, unpublished). It is more likely that the altered emission is due to the dissociation of a nonprotein chromophore,

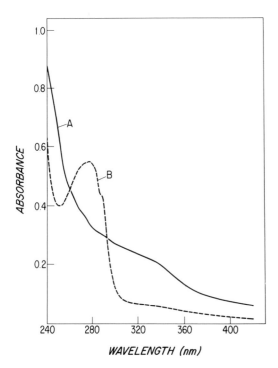

Fig. 4. Methanol extraction of neocarzinostatin. Absorption spectra of methanol-soluble fraction A, in methanol (——) and methanol-insoluble fraction B in water (– – –). The fractions were diluted to a volume approximately equivalent to 1 mg/ml neocarzinostatin (from Napier et al. 1979)

which quenches the tryptophan emission by competitively absorbing excitation energy and by serving as an energy acceptor.

Separation of a nonprotein UV/visible absorbing material from the NCS protein is possible by several procedures. Sephadex G-50 chromatography of guanidine-treated, of β-mercaptoethanol-treated, or of base-treated (0.1 N NaOH), but not of acid-treated (0.1 N HCl), NCS, separates the protein and nonprotein components. The most convenient separation procedure, however, is methanol extraction of the lyophilized drug. The methanol fraction (A) (Fig. 4), free of common amino acids, has two broad absorption shoulders near 270 nm and 340 nm, CD and MCD activity above and below 300 nm, and fluorescence emission at 420 nm and 490 nm (Fig. 5A). The methanol-insoluble fraction (B) with an amino acid composition identical to NCS, exhibits typical protein absorbance (Fig. 4), with λmax at 277 nm (ϵ 14,000). The absorption at 250 nm, seen for NCS, is reduced. Absorption, fluorescence excitation, and optical activity above 300 nm are virtually absent. Also, there are no changes in the tryptophan residues as determined by MCD. The only significant fluorescence remaining is the tryptophan emission at 345 nm (Fig. 5B). The spectral properties of fraction B are similar to those found for S-carboxymethylated NCS which has lost the nonprotein chromophore probably due to reduction and methanol extraction procedures used in its preparation (Maeda et al. 1974).

Thus, NCS purified by chromatography is an acid-stable complex of a protein and a nonprotein chromophore which can be dissociated and separated by a number of procedures. It is of interest that carzinostatin, an uncharacterized antibiotic isolated from the parent strain of *S. carzinostaticus,* has been described as a complex consisting

EMISSION

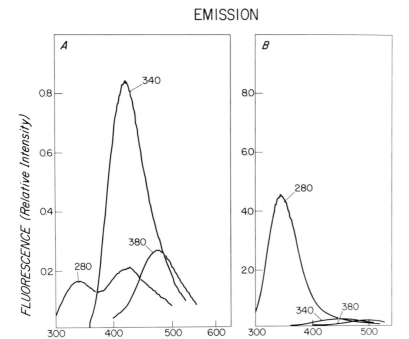

Fig. 5. Fluorescence emission spectra (excitation wavelength = 280 nm, 340 nm, 380 nm) of neo-carzinostatin fraction A and fraction B (from Napier et al. 1979)

of a methanol-soluble, low molecular weight fraction and a methanol-insoluble, water-soluble, high molecular weight fraction (Shoji 1961). Neither fraction had significant biological activity alone, but recombination restored activity. NCS is a more stable antibiotic isolated from variant 41 of *S. carzinostatin,* with biological properties similar to carzinostatin.

C. Evidence for DNA as a Target

I. In vivo Effects on DNA Synthesis and DNA Strand Breakage

There is considerable evidence that DNA is an important target in the action of NCS. Thus, low levels of NCS selectively inhibit DNA in sensitive bacteria and mammalian cells (Ono et al. 1966; Homma et al. 1970; Sawada et al. 1974; Beerman and Goldberg 1977; Hatayama and Boldberg 1979), induce degradation of existing DNA in bacteria (Ono et al. 1966; Ohtsuki and Ishida 1975a), and produce breaks in the DNA of mammalian cells (Beerman and Goldberg 1974; Tatsumi et al. 1974; Sawada et al. 1974; Ohtsuki and Ishida 1975b; Beerman and Goldberg 1977; Beerman et al. 1977; Hatayama and Goldberg 1979). A correlation exists between the ability of NCS to induce breakage of cellular DNA in HeLa cells and its inhibition of DNA replication and cell growth

(Beerman and Goldberg 1977; Beerman et al. 1977). NCS-induced single-strand breaks in HeLa cells can be repaired to a considerable extent but repair of double-strand breaks is less efficient (Hatayama and Goldberg 1979) and the latter may account for the cell-killing activity of the antibiotic. We find that there are about 55–60 double-strand breaks per cell at the mean lethal dose (at which there is 37% survival of cells) of 0.01 μg/ml NCS (Hatayama and Goldberg 1979). Thus it appears that a small number of such breaks in a critical region of the genome may be the lethal event. Further, the so-called DNA complex consisting of DNA, lipid, and protein is disrupted by NCS treatment of HeLa cells and this is accompanied by the release within the cell of free DNA (Hatayama and Goldberg 1979). DNA repair synthesis is activated by NCS-induced DNA damage a as revealed by the induction of unscheduled DNA synthesis in treated lymphocytes (Tatsumi et al. 1975) and by the marked stimulation of thymidine incorporation into parental (but not newly made replicative) DNA in HeLa cells and isolated nuclei (Kappen and Goldberg 1978a). Further, low levels of NCS cause a block in the G_2 phase of the cell cycle (Ebina et al. 1975; Rao and Rao 1976) and chromosomal abberrations in mammalian cells (Kumagai et al. 1966; Rao and Rao 1976).

II. Mutagenic Action and Effect of DNA Repair Systems

Strong evidence for the involvement of DNA in NCS action comes from experiments showing that NCS is a mutagen for *Escherichia coli* and that mutagenicity and cell killing are affected in inverse ways by a recA mutation (Tatsumi and Nishioka 1977). We have also found NCS to be mutagenic for *Salmonella typhimurium*, especially in strains bearing the plasmid pKM101 which has been implicated in error-prone repair (Eisenstadt et al. 1979). Finally, the lethal effect of NCS in L-1210 cells has been found to be potentiated by caffeine, an inhibitor of postreplication repair (Sasada et al. 1976; Tatsumi et al. 1979).

III. In vitro Scission of DNA

NCS causes single-strand nicks in superhelical (Fig. 6) and linear duplex DNA in vitro and this reaction is stimulated at least 1000-fold by the presence of a mercaptan (Beerman and Goldberg 1974; Tatsumi et al. 1974; Ishida and Takahashi 1976; Beerman et al. 1977; Poon et al. 1977; Kappen and Goldberg 1977; Kappen and Goldberg 1978b; Sim and Lown 1978; Ishida and Takahashi 1978). At higher doses of NCS do double-strand breaks are produced, presumably as the result of the random placement of single-strand breaks within a small number of base-apirs of one another.

NCS-treated DNA is a much poorer primer for *E. coli* DNA polymerase I than untreated DNA (Fig. 7) (Tsuruo et al. 1971; Kappen and Goldberg 1977), and the enzyme has been shown to bind to the site of DNA damage in a nonfunctional form (Kappen and Goldberg 1977). The degree of inhibition of DNA synthesis by NCS depends on the concentrations of NCS and of 2-mercaptoethanol during preincubation (Fig. 7A). In the control containing no drug, [^{3}H]dTMP incorporation is not affected to any significant extent by concentrations of 2-mercaptoethanol up to 10 mM in the preincubation reaction (4 mM in the final polymerization reaction), whereas in the presence of

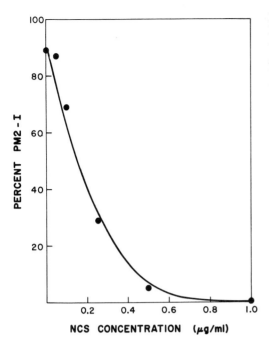

Fig. 6. Single strand scission of superhelical PM2 (Form I) DNA by NCS. The conversion of suberhelical PM2 DNA (Form I) to the singly nicked circular duplex form (Form II) was followed on alkaline sucrose gradients (from Beerman et al. 1977)

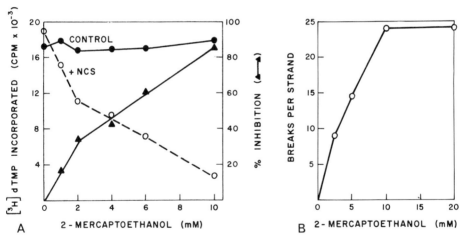

Fig. 7A, B. Influence of 2-mercaptoethanol on NCS-induced inhibition of DNA polymerase I and breakage of DNA. **A** DNA synthesis by DNA polymerase I. **B** Quantitation of NCS-induced strand breaks in λDNA by alkaline sucrose gradient analysis (from Kappen and Goldberg 1977)

NCS the inhibition increases with increasing concentrations of 2-mercaptoethanol with maximal inhibition being obtained at 10 mM. The NCS-induced inhibition of DNA polymerase I activity, as described by the concentration of 2-mercaptoethanol, is related to the number of single-strand breaks in the DNA. As shown in Fig. 7A, the maximal inhibition of dTMP incorporation by NCS occurs under conditions of maximal strand scission (Fig. 7B), suggesting that there is a connection between the two effects of the drug. That NCS-induced strand breakage generates inactive binding sites for DNA polymerase I is shown in experiments in which either untreated DNA or excess enzyme are able to overcome the inhibition in a competitive way. Further, [²⁰³Hg]DNA plymerase I bound to NCS-treated DNA and binding correlates with the extent of nicking (Kappen and Goldberg 1977). It appears that one molecule of enzyme binds to each break introduced by NCS.

IV. Chemical Mature of the Break

The breaks possess both 3'- and 5'-phosphoryl termini, indicative of the existence of a gap of one or more nucleotides (Poon et al. 1977; Kappen and Goldberg 1978c). As expected, generation of a 3'-OH group by treatment of NCS-nicked DNA with alkaline phosphatase converts the DNA into a much better primer for DNA polymerase I than DNA unexposed to the antibiotic. Recent studies have shown that a damaged sugar fragment (carbons 1', 2', and 3') can be released from NCS-treated DNA in the form of a material behaving like malondialdehyde, especially after alkaline treatment (Hatayama and Goldberg, unpublished data), suggesting that sugar fragment release is incomplete without additional treatment. DNA strand cleavage is also accompanied by the release of the bases thymine (Ishida and Takahashi 1976; Poon et al. 1977) and adenine (Poon et al. 1977). Only about 15% as much adenine is released as is thymine. The requirement for thymidylic and deoxyadenylic acids in the scission reaction was also revealed by experiments in which various synthetic and natural DNA's of different base composition were tested for their ability to protect against the cutting of radioactive lambda virus DNA (Poon et al. 1977). The amount of thymine released correlated well with the number of strand scissions. Further, using the DNA sequencing technique of Maxam and Gilbert (1977), NCS was shown in the presence of 2-mercaptoethanol to cleave double-stranded φX174 DNA restriction fragments almost exclusively at deoxythymidylic and deoxyadenylic acid residues (Hatayama et al. 1978). Overall, deoxythymidylic acid residues are attacked much more frequently than are deoxyadenylic acid residues (Fig. 8), although there is variability in the attack rate for both nucleotides at different locations in the DNA molecule. While all deoxythymidylic acid residues are sites of scission, not all deoxyadenylic acid residues are cleavage sites, but there appears to be no clear-cut nucleotide sequence specificity in determining cleavage frequency. In agreement with the protection experiments, single-stranded DNA is a very poor substrate for NCS-induced scission. It is of interest that both members of a base pair, deoxythymidylic and deoxyadenylic acids, are the main targets for NCS, for the changes of producing double-strand breaks in AT-poor or homopolymeric regions will be significantly increased over that caused by the otherwise random placement of single-strand scissions. Since double- and not single-strand breaks are considered to be lethal events (Freifelder 1968; Leenhouts and Chadwick 1978), this action of NCS may be important in determining its cytotoxicity. NCS attack sites at thymidylic and deoxyadenylic acid residues in DNA have also been found by D'Andrea and Haseltine (1978). Further, these results are in agreement with the earlier finding with DNA polymerase I that DNA synthesis directed by poly-[d(A-T)] is much more sensitive to NCS than that by poly[d(G-C)] (Kappen and Goldberg 1977).

Z_8—Hinf I— long fragment

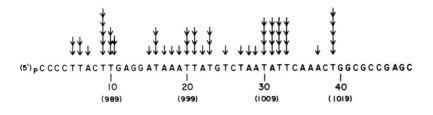

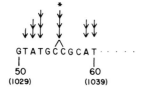

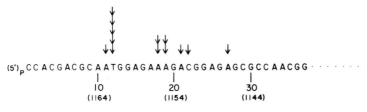

Z_8—Hinf I— short fragment

Fig. 8. NCS cleavage sites in two restriction enzyme fragments of duplex ϕX174 DNA as determined by the DNA sequencing technique of Maxam and Gilbert (1977). The number of *arrows* indicates the relative intensity of the bands on the autoradiograms. Very faint bands deserving significantly less than one *arrow* are not shown. (*) Position attacked by neocarzinostatin in the single-stranded fragment. The nucleotide position in the Sanger map is indicated within parentheses. The sequence of the short fragment is the complement of that in the Sanger map (from Hatayama et al. 1978)

D. Drug Activation by Reducing Agent and Need for O_2

While NCS-induced cutting of DNA in vitro is markedly stimulated by the presence of a mercaptan, high levels of the mercaptan inhibit the reaction (Fig. 9) (Kappen and Goldberg 1978b). This latter effect is seen especially clearly with the radiation-protector (and free radical scavenger) S,2-aminoethylisothiuronium bromide·HBr (AET). Since AET rearranges rapidly when neutralized to form 2-mercaptoethylguanidine, it is likely that the latter is the active form. While less active than a mercaptan, we have found that sodium borohydride can also activate the reaction. In addition, preincubation of NCS with a mercaptan results in its rapid inactivation (Kappen and Goldberg 1978b; Ishida and Takahashi 1978). These data suggested that reduction of the disulfide(s) of NCS was necessary to generate an active form of the drug and that a free radical mechanism

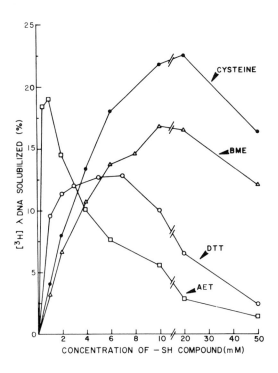

Fig. 9. Effect of different sulfhydryl compounds on the activity of NCS (100 μg/ml). The concentrations of the -SH agents (BME, β-mercapto-ethanol; DTT, dithiothreitol) were varied as indicated. The trichloroacetic acid-solubility of [³H]λDNA was measured at 20 min (from Kappen and Goldberg 1978b)

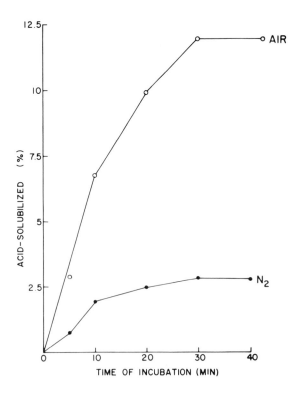

Fig. 10. Time course of activity of NCS in air or in N₂. Air or N₂ was bubbled through the reaction mixture containing 100 μg/ml NCS and 7.5 mM β-mercaptoethanol. The acid-solubilization of [³H]λDNA was measured (from Kappen and Goldberg 1978b)

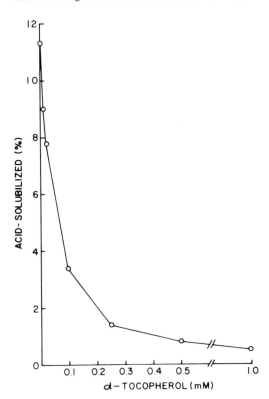

Fig. 11. Inhibition of NCS activity by α-tocopherol. NCS (100 μg/ml)-induced acid-solubilization of [³H]DNA was measured at 30 min in a reaction similar to that in Fig. 10 (from Kappen and Goldberg 1978b)

may be involved in the DNA cutting reaction. The former was also suggested by experiments in which NCS activation was obtained by photoreduction of the disulfides in the presence of sodium hypophosphite (Sim and Lown 1978). Support for involvement of a radical mechanism comes from the requirement for O_2 in the DNA scission reaction (Fig. 10) (Kappen and Goldberg 1978b; Sim and Lown 1978; Burger et al. 1978) and its inhibition by low concentrations of the potent peroxyl free radical scavenger α-tocopherol (Fig. 11) (Kappen and Goldberg 1978b). Acid solubilization of DNA by NCS was blocked 50% at 50 μM α-tocopherol. On the other hand, exogenous metals have yet to be implicated in the reaction since various metal chelators (EDTA, 8-hydroxy-quinoline, deferoxamine, diethyldithiocarbamate, neocuproine, cuprizone) do not block the reaction (Kappen and Goldberg 1978b and unpublished data). Similarly, superoxide dismutase does not inhibit the reaction. Unexpectedly, various alcohols, organic solvents, and other agents, such as urea, that affect protein conformation have been found to stimulate the mercaptan-dependent reaction significantly (Kappen and Goldberg 1978b). It is of interest that, like the mercaptans, these agents inactivate NCS in a preincubation; it is possible that they facilitate the generation of an active, labile (presumably containing free sulfhydryls) form of the antibiotic, possibly involving activation of the chromophore. As of this writing, the precise mechanism whereby NCS cuts DNA is not known. The generation of sulfhydryls in NCS and their possible oxidation to free radical forms may be responsible but additional experimentation is required. Further, experiments are in progess to identify the nonprotein chromophore in NCS and to determine its contribution to the biological activity of the antibiotic.

E. Mechanism in vivo

It also remains to be demonstrated conclusively that the mechanism for DNA damage discovered in the in vitro system pertains in the intact cell. Mechanisms for cell killing not involving direct effects on DNA have also been proposed. Thus, studies using NCS covalently bound to Agarose or Sepharose suggest the possibility that the cytotoxic effect may be a membrane-mediated event (Nakamura and Ono 1974; Lazarus et al. 1977). Also, very high levels of NCS inhibit the vinblastine-induced formation of micro-tubular paracrystals (Ebina and Ishida 1975) and cap formation and cell spreading in mammalian cells (Ebina et al. 1977), but these effects are probably not responsible for cell killing. Definitive evidence for DNA as a primary target for NCS in cells and for the base specificity existing under in vitro conditions being involved in cellular DNA damage must come from genetic experiments and DNA-sequencing studies in intact cells.

Acknowledgment. This work was supported by U.S. Public Health Service Research Grant GM 12573 from the National Institutes of Health

References

Beerman TA, Goldberg IH (1974) DNA strand scission by the antitumor protein neocarzinostatin. Biochem Biophys Res Commun 59: 1254–1261

Beerman TA, Goldberg IH (1977) The relationship between DNA strandscission and DNA synthesis inhibition in HeLa cells treated with neocarzinostatin. Biochim Biophys Acta 475: 281–293

Beerman TA, Poon R, Goldberg IH (1977) Single-strand nicking of DNA in vitro by neocarzino-statin and its possible relationship to the mechanism of drug action. Biochim Biophys Acta 475: 294–306

Burger RM, Peisach J, Horwitz SB (1978) Effect of light and oxygen on neocarzinostatin stability and DNA-cleaving activity. J Biol Chem 253: 4830–4832

D'Andrea AD, Haseltine WA (1978) Sequence specific cleavage of DNA by the antitumor antibiotics neocarzinosatin and bleomycin. Proc Natl Acad Sci USA 75: 3608–3612

Ebina T, Ishida N (1975) Inhibition of formation of microtubular paracrystals in HeLa-S$_3$ cells by neocarzinostatin. Cancer Res 35: 3705–3709

Ebina T, Ohtsuki K, Seto M, Ishida N (1975) Specific G2 block in HeLa-S$_3$ cells by neocarzinostatin. Eur J Cancer 11: 155–158

Ebina T, Satski M, Ishida N (1977) Inhibition of surface immunoglobulin central capping of Daudi cells and cell spreading of HeLa-S$_3$ cells by neocarzinostatin. Cancer Res 37: 4423–4429

Eisenstadt E, Wolf M, Goldberg IH (1979) Mutagenicity of neocarzinostatin in *Escherichia coli* and *Salmonella typhimurium.* Abstract for the Tenth Annual Meeting of the Environmental Mutagen Society. Environ Mutagenesis 1: 139

Freifelder D (1968) Rate of Production of single-strand breaks in DNA by X-irradiation *in situ.* J Mol biol 35: 303–309

Hatayama T, Goldberg IH (1979) DNA damage and repair in relation to cell killing in neocarzino-statin-treated HeLa cells. Biochim Biophys Acta 563: 59–71

Hatayama T, Goldberg IH, Takeshita M, Grollman AP (1978) Nucleotide specificity in DNA scission by neocarzinostatin. Proc Natl Acad Sci USA 75: 3603–3607

Homma M, Koida T, Saito-Koide T, Kamo I, Seto M, Kumagai K, Ishida N (1970) Specific inhibition of the initiation of DNA synthesis in HeLa cells by neocarzinostatin. Prog Antimicrob Anticancer Chemother Proc IV Internat Congr Chemother 2: 410–415

Ishida N, Miyazaki K, Kumagai K, Rikimaru M (1965) Neocarzinostatin, an antitumor antibiotic of high molecular weight. J Antibiot 18: 29–37

Ishida R, Takahashi T (1976) In vitro release of thymine from DNA by neocarzinostatin. Biochim Biophys Res Commun 68: 256–261

Ishida R, Takahashi T (1978) Role of mercaptoethanol in in vitro DNA degradation by neocarzino-statin. Cancer Res 38: 2617–2620

Kappen LS, Goldberg IH (1977) Effect of neocarzinostatin-induced strand scission on the template activity of DNA for DNA polymerase I. Biochemistry 16: 479–485

Kappen LS, Goldberg IH (1978a) Neocarzinostatin induction of DNA repair synthesis in HeLa cells and isolated nuclei. Biochim Biophys Acta 520: 481–489

Kappen LS, Goldberg IH (1978b) Activation and inactivation of neocarzinostatin-induced cleavage of DNA. Nucleic Acids Res 5: 2959–2967

Kappen LS, Goldberg IH (1978c) Gaps in DNA inruced by neocarzinostatin bear 3'- and 5'-phosphoryl termini. Biochemistry 17: 729–733

Khokhlov AS, Cherches BZ, Reshetov PD, Smirnova GM, Sorokina IB, Prokoptzeva TA, Koloditskaya TA, Smirnov VV, Navashin SM, Fomina IP (1969) Physico-chemical and biological studies on actinoxanthin, an antibiotic from *Actinomyces globisporus* 1131. J Antibiotic 22: 541–544

Khokhlov AS, Reshetov PD, Chupova LA, Cherches BZ, Zhigis LS, Stoyachenko IA (1976) Chemical studies on actinoxanthin. J Antibiot 29: 1026–1034

Kohno M, Haneda I, Koyama Y, Kikuchi M (1974) Studies on the stability of antitumor protein, neocarzinostatin. I. Stability of solution of neocarzinostatin. Jap J Antibiot 27: 707–714

Kikuchi M, Shoji M, Ishida N (1974) Pre-neocarzinostatin, a specific antagonist of neocarzinostatin. J Antibiot 27: 766–774

Kumagai K, Ono Y, Nishikawa T, Ishida N (1966) Cytological studies on the effect of neocarzino-statin on HeLa cells. J Antibiot 19: 69–74

Lazarus H, Raso V, Samy TSA (1977) In vitro inhibition of human leukemic cells (CCRF-CEM) by agarose-immobilized neocarzonostatin. Cancer Res 37: 3731–3736

Leenhouts HP, Chadwick KH (1978) The crucial role of DNA double-strand breaks in cellular radiobiological effects. Adv Radiat Biol 7: 55–101

Maeda H (1974) Chemical and biological characterization of succinyl neocarzinostatin. J Antibiot 27: 303–311

Maeda H, Kuromizu K (1977) Spontaneous deamidation of a protein antibiotic, neocarzinostatin, at weakly acidic pH. J Biochem (Tokyo) 81: 25–35

Maeda H, Shiraishi H, Onodera S, Ishida N (1973) Conformation of antibiotic protein, neocarzino-statin, studied by plane polarized infrared spectroscopy, circular dichroism and optical rotatory dispersion. Int J Pept Protein Res 5: 19–26

Maeda H, Glaser CB, Czombos J, Meienhofer J (1974) Structure of the antitumor protein neocarzino-statin. Purification, amino acid composition, disulfide reduction, and isolation and composition of tryptic peptides. Arch Biochem Biophys 164: 369–378

Maxam AM, Gilbert W (1977) A new method for sequencing DNA. Proc Natl Acad Sci USA 74: 560–564

Meienhofer J, Maeda H, Glaser CB, Czombos J, Kuromizu K (1972a) Primary structure of neocarzino-statin, an antitumor antibiotic. Science 178: 875–876

Meienhofer J, Maeda H, Glaser CB, Czombos J (1972b) Structural studies on neocarzinostatin, an antitumor polypeptide antibiotic. In: Lande S (ed) Progress in peptide research, vol 2, pp 295–306. Gordon and Breach, New York

Nakamura H, Ono K (1974) Growth inhibition of leukemic cells by neocarzinostatin-sepharose. Proc Jap Cancer Assoc 33rd Annual Meeting, p 112

Napier MA, Holmquist B, Strydom DJ, Goldberg IH (1979) Neocarzinostatin: Spectral characterization and separation of a nonprotein chromophore. Biochem Biophys Res Commun 89: 635–642

Ohtsuki K, Ishida N (1975a) Mechanism of DNA degradation induced by neocarzinostatin in *Bacillus subtilis*. J Antibiot 28: 229–236

Ohtsuki K, Ishida N (1975b) Neocarzinostatin-induced breakdown of deoxyribonucleic acid in HeLa-S_3 cells. J Antibiot 28: 143–148

Ono Y, Watanabe Y, Ishida N (1966) Mode of action of neocarzinostatin: Inhibition of DNA synthesis and degradation of DNA in *Saccina lutea*. Biochim Biophys Acta 119: 46–58

Poon R, Beerman TA, Goldberg IH (1977) Characterization of DNA strand breakage in vitro by the antitumor protein neocarzinostatin. Biochemistry 16: 486–492

Rao AP, Rao PN (1976) The cause of G2 arrest in CHO cells treated with anticancer drugs. J Natl Cancer Inst 57: 1139–1143

Samy TSA (1977) Neocarzinostatin: effect of modification of side chain amino and carboxyl groups on chemical and biological properties. Biochemistry 16: 5573–5578

Samy TSA, Meienhofer J (1974) Chemical modification and physical studies of the antitumor protein neocarzinostatin. In: Bewley TA, Lin M, Ramachandran J (eds) Proceedings of the international workshop on hormones and proteins, San Francisco, 1974, pp 143–156. The Chinese Univ of Hon Kong, Hong Kong

Samy TSA, Atreyi M, Maeda H, Meienhofer J (1974) Selective tryptophan oxidation in the antitumor protein neocarzinostatin and effects on conformation and biological activity. Biochemistry 15: 1007–1013

Samy TSA, Hu, J-M, Meienhofer J, Lazarus H, Johnson RK (1977) A facile method of purification of neocarzinostatin, an antitumor protein. J Natl Cancer Inst 58: 1765–1768

Sawada H, Tatsumi K, Sasada M, Shirakawa S, Nakamura T, Wakisaka G (1974) Effect of neocarzinostatin on DNA synthesis in L1210 cell. Cancer Res 34: 3341–3346

Sasada M, Sawada H, Nakamura T, Uchino H (1976) Caffeine potentiation of lethality of L-1210 cells treated with neocarzinostatin. Gann 67: 447–449

Sawyer TH, Guetzow K, Olson MOJ, Busch H, Prestayko AW, Crooke ST (1979) Amino terminal amino acid sequence of macromomycin, a protein antitumor antibiotic. Biochem Biophys Res Commun 86: 1133–1138

Shoji J (1961) Preliminary studies on the isolation of carzinostatin complex and its characteristics. J Antibiotic 14: 27–33

Sim S-K, Lown JW (1978) The mechanism of the neocarzinostatin-induced cleavage of DNA. Biochem Biophys Res Commun 81: 99–105

Tatsumi K, Nishioka H (1977) Effect of DNA repair systems on antibacterial and mutagenic activity of an antitumor protein, neocarzinostatin. Mutat Res 48: 195–204

Tatsumi K, Nakamura T, Wakisaka G (1974) Damage of mammalian cell DNA in vivo and in vitro induced by neocarzinostatin. Gann 65: 459–461

Tatsumi K, Sakane T, Sawada H, Shirakawa S, Nakamura T, Wakisaka G (1975) Unscheduled DNA synthesis in human lymphocytes treated with neocarzinostatin. Gann 66: 441–444

Tatsumi K, Tashima M, Shirakawa S, Nakamura T, Uchino H (1979) Enhancement by caffeine of neocarzinostatin cytotoxicity in murine leukemia L1210 cells. Cancer Res 39: 1623–1627

Tsuruo T, Satoh H, Ukita T (1971) Effect of the antitumor antibiotic neocarzinostatin on DNA synthesis in vitro. J Antibiot 24: 423–429

Note added in proof. Our recent work [Kappen LS, Napier MA, Goldberg IH (1980) Roles of chromophore and apoprotein in neocarzinostatin action. Proc Natl Acad Sci USA 77: 1970–1974] has shown that the non-protein chromophore is responsible for the biological effects of NCS in vivo and in vitro; the NCS apoprotein acts to stabilize the labile chromophore and to control its release for interaction, by an intercalative mechanism [Povirk LF, Goldberg IH (1980) Binding of the non-protein chromophore of neocarzinostatin to DNA. Biochemistry, in press]. A partial structure of the active chromophore has been reported [Albers-Schönberg G, Dewey RS, Hensens OD, Liesch JM, Napier MA, Goldberg IH (1980) Neocarzinostatin: Chemical characterization and partial structure of the non-protein chromophore. Biochem Biophys Res Commun, in press].

7. Site-Specific Mutagenesis in the Analysis of a Viral Replicon

D.NATHANS

A. Introduction

The small circular genome of Simian Virus 40 (SV40) is a relatively simple mammalian replicon, a DNA molecule that contains a signal for the origin of DNA replication and gene(s) for protein(s) involved in initiating DNA synthesis at the origin (Jacob and Brenner 1963). SV40 DNA replicates in the nucleus of infected monkey cells, replication commencing at a specific site in the DNA located in a regulatory segment of the viral genome between the start of early and late genes, and proceeding bidirectionally from that site (see Fig. 1) (Danna and Nathans 1972; Fareed et al. 1972). That SV40 DNA codes for a protein involved in the initiation of viral DNA replication was inferred from the observation that tsA mutants of SV40 are defective in initiation of DNA replication at the nonpermissive temperature (Tegtmeyer 1972). The mutational alterations in tsA mutants map in the gene for the SV40 T antigen (Lai and Nathans 1975). Direct interaction in vitro of T antigen with the SV40 replication origin region (Tjian 1978a, b) suggests that specific T antigen binding to the origin signal may be the initial event in SV40 DNA replication.

My colleagues, David Shortle and Robert Margolskee, and I have recently undertaken a mutational analysis of the SV40 replicon in an attempt to define the replication origin (*ori*) at the nucleotide sequence level, and to explore the interaction of virus-encoded proteins with the *ori* signal (Shortle and Nathans 1979). For this purpose mutants with single base-pair changes in the *ori* sequence or in the relevant parts of the T antigen gene would be especially valuable. The clasical method for obtaining such mutants is by random mutagenesis of the virus followed by selection of desired mutant phenotypes. However, a more direct approach is now possible, namely site-specific mutagenesis, based on the availability of restriction endonucleases and a physical and functional map of the SV40 genome, including its total nucleotide sequence (Reddy et al. 1978; Fiers et al. 1978). With this objective in mind, Shortle developed a "local mutagenesis" method that efficiently generates viral mutants with base substitutions at preselected sites in the viral DNA (Shortle and Nathans 1978). By means of this method, mutants were constructed with single base-pair substitutions within a region of SV 40 DNA corresponding to the previously mapped replication origin. Several of these mutants were partially or conditionally defective in viral DNA replication. By

Dedicated to Fritz Lipmann, with admiration and affection.

Department of Microbiology, The Johns Hopkins University, School of Medicine, Baltimore, Maryland 21205, USA

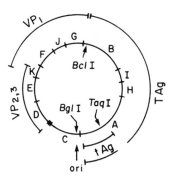

Fig. 1. A map of the 5226 base-pair, circular genome of SV40, showing *Bgl I, Taq* I, and *Bcl* I cleavage sites; *Hin* dII + III restriction fragments (*A* to *K*); coding sequences for T antigens and viral structural proteins (*VP 1, 2, 3*): and the origin of DNA replication (*ori*). *Below* is shown the palindromic sequence described later in the text; *numbers* refer to nucleotide positions in the SV40 DNA (Reddy et al. 1978)

```
5'  CAGAGGCCGAGGC  G  GCCTCGGCCTCTG
3'  GTCTCCGGCTCCG  C  CGGAGCCGGAGAC
```

5148 –   5161 –   5174 –

correlating replication defects with nucleotide sequence changes we could operationally define the *ori* signal at the sequence level. Subsequently pseudorevertants of defective *ori* mutants were isolated and found to map in the gene for T antigen (Shortle et al. 1979), thus establishing the relevance of T antigen-origin interaction to the regulation of viral DNA replication and providing material for exploring this interaction at the biochemical level. In this communication I outline the methods used to construct the desired replicon mutants by local mutagenesis and summarize some of their properties. A more detailed account can be found in the original publications (Shortle and Nathans 1978; Shortle and Nathans 1979; Shortle et al. 1979).

B. Local Mutagenesis

The local mutagenesis procedure for producing base-pair substitutions at preselected sites in duplex viral DNA involves the creation of a localized target for mutagenesis, namely a small single-stranded region, followed by treatment of the DNA with a mutagen specific for bases in single-stranded polynucleotides. An especially suitable mutagen is bisulfite, which deaminates cytosine residues only in unpaired regions of DNA or RNA (Shapiro et al. 1973; Hayatsu 1976), thus creating the natural base uracil and hence a clear change in base-pairing properties. Since a base substitution occurs at an unpaired position, it is equivalent to a base-pair substitution that cannot be reversed by cellular repair systems. Positioning of the gap can be accomplished in various ways, for example by restriction endonuclease scission of a single strand of duplex DNA at a given restriction site, followed by controlled exonucleolytic extension of the nick. After treatment of gapped DNA with the mutagen, the DNA can be used directly to generate plaques in monolayers of transfected cells. If there is only one site in the

DNA for the restriction enzyme used to position the initial single-strand break, one can close the mutagenized gap in vitro and then cut with the same restriction enzyme to eliminate all molecules that have escaped mutagenesis within the restriction site, thereby enriching for the desired mutant DNA. In practice, viral mutants have been generated by this procedure at a frequency of about 80%, i.e., about 80% of viral plaques tested contain mutants with nucleotide changes at the given restriction site (Shortle and Nathans 1978).

C. Construction of *ori* Mutants by Local Mutagenesis

The procedure just outlined was used to construct SV40 mutants with base-pair substitutions within the regulatory segment of DNA that contains the origin of DNA replication (Shortle and Nathans 1979). Examination of the nucleotide sequence in this region reveals many symmetrical sequences (Subramanian et al. 1976), including a 27 base-pair palindrome, the twofold symmetry of which could be related to the known bidirectionality of SV40 DNA replication (Fig. 1). Furthermore, restriction endonuclease *Bgl* I, which cuts SV40 DNA once, cleaves within this palindrome (Zain and Roberts, pers. comm.; Shortle and Nathans 1979). We therefore set out to deaminate cytosine residues in and around the *Bgl* I site by local bisulfite mutagenesis, as outlined in Fig. 2, and to isolate *Bgl* I-resistant (*Bgl* I^r) mutants from DNA thus modified.

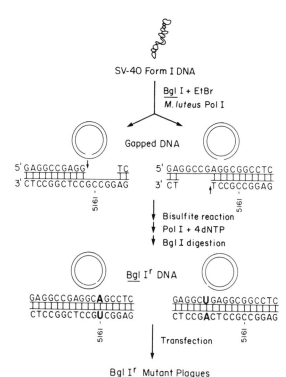

Fig. 2. Construction of *Bgl* I-resistant, *ori* mutants by local mutagenesis. Eth Br, ethidium bromide, which limits digestion by *Bgl* I to a single nick; *M. luteus* Pol I is DNA polymerase, used for its 5′ → 3′ exonuclease activity at the nick

```
                         5154          5161
                         5155          5162
wild type SV-40   5'  TCAGAGGCCGAGGCGGCCTCGGCCTCTG
Class I  (ar-1026)    TCAGAGGCCGAGGCAGCCTCGGCCTCTG
Class II (shp-1027)   TCAGAGGCAGAGGCGGCCTCGGCCTCTG
Class III (sp-1030)   TCAGAGGCCGAGGCGACCTCGGCCTCTG
Class IV (cs-1031)    TCAGAGGTCGAGGCGGCCTCGGCCTCTG
```

Fig. 3. Base substitutions in the DNA of *ori* mutants described in the text (Shortle and Nathans 1979). Nucleotide sequence analysis was carried out by the method of Maxam and Gilbert (1977)

When the *Bgl* I^r, mutagenized DNA (diagrammed in Fig. 2) was used to generate virus plaques in cell monolayers, four types of mutant plaques were produced: *ar* ("altered restriction"), which were indistinguishable from wild-type plaques (but mutant DNA was *Bgl* I^r); *shp* ("sharp plaques"); *sp* ("small plaques"); and *cs* ("cold-sensitive"), which were small at 32°C but wild-type at 37°C and 40°C. Each of these classes of *Bgl* I^r mutants corresponds to a change in a different C·G base pair within the *Bgl* I recognition sequence, as indicated in Fig. 3, i.e., within the palindromic sequence of the regulatory segment of the viral genome at precisely the positions expected by the procedure used to generate the mutants.

D. DNA Replication of *ori* Mutants

To determine the effect of each of the mutations shown in Fig. 3 on viral DNA replication, rates of viral DNA synthesis were measured in cultured cells infected by each mutant or by wild-type SV40. *ar* mutants had wild-type rates of viral DNA replication; *shp* mutants had elevated rates of viral DNA replication; *sp* mutants had markedly depressed rates; and *cs* mutants showed cold-sensitive DNA replication (Table 1). In all cases tested replication was still bidirectional, even where the symmetry of the palindromic sequence was disturbed. To learn whether defective mutants had a *cis-*dominant defect in DNA replication, as expected if their mutations were in the *ori* signal, viral DNA replication was measured in cells coinfected with a mutant and with wild-type SV40. (Mutant and wild-type DNA could be separated after *Bgl* I digestion.) In this experiment we were asking whether wild-type gene products can correct the defect in mutant DNA replication. In no case was the mutant pattern of DNA replication corrected. We conclude that single base-pair changes at positions 5154,

Table 1. Replication origin mutants of SV40

Base-pair change	Plaque morphology	DNA replication
5161 G/C → A/T	wild type	normal
5155 C/G → A/T	small, sharp	increased
5162 G/C → A/T	small	decreased
5154 C/G → T/A	small at 32°C	decreased at 32°C
	wild type at 40°C	normal at 40°C

5155, and 5162 have affected a DNA element that regulates the rate of viral DNA replication, and thus these mutations operationally identify at least part of the *ori* signal of the SV40 replicon.

E. Second-Site Revertants or *ori* Mutants

As noted in the introduction, SV40 T antigen has been shown to bind to the replication origin region in vitro. If this interaction occurs in infected cells as part of the initiation of DNA replication, as inferred from the properties of tsA mutants, one might
expect that appropriate mutations in the gene for T antigen would correct or suppress the defect caused by mutations in the T antigen binding site. To test this possibility Shortle and Margolskee isolated and characerrized second-site revertants (pseudorevertants) of defective *ori* mutants, i.e., viruses that produce wild-type plaques as a
result of a second mutation in the genome (Shortle et al. 1979). For this purpose *sp*
or *cs ori* mutant DNA was mutagenized with bisulfite by *random* local mutagenesis.
Bisulfite targets were single-stranded regions in a population of DNA molecules, each
of which contained a single small gap initiated by nicking with pancreatic DNAase
(Parker et al. 1977). This novel procedure was used in order to avoid true reversion at

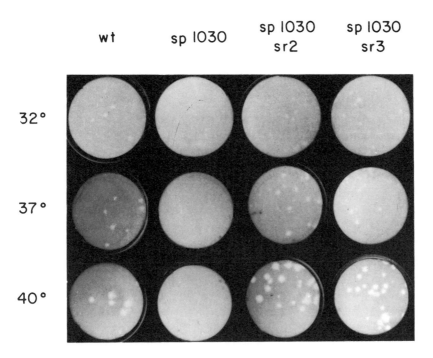

Fig. 4. Plaques of second-site revertants (*sr2* and *sr3*) or mutant *sp* 1030, compared with those of
wild-type SV40 and *sp* 1030

Mapping Pseudorevertants by
in vitro Recombination

Fig. 5. Mapping second site mutations by
in vitro recombination of viral DNA fragments

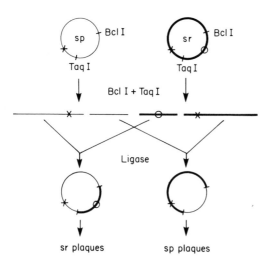

the *ori* site (since bisulfite attacks only cytosine) and to minimize multisite mutations. Examples of the plaque morphology of such second-site revertants (sr) are illustrated in Fig. 4.

To determine the map positions of suppressing mutations present in pseudorevertants, we used in vitro recombination to substitute fragments of DNA from pseudorevertants for those of the *ori* mutant from which the revertant was derived, as diagrammed in Fig. 5, and scored the morphology of plaques produced by such recombinant DNA (Shortle et al. 1979). Results of one set of experiments with *sp* 1030
and two of its revertants are presented in Table 2. As seen in the table, the suppressing mutations of *sp* 1030-sr2 and -sr3 map in a fragment derived from the T antigen
genes. Additional experiments have localized the *sr* mutations to an even smaller segment of the large T antigen gene. From these results we infer that the T antigen interacts with the *ori* signal in infected cells and that this interaction controls the rate of
viral DNA replication.

F. Conclusions

The availability of detailed functional and chemical maps of DNA molecules together
with methods for site-specific mutagenesis allow a new approach to mutational analysis of genetic elements, namely the construction of mutants with sequence changes
at preselected sites in a DNA genome followed by phenotypic analysis. We have applied this approach to a mutational analysis of the elements that have been hypothesized to regulate the rate of replication of SV40 DNA in infected cells: the origin of
replication and the SV40 T antigen. Single base-pair substitutions in the *ori* signal

Table 2. Mapping by in vitro recombination

Bcl/Taq Fragment		Plaque type	
Large	Small	sp	sr
sp 1030	sp 1030	+	−
sr2	sr2	−	+
sr3	sr3	−	+
sp 1030	sr2	−	+
sr2	sp 1030	+	−
sp 1030	sr3	−	+
sr3	sp 1030	+	−
none	sp 1030	no plaques	
none	sr2	no plaques	
none	sr3	no plaques	

produced by local mutagenesis depress or enhance the rate of viral DNA replication or cause DNA replication to be cold-sensitive. Compensating mutations in the T antigen gene can suppress the effect of a defective *ori* signal. We infer that T antigen-*ori* interaction regulates the rate of viral DNA replication. More direct analysis of this interaction, utilizing various combinations of mutant T antigens and *ori* sites, may provide some insight into the molecular interactions involved.

It should be noted that pseudorevertants of *ori* mutants of the type just described represent new viral replicons. Both components of the replicon have been modified coordinately, the *ori* signal and the protein that interacts with it. Additional changes can be brought about by serial mutation or by recombining various *ori* and T antigen mutants, thus providing interesting material for studying structure-function relationships in this important regulatory interaction. Moreover, since T antigen is also essential for cell transformation to tumorigenicity by SV40, these new SV40 replicons may also be useful in studies of the mechanism of SV40 tumorigenesis. Finally, I point out the general applicability of the approach we are taking to analyze the SV40 replicon to other viral regulatory elements and the proteins that interact with them, including cellular proteins.

Acknowledgments. This research was supported by grant number 5 P01 CA16519 from the National Cancer Institute, U.S. Public Health Service, and by the Whitehall Foundation.

References

Danna KJ, Nathans D (1972) Bidirectional replication of simian virus 40 DNA. Proc Natl Acad Sci USA 69: 3097–3100

Fareed GC, Garon CF, Salzman NP (1972) Origin and directionof simian virus 40 deoxyribonucleic acid replication. J Virol 10: 484–491

Fiers W, Contreras R, Haegeman G, Rogiers R, Van de Voorde A, Van Heuverswyn H, Van Herreweghe J, Volckaert G, Ysebaert M (1978) Complete nucleotide sequence of SV40 DNA. Nature 273: 113–120

Hayatsu H (1976) Bisulfite modification of nucleic acids and their constituents. Prog Nucleic Acid Res Mol Biol 16: 75–124

Jacob F, Brenner S (1963) Sur la regulation de la synthese du DNA chez les bacteries: l'hyopthese du replicon. C R Acad Sci, Ser D 256: 298–300

Lai C-J, Nathans D (1975) A map of temperature-sensitive mutants of simian virus 40. Virology 66: 70–81

Maxam AM, Gilbert W (1977) A new method for sequencing DNA. Proc Natl Acad Sci USA 74: 560–564

Parker RC, Watson RM, Vinograd J (1977) Mapping of closed circular DNAs by cleavage with restriction endonucleases and calibration by agarogse gel electrophoresis. Proc Natl Acad Sci USA 74: 851–855

Reddy VB, Thimmappaya B, Shar R, Subramanian KN, Zain BS, Pan J, Ghosh PK, Celma ML, Weissman SM (1978) The genome of simian virus 40. Science 200: 494–502

Shapiro R, Braverman B, Louis JB, Servis RE (1973) Nucleic acid reactivity and conformation. II. Reaction of cytosine and uracil with sodium bisulfite. J Biol Chem 248: 4060–4064

Shortle D, Nathans D (1978) Local mutagenesis: A method for generating viral mutants with base substitutions in preselected regions of the viral genome. Proc Natl Acad Sci USA 75: 2174

Shortle D, Nathans D (1979) Regulatory mutants of simian virus 40: constructed mutants with base substitutions at the origin of DNA replication. J Mol Biol 131: 801–817

Shortle D, Margolskee RF, Nathans D (1979) Mutational analysis of the simian virus 40 replicon: Pseudorevertants of mutants with a defective replication origin. Proc Natl Acad Sci USA 76: 6128–6131

Subramanian KN, Dhar R, Weissman SM (1976) Nucleotide sequence of a fragment of SV40 DNA that contains the origin of DNA replication and specifies the 5' ends of "early" and "late" viral RNA. J Biol Chem 252: 355–367

Tegtmeyer P (1972) Simian virus 40 DNA synthesis: The viral replicon. J Virol 10: 591–598

Tjian R (1978a) The binding site on SV40 DNA for a T-antigen related protein. Cell 13: 179

Tjian R (1978b) Protein-DNA interactions at the origin of simian virus 40 DNA replication. Cold Spring Harbor Symp Quant Biol 43: 655–662

D. Protein Biosynthesis

1. Molecular Mechanism of Protein Biosynthesis and an Approach to the Mechanism of Energy Transduction

Y. KAZIRO

A. Introduction

In 1941 Fritz Lipmann wrote an article in *Advances in Enzymology* which is now regarded as a landmark in the history of biochemistry. In this review he contemplated his new concept that the phosphate bonds serve as general carriers in energy transformation and in biosynthesis. The concept of "phosphate bond energy" or "energy-rich phosphate bond" has promptly prevailed among biochemists, and since then much effort has been focused on the mechanism of "group transfer" reactions in which the phosphate bond energy is very often utilized for the synthesis of covalent bonds.

Among them was the activation of amino acid to form aminoacyl-tRNA, an immediate precursor for protein biosynthesis. In the reaction catalyzed by aminoacyl-tRNA synthetase, amino acid is first activated by ATP to yield an enzyme-bound aminoacyladenylate, from which the aminoacyl moiety is subsequently transferred to tRNA to yield aminoacyl-tRNA.

The utilization of the energy of ATP for the "mechanochemical coupling" was first noted in 1942, when Engelhardt discovered that myosin is actually an ATPase for muscle contraction. Later, the process of transport of cations across biomembranes was found to be dependent on ATP (Skou 1957). Thus, the functional role of Na^+, K^+ ATPase in cytoplasmic membrane, and Ca^{2+}, Mg^{2+} ATPase in sarcoplasmic reticulum, was assigned as the vectorial transport of cations across the membrane barrier and, very often, against the concentration gradient.

The finding that the energy of ATP could be utilized for formation of an electrochemical gradient was further extended by Mitchell (1968) to the formulation of the "chemiosmotic theory" of oxidative phosphorylation. He claimed that the energy released during electron transfer in mitochondrial membrane is stored in the form of an electrochemical potential of proton, and then transformed to "phosphate potential" of ATP by the action of F_1 ATPase.

I have been interested in the molecular mechanism of energy transduction. Namely, the mechanism by which the phosphate bond energy could be utilized to drive various "mechanochemical reactions". For this purpose, I chose a rather classical problem, i.e., the role of GTP in protein biosynthesis. The rationale of working on this problem and the experimental details have already been published in several review articles (Kaziro 1973, 1976, 1978), and will not be given here in too much detail.

Institute of Medical Science, University of Tokyo, Minatoku, Takanawa, Tokyo 108, Japan

However, I just want to add briefly that back in the early 1960's Lipmann's laboratory at Rockefeller Institute made a great contribution in resolving the "transfer enzyme" into two, and later into three, components which are now designated as EF-Tu, EF-Ts, and EF-G (see for a review, Lukas-Lenard and Lipmann 1971). Later, the properties as well as the basic function of these factors have been worked out in his laboratory, and several other laboratories including our own. The early phase of these works was presented at Cold Spring Harbor Symposium in 1969 when F. Lipmann celebrated his 70th birthday (cf. Cold Spring Harbor Symposia on Quantitative Biology, Vol. 34, 1969).

The work on elongation factors was further advanced in the last 10 years, and at present the structure, function, genetics, and biosynthesis of these factors have been clarified in considerable detail (see for recent reviews, Miller and Weissbach 1977; Kaziro 1978). We have been very fortunate to be able to solve the basic mechanism by which the energy of 2 GTP molecules are utilized for the elongation of a peptide bond. Shortly, our studies on this aspect have revealed that conformational transitions of EF-Tu and EF-G in association with the change of their nucleotide ligands, i.e., from GDP to GTP or vice-versa, are of primary importance for this type of reaction. I believe that all the mechanochemical reactions utilizing the energy of nucleoside triphosphate are mediated by a similar and common mechanism.

In this paper I shall describe first the function of GTP in polypeptide chain elongation, and then some of the more recent work on *E. coli* EF-Tu including cloning of two genes for EF-Tu, i.e., *tufA* and *tufB*, determination of the partial nucleotide sequences, and regulation of the expression of these genes.

Finally, I would like to come back again to the bioenergetic aspect of protein biosynthesis, and to try to extrapolate the findings on protein synthesizing systems to other mechanochemical type reactions. A generalized concept which may be applicable to all the energy-transducing reactions involved in a variety of cellular functions will be proposed.

B. Role of GTP in Polypeptide Chain Elongation

Figure 1 illustrates the polypeptide chain elongation cycle in *Escherichia coli* (for references, see Miller and Weissbach 1977; Kaziro 1978). Three complementary factors, EF-Tu, EF-Ts and EF-G, are involved in these reactions. EF-Tu·GTP interacts with aminoacyl-tRNA to form a ternary complex, aminoacyl-tRNA·EF-Tu·GTP. The complex is then transferred to the A site of a ribosome having a peptidyl-tRNA prebound to its P site, GTP is hydrolyzed and EF-Tu·GDP is released. EF-Tu·GDP is subsequently converted to EF-Tu·GTP through exchange of its GDP moiety with GTP in the presence of EF-Ts.

Then, a new peptide bond is formed by the transfer of the carboxyl group of peptidyl-tRNA to the adjacent amino group of a newly bound aminoacyl-tRNA. The resulting ribosome possesses a deacylated tRNA in its P site, and a peptidyl-tRNA, having a chain one amino acid longer in its A site. EF-G catalyzes the translocation of peptidyl-tRNA from the A site to the P site with concomitant release of deacylated

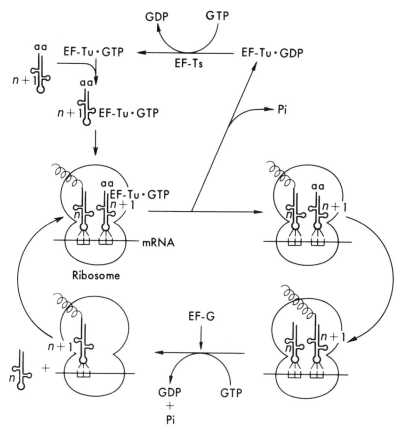

Fig. 1. Polypeptide chain elongation cycle in *E. coli.* The initial substrate (not shown) is a complex having peptidyl-tRNA$_n$ at the P site. The *leftmost bottom structure* (with tRNA$_{n+1}$ at the P site) denotes the product of one turnover of the elongation cycle. From Kaziro (1973)

tRNA from the P site and the movement of mRNA by a three-nucleotide distance on the 30S ribosomal subunit. One mol of GTP is hydrolyzed in the translocation reaction catalyzed by EF-G. The result of the above sequences of reactions is the elongation by one peptide at the expense of two molecules of GTP. One molecule of GTP is consumed for the entry of an aminoacyl-tRNA, while another molecule of GTP is utilized for the exit of a deacylated tRNA.

Less was known about polypeptide chain elongation in the eukaryotic system than in prokaryotes. However, recent studies in our laboratory, especially the isolation of EF-1α (Nagata et al. 1977) and EF-1βγ (Motoyoshi et al. 1977) from mammalian tissues revealed that the process in these two cell types is essentially analogous. The function of eukaryotic EF-1α, EF-1βγ, and EF-2 has been shown to be practically identical with that of prokaryotic EF-Tu, EF-Ts, and EF-G, respectively (see Kaziro 1978; Iwasaki 1979; Iwasaki and Kaziro 1979).

The EF-Tu-promoted, GTP-dependent binding of aminoacyl-tRNA may be considered as a process analogous to active transport. In this reaction EF-Tu functions as a carrier protein which promotes the energy-dependent, unidirectional transport of aminoacyl-tRNA to ribosomes. On the other hand, the translocation reaction has been compared to muscle contraction since mRNA moves or slides on the 30S ribosomal subunit. Below, I would like to show how both reactions are mediated through essentially similar mechanisms and that the role of GTP in these two reactions is exactly comparable.

In order to investigate the role of GTP in the above two reactions and to understand the molecular mechanism by which the energy derived from GTP is utilized for each reaction, it was essential to investigate the mechanism of each partial reaction using the highly purified factors in a substrate quantity. Table 1 lists the properties of the elongation factors from *E. coli, Thermus thermophilus*, and pig liver, purified in our laboratory.

To cut a long story short, the molecular mechanism of the process of polypeptide chain elongation has been clarified as illustrated in Fig. 2 (Kaziro 1978). The most remarkable feature of the EF-Tu-promoted aminoacyl-tRNA binding was that only EF-Tu·GTP and not EF-Tu·GDP could interact with aminoacyl-tRNA to form a ternary complex, aminoacyl-tRNA·EF-Tu·GTP. EF-Tu·GTP binds selectively with an aminoacylated form of tRNA in preference to its deacylated form, whereas EF-Tu·GDP interacts with neither aminoacyl-tRNA nor deacylated tRNA. It was further shown that in the absence of aminoacyl-tRNA, EF-Tu·GTP and not EF-Tu·GDP interacts with ribosomes. Therefore, it was suggested that the binding of GTP to EF-Tu alters its conformation in a manner that facilitates its interaction with aminoacyl-tRNA

Table 1. Properties of the elongation factors from *E. coli, T. thermophilus,* and pig liver (from Kaziro 1978)

	Elongation factors	State	M.W.	SH required	Stabilizer	K_d for GTP(M)	K_d for GDP(M)
E. coli:							
	EF-Tu	Crystalline	47,000	Yes	GDP	3.6×10^{-7}	4.9×10^{-9}
	EF-Ts	Homogeneous	36,000	Yes	$-^c$	$-^d$	$-^d$
	EF-G	Crystalline	83,000	Yes	sucrose	1.4×10^{-5}	1.1×10^{-5}
T. thermophilus:							
	EF-Tu	Crystalline	49,000	No	$-^c$	5.8×10^{-8}	1.1×10^{-9}
	EF-Ts	Homogeneous	$27,000^a$	No	$-^c$	$-^d$	$-^d$
	EF-G	Crystalline	85,000	Yes	$-^c$	1.2×10^{-5}	6.7×10^{-7}
Pig liver:							
	EF-1α	Crystalline	53,000	Yes	glycerol	1.9×10^{-7}	2.4×10^{-8}
	EF-1β	Homogeneous	30,000	Yes	phosphate	$-^d$	$-^d$
	EF-1γ	Homogeneous	55,000	$-^b$		$-^d$	$-^d$
	EF-2	Homogeneous	100,000	Yes	sucrose	1.4×10^{-5}	4.1×10^{-7}

[a] Molecular weight as a monomer   [b] Not known   [c] Not required
[d] No interaction

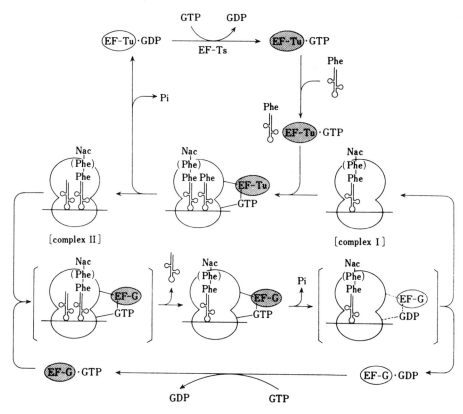

Fig. 2. Mechanism of polypeptide chain elongation. *Complex I* and *Complex II* are the post- and pre-translocational complexes, respectively. *Shaded* and *open ellipses* denote the GTP- and GDP-induced conformations of elongation factors, respectively. The *complexes in brackets* are the hypothetical transient intermediates. From Kaziro (1978)

and ribosomes. The aminoacyl-tRNA·EF-Tu·GTP complex binds to the A site of ribosomes, and then the GTP moiety of the complex was hydrolyzed to GDP and inorganic phosphate. The hydrolysis of GTP is apparently required to release EF-Tu·GDP from ribosomes by changing the conformation of EF-Tu to one which is no longer favorable for interaction with ribosomes. And then, the exogenous energy is fed into this system simply through displacement of GDP bound to EF-Tu with external GTP, to regain the original conformation of EF-Tu·GTP, Thus, it appears that the reactivity of the EF-Tu molecule is modulated through the sequential transitions of its conformation.

The intermediate of the EF-Tu-promoted aminoacyl-tRNA binding reaction could be isolated with the use of nonhydrolyzable analogs of GTP, GMP-P(NH)P and GMP-P(CH$_2$)P, and a substrate amount of EF-Tu. In this intermediate, the newly bound Phe-tRNA had not been reacted with the prebound N-acetyl-Phe-tRNA. However, when the intermediate complex was centrifuged through 10% sucrose solution containing 0.2 mM GDP, EF-Tu was released from the complex, and the peptidyl transfer reaction took place (Yokosawa et al. 1973, 1975). This indicates that the hydrolysis of GTP is not directly required for the peptidyl transfer reaction itself, but for the removal of EF-Tu, the presence of which could be inhibitory to the subsequent peptidyl transfer.

An essentially analogous mechanism was found for the EF-G-promoted translocation reaction with respect to the utilization of the GTP energy. Although bound less tightly than EF-Tu, EF-G can bind with guanine nucleotides to form a binary complex, EF-G·GTP or EF-G·GDP (N. Arai et al. 1975, 1977a). The affinity of EF-G toward ribosomes is again modulated by the species of guanine nucleotide ligand; in the presence of GTP, EF-G can interact with ribosomes to form a stable ternary complex, whereas in the presence of GDP it has practically no affinity to ribosomes.

The intermediate of the translocation was again isolated with the use of unhydrolyzable analogs of GTP and a substrate amount of EF-G (Inoue-Yokosawa et al. 1974). In this intermediate, peptidyl-tRNA had been shifted from the A to the P site (i.e., from the puromysin-unreactive to the puromysin-reactive state), and the deacylated-tRNA had been released from ribosomes. Thus, it appears that not the hydrolysis of GTP, but the binding of EF-G·GTP to ribosomes is the trigger for translocation. The hydrolysis of GTP is required for the release of EF-G from ribosomes, in order to proceed to the subsequent step of polypeptide chain elongation, and to utilize EF-G for a new cycle of reactions.

Above results indicate that the precise mechanism of the utilization of GTP energy in EF-Tu-promoted aminoacyl-tRNA binding and EF-G-promoted translocation is essentially analogous. In both cases, it is the conformation and reactivity of the protein factors associated with GTP which is utilized to drive the reaction. Afterwards, bound GTP is hydrolyzed to GDP to shift the conformation as well as the reactivity of the proteins to the alternate state. The intriguing feature of the reaction mechanism depicted in Fig. 2 is that the reactivity of both EF-Tu and EF-G is converted qualitatively and reversibly by the "phosphate-energy level" of their ligands. This has led us to the proposal that the transformation of the energy is achieved by the conformational transition induced by "high-energy phosphates" (Kaziro 1976, 1978).

The nature of the conformational change appears to be a local one, since we could not detect any gross conformational change by optical rotatory dispersion measurements or by slow hydrogen-tritium exchanges (K. Arai et al. 1977; N. Arai et al. 1977a). The functionally important conformational change between EF-Tu·GTP and EF-Tu·GDP, or EF-G·GTP and EF-G·GDP, appears to be rather restricted to the neighborhood of the active site of the proteins. Below, I shall discuss mainly the nature of the conformational change of EF-Tu. Essentially similar observations have been made also on EF-G (N. Arai et al. 1975, 1976, 1977b).

The chemical studies on the structure of EF-Tu have revealed that the protein contains two reactive and one unreactive sulfhydryl group. One reactive sulfhydryl (SH_1) is essential for the binding of GDP or GTP, and is protected by the guanine nucleotides, while the other one (SH_2) is essential for aminoacyl-tRNA binding and is protected by aminoacyl-tRNA only when EF-Tu exists as EF-Tu·GTP (K. Arai et al. 1974b). The third sulfhydryl group in EF-Tu is nonreactive, and can be titrated only after complete denaturation of the protein. These results suggest that the binding of GTP to the neighborhood of SH_1 induces a conformational transition around SH_2, and facilitates the interaction with aminoacyl-tRNA.

The proposed conformational change has been verified by several means. First, the reactivity of SH_2 toward various sulfhydryl reagents was found to be markedly increased in EF-Tu·GTP compared with EF-Tu·GDP (K. Arai et al. 1974b). Second, the introduction of the spin-labeled analogs of N-ethylmaleimide into this particular sulfhydryl group revealed the differential electron spin resonance spectra between the spin-labeled EF-Tu·GTP and EF-Tu·GDP (K. Arai et al. 1974a, 1976b). Third, fluorescent probe using l-anilino-8-naphthalenesulfonate (ANS) indicated the increase in the fluorescence emission when EF-Tu·GDP·ANS complex was converted to EF-Tu·GTP·ANS complex (Crane and Miller 1974; K. Arai et al. 1975). Equilibrium studies revealed that EF-Tu·GTP binds three mols of ANS, whereas EF-Tu·GDP binds two mols of ANS per mol of the protein, suggesting that one additional hydrophobic site is formed when EF-Tu·GDP is converted to EF-Tu·GTP. Fourth, the motility of EF-Tu·GDP and EF-Tu·GTP is found to be different when measured by the rapid tritium-hydrogen (Printz and Miller 1973) and deuterium-hydrogen (Ohta et al. 1977) exchanges. And fifth, the intrinsic fluorescence of tryptophan increased on conversion of EF-Tu·GDP to EF-Tu·GTP (K. Arai et al. 1977).

EF-Tu may be regarded as a type of allosteric protein in that its ability to interact with aminoacyl-tRNA at one site is turned on or off by a ligand change, GDP to GTP, or GTP to GDP, at a different site. It is intriguing that, in contrast to other proteins, which are usually composed of several identical or nonidentical subunits, EF-Tu is a monomeric protein consisting of a single polypeptide chain with a molecular weight of approximately 47,000. We hope that the X-ray analysis of crystals of EF-Tu·GDP and EF-Tu·GTP may elucidate the precise nature of the conformational transitions between these two forms. Studies along this line may throw more light on the molecular basis of allostery as well as on the molecular mechanism of energy transduction.

C. Primary Structure of *E. coli* EF-Tu

The studies on the limited hydrolysis of EF-Tu·GDP by trypsin have revealed that the molecule was first cleaved into two fragments, Fragment D and Fragment A, and the latter was subsequently hydrolyzed into Fragments B and C. The molecular weight of Fragments A, B, C, and D were 39,000, 22,000, 12,000, and 8000, respectively (K. Arai et al. 1976a; Gast et al. 1976; Nakamura et al. 1977).

Fragments A and D, as well as Fragments B, C, and D are tightly associated and form a complex which migrates as a single peak on a Sephadex G-75 column. It was remarkable that the complexes still retained the ability to interact with guanine nucleotides as well as the ability to change conformation upon interaction with the nucleotides (K. Arai et al. 1976a). On the other hand, neither the complex of Fragments A and D nor that of Fragments B, C, and D showed any interaction with aminoacyl-tRNA. It appears that the binding properties of EF-Tu with aminoacyl-tRNA and ribosomes were destroyed in the early phase of tryptic digestion.

The analysis of N-terminal and C-terminal sequences of these tryptic fragments revealed that Fragment D and Fragment C were derived from the N-terminus and C-terminus of EF-Tu, respectively. The N-terminal sequences of Fragment B (NH_2-Gly-Ile-Thr-) were identical to those of Fragment A. Taking all these results together, the alignment of the tryptic fragments in the EF-Tu molecule was deduced as: X-NH-[Fragment D]-[Fragment B]-[Fragment C]-COOH (Nakamura et al. 1977). From the sequence studies, the three sulfhydryl groups of EF-Tu were located in Fragment B in the order SH_2, SH_1, and SH_3 (Nakamura et al. 1977). The distance between the two active sulfhydryl groups of EF-Tu, i.e., SH_2 and SH_1, is 55 amino acid residues. The assignment of the individual sulfhydryl group is based on the previous studies in which the three sulfhydryl-containing tryptic peptides were purified and sequenced, and their chromatographic behaviors were compared to those of the tryptic digests of EF-Tu differentially labeled at each sulfhydryl group with radioactive iodoacetic acid (Nakamura et al. 1975).

The results of the amino acid sequencing by Laursen's group (Wade et al. 1975; Laursen et al. 1977) and ourselves (Nakamura et al. 1975, 1977) together with those of the determination of the sequence of *tufA* DNA by Yokota et al. (unpublished results) cover at present as much as 90% of the total amino acid sequence of *E. coli* EF-Tu.

D. Two Genes for *E. coli* EF-Tu, tufA, and tufB

Nomura and his collaborators made a remarkable observation that two distinct structural genes for EF-Tu are present in *E. coli* chromosome; one located at 72 min (*tufA*) and the other at 88 min (*tufB*) on a recalibrated map of *E. coli* (Jaskunas et al. 1975). The transducing λ phages containing *tufA* (λ*fus3*) and *tufB* (λ*rif*^d18) have been isolated and their fine genetic organizations have been established (Nomura et al. 1977).

The purpose of our investigation on the genetic elements of EF-Tu is twofold. First, we wish to determine the nucleotide sequence of both *tufA* and *tufB*. Although the products of the two genes appear to be almost identical in terms of their physical, chemical, and catalytic properties (Furano 1977; Miller et al. 1978), it would be of interest to compare, in a more direct manner, the identity of these two genes by analyzing their nucleotide sequences. Second, we are interested in studying the expression and its regulation of *tufA* and *tufB* genes. Although *tufA* is cotranscribed with its upstream genes *rpsL, rpsG*, and *fus* in the *str* operon, the level of the *tufA* product is about three to four times higher than that of the other proteins. Therefore, it is expected that a sequence could be found between *fus* and *tufA* which increases the translational frequency of *tufA*. On the other hand, transcription of *tufB* with its adjacent putative promoter is supposed to be under the stringent response. Therefore, the studies on the transcription and translation of *tufB* gene in a well-defined cell-free system may lead to the solution of the molecular mechanism of stringent control.

By cloning the *EcoRI* fragments containing *tufA* and *tufB*, respectively, on a ColEl derivative plasmid RSF2124 (ColEl-Ap^r), we isolated three new plasmids pTUAl (Shibuya et al. 1979), and pTUBl and pTUB2 (Miyajima et al. 1979). The latter two plasmids contained the DNA fragment carrying *tufB* in an opposite orientation with respect to the vector DNA. Below, I should like to give a brief summary of the results which have so far been obtained using these plasmids: (1) Most of the nucleotide sequence of the *tufA* region coding for EF-Tu(A) has been determined (Yokota et al., in preparation). The amino acid sequence of EF-Tu(A) deduced from the *tufA* DNA sequence determined so far was in complete agreement with the amino acid sequence obtained with the mixture of products from *tufA* and *tufB* genes. (2) Codon usage of *tufA* is very similar to that of *rplA, rplJ, rplK,* and *rplL* which was recently reported by Post et al. (1979). (3) The DNA sequence of the *fus-tufA* junction has been determined (Yokota et al., in preparation). The gap between the codon for C-terminal Lys of EF-G and the GUG codon for the initiation of EF-Tu(A) was 73 nucleotide base pairs. The two Shine-Dalgano sequences and/or one twofold symmetry in this region may be responsible for the increased translational frequency of the *tufA* mRNA. (4) Using a kirromycin-resistant mutant, *E. coli* LBE2012 (Van de Klundert et al. 1977), it was shown that *tufB* in pTUBl or pTUB2, but not *tufA* in pTUAl, is expressed in *E. coli* cells (Miyajima et al., unpublished). (5) The synthesis of EF-Tu(B) directed by λ*rif*^{d}18 DNA or pTUB DNA in the cell-free transcription-translation coupled system, is inhibited by ppGpp (Shibuya and Kaziro 1979). (6) The RNA transcripts from pTUBl DNA and λ*rif*^{d}18 DNA were assayed by RNA·DNA hybridization using λ*fus*3 DNA and pTUAl DNA, respectively. It was shown that the transcription of *tufB* DNA in the cell-free system consisting of purified RNA polymerase holoenzyme is sensitive to ppGpp (Shibuya et al., unpublished). The more detailed mechanism of this inhibition is now under investigation.

E. Generalized Concept for Energy Transduction

As has been described above (see Section B) and also elsewhere (Kaziro 1973, 1976, 1978), the role of GTP in EF-Tu- and EF-G-promoted reactions appears to be quite analogous. In both cases, the conformation as well as reactivity of the protein molecules are reversibly and qualitatively altered by the change of their nucleotide ligands. A single turnover reaction is accomplished by the specific conformation induced by GTP, and the splitting of GTP is required to shift the protein to another conformation (Fig. 2). An analogous mechanism of GTP cleavage was also found in IF-2-promoted binding of fMet-tRNA to ribosomes (Benne and Voorma 1972; Dubnoff et al. 1972) and, I believe, will be found in the release factor-promoted termination reaction. It is thus conceivable that the four GTP-utilizing reactions in protein synthesis, catalyzed by IF-2, EF-Tu, EF-G, and RF, respectively, are probably mediated through similar and common reaction mechanisms. Furthermore, various ATP-utilizing mechanochemical reactions including muscle contraction, active transport, and oxidative phosphorylation may also share common characteristics with the GTP-utilizing reactions of protein biosynthesis.

Table 2 lists various energy-transducing nucleoside triphosphatases involved in (1) macromolecular biosyntheses, (b) biological movements and self-assembly systems, and (3) some cellular regulatory processes. Among them, only a few examples will be discussed below.

In macromolecular biosyntheses both replication and recombination of DNA are dependent on ATP hydrolysis, and several DNA-dependent ATPases are found in these processes. For example, Scott et al. (1977) found that a *rep* protein which is required

Table 2. Energy-transducing nucleoside triphosphatases

Class	Function	Designation
I. Macromolecular biosynthesis	DNA replication and repairs	*rep* ATPase DNA gyrase *dnaB* ATPase DNA unwinding enzyme (ATPase) *recA* ATPase *recBC* nuclease (ATPase)
	Transcription	*rho* factor (ATPase)
	Translation	IF-2, eIF-2 (GTPase) EF-Tu, EF-1 (GTPase) EF-G, EF-2 (GTPase) eRF (GTPase)
II. Biological movements and self assembly	Muscle contration	Myosin ATPase
	Active transport	Na^+, K^+ ATPase Ca^{2+}, Mg^{2+} ATPase
	Oxidative phosphorylation	F_1 ATPase
	Microtubules	Dynein ATPase Assembly GTPase
III. Cellular regulation	Hormonal response	catecholamine-dependent GTPase
	Photo-reception	light-dependent GTPase

for in vitro replication of phage ϕX174 exhibits a single-stranded DNA-dependent ATPase activity. Protein *rep* appears to function for DNA strand separation during replication, for it was shown that, in a reaction uncoupled from replication, *cisA, rep,* DNA-binding protein, and ATP were able to separate the supercoiled replicative form I of phage ϕX174 into two single strands. Furthermore, in a reaction system coupled with DNA synthesis, it was observed that two ATP molecules were split per base pair melted and nucleotide residue incorporated (Kornberg et al. 1978). Therefore, it is evident that just as the elongation of polypeptide chain in protein synthesis requires the energy of GTP, the elongation of polynuceotide chain in DNA replication is dependent on the hydrolysis of ATP. I would assume that the basic mechanism by which the energy of nucleoside triphosphate is utilized for these two reactions is again analogous.

We have studied the role of GTP in the energy-dependent self-assembly system of brain microtubules (T. Arai and Kaziro 1976, 1977). Polymerization of brain tubulin, a dimeric protein with a molecular weight of approximately 110,000, into microtubules requires the presence of GTP, and the amount of GTP hydrolyzed to GDP and P_i was stoichiometric with that of tubulin incorporated into microtubules. We have demonstrated that GMP-P(NH)P could replace GTP in supporting the polymerization of microtubules. The microtubules formed in the presence of GMP-P(NH)P were indistinguishable from those formed in the presence of GTP under electron microscopy but they had a markedly decreased sensitivity toward Ca^{2+}. A similar observation was reported independently by Weisenberg et al. (1976). Based on these observations, we suggested that the role of GTP in the assembly of microtubules is to confer upon tubulin a conformation favorable for polymerization. On the other hand, the conformation of tubulin bound to GDP was the one which was unfavorable for polymerization. The equilibrium of the assembly reaction was then shifted irreversibly toward the formation of microtubules through hydrolysis of tubulin-bound GTP. The cleavage of GTP appears to be also necessary for subsequent disassembly of microtubules, i.e., for the recycling of tubulin, since GTP-assembled microtubules, but not GMP-P(NH)P-assembled microtubules, are sensitive to calcium-induced depolymerization (T. Arai and Kaziro 1976, 1977).

An extremely interesting analogy has recently been found in hormone-activated adenylate cyclase system in avian erythrocytes with respect to the role of GTP. It has been known that hormonal stimulation of adenylate cyclase requires GTP (Rodbell et al. 1971) and, more recently, the presence of hormone-dependent GTPase activity was demonstrated in turkey erythrocyte membranes (Cassel and Selinger 1976). Pfeuffer (1977, 1979) succeeded in separating from pigeon erythrocyte membranes a guanine nucleotide binding protein (G-protein) which, in association with GTP but not with GDP, can activate the adenylate cyclase catalytic unit. The hydrolysis of GTP bound to G-protein appears to be required for the shut-off of the hormonal stimulation. Thus, in the presence of cholera toxin which inhibits the hormone-induced GTPase activity (Cassel and Selinger 1977) through ADP-ribosylation of G-protein (Cassel and Pfeuffer 1978), adenylate cyclase becomes persistently activated. More detailed studies on the reaction mechanism of this system have revealed that the binding of hormone to the receptor protein stimulates the exchange of GDP tightly bound to G-protein with external GTP (Cassel and Selinger 1978). This process is obviously analogous to the EF-Ts-promoted nucleotide exchange reaction; EF-Ts induces the dissociation of the tightly-bound GDP from EF-Tu and thereby facilitates its displacement with external GTP.

The resulting G-protein·GTP complex can interact and activate adenylate cyclase through formation of the adenylate cyclase·G-protein·GTP complex (holoenzyme). It has been shown that G-protein can form a complex with the catalytic unit of adenylate cyclase in the presence of GTPγS or GMP-P(NH)P, whereas no complex formation was detected in the presence of GDP (Pfeuffer 1979).

In the foregoing discussion I have emphasized the importance of the conformational transitions induced by nucleoside diphosphate and triphosphate for various energy-transducing reactions. It must be noted that, in the cation transport systems, the intermediate phospho-enzyme also plays a significant role. It is conceivable that the conformation of ADP-sensitive E~P and of cation-sensitive E-P, possibly together with that of the enzyme·ATP complex and of free enzyme, may be implicated in the transport and countertransport of cations. In this case also, the reaction is driven by the conformational transitions induced by the "phosphate potential", except that the phosphate group is covalently linked to the protein.

I should like further to extend the above contention to include the conformational transitions of electron transferring proteins between oxidized and reduced forms, and of the photoreceptor protein between illuminated and nonilluminated forms. In the former case, the conformational change which is induced by the "oxidation-reduction potentials" of the prosthetic groups could be utilized for generation of a chemiosmotic gradient across the inner membrane of mitochondria and, in the latter case, a light-driven proton pump in *Halobacterium halobium* membranes could function through the conformational transitions of bacteriorhodopsin between its photoactivated and deactivated forms.

In conclusion, the reversible and energy-dependent conformational transitions of the functional protein are of prime importance for energy transduction in biological systems. The conformational transitions in a protein molecule are achieved again by reversible and energy-dependent changes in ligands covalently or noncovalently associated with the protein. These could be the general characteristics of the mechanism by which various energy-transducing reactions are catalyzed.

References

Arai K, Kawakita M, Kaziro Y, Maeda T, Ohnishi S (1974a) Conformational transition in polypeptide elongation factor Tu as revealed by electron spin resonance. J Biol Chem 249: 3311–3313

Arai K, Kawakita M, Nakamura S, Ishikawa I, Kaziro Y (1974b) Studies on the polypeptide elongation factors from *E. coli*. VI. Characterization of sulfhydryl groups in EF-Tu and EF-Ts. J Biochem (Tokyo) 76: 523–534

Arai K, Arai T, Kawakita M, Kaziro Y (1975) Conformational transitions of polypeptide chain elongation factor Tu. I. Studies with hydrophobic probes. J Biochem (Tokyo) 77: 1095–1106

Arai K, Nakamura S, Arai T, Kawakita M, Kaziro Y (1976a) Limited hydrolysis of the polypeptide chain elongation factor Tu by trypsin: Isolation and characterization of the polypeptide fragments. J Biochem (Tokyo) 79: 69–83

Arai K, Maeda T, Kawakita M, Ohnishi S, Kaziro Y (1976b) Conformational transitions of polypeptide chain elongation factor Tu. II. Further studies by electron spin resonance. J Biochem (Tokyo) 80: 1047–1055

Arai K, Arai T, Kawakita M, Kaziro Y (1977) Further studies on the properties of the polypeptide chain elongation factors Tu and Ts: Hydrogen-Tritium exchange, optical rotatory dispersion, and intrinsic fluorescence. J Biochem (Tokyo) 81: 1335–1346

Arai N, Arai K, Kaziro Y (1975) Formation of a binary complex between elongation factor G and guanine nucleotides. J Biochem (Tokyo) 78: 243–246

Arai N, Arai K, Maeda T, Ohnishi S, Kaziro Y (1976) Conformational transitions of polypeptide chain elongation factor G as determined by electron spin resonance. J Biochem (Tokyo) 80: 1057–1065

Arai N, Arai K, Kaziro Y (1977a) Further studies on the interaction of the polypeptide chain elongation factor G with guanine nucleotides. J Biochem (Tokyo) 82: 687–694

Arai N, Arai K, Nakamura S, Kaziro Y (1977b) Properties and function of the sulfhydryl group in the polypeptide chain elongation factor G from *E. coli*. J Biochem (Tokyo) 82: 695–702

Arai T, Kaziro Y (1976) Effect of guanine nucleotides on the assembly of brain microtubules: Ability of 5'-guanylyl imidodiphosphate to replace GTP in promoting the polymerization of microtubules in vitro. Biochem Biophys Res Commun 69: 369–376

Arai T, Kaziro Y (1977) Role of GTP in the assembly of microtubules. J Biochem (Tokyo) 82: 1063–1071

Benne R, Voorma HO (1972) Entry site of formylmethionyl-tRNA. FEBS Lett 20: 347–351

Cassel D, Pfeuffer T (1978) Mechanism of cholera toxin action: Covalent modification of the adenylate cyclase system. Proc Natl Acad Sci USA 75: 2669–2673

Cassel D, Selinger Z (1976) Catacholamine-stimulated GTPase activity in turkey erythrocyte membranes. Biochim Biophys Acta 452: 538–551

Cassel D, Selinger Z (1977) Mechanism of adenylate cyclase activation by cholera toxin: Inhibition of GTP hydrolysis at the regulatory site. Proc Natl Acad Sci USA 74: 3307–3311

Cassel D, Selinger Z (1978) Mechanism of adenylate cyclase activation through the β-adrenergic receptor: Catecholamine-induced displacement of bound GDP by GTP. Proc Natl Acad Sci USA 75: 4155–4159

Crane LJ, Miller DL (1974) Guanosine triphosphate and guanosine diphosphate as conformation determining molecules. Differential interaction of a fluorescent probe with the guanosine nucleotide complexes of bacterial elongation factor Tu. Biochemistry 13: 933–939

Dubnoff JS, Lockwood AH, Maitra U (1972) Studies on the role of guanosine triphosphate in polypeptide chain initiation in *Escherichia coli*. J Biol Chem 247: 2884–2894

Furano AV (1977) The elongation factor Tu coded by the *tufA* gene of *Escherichia coli* K-12 is almost identical to that coded by the *tufB* gene. J Biol Chem 252: 2154–2157

Gast WH, Leberman R, Schulz GE, Wittinghofer A (1976) Crystals of partially trypsin-digested elongation factor Tu. J Mol Biol 106: 943–950

Inoue-Yokosawa N, Ishikawa C, Kaziro Y (1974) The role of guanosine triphosphate in translocation reaction catalyzed by elongation factor G. J Biol Chem 249: 4321–4323

Iwasaki K (1979) Peptide chain elongation in eukaryotic cells. Seikagaku (in Japanese) 51: 1–18

Iwasaki K, Kaziro Y (1979) Polypeptide chain elongation factors from pig liver. In: Moldave K, Grossman L (eds) Methods in Enzymol, vol LX, pp 657–676. Academic Press, New York

Jaskunas SR, Lindahl L, Nomura M, Burgess RR (1975) Identification of two copies of the gene for the elongation factor EF-Tu in *E. coli*. Nature 257: 458–462

Kaziro Y (1973) The role of GTP in the polypeptide elongation reaction in *E. coli*. In: Nakao M, Packer L (eds) Organization of energy transducing membranes, pp 187–200. University of Tokyo Press, Tokyo

Kaziro Y (1976) Studies on high-energy phosphate bonds: from biosynthetic to mechanochemical reactions. In: Kornberg A, Horecker BL, Cornudella L, Oro J (eds) Reflections on biochemistry in honour of Severo Ochoa, pp 85–94. Pergamon Press, Oxford New York

Kaziro Y (1978) The role of guanosine 5'-triphosphate in polypeptide chain elongation. Biochim Biophys Acta 505: 95–127

Kornberg A, Scott JF, Bertsch LL (1978) ATP utilization by *rep* protein in the catalytic separation of DNA strands at a replicating fork. J Biol Chem 253: 3298–3304

Laursen RA, Nagarkatti S, Miller DL (1977) Amino acid sequence of elongation factor Tu. Characterization and alignment of the cyanogen bromide fragments and location of the cysteine residues. FEBS Lett 80: 103–106

Lipmann F (1941) Metabolic generation and utilization of phosphate bond energy. Adv Enzymol 1: 99–162

Lucas-Lenard J, Lipmann F (1971) Protein biosynthesis. Annu Rev Biochem 40: 409–448

Miller DL, Weissbach H (1977) Factors involved in the transfer of aminoacyl-tRNA to the ribosome. In: Weissbach H, Pestka S (eds) Molecular mechanisms of protein biosynthesis, pp 323–373. Academic Press, New York

Miller DL, Nagarkatti S, Laursen RA, Parker J, Friesen JD (1978) A comparison of the activities of the products of the two genes for elongation factor Tu. Mol Gen Genet 159: 57–62

Mitchell P (1968) Chemiosmotic coupling and energy transduction. Glynn Research, Bodmin

Miyajima A, Shibuya M, Kaziro Y (1979) Construction and characterization of the two hybrid ColEl plasmids carrying *Escherichia coli tufB* gene. FEBS Lett 102: 207–210

Motoyoshi K, Iwasaki K, Kaziro Y (1977) Purification and properties of polypeptide chain elongation factor-$1\beta\gamma$ from pig liver. J Biochem 82: 145–155

Nagata S, Iwasaki K, Kaziro Y (1977) Purification and properties of polypeptide chain elongation factor-1α from pig liver. J Biochem 82: 1633–1646

Nakamura S, Arai K, Takahashi K, Kaziro Y (1975) Amino acid sequences of two sulfhydryl-containing tryptic peptides of the polypeptide chain elongation factor Tu. Biochem Biophys Res Commun 66: 1069–1077

Nakamura S, Arai K, Takahashi K, Kaziro Y (1977) Alignment of the tryptic fragments and location of sulfhydryl groups of the polypeptide chain elongation factor Tu. Biochem Biophys Res Commun 77: 1418–1424

Nomura M, Morgan EA, Jaskunas SR (1977) Genetics of bacterial ribosomes. Annu Rev. Genet 11: 297–347

Ohta S, Nakanishi M, Tsuboi M, Arai K, Kaziro Y (1977) Structural fluctuation of the polypeptide-chain elongation factor Tu. Eur J Biochem 78: 599–608

Post LE, Strycharz GD, Nomura M, Lewis H, Dennis PP (1979) Nucleotide sequence of the ribosomal protein gene cluster adjacent to the gene for RNA polymerase subunit β in *Escherichia coli*. Proc Natl Acad Sci USA 76: 1697–1701

Pfeuffer T (1977) GTP-binding proteins in membranes and the control of adenylate cyclase activity. J Biol Chem 252: 7224–7234

Pfeuffer T (1979) Guanine nucleotide-controlled interactions between components of adenylate cyclase. FEBS Lett 101: 85–89

Printz MP, Miller DL (1973) Evidence for conformational changes in elongation factor Tu induced by GTP and GDP. Biochem Biophys Res Commun 53: 149–156

Rodbell M, Birnbaumer L, Pohl SL, Krans HMJ (1971) The glucagon-sensitive adenyl cyclase system in plasma membrane of rat liver. V. An obligatory role of guanyl nucleotides in glucagon action. J Biol Chem 246: 1877–1882

Scott JF, Eisenberg S, Bertsch LL, Kornberg A (1977) A mechanism of duplex DNA replication revealed by enzymatic studies of phage φX174: Catalytic strand separation in advance of replication. Proc Natl Acad Sci USA 74: 193–197

Shibuya M, Kaziro Y (1979) Studies on stringent control in a cell-free system. J Biochem (Tokyo) 86: 403–411

Shibuya M, Nashimoto H, Kaziro Y (1979) Cloning of an *Eco*RI fragment carrying *E. coli tufA* gene. Molec Gen Genet170: 231–234

Skou JC (1957) The influence of some cations on an adenosine triphosphatase from peripheral nerves. Biochim Biophys Acta 23: 394–401

Van de Klundert JAM, den Turk E, Borman AH, van der Meide PH, Bosch L (1977) Isolation and characterization of a mocimycin resistant mutant of *Escherichia coli* with an altered elongation factor EF-Tu. FEBS Lett 81: 303–307

Wade M, Laursen A, Miller DL (1975) Amino acid sequence of elongation factor Tu. Sequence of a region containing the thiol group essential for GTP binding. FEBS Lett 53: 37–39

Weisenberg RC, Deery WJ, Dickinson PJ (1976) Tubulin-nucleotide interactions during the polymerization and depolymerization of microtubules. Biochemistry 15: 4248–4254

Yokosawa H, Inoue-Yokosawa N, Arai K, Kawakita M, Kaziro Y (1973) The role of GTP hydrolysis in elongation factor Tu promoted binding of aminoacyl-tRNA to ribosomes. J Biol Chem 248: 375–377

Yokosawa H, Kawakita M, Arai K, Inoue-Yokosawa N, Kaziro Y (1975) Binding of aminoacyl-tRNA to ribosomes promoted by elongation factor Tu: Further studies on the role of GTP hydrolysis. J Biochem (Tokyo) 77: 719–728

2. On Codon-Anticodon Interactions

H. GROSJEAN and H. CHANTRENNE

A. Introduction

One of the milestones on the road to the discovery of protein biosynthesis pathway was the demonstration by Lipmann's group that the tRNA's bring the aminoacids at the proper place on the template, that genetic information is read by tRNA's (Chapeville et al. 1962).

Accuracy of translation therefore rests on correct charging of tRNA's and on faultless selection on the ribosome of charged tRNA's by the codons of the messenger.

Recognition of a codon by a complementary triplet (the anticodon of the tRNA) looks straightforward enough and at first it did not seem to raise any problem. However, when the molecular process and its selectivity are considered more closely, serious difficulties arise.

The stability of GC pairs is greater than that of AU pairs in a regular double helix. For instance, the melting temperature of a double strand formed by self-complementary A_3U_3 is $-8°$, that of a similar helix made of two G_3C_3 is $+94°$ under identical conditions (Borer et al. 1974).

One would expect codons made of Gs and (or) Cs to bind their complementary tRNA much more strongly than codons made of As and (or) Us; if so, the time of residence of the different tRNA's on their respective complementary codon would be quite different. Could such a situation be compatible with smooth translation of the message? (Eigen 1971, 1978)

Protein synthesis is very accurate: substitution of valine for isoleucine, or cystein for arginine at certain positions of a protein were estimated to occur with a frequency less than 4×10^{-4} per codon (Loftfield and Vanderjagt 1972; Edelmann and Gallant 1977). We might expect that selective binding and complete processing of the aminoacyl-tRNA must be at least as accurate as that, and actually more accurate since a correct aminoacyl-tRNA competes for each codon with more than twenty incorrect ones at variable concentrations. If the selection of tRNA's rests only on the difference in free energy of complex formation, at equilibrium a difference of ΔG in the range of 5 to 7 kcal between wrong and right association is required. Could this be accounted for by a difference of two H bonds in an aqueous environment? For instance, the ΔG difference between a correct AU pair and an incorrect GU pair is much lower than that

This paper is dedicated to Prof. F. Lipmann

Department of Molecular Biology, University of Brussels, 1640 Rhode St-Genese, Belgium

(Fink and Crothers 1972; Tinoco et al. 1973; Ninio 1979a). Nevertheless, replacement of A by G in the first two positions of any codon changes its meaning completely.

Absolute values of pairing energy derived from measurements on homogeneous complementary oligonucleotides show that the complex formed by two complementary trinucleotides must be very labile indeed; association is practically undetectable in dilute solution (Jaskunas et al. 1968). How can the tRNA bind specifically and strongly enough to its complementary codon to insure correct translation?

In the presence of 30S ribosomal subunits or 70S ribosomes most tRNA's do bind specifically to their complementary trinucleotide; they form at $0°$ a complex stable enough for being isolated easily together with the 30S or 70S particle (Glukhova et al. 1975). This is the basis of one of the methods which contributed to solving the genetic code (Nirenberg and Leder 1964). In one way or another, the 30S particles of the 70S ribosomes play a part in the accuracy of translation: mistakes are caused by streptomycin and other antibiotics (Davies et al. 1965; Tai et al. 1978); certain mutants affected in 30S ribosomal proteins resist streptomycin, others require the antibiotic for translating the messages correctly (Gorini et al. 1966; Apirion and Schlessinger 1967; Bollen et al. 1969; Topisirovic et al. 1977). Accuracy of translation must depend on subtle structural details of certain ribosomal proteins to which the messenger is associated. The exact process escapes us completely, but it is clear that built-in features of ribosomal proteins are decisive for accuracy (Kurland 1977; de Wilde et al. 1977).

B. Pecularities of the Anticodon Loop of tRNA

Built-in features of the anticodon loop are also essential for its correct association with a codon; their study gives a first glimpse of what happens on the ribosome when a codon selects a tRNA.

To approach this problem, an amenable system is the interaction of pairs of tRNA's having complementary anticodons.

At $0°C$ two complementary trinucleotides in solution do not associate to any measurable extent (Jaskunas et al. 1968); but the association of the anticodon of a tRNA with its complementary trinucleotide can be detected and studied quantitatively (Högenauer 1970; Uhlenbeck 1972; Pongs et al. 1973). For instance, (see Fig. 1) trinucleotide UUC binds to anticodon of tRNA^{Phe} one thousand times more strongly than to the free trinucleotide GAA (Eisinger et al. 1971; Yoon et al. 1975). The interaction between tRNA^{Phe} (anticodon GmAA) and tRNA^{Glu} (anticodons s^2U*UC) (Fig. 1c) is still more than one thousand times stronger (Eisinger 1971; Eisinger and Gross 1975). Strong association of this sort is observed with other pairs of tRNA's as well (Eisinger and Gross 1974; Grosjean et al. 1978a); it can even be used for isolating tRNA's or tRNA precursors by affinity chromatography on columns to which the complementary tRNA is covalently linked (Grosjean et al. 1973; Vögeli et al. 1975; Buckingham 1976). This dramatic increase of stability by more than six orders of magnitude at $0°$, observed when the two trinucleotides are within the anticodon loop, is as unexpected and puzzling as the stabilization exerted by the ribosome on the tRNA-codon complex. But the system is much simpler, since it involves only two well-known molecules in free solution.

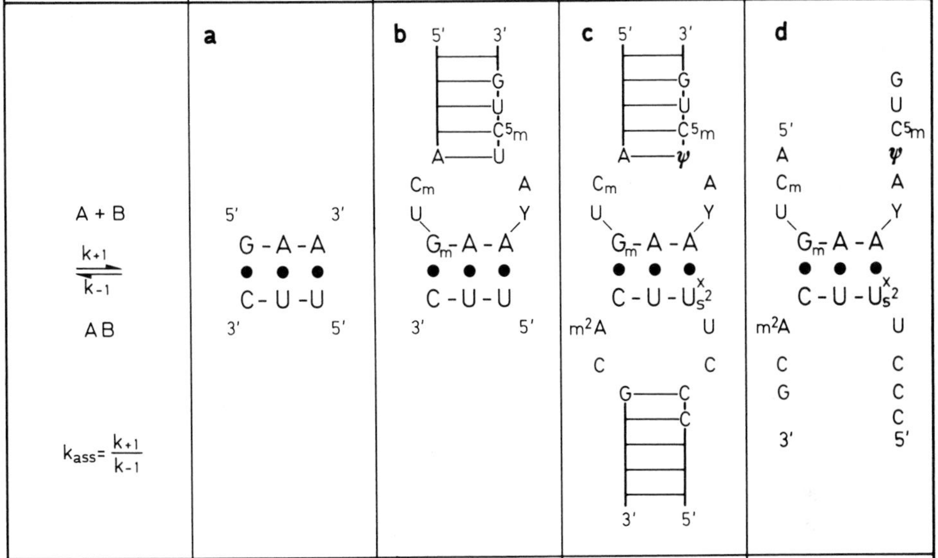

Type of complex	(Codon)^2	(Anticodon - Codon)	(Anticodon)^2	(Anticodon Fragment)^2
	a	**b**	**c**	**d**

The body of the comparison table (schematic structures):

- **a** (Codon)^2:
  5'   3'
  G - A - A
  • • •
  C - U - U
  3'   5'

- **b** (Anticodon - Codon):
  G_m - A - A
  • • •
  C - U - U

- **c** (Anticodon)^2:
  G_m - A - A
  • • •$_x$
  C - U - U_s^2

- **d** (Anticodon Fragment)^2:
  G_m - A - A
  • • •$_x$
  C - U - U_s^2

Left column reaction:

A + B $\underset{k_{-1}}{\overset{k_{+1}}{\rightleftharpoons}}$ AB

$k_{ass} = \dfrac{k_{+1}}{k_{-1}}$

$K_{ass} \left[\begin{array}{c}0°C\\20°C\end{array}\right] M^{-1}$	$\geqslant 1$ —	1.8×10^3 —	4.4×10^7 2.1×10^6	9.8×10^5 7.2×10^4
$\Delta H : kcal \cdot M^{-1}$	−14 (calculated)	−14 ± 4	−24 ± 2	−23 ± 2
k_{+1} (0') $M^{-1} sec^{-1}$ k_{-1} (0') sec^{-1}	— —	1.8×10^5 100.0	5.3×10^6 0.12	3.3×10^6 3.4
Références	JASKUNAS et al 1968	YOON et al. 1975	GROSJEAN, SÖLL, CROTHERS 1976	

Fig. 1a−d. Comparison of thermodynamic and kinetic parameters for the association of two complementary sequences in polynucleotides. The data are taken from three different papers but they were obtained under closely comparable experimental conditions (ionic strength 0.1 or 0.2, 10 mM Mg^{++}, buffer at pH 6.8−7.0). *Abbreviations:* Cm: 2'-O-Methylcytidine; Gm: 2'-O-Methylguanosine; ψ: Pseudouridine; C^{5}m: 5-Methylcytidine; m^{2}A: 2-Methyladenosine; s^{2}U: 5-methylaminomethyl-2-thiourine; Y: a tricyclic derivative of Guanosine. For simplicity, in **b** and **c** bases in anticodon stem and the rest of tRNA molecule are not mentioned

A first tentative explanation which comes to mind in the case of this simpler system is as follows. Trinucleotides in solution enjoy many degrees of freedom; if in tRNA their conformation was frozen in a geometry adequate for pairing as a segment of double helix, the unfavorable entropy term might be considerably reduced, association made more probable, and the complex more stable (Eisinger 1971).

Grosjean et al. (1976) determined kinetic and thermodynamic paramters of the tRNA^{Phe}-tRNA^{Glu} association using the temperature jump method and the highly sensitive and accurate spectrophotometer described by Crothers (1971) and Rigler et al. (1974). It was found that stabilization is not entropic, it is almost entirely due to an extra *enthalpy* change of about −10 kcal over that expected for GAA-UUC pairing. Base stacking being the major source of enthalpy production in nucleic acid pairing (Martin et al. 1971; Pörschke 1977), this large enthalpy change is assumed to be due to some base stacking which occurs when the complementary tRNA associate, and does not occur when the free trinucleotides are confronted. The difference in behavior of the two systems must be looked for in a different geometry, or in base stacking inside and outside the nucleotide triplets, or in some effect pertaining to the loop constraints, or in the tertiary structure of tRNA's.

Pieces of tRNA's 10 and 12 nucleotides long and containing the anticodons were confronted. None of them could form a closed anticodon loop or pair, except by the complementary anticodon triplets (Fig. 1d). Their interaction is weaker by a factor of 44 than that of the intact tRNA's; nevertheless, at 0° it remains more than 10^5 times stronger than that of the free trinucleotides (Grosjean et al. 1976). Unexpected high affinity for the binding of the UUCA tetranucleotide on the same fragment of yeast tRNA^{Phe} was also reported by Yoon et al. (1976). Therefore, although the loop constraints may contribute to stability by two orders of magnitude, the essential enthalpy contribution must come from the unpaired bases flanking the paired triplets and from the modified bases at the first position of anticodons. Nonpaired bases at the ends of a double helical segment are indeed known to stabilize such a structure, probably by stacking on the ordered helix (Martin et al. 1971; Pörschke 1977). Stabilization of tRNA-tRNA pairing must be due to a large extent to stacking of dangling bases on both sides of the anticodons. This conclusion is supported by direct experiment.

In tRNA^{Phe} from yeast, the base on the 3' side of the anticodon is a tricyclic molecule derived from guanine, and named base Y; it can be split selectively under mild conditions (Thiebe and Zachau 1968). Removal of Y decreases the stability of tRNA^{Phe}-tRNA^{Glu} considerably with a loss of −7 kcal in ΔH. Also, tRNA^{Phe} from Mycoplasma has a m^{1}G instead of Y base next to the 3' side of anticodon: it forms a strong complex with tRNA^{Glu} and the ΔH for this reaction is −20 kcal M^{-1} instead of −24 kcal M^{-1} for yeast tRNA^{Phe}-RNA^{Glu} (Grosjean et al. 1976).

No direct proof for the effects of modified bases in the first position of anticodon on stability of tRNA^{Phe}-tRNA^{Glu} is yet available, but physicochemical studies on oligonucleotides containing 2-thio-Uridine (Mazumdar et al. 1974; Plesiewicz et al. 1976) or 2-O' methyl on the ribose (Drake et al. 1974) clearly demonstrate that these modified nucleotides stack much more strongly with their neighboring bases than the corresponding nonmodified ones.

The picture one can form from these results is as follows: the anticodon loop in free tRNA's in solution is rather flexible (Langlois et al. 1975; Turner et al. 1975) even in a crystal (Holbrook et al. 1978; Schevitz et al. 1979), and the triplet is not quite in the conformation it would adopt when paired in a double helix segment (Sundaralingam et al. 1975; Rich and Rajbhandary 1976). On both sides there are unpaired dangling bases, e.g., in tRNA^{Phe} base Y on the 3' side and two pyrimidines on the 5' side of the anticodon. When complementary tRNA's associate, the bases of both anti-

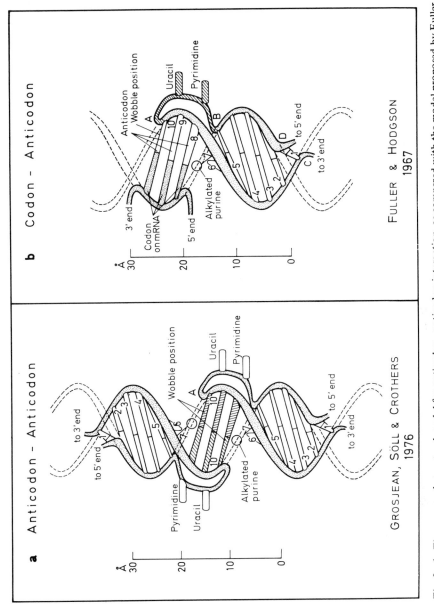

Fig. 2a, b. The proposed structural model for anticodon–anticodon interaction, as compared with the model proposed by Fuller and Hodgson (1967) for codon–anticodon interaction on the ribosome

codons stack better and the dangling bases stack on the ends of the short helix segment formed (Fig. 2a). This accounts for an extra enthalpy production which stabithe complex; it fits with an earlier model for codon–anticodon interaction intially proposed by Fuller and Hodgson (1967) (Fig. 2b).

Compilation of 123 sequences of tRNA :

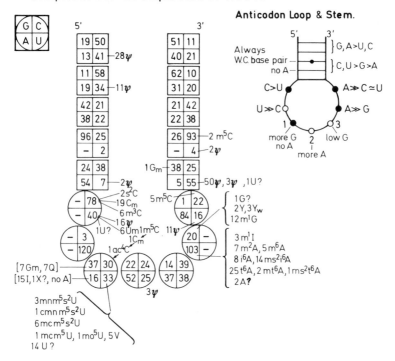

Fig. 3. Compilation of all tRNA sequences known up to April, 1979. Most tRNA sequences are taken from Gauss et al. (1979) and Dirheimer et al. (1979); several were in press. Each position of the anticodon loop is represented by a *circle*, or by a *square* in the anticodon stem. The frequency of each of the four bases are mentioned in these *boxes* according to the *pattern* indicated in the small insert. The fraction of these numbers which corresponds to a modified base is indicated outside the corresponding *box. Minus* means that no corresponding base was ever found at this position. Abbreviations for the modified bases are those used by Gauss et al. (1979). Inosine (I) being a post-transcriptional derivative of Adenosine (Durge and Cedergren 1974), is counted as Adenosine. The computer facilities of the University of Montreal (Cedergren and Sankoff) were used for this compilation

Compilation of data concerning 58 prokaryotic and 65 eukaryotic tRNA's (Fig. 3) shows that the anticodon is always flanked on its 5' side by two pyrimidines; the one immediately adjacent to the anticodon is uracil in 120 out of 123 sequences examined. The base next to the anticodon on the 3' side is always a purine, which is modified in many different manners in 85 of the 123 sequences tRNA's. These common features must reflect some common requirement for tRNA function (see McCloskey and Nishimura 1977; Feldman 1977). The data described above suggest that they are related at least to stabilization of anticodon pairing.

Inasmuch as the association of complementary tRNA's can be regarded as a model of what happens on the ribosome (Fig. 2), we shall assume that the features of the anticodon loop which stabilize the complex of complementary anticodons play the same role in binding of tRNA to messenger on the ribosome. There is evidence supporting this. Removal of the Y base from tRNA^{Phe} results in loss of codon recognition in the classical in vitro test systems (Thiebe and Zachau 1968); the same applies to its ability to bind UUU and UUC triplets in the absence of ribosome (Pong et al. 1973). Similar observations were reported on tRNA's lacking the normal modification of the adenine next to the anticodon (for example: Fittler and Hall 1966; Gefter and Russell 1969; Furuichi et al. 1970; Weissenbach and Grosjean 1980). More recently, a mutant was found in which the isopentenyl adenosine normally present in that same position in a suppressor tRNA is replaced by adenosine; the efficiency of suppression in vivo (i.e., the effiency of that tRNA in protein synthesis) is much reduced (Laten et al. 1978). This is what one would expect if base modification resulted in stabilization of tRNA binding to messenger on the ribosome.

It will be noticed in the models of Fig. 2 that the two pyrimidines on 5' side of the anticodon are "looped out" and exposed; they might interact with the 30s ribosomal RNA or proteins at some stages of the elongation process.

Interestingly enough, the strong binding that characterizes the complex between complementary anticodons is restricted to the three central bases of a loop of seven bases: the four-bases pair complex involving the three anticodon bases and the additional U residue on the 5' side was found to be weaker by more than two orders of magnitude (Grosjean et al. 1976). This is just the opposite of what was expected on the basis of results obtained for the binding of the tetra- or pentanucleotides (UUCA or UUCAG) on the anticodon loop of yeast tRNA^{Phe} (Uhlenbeck 1972; Eisinger and Spahr 1973; Yoon et al. 1976). It thus appears that the preferred complex for two loops of seven bases has a pseudo twofold axis of symmetry at the bond between the center bases of the two anticodons. On the other hand, one of the frameshift suppressor tRNA's, which uses a space of four bases at a time on the mRNA, has eight bases in its anticodon loop instead of seven (Riddle and Carbon 1973).

All these observations taken together suggest that the triplet nature of the genetic code is fundamentally related to the special architecture of the anticodon loop; they suggest also that the function of the ribosomal proteins implicated in the accuracy of translation might conceivably be to confer to the codon at the reading site a geometry and an environment comparables to those of the anticodon in a tRNA. Such an effect of the ribosomal protein might not be possible with small oligonucleotides or with certain synthetic polynucleotides (poly U for example) and great care should be taken when one has to interpret results obtained with such nonnatural templates. For example, this might explain why the rate of exchange of a trinucleotide on the ribosome is much faster than the rate of exchange of the complementary aminoacyl-tRNA (Hatfield and Nirenberg 1971), a property incompatible with a template function of mRNA; also this might explain why some oligonucleotides complementary to an anticodon loop do not stimulate at all the tRNA binding to 30S ribosomes (Möller et al. 1978).

From the above assumption it follows that the "architecture" of the successive codons in the mRNA (taken by "5 bases out of 3") must be very important and may be symmetrical to that of an anticodon loop. This is exactly what appears from statistical analysis of the codon neighboring bases in most mRNA sequenced up to now (Crick et al. 1976; Eigen 1978; Grantham, Gauthier and Grosjean, unpublished results). Accordingly, the bases flanking the codon in the messenger (and belonging to the neighbor codons) should stabilize tRNA binding in the same manner as the dangling bases of the anticodon loop. This may explain the puzzling observation that the efficiency of translation of individual codons in vivo depends on the nature of neighboring triplets (Colby et al. 1976; Akaboshi et al. 1976; Fluck et al. 1977; Feinstein and Altman 1978). The stacking contributions of a dangling adenine or a dangling cytosine would indeed be quantitatively different. A more farfetched but intriguing possibility is that the effect of neighboring triplets may account in part for the nonrandom use of the different codons of individual aminoacids in natural messengers (Grosjean et al. 1978b; Grantham 1978; Fiers and Grosjean 1979; Grosjean 1979). A special situation may exist in the formation of the initiation complex; apparently four bases might be involved in the recognition of the initiator tRNA (Taniguchi and Weissmann 1978; Manderscheid et al. 1978; Ganoza et al. 1978; Belin et al. 1979).

Another interesting result of the study of association of complementary tRNA's concerns kinetic parameters. The equilibrium constant of tRNA^{Phe}-tRNA^{Glu} association is 10^6 times smaller than that of the corresponding free trinucleotides GAA-UUC, as noted above. This could in principle result from an increased association velocity and/or from a reduced dissociation rate of the complex. Temperature jump experiments show that the change in equilibrium constant is accounted for entirely by a decrease of the dissociation rate by a factor 10^6. Complex formation is not more frequent, but once formed the complex has a longer life (Grosjean et al. 1976). This type of stabilization, according to Ninio (1974), may be important for accuracy.

The above considerations rest on the study of one single pair of complementary tRNA's, namely tRNA^{Phe}-tRNA^{Glu}. Information about the specificity of pairing was derived from experiments on some 60 pairs of tRNA with either complementary or partly complementary anticodons. A first striking result of this study (Grosjean et al. 1978a) was evidence that the lifetime of a pair (which is directly related to binding energy as we have just seen) is not correlated to the number of GC or AU pairs involved. For instance, from data on double helical ribonucleotides (Pörschke et al. 1973; Borer et al. 1974; Ninio 1979a) one would expect an association involving 3 GC pairs to live at least 100 times longer than one with 1 GC and 2 AU pairs. Temperature jump experiments show that paradoxically the lifetime of the complex tRNA^{Ala}-tRNA^{Gly} (3 GC pairs) is just a little *shorter* than that of the complex tRNA^{Phe}-tRNA^{Glu} (1 GC and 2 AU). The lifetime of all tested correct pairs of complementary anticodons fall actually within a narrow range differing by less than one order of magnitude irrespective of their GC or AU content (Fig. 4). How can this be explained?

It is known that sequence in influences the interaction energy due to differences in overlap and relative orientation of the bases (Borer et al. 1974). But this cannot account for the close value of the lifetime found for all the correct pairs. It turns out (Fig. 5) that anticodons containing As or Us are always flanked on the 3' side by a much modified purine (base Y in the case of tRNA^{Phe}, i^6A or ms^2i^6A, t^6A, mt^6A or ms^2t^6A in other systems) which stabilizes the complex, as emphasized above. On the contrary, a normal purine A is found on the 3' side of anticodons made of only Gs and/or Cs (such as in tRNA^{Gly} and tRNA^{Ala}), as if they did not need as much stabilization. A remarkable illustration of this correlation is that a mutation resulting in the replacement of C by A at the third position within the anticodon is accompanied by a

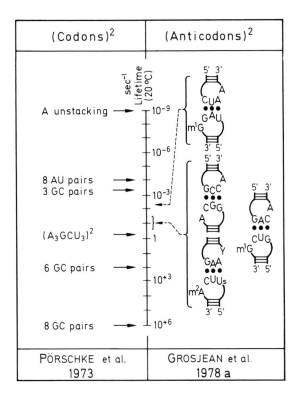

Fig. 4. Comparison of relaxation time constants (lifetimes) of different types of nucleic acid associations at 20°C. The time scale is logarithmic. Relaxation time constants of the bimolecular reaction are given at zero concentration and correspond to the reverse of the kinetic rate constants of dissociation (k_{-1}) of the complex. The data on the interaction between oligonucleotides were kindly provided by Dr. D. Pörschke

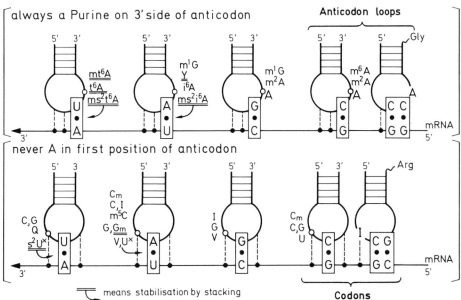

Fig. 5. Correlation between modified bases located next to the 3' side of anticodon or at the first position of anticodon, with its adjacent anticodon base. Doubly underlined bases are thought to stabilize the helical structure by stacking more efficiently on the neighboring base

base, modification which did not occur in the wild-type tRNA, namely addition of a bulky isopentyl thiomethyl group on the NH_2 of the normally unmodified adenine flanking the anticodon on the 3' side (Carbon and Fleck 1974). Similarly, a mutation which replaces C by U at the third position of the anticodon is accompanied by the addition of a carbamoylthreonyl to the normally unmodified A next to the anticodon (Roberts and Carbon 1974). The modification enzymes react to the presence of A or U at the third position of the anticodon in modifying the flanking base, but they do not modify it when the third base of the anticodon is C.

In the first position of anticodon, substitution of a modified base for a regular one might be also related to the change of the interaction energy and/or specificity not only with the third base of the codon as it is usually thought, but also with the base located *next* to it at the middle position of anticodon. Indeed, 2-thio derivatives or Uridine, which stack better than U, are always found next to U or to A occupying the middle position of an anticodon. The systematic absence of Adenosine at the first position of anticodon might be related also to its intrinsically weak potential of AU pairing but it is most probable that G or 2-o' methyl G are used because of better stacking at this position, which insures a correct reading of the middle anticodon base. Situation with Inosine at the first position of anticodon is particular since IC pairs are less stable than GC, but nevertheless in the same range of stability as a GU or IU base pairs (Unger and Takemura 1973; Grosjean et al. 1978a). Again, this might be related to a more global property of the codon—anticodon interaction, such as an optimization of its energy of association (Ninio 1971).

This does not exclude other essential roles of the bulky substituents found in anticodon loops such as the restriction and the amplification of the wobble recognition (Hillen et al. 1978), the possible interaction with some part of the ribosomal RNA or ribosomal proteins (Feldman 1977), and their possible incidences on the regulation (Cortese 1979) and the development of the cellular processes (Vold 1978; Hoburg et al. 1979).

From these data it is obvious, however, that the choice of modified bases within the anticodon or next to it results in a modulation of the interaction energy between tRNA's with complementary anticodons. Transposing this conclusion to the case of messenger tRNA binding on the ribosome one may surmise that a function of base modifications in the anticodon loop, as well as a proper choice of the successive codons in mRNA, is to adjust the association energy so as to bring the residence time of all tRNA's irrespective of the codon composition, within a narrow range and in the correct frame, and thus to allow a smooth in phase translation. In this perspective, the molecular tinkering which takes place in tRNA formation and the amazing diversity of base modifications observed begin to make sense; they are part of the *fine tuning* of the translation machinery (see also Weiss 1973; Chantrenne 1978; Ninio 1979b).

C. What Happens After Codon-Anticodon Association?

Grosjean et al. (1978a), Labuda and Pörschke (unpublished results), Grosjean and Houssier (unpublished results) also studied the behavior of several pairs of tRNA's

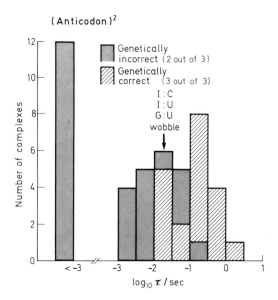

Fig. 6. Frequency of observation of anticodon–anticodon lifetimes falling between τ and $\sqrt{10}\tau$. This Figure summarizes all the experiments concerning 60 pairs of tRNA's (Grosjean et al. 1978a). The 60 pairs analyzed are arranged according to whether or not their pairing is in agreement with the "wobble" hypothesis of Crick (1966)

with anticodons which are partly complementary, i.e., which can make only two correct base pairs out of three. In most cases, the tRNA's do associate through their anticodons, but these incorrect associations are short-lived (unstable) as compared to correct pairs. However, for some of them the differences in stability observed are not large enough for ensuring a selectivity comparable to that of the actual process of protein synthesis: clearly, there exists an overlap in the lifetime of the pairs which are correct according the genetic code and the lifetime of some unexpected pairs which are genetically incorrect (Fig. 6). These misbindings, therefore, reflect an unexpectedly high intrinsic affinity between particular sequences of two nucleic acids; it is not limited to anticodon–anticodon interactions, it occurs also in codon–anticodon interactions on the ribosome. For example (Fig. 7a), *E. coli* tRNA$_V^{leu}$, a minor species of tRNA^{Leu} having an anticodon XAA (where X is still unidentified) do form a strong complex with *E. coli* tRNA^{Lys} (anticodon s^{2}U*UU). The lifetime (stability) of this complex is about the same as for the expected complex formed between tRNA^{Lys}-tRNA^{Phe}, involving a GU wobble pair. Thus tne anticodon of tRNA$_V^{leu}$ has an intrinsic high affinity for UUU sequence; this tRNA is also able to misread poly U on the ribosome only half as efficient as tRNA^{Phe} (Fig. 7b). *E. coli* tRNA^{Leu} species II, III and IV do not bind significantly with tRNA^{Lys} in an "anticodon–anticodon binding test". Nevertheless, they are able to misread poly U in the in vitro system, but only in the presence of streptomycin and at low efficiency. Moreover, tRNA$_f^{Leu}$, the major species of *E. coli* tRNA^{Leu}, does not bind at all to tRNA^{Lys} nor does it misread poly U on the ribosome, even in the presence of streptomycin or with the *ram* A or *ram* C ribosomes (Grosjean et al. 1979).

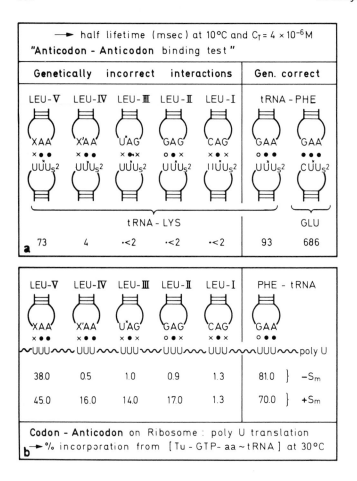

Fig. 7a, b. Correlation between results obtained in the "anticodon–anticodon binding test" and those obtained with the same pure tRNA's in poly U programmed in vitro protein synthesis. In **a** the half lifetimes (= 069 τ) in msec for different pairs of anticodon–anticodon were measured under strictly comparable conditions, they characterize the relative stability of these different complexes. In **b** the *numbers* correspond to the fraction (in %) of aminoacyl-tRNA and of peptidyl-tRNA bound to the poly U programmed ribosome after 30 min incubation at 30°C. They characterize the relative efficiency of these pure tRNA's for decoding poly U on the ribosome. When streptomycin was added, the drug/ribosome ratio was 10. All pure tRNA's used are from *E. coli*. Other experimental conditions were according to Petska et al. (1965)

These results demonstrate two important things: first, only the weak codon–anticodon interactions are the most sensitive to change in the structure of the ribosome, yet with certain selectivity; second the system involving a standardization of a high binding energy, intrinsic to codon–anticodon interaction as discussed above, does not produce necessarily a highly accurate system for reading the genetic code. Evidently, the real pattern of codon–anticodon interaction, as we know it from the genetic code, depends also on other factors. When tRNA's interact with the messenger on the ribosome, they are not in pure solution; the entity which binds is not free tRNA, it is a complex of aminoacyl-tRNA with GTP and elongation factor Tu — another child from Lipmann's laboratory (Lucas-Lenard and Lipmann 1966) — and there are some 60 different GTP-Tu-aa-tRNA complexes around, competing for the acceptor site. What really happens is indeed difficult to visualize. Does the high accuracy rest mostly on the difference of binding energy of tRNA anticodon loop with correct versus incorrect codons, or is there another checking process which rejects incorrect tRNA's, or which reacts to correct pairing by further stabilizing the association and by making it practically irreversible until peptide bond is formed? There is evidence that something of that sort does occur.

The common TψCG sequence which happens to be complementary to a region of 5S ribosomal RNA is not available for pairing with its complementary tetranucleotide in vitro (Uhlenbeck 1972); in crystalline tRNA it is internally bound to another loop of the molecule (Robertus et al. 1974; Kim et al. 1974; Stout et al. 1976). However, when tRNA binds its corresponding codon, then and only then, its TψCG sequence becomes available for pairing, with high affinity, with its complementary CGAA tetranucleotide (Schwartz et al. 1976; Möller et al. 1979). Correct association of tRNA with the corresponding codon indeed triggers a transconformation (Kurland et al. 1975; Robertson et al. 1977; Crothers and Cole 1978) which might conceivably make TψCG available for pairing with 5S ribosomal RNA. Alternatively, or in addition, there might be in tRNA other structural features which are also essential for its correct and tight binding on the mRNA-ribosome complex: modification of codon specificity by a mutation in the dihydrouridine stem of $E.$ $coli$ tRNA^{Trp} may be one of the most striking examples (Hirsh 1971; Buckingham and Kurland 1977). This may, in turn, be a condition for peptide bond formation. Data compatible with a proof reading process (Hopfield 1974; Ninio 1975; Blomberg 1977) involving GTP hydrolysis for rejecting incorrectly bound tRNA's have also been reported (Thompson and Stone 1977). A controversy exists on the exact function of GTP hydrolysis for accurate protein synthesis (Gavrilova et al. 1976; Kurland 1978); nevertheless, all agree on the possible importance of this sort of nonspecific reaction for improving the fidelity and for driving proper codon—anticodon selection in an irreversible path.

In vivo, the iso-tRNA concentrations are correlated to codon frequency of the mRNA of the cell type considered, at least in eukaryotic organisms (Garel 1974; Chavancy et al. 1979); whether such a correlation is important to optimize translation accuracy is difficult to establish, but some in vitro experiments are in agreement with such an idea (Hunter and Jackson 1970; Atkins et al. 1979). What is clear also is that in drastic starvation of certain amino acids incorrect amino acids are incorporated, and the level of mistranslation in vivo is at least two orders of magnitude higher than that estimated for normal protein synthesis (O'Farrell 1978; Parker et al. 1978). This shows that the triggering or the proofreading mechanisms, whatever they may be, are not foolproof. Under conditions of extreme amino acid starvation, protein synthesis is very slow and inefficient: elongation is delayed for a long time at each step where the lacking amino acid should normally be incorporated. The peptidyl-tRNA may have a nonnegligible probability to leave the ribosome (Menninger 1977; Manley 1978), unless a misincorporation occurs and allows the polypeptide elongation to proceed (Holmes 1978; Tai et al. 1978). Under normal conditions, such situations are avoided because the correct aminoacyl-tRNA always outcompetes the others (Goldman et al. 1979). This might be regarded as a case of kinetic gain in accuracy under normal conditions versus starvation, which depends on the whole collection of tRNA's and the balance between them (Ninio 1971, 1974). A related problem is the competition between isoacceptor tRNA's for the different codons of their cognate amino acid.

In vitro, one single species of tRNA^{Leu} from yeast is able to read the six codons of leucine (Weissenbach et al. 1977) but in extracts of mammalian cells, the translation of certain mRNA's is strictly dependent on the presence of one minor species of tRNA^{Leu} (Zilberstein et al. 1976). In *E. coli,* different species of tRNA^{Leu} are able to read two or four codons (Holmes et al. 1978; Goldman et al. 1979). Each isoacceptor tRNA^{Val} from *E. coli* or yeast can read the four codons of valine (Mitra et al. 1977), yet at different efficiency levels (Mitra et al. 1979). A tRNA^{Lys} from yeast apparently reads only the two lysine codons (Mitra 1978). Lagerkvist (1978) assumes that the probability of such misreading is governed by the base-pairing strength; in the first two positions of the codon, indeed, those codons which are read by an apparent "two-out-of-three" mechanism correspond to families of anticodons that bind the most strongly. The difference must be very subtle since, as pointed out above, the whole anticodon loop of a tRNA as well as the choice of successive codons on mRNA may have evolved so as to reduce such differences.

When comparing the sequences of all tRNA available up to now, it is possible to establish a sort of phylogenic tree (Cedergren et al. 1972, 1980). The interesting observation is that the three present-day tRNA^{Gly} are clearly derived from two very ancient protosequences: one of them has given the present-day tRNA with GCC anticodons that still read the two codons ending with C or U; the other one has given, later in evolution, the other two tRNA^{Gly} (anticodon CCC and U*CC, where U* is an unknown modified Uridine) that read each one of the two codons ending by a purine (Cedergren et al. 1980; Larue et al. 1979). Thus, in the early days there was an ancient tRNA^{Gly} that read the two codons ending by a purine. At a primitive stage, therefore, the genetic code might have consisted of doublets separated by "comma-bases" (Crick 1968) read by a more systematic two-out-of-three mechanism.

The affinity of the three codons that correspond to the termination signals with the complementary anticodons is still very low at the present time; this may be the reason why they became terminators. The efficiency of suppression of such codons by suppressor tRNA carrying the complementary anticodon never reaches 100%. They form the pairs that are the most dependent on several factors (Steege and Söll 1979); presence of a modified purine next to the 3' end of anticodon (Gefter and Russell 1969; Laten et al. 1978), presence of modified Uridine in the first position of the anticodon (Comer et al. 1975; Altman 1976; Colby et al. 1976; Pope and Reeves 1978) as well as outside the anticodon loop (Hirsh 1971; Leon et al. 1977), ribosomal mutations (Yates et al. 1977; Topisirovic et al. 1977), drugs binding to the ribosomes (Rosset and Gorini 1969; Strigini and Brickman 1973), in vitro experimental conditions such as temperature (Manley and Gesteland 1978), and finally messenger context.

In the "anticodon-anticodon binding test", *E. coli* tRNA$_{su}^{Tyr+}$ and yeast tRNA^{Leu} (anticodon UAG) form also the only complementary pair of tRNA's that does not fall within the common high affinity range (Fig. 4); it is weaker by more than one order of magnitude (Grosjean et al. 1978a).

D. What Next?

In order to fully understand the physical basis of high accuracy in protein synthesis, it would be essential to know to what extent each tRNA shows preferences for individual codons under normal conditions in vivo, where the isoacceptors are submitted to competition with one another, and to elucidate more elaborate processes involved in eliminating the aminoacyl-tRNA or even the peptidyl-tRNA incorrectly bound on the ribosome. Another puzzling process is the proteolytic degradation of the "abnormal" proteins that is observed in vitro (Pinkett and Brownstein 1974; Ballard 1977); here we are dealing with a typical protein–protein recognition that bears some resemblance to the discrimination between self and non-self by the immunological system.

In our laboratory, we have undertaken various experiments with the aim of determining the accuracy of each step of protein synthesis within a living cell. Our approach rests on the translation of an eukaryotic mRNA of known sequence injected into frog oocytes (Chantrenne and Marbaix 1972), a technique initially developed by Gurdon and co-workers (1971).

Acknowledgments. The temperature-jump experiments on anticodon-anticodon interactions mentioned in this paper were performed mainly in the laboratories of Profs. D. Crothers and D. Söll at Yale University but also in the laboratories of Prof. G. Maass and Dr. C. Urbanke at the Medizinische Hochschule of Hannover, of Prof. M. Eigen, Drs. D. Pörschke and D. Labuda (Germany), and of Profs. V. Dereux, E. Fredericq, and Dr. C. Houssier at the University of Liege. We thank N.I.H. (USA), N.A.T.O. (USA), D.A.A.D. (Germany), the "PATRIMOINE of University of Liege", the "ACTIONS CONCERTEES", and the "FONDS NATIONAL DE LA RECHERCHE SCIENTIFIQUE" in Belgium for financial support.

References

Akaboshi E, Inouye M, Tsugita A (1976) Effect of neighboring nucleotide sequences on suppression efficiency in amber mutants of T4 phage lysozyme. Mol Gen Genet 149: 1–4

Altman S (1976) A modified uridine in the anticodon of *E. coli* tRNA$_1^{Tyr}$ su$_{co}^+$. Nucleic Acids Res 3: 441–448

Apirion D, Schlessinger D (1967) Reversion from streptomycin dependence in *E. coli* by a further change in the ribosome. J Bacteriol 94: 1275–1280

Atkins JK, Gesteland RF, Reid RR, Anderson CW (1979) Normal transfer RNA promote ribosomal frameshifting. Cell 18: 1119–1131

Ballard FJ (1977) Intracellular protein degradation. In: Campbell PN, Aldridge WN (eds) Assays in biochemistry, vol XIII, pp 1–31. Academic Press, New York

Belin D, Hedgpeth J, Selzer GB, Epstein RH (1979) Temperaturesensitive mutation in the initiation codon of the *rIIB* gene of bacteriophage T4. Proc Natl Acad Sci USA 76: 700–704

Blomberg C (1977) A kinetic recognition process for tRNA at the ribosome. J Theoret Biol 66: 301–325

Bollen A, Helser T, Yamada T, Davies J (1969) Altered ribosomes in antibiotic-resistant mutants of *E. coli*. Cold Spring Harbor Symp Quant Biol 34: 95–100

Borer PN, Dengler B, Tinoco I, Uhlenbeck OC (1974) Stability of ribonucleic acid double-stranded helices. J Mol Biol 86: 843–853

Buckingham RH (1976) Anticodon conformations and accessibility in wild type and suppressor tryptophan tRNA from *E. coli*. Nucleic Acids Res 3: 965–975

Buckingham RH, Kurland CG (1977) Codon specificity of UGA suppressor tRNA^{TRP} from *E. coli*. Proc Natl Acad Sci USA 74: 5496–5498

Carbon J, Fleck E (1974) Genetic alterations of structure and function in glycine tRNA of *E. coli*: mechanism of suppression of the tryptophan synthetase A78. J Mol Biol 85: 371–391

Cedergren RJ, Cordeau JR, Robillard P (1972) On the phylogeny of tRNAs. J Theoret Biol 37: 209–220

Cedergren RJ, Larue B, Sankoff D, Grosjean H (1980) The evolving transfer RNA molecule. In: CRC critical reviews in biochem, vol VIII. CRC Press, Cleveland, Ohio, USA

Chantrenne H (1978) Why should tRNAs be so elaborate? Molec Cell Biochem 21: 3–8

Chantrenne H, Marbaix G (1972) Traduction des RNA 9S, messagers de l'hémoglobine, dans des systèmes hétérologues. Biochimie 54: 1–5

Chapeville F, Lipmann F, Ehrenstein G von, Weisblum B, Ray W, Benzer S (1962) On the role of soluble RNA in coding for aminoacids. Proc Natl Acad Sci USA 48: 1086–1092

Chavancy G, Chevallier A, Fournier A, Garel JP (1979) Adaptation of iso-tRNA concentrations to mRNA codon frequency in the eukaryote cell. Biochimie 61: 71–78

Colby DS, Schedel P, Guthrie C (1976) A functional requirement for modification of the wobble nucleotide in the anticodon of a T4 suppressor tRNA. Cell 9: 449–463

Comer MM, Foss K, McClain WH (1975) A mutation of the wobble nucleotide of a bacteriophage T4 tRNA. J Mol Biol 99: 283–293

Cortese R (1979) The role of tRNA in regulation. In: Goldberger RF (ed) Regulation and development, vol I, pp 401–432. Plenum Press, New York

Crick FHC (1966) Codon-anticodon pairing: the wobble hypothesis. J Mol Biol 19: 548–555

Crick FHC (1968) The origin of the genetic code. J Mol Biol 38: 367–379

Crick FHC, Brenner S, Klug A, Pieczenik G (1976) A speculation on the origin of protein synthesis. Origins of life 7: 389–397

Crothers DM (1971) Temperature-jump methods. In: Cantoni G, Davies D (eds) Procedures in nucleic acid research, vol 2, pp 369–388.

Crothers PM, Cole PE (1978) Conformational changes of tRNA. In: Altman S (ed) Transfer RNA. Cell monograph series 2. The Massachussetts Institute of Technology, Cambridge/Mass.

Davies J, Gorini L, Davies BD (1965) Misreading of RNA codewords induced by aminoglycoside antibiotics. Mol Pharmacol 1: 93–106

De Wilde M, Cabezon T, Herzog A, Bollen A (1977) Apport de la génétique a la connaissance du ribosome bactérien. Biochimie 59: 125–140

Dirheimer G, Keith G, Sibler A, Martin R (1979) In: Schimmel P, Söll D, Abelson J (eds) Transfer RNA, part I: Structure, properties and recognition. Cold Spring Harbor Monograph 9A, 19–41

Drake AF, Mason SF, Trim AR (1974) Optical studies of the base stacking properties of 2'-O methylated dinucleoside monophosphates. J Mol Biol 86: 727–739

Dugre M, Cedergren RJ (1974) Origine de l'Inosine dans les tRNA de levure. Can J Biochem 52: 417–422

Edelmann NP, Gallant J (1977) Mistranslation in *E. coli*. Cell 10: 131–137

Eigen M (1971) Selforganization of matter and evolution of biological macromolecules. Naturwissenschaften 58: 465–523

Eigen M (1978) The hypercycle: principle of natural self-organization; part C; the realistic hypercycle. Naturwissenschaften 65: 341–369

Eisinger J (1971) Visible gel electrophoresis and the determination of association constants. Biochem Biophys Res Commun 43: 854–861

Eisinger J, Gross N (1974) The anticodon-anticodon complex. J Mol Biol 88: 165–174

Eisinger J, Gross N (1975) Conformers, dimers and anticodons complexes of tRNA$_2^{Glu}$ of *E. coli*. Biochem 14: 4031–4041

Eisinger J, Spahr PF (1973) Binding of complementary pentanucleotides to the anticodon loop of tRNA. J Mol Biol 73: 131–137

Eisinger J, Feuer B, Yamane T (1971) Codon-anticodon binding in tRNA^{Phe}. Nature New Biol 231: 126–12

Feinstein SI, Altman S (1978) Context effects on nonsense codon suppression in *E. coli*. Genetics 88: 201–219

Feldman MY (1977) Minor components in tRNA: the location-function relationships. Prog Biophys Mol Biol 32: 83–102

Fluck MM, Salser W, Epstein RH (1977) The influence of the reading context upon the suppression of nonsense codon. Mol Gen Genet 151: 137–149

Fiers W, Grosjean H (1979) On codon usage. Nature 277: 328

Fink TR, Crothers DM (1972) Free energy of imperfect nucleic acid helices: I. The bulge defect. J Mol Biol 66: 1–12

Fittler F, Hall RH (1966) Selective modification of yeast tRNA^{Ser} and its effect on the acceptance and binding functions. Biochem Biophys Res Commun 25: 441–446

Fuller W, Hodgson A (1967) Conformation of the anticodon loop in tRNA. Nature 215: 817–821

Furuchui Y, Wataya Y, Hayatsu H, Yukita T (1970) Chemical modification of tRNA^{Tyr} of yeast with bisulfite: a new method to modify isopentenyladenosine residue. Biochem Biophys Res Commun 41: 1185–1191

Ganoza MC, Fraser AR, Neilson T (1978) Nucleotides contiguous to AUG effect translational initiation. Biochemistry 17: 2769–2775

Garel JP (1974) Functional adaptation of tRNA population. J Theoret Biol 43: 211-225

Gauss DH, Grüter F, Sprinzl M (1979) Compilation of tRNA sequences. Nucleic Acids Res 6: r1–r20

Gavrilova LP, Kostiashkina OE, Koteliansky VG, Rutkevitch M, Spirin AS (1976) Factor-free (non-enzymic) and factor-dependent systems of translation of poly U by *E. coli* ribosomes. J Mol Biol 101: 537–552

Gefter ML, Russell RL (1969) Role of modification in Tyrosine tRNA: a modified base affecting ribosome binding. J Mol Biol 39: 145–157

Glukhova MA, Belitsina NV, Spirin AS (1975) A study of codondependent binding of aminoacyl-tRNA with the ribosomal 30S-subparticle of *E. coli*: determination of the active-particle fractions and binding constants in different media. Eur J Biochem 52: 197–202

Goldman E, Holmes WM, Hatfield GW (1979) Specificity of codon recognition by *E. coli* tRNA^{Leu} isoaccepting species determined by proteins synthesis *in vitro* directed by phage RNA. J Mol Biol 129: 567–785

Gorini L, Jacoby GA, Breckenridge L (1966) Ribosomal ambiguity. Cold Spring Harbor Symp 31: 657–664

Grantham R (1978) Viral, prokaryote and eukaryote genes contrasted by mRNA sequence FEBS Lett 95: 1–11

Grosjean H (1979) Codons usage in several organisms. In: Söll D, Abelson J, Schimmel P.(eds) Transfer RNA, part 2: Biological aspects. Cold Spring Harbor Monograph 9B (in press)

Grosjean H, Takada C, Petre J (1973) Complex formation between tRNAs with complementary anticodons: use of matrix bound tRNA. Biochem Biophys Res Commun 53: 882–893

Grosjean H, Söll DG, Crothers DM (1976) Studies of the complex between tRNA with complementary anticodons. I. Origins of enhanced affinity between complementary triplets. J Mol Biol 103: 499–519

Grosjean H, Söll D, Crothers D (1977) Studies of the complex between tRNAs with complementary anticodons: a direct approach to the "wobble" problem. Arch Int Physiol Biochim 85: 414–415

Grosjean H, de Henau S, Crothers D (1978a) On the physical basis for ambiguity in genetic coding. Proc Natl Acad Sci USA 75: 610–614

Grosjean H, Sankoff D, Min Jou W, Fiers W, Cedergren RJ (1978b) Bacteriophage MS_2-RNA: a correlation between the stability of the codon-anticodon interaction and the choice of code words. J Mol Evol 12: 113–119

Grosjean H, Ballivian L, de Henau S, Bollen A (1979) A minor species of *E. coli* tRNA^{Leu} is able to decode poly-U in a cell-free system because of a high affinity of its anticodon for UUU sequence. Arch Int Physiol Biochim 87: 415–417

Gurdon JB, Lane CD, Woodland HR, Marbaix G (1971) Use of frog eggs and oocytes for the study of messenger RNA and its translation in living cells. Nature 233: 177–182

Hatfield D, Nirenberg M (1971) Binding of radioactive oligonucleotides to ribosomes. Biochemistry 10: 4318–4323

Hillen W, Egert E, Lindner HJ, Gassen HG (1978) Restriction or amplification of wobble recognition: the structure of 2-thio-5-methyl-aminomethyluridine and the interaction of odd uridines with the codon loop backbone. FEBS Lett 94: 361–364

Hirsh D (1971) Tryptophan tRNA as the UGA suppressor. J Mol Biol 58: 439–458

Hoburg A, Aschhoff HJ, Kersten H, Manderschied FR, Gassen HG (1979) On the function of the modified nucleosides m7G, rT and ms^2i^6A in prokaryotic transfer RNA. J Bacteriol 140: 408–414

Högenauer F (1970) The stability of a codon tRNA complex. Eur J Biochem 12: 527–532

Holbrook SR, Sussman JL, Warrant RW, Kim SH (1978) Cristal structure of yeast tRNA^{Phe}. II. Structural features and functional implications. J Mol Biol 123: 631–660

Holmes M, Hatfield GW, Goldman E (1978) Existence for misreading during tRNA-dependent protein synthesis in vitro. J Biol Chem 253: 3482–3486

Hopfield J (1974) Kinetic proofreading: a new mechanism for reducing errors in biosynthetic processes requiring high specificity. Proc Natl Acad Sci USA 71: 4135–4139

Hunter AR, Jackson RJ (1970) Miscoding by *E. coli* tRNAs for Methionine, Cysteine and Valine in the synthesis of rabbit globin. Eur J Biochem 15: 381–390

Jaskunas SR, Cantor CR, Tinoco I (1968) Association of complementary oligoribonucleotides in aqueous solution. Biochemistry 7: 3164–3178

Kim SH, Sudath GJ, McPherson A, Sussman J, Wang A, Seeman N, Rich A (1974) Three dimensional tertiary structure of yeast tRNA^{Phe}. Science 185: 435–440

Kurland CG (1977) Ribosomes: structure and function. In: Weissbach H, Pestka S (eds) Molecular mechanisms of protein biosynthesis, pp 81–113. Academic Press, New York

Kurland CG (1978) The role of Guanine nucleotides in protein biosynthesis. Biophys J 22: 373–392

Kurland CG, Rigler A, Ehrenberg M, Blomberg C (1975) Allosteric mechanism for codon-dependent tRNA selection of ribosomes. Proc Natl Acad Sci USA 72: 4248–4251

Lagerkvist U (1978) Two out of three: an alternative method for codon reading. Proc Natl Acad Sci USA 75: 1759–1762

Langlois R, Kim SH, Cantor CR (1975) A comparison of the fluorescence of the Y base of yeast tRNA^{Phe} in solution and in crystal. Biochemistry 14: 2554–2558

Larue B, Cedergren RJ, Sankoff D, Grosjean H (1979) Evolution of methionine initiator and phenylalanine tRNA. J Mol Evol 14: 287–300

Laten H, Gorman J, Bock R (1978) Isopentenyladenosine deficient tRNA from an antisuppressor mutant of *Saccharomyces cerevisiae*. Nucleic Acids Res 5: 4329–4342

Leon V, Altman S, Crothers DM (1977) Influence of the A15 mutation on the conformational energy balance in *E. coli* tRNA^{Tyr}. J Mol Biol 113: 253–265

Loftfield R, Vanderjagt D (1972) The frequency of errors in protein biosynthesis. Biochem J 128: 1363–1356

Lomant AJ, Fresco JR (1975) Structural and energetic consequences of non-complementary base oppositions in nucleic acid helices. In: Davidson JN, Cohen WE (eds) Progress in nucleic acid research and molecular biology, vol XV, pp 185–218. Academic Press, New York

Lucas-Lenard J, Lipmann F (1966) Separation of three microbial aminoacid polymerization factors. Proc Natl Acad Sci USA 55: 1562–1566

Manderschied U, Bertram S, Gassen HG (1978) Initiator-tRNA recognizes a tetranucleotide codon during the 30S initiation complex formation. FEBS Lett 90: 162–166

Manley JL (1978) Synthesis and degradation of termination and premature termination fragments of β-galactosidase in vitro and in vivo. J Mol Biol 125: 407–432

Manley JL, Gesteland F (1978) Suppression of amber mutants in vitro induced by low temperature. J Mol Biol 125: 433–447

Martin FH, Uhlenbeck OC, Doty P (1971) Self complementary oligo-ribonucleotides: adenylic acid-uridylic black copolymers. J Mol Biol 57: 201–215

Mazumdar SK, Saenger W, Scheit KH (1974) Molecular structure of poly-2-thiouridylic acid, a double helix with non-equivalent poly-nucleotide chains. J Mol Biol 85: 213–229

McCloskey JA, Nishimura S (1977) Modified nucleosides in tRNA. Acc Chem Res 10: 403–410

Menninger JR (1977) Ribosome editing and the error catastrophe hypothesis of cellular aging. Mech Ageing Dev 6: 131–142

Mitra SK (1978) Yeast tRNA^{Lys} (anticodon CUU) translates AAA codon. FEBS Lett 91: 78–80

Mitra SK, Lustig F, Akesson B, Lagerkvist U (1977) Codon-anticodon recognition in the valine codon family. J Biol Chem 252: 471–478

Mitra SK, Lustig F, Akesson B, Axberg T, Elias P, Lagerkvist U (1979) Relative efficiency of anticodons in reading the valine codons. J Biol Chem 254: 6397–6401

Möller A, Schwarz U, Lipecky R, Gassen HG (1978) Effective binding of oligonucleotides to the anticodon of a tRNA without stimulation of tRNA binding to 30S ribosomes. FEBS Lett 89: 263–266

Möller A, Wild U, Riesner D, Gassen HG (1979) Evidence for U.V. absorption measurements for a codon induced conformational change in Lysine tRNA from E. coli. Proc Natl Acad Sci USA 76: 3266–3270

Ninio J (1971) Codon-anticodon recognition: the missing triplet hypothesis. J Mol Biol 6: 63–82

Ninio J (1974) A semi-quantitative treatment of missense and nonsense suppression in the str A and ram ribosomal mutant of E. coli: evaluation of some molecular parameters of translation in nitro. J Mol Biol 84: 297–313

Ninio J (1975) Kinetic amplification of enzyme discrimination. Biochimie 57: 587–595

Ninio J (1979a) Prediction of pairing schemes in RNA molecules: loop contribution and energy of wobble and nono wobble pairs. Biochimie 61: 1133–1150

Ninio J (1979b) The origin of the code and molecular defenses. In: Approches moléculaires de l'evolution, pp 62–88, 102–115. Masson, Paris

Nirenberg M, Leder P (1964) RNA codewords in protein synthesis: the effect of trinucleotides upon the binding of sRNA to ribosomes. Science 145: 1399–1407

O'Farrell PH (1978) The suppression of defective translation by ppGpp and its role in the stringent response. Cell 14: 545–557

Parker J, Pollard JW, Friesen JD, Stanners CP (1978) Stuttering: high-level mistranslation in animal and bacterial cells. Proc Natl Acad Sci USA 75: 1091–1095

Pestka S, Marchal R, Nirenberg MW (1965) RNA codewords and protein synthesis: V Effects of streptomycin on the formation of ribosome-sRNA complexes. Proc Natl Acad Sci USA 53: 639–646

Pinkett MO, Brownstein BL (1974) Streptomycin-induced synthesis of abnormal protein in an E. coli mutant. J Bacteriol 119: 345–350

Plesiewicz E, Spepien E, Bolewska K, Wirzchowski KL (1976) Stacking self-association of pyrimidine nucleosides and cytosines: effects of methylation and thiolation. Nucleic Acids Res 3: 1295–1305

Pörschke D (1977) Elementary steps of base recognition and helix-coil transition in nucleic acids. In: Pecht I, Rigler R (eds) Molecular biology, biochemistry and biophysics, vol 24, pp 191–218. Springer, Berlin Heidelberg New York

Pörschke D, Uhlenbeck OC, Martin FA (1973) Thermodynamics and kinetics of the helix-coil transition of oligomers containing GC base pairing. Biopolymers 12: 1313–1335

Pongs O, Bald R, Reinwald E (1973) On the structure of yeast tRNA^{Phe} complementary-oligonucleotide binding studies. Eur J Biochem 32: 117–125

Pope WT, Reeves RH (1978) The identification of the tRNA substrates for the *sup K* tRNA methylase. Nucleic Acids Res 5: 1041–1057

Rich A, Rajbandhary U (1976) Transfer RNA: molecular structure, sequence, and properties. Ann Rev Biochem 45: 806–852

Riddle DL, Carbon J (1973) Frameshift suppression: a nucleotide addition in the anticodon of a glycine tRNA. Nature New Biol 242: 230–234

Rigler R, Rabl CR, Jovin T (1974) A temperature-jump apparatus for fluorescence measurements. Rev Sci Instruments 45: 550–588

Roberts JW, Carbon J (1974) Molecular mechanism for misense suppression in *E. coli*. Nature 250: 412–414

Robertson JM, Kahan M, Wintermeyer W, Zachau HG (1977) Interaction of yeast tRNA^{Phe} with ribosomes from yeast and *E. coli*: a fluorescence spectroscopic study. Eur J Biochem 72: 117–125

Robertus JD, Ladner JE, Finch JT, Rhoades D, Brown RS, Clark BF, Klug A (1974) Structure of yeast phenylalanine tRNA at 3 Å resolution. Nature 250: 546–551

Rosset R, Gorini L (1969) A ribosomal ambiguity. J Mol Biol 39: 95–112

Schevitz RW, Podjarny AD, Krishnamachari N, Hughes JJ, Sigler PB, Sussman JL (1979) Crystal structure of a eukaryotic initiator tRNA. Nature 278: 188–190

Schwarz U, Menzel HM, Gassen HG (1976) Codon-dependent rearrangement of the three-dimensional structure of phenylalanine tRNA exposing the TψCG sequence for binding to the 50S ribosomal subunits. Biochemistry 15: 2484–2490

Steege DA, Söll DG (1979) Suppression. In: Goldberger RF (ed) Regulation and development, vol I, pp 435–475. Plenum Press, New York

Stout CD, Mizuno H, Rubin J, Brennan T, Rao ST, Sundaralingam M (1976) Atomic coordinates and molecular conformation of yeast tRNA^{Phe}: an independent investigation. Nucleic Acids Res 3: 1111–1124

Strigini P, Brickman E (1973) Analysis of specific misreading in *E. coli*. J Mol Biol 75: 659–672

Sundaralingam M, Brennan T, Yathindra N, Ichikawa T (1975) Stereochemistry of mRNA (codon) tRNA (anticodon) interaction on the ribosome during peptide bond formation. In: Sundaralingam M, Rao ST (ed) Structure and conformation of nucleic acids and protein-nucleic acid interactions. University Park Press, Baltimore, pp 101–115

Tai PC, Wallace BJ, Davis BD (1978) Streptomycin causes misreading of natural messenger by interacting with ribosomes after initiation. Proc Natl Acad Sci USA 75: 275–279

Taniguchi T, Weissmann C (1978) Site-directed mutations in the initiator region of the bacteriophage Q β coat cistron and their effect on ribosome binding. J Mol Biol 118: 533–565

Thiebe R, Zachau HG (1968) A specific modification next to the anticodon of phenylalanine tRNA. Eur J Biochem 5: 546–555

Thompson RC, Stone PJ (1977) Proofreading of the codon-anticodon interaction on ribosomes. Proc Natl Acad Sci USA 74: 198–202

Tinoco I, Borer PN, Dengler B, Levine MD, Uhlenbeck OC, Crothers DM, Gralla J (1973) Improved estimation of secondary structure in RNA. Nature New Biol 246: 40–41

Topisirovic L, Villarroel R, de Wilde M, Herzog A, Cabezon T, Bollen A (1977) Translation fidelity in *E. coli*: contrasting role of *nea A* and *ram A* gene products in the ribosome functioning. Mol Gen Genet 15: 89–94

Turner DH, Tinoco I, Maestre JF (1975) Fluorescence detected circular dischroism study of the anticodon loop of yeast tRHA^{Phe}. Biochemistry 14: 3795–3799

Uhlenbeck OC (1972) Complementary oligonucleotide binding to tRNA. J Mol Biol 65: 25–41

Unger FM, Takemura S (1973) A comparison between inosine and guanosine containing anticodon in ribosome-free codon-anticodon binding. Biochem Biophys Res Commun 52: 1141–1147

Vögeli G, Grosjean H, Söll D (1975) A method for the isolation of specific tRNA precursor. Proc Natl Acad Sci USA 72: 4790–4794

Vold BD (1978) Post-transcriptional modifications of the anticodon loop regions: alterations in isoaccepting species of tRNAs during development in *B. subtilis*. J Bacteriol 135: 124–132

Weiss GB (1973) Translational control of protein synthesis by tRNA unrelated to changes in tRNA concentration. J Mol Evol 2: 199–204

Weissenbach J, Grosjean H (1980) Effect of threonylcarbamoyl modification in yeast tRNA$_{III}^{Arg}$ on codon-anticodon and anticodon-anticodon binding: a thermodynamic and kinetic evaluation. Eur J Biochem (in press)

Weissenbach J, Dirheimer G, Falcoff R, Sanceau J, Falcoff E (1977) Yeast tRNA^{Leu} (anticodon UAG) translates all six leucine codons in extracts from interferon treated cells. FEBS Lett 82: 71–76

Yates JL, Gette WR, Furth ME, Nomura M (1977) Effects of ribosomal mutations on the read-through of a chain termination signal: studies on the synthesis of bacteriophage λ o-gene protein in vitro. Proc Natl Acad Sci USA 74: 689–693

Yoon K, Turner DH, Tinoco I (1975) The kinetics of codon-anticodon interaction in yeast phenyl-alanine tRNA. J Mol Biol 99: 507–518

Yoon K, Turner DH, Tinoco I, von der Haar F, Cramer F (1976) The kinetics of binding UUCA to a dodecanucleotide anticodon fragment from yeast tRNA^{Phe}. Nucleic Acids Res 3: 2233–2241

Zilberstein A, Dudock B, Berissi H, Revel M (1976) Control of mRNA translation by minor species of leucyl-tRNA in extracts from interferontreated L cells. J Mol Biol 108: 43–54

3. Fluorescent tRNA Derivatives and Ribosome Recognition

W. WINTERMEYER, J. M. ROBERTSON and H. G. ZACHAU

A. Introduction

The interactions of tRNA's with aminoacyl-tRNA synthetases and ribosomes have
been studied rather extensively using a variety of approaches (Rich and RajBhandary
1976; Altman 1978; Schimmel et al. 1979). For a number of years we have investigated the chemistry of introducing fluorescent groups into tRNA in order to use the fluorescent tRNA's as tools in the study of aminoacyl-tRNA synthetase and ribosome recognition (Wintermeyer and Zachau 1971, 1979). The procedure eventually applied
allows the specific replacement of odd bases in internal positions of the tRNA molecule with fluorescent dyes. The replacement basically involves a two-step reaction: a
ribosylic aldehyde group is created in the tRNA by selective excision of a base which
subsequently can be condensed with a fluorophor possessing either a primary amino
or a hydrazino group (Fig. 1). For the experiments described below fluorescent derivatives of tRNA^{Phe} (yeast) have been used in which wybutine (position 37, next to the
anticodon) or dihydrouracil (positions 16 or 17 in the D-loop) had been replaced with
ethidium (Etd) or proflavin (Prf). Upon modification the tRNA^{Phe} retained its activity
in the partial reactions of protein synthesis, indicating that the essential elements of
the native structure are not disturbed by insertion of the dyes. In support of this conclusion, the direct structural investigation of tRNA$^{Phe}_{Prf}$ 37 by NMR spectroscopy has
shown that the insertion of Prf reverses the substantial changes of the spectrum which
are introduced by the excision of wybutine from tRNA^{Phe} (Wong et al. 1975).

The fluorescence spectroscopic properties of the tRNA^{Phe}-dye compounds, which
are determined to a large extent by interactions of the dyes with the neighboring bases
(Wintermeyer and Zachau 1979), proved to be very useful for studies on the conformation of the tRNA molecule (Ehrenberg et al. 1979; Wintermeyer et al. 1979) and on

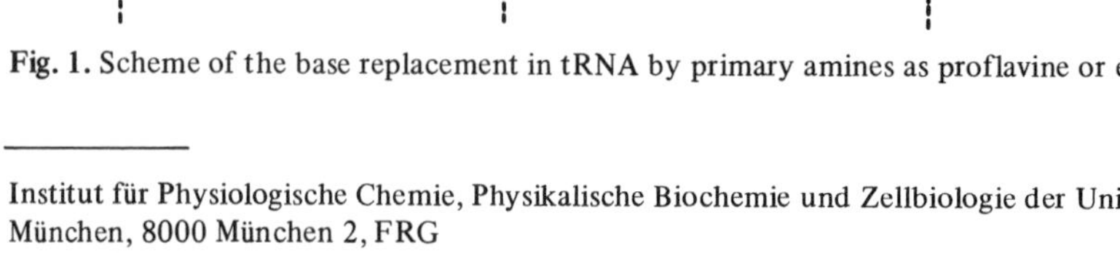

Fig. 1. Scheme of the base replacement in tRNA by primary amines as proflavine or ethidium

Institut für Physiologische Chemie, Physikalische Biochemie und Zellbiologie der Universität
München, 8000 München 2, FRG

its interactions with phenylalanyl-tRNA synthetase (Pachmann et al. 1973) and ribosomes (Robertson et al. 1977). In the present contribution we wish to concentrate on two aspects of this work and discuss (1) conformational changes of the tRNA^{Phe} molecule and their possible functional significance, and (2) the interaction of tRNA^{Phe} with ribosomes as measured in equilibrium and kinetic studies.

B. Conformational Changes in Fluorescent tRNA^{Phe}

Since the determination of the crystal structure of tRNA^{Phe} (yeast) (Ladner et al. 1975; Quigley et al. 1975), much attention has been given to the question of whether the conformation found in the crystal represents the conformation of the tRNA^{Phe} molecule found in solution. There are indications from enzymatic degradation (Zachau et al. 1972) and chemical modification (Wintermeyer and Zachau 1975a) experiments, as well as from spectroscopic studies using dynamic light scattering (Olson et al. 1976) or NMR (Johnston and Redfield 1977), which suggest that the tRNA molecule in solution exists in more than one conformation, the proportions of which depend on the solution conditions.

More detailed studies on the solution structure of tRNA^{Phe} have been possible by the use of tRNA^{Phe}-Etd derivatives. It turned out that the fluorescence properties of these compounds are rather sensitive to changes of the solution conditions. In addition, the fluorescence of wybutine in the anticodon loop was measured, which is rather sensitive to changes of the Mg^{2+} concentration and has been used previously for structural studies on tRNA^{Phe} (Beardsley et al. 1970). We have utilized this parameter for a comparison of tRNA$^{Phe}_{Etd\ 16/17}$ and unmodified tRNA^{Phe} (Fig. 2). A similar response was observed in both tRNA's, the difference in the extent of the fluorescence enhancement being due to energy transfer from wybutine to Etd. When the fluorescence of Etd in the D-loop was monitored, a substantial quenching was observed upon additon of Mg^{2+} to tRNA$^{Phe}_{Etd\ 16/17}$ (Fig. 2). The same result was obtained when the fluorescence of Etd in the anticodon loop was monitored (Rigler et al. 1977). No comparable Mg^{2+} effect on the Etd-fluorescence was observed when analogous experiments were performed with Etd-containing fragments of tRNA^{Phe} including a half molecule carrying Etd in the D-loop (Ehrenberg et al. 1979; Rigler et al. 1977). From these observations we conclude that the binding of Mg^{2+} induces structural transitions in both the anticodon loop and the D-loop which depend on the presence of the intact tertiary structure of the tRNA molecule. Although independent Mg^{2+}-induced conformational changes in the two loop regions cannot be excluded, the striking similarity of the effects suggests that the conformational changes are coupled and represent a transition of the tertiary structure of the whole tRNA molecule.

The Mg^{2+}-induced conformational changes of the tRNA^{Phe}-Etd derivatives have been investigated in greater detail by fluorescence decay measurements and chemical relaxation experiments (Rigler et al. 1977; Ehrenberg et al. 1979). The analysis has shown that there is a Mg^{2+}-dependent equilibrium of at least two conformations of the tRNA molecule, which are characterized by different fluorescence lifetimes of the Etd label. The equilibirum is shifted with increasing Mg^{2+} concentration towards the

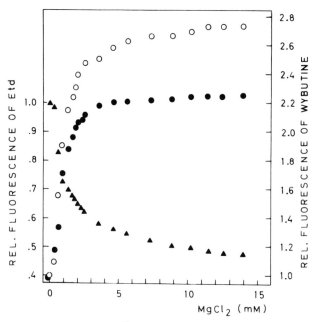

Fig. 2. Influence of the Mg^{2+} concentration on the fluorescence of tRNA^{Phe} (O) and tRNA$^{Phe}_{Etd}$ 16/17 (*black spots*). 0.25 μM solutions of the tRNA's in 10 mM Tris-HCl, pH 7.5, 100 mM KCl, 0.5 mM EDTA were titrated at 24°C by adding small volumes of the same solutions containing 100 mM MgCl$_2$. The fluorescence of wybutine (*open rings*) and Etd (*black triangles*) was excited at 312 nm and measured with two photomultipliers after passing a bandpass filter (Schott SFK 436) or a cut-off filter (Schott OG 590), respectively

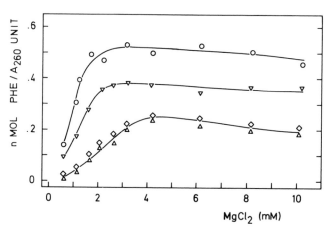

Fig. 3. Mg^{2+} dependence of the aminoacylation rate of tRNA^{Phe} (*open rings*), tRNA$^{Phe}_{Etd}$ 16/17 (*downward triangles*), and two isomeric species of tRNA$_{Etd}$ 37 (*diamonds, triangles*). The aminoacylation by phenylalanyl-tRNA synthetase from yeast was measured in the buffer of Fig. 2, containing 0.1 mM ATP and 0.05 mM [^{14}C]L-phenylalanine in addition. Figure reproduced from Wintermeyer and Zachau (1979) with permission

shorter lifetime state; thus the latter state probably represents the conformation of the tRNA in the crystal.

It is well known that Mg^{2+} strongly influences the folding of the tertiary structure of tRNA's from the denatured or melted forms. It should be emphasized, however, that the effects described here are observed at Mg^{2+} concentrations in the mM range, i.e., under conditions in which the tRNA can be regarded as being in the native state. This point is documented in Fig. 3, which shows that the rate of the aminoacylation reaction reaches a plateau at 2–4 mM Mg^{2+}. The observation that modification of the tRNA slightly changes the Mg^{2+} optimum of the reaction indicates that it is influenced by Mg^{2+} binding to the tRNA. Thus the simultaneous occurrence of high flexibility (Fig. 2) and optimal charging (Fig. 3) suggests that the ability to exist in alternative conformations is an important feature of the functional design of the tRNA molecule.

C. Ribosome Binding of Fluorescent tRNA^{Phe}

In order to explain the high specificity of aminoacyl-tRNA selection by the programmed ribosome, a number of proofreading models have been proposed. Some of these models involve conformational transitions of an initially unspecific tRNA-ribosome complex and lead to the desired preference for the correct tRNA (Ninio 1973; Kurland et al. 1975). In fact, codon-dependent conformational changes of the tRNA molecule have been proposed on the basis of oligonucleotide binding data (Schwarz and Gassen 1977). However, there are only very few thermodynamic and essentially no kinetic data available which would allow one to decide whether the codon-dependent binding of tRNA to the decoding site on the ribosome proceeds by such a mechanism.

Fluorescent tRNA dervatives seem to be well suited to study tRNA-ribosome complex formation and, beyond that, to investigate the question of whether changes of the tRNA conformation are involved in the recognition mechanism. An important prerequisite for such studies was to show that the efficiency of ribosome binding was not appreciably impaired by insertion of the dye into tRNA^{Phe} (Wintermeyer and Zachau 1975b; Robertson et al. 1977; Schleich et al. 1978; Odom et al. 1975). The initial fluorescence studies have been performed with the deacylated tRNA^{Phe}-dye derivatives under conditions in which the tRNA is bound to the P-site of the poly(U)-programmed ribosomes. Both 70S ribosomes from *E. coli* and 80S ribosomes from yeast have been used, yielding rather similar results.

Ribosome binding of the tRNA^{Phe}-dye compounds considerably altered the fluorescence properties of the probes in both positions of the tRNA molecule (Robertson et al. 1977; Fairclough et al. 1979). As an example, Fig. 4A shows titrations of tRNA$^{Phe}_{Etd}$ 37 with yeast ribosomes, as monitored by the increase in the fluorescence polarization, reflecting the immobilization of the fluorophor upon complex formation. In the presence of poly(U) the Etd in the complex was found to be rigidly fixed, whereas in the absence of poly(U) the probe retained considerable mobility. The quantitative evaluation of the saturation curves has revealed comparable apparent binding constants fot the complexes formed in the presence or absence of poly(U). On the other hand, the competition experiment (Fig. 4B) shows that dissociation of the com-

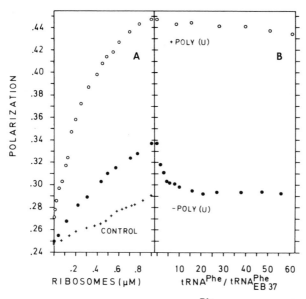

Fig. 4. Stability of the complexes of tRNA$_{Etd}^{Phe}$ 37 with yeast ribosomes. (**A**) tRNA$_{Etd}^{Phe}$ 37 was titrated at 20 mM Mg^{2+} with ribosomes to near saturation in the presence (*open rings*) or absence (*black spots*) of poly(U), as well as in the presence of tRNA^{Phe} in 60-fold excess without poly(U) (*crosses*; this control shows the effect of the increasing viscosity on the polarization exhibited by free tRNA$_{Etd}^{Phe}$ 37). (**B**) To the complexes formed in the presence or absence of poly(U) increasing amounts of tRNA^{Phe} were added. For experimental details see Robertson et al. (1977) from which the figure is reproduced with permission

plex was extremely slow in the presence of poly(U), whereas the tRNA was readily released from the complex formed in the absence of poly(U). These results illustrate the potential of the fluorescence technique since, unlike the filter binding assay usually employed for studies of tRNA-ribosome complexes, it allows the detection of weak or readily dissociating complexes.

The emission spectra of Prf in both the anticodon and the D-loop of tRNA^{Phe} were changed substantially when the tRNA bound to poly(U)-programmed ribosomes (Fig. 5). The blue shift of 10 nm and the concomitant increase of the quantum yield, which was not observed in the absence of poly(U) (Fairclough et al. 1979), has to be interpreted with caution. On the basis of the present results we cannot distinguish between environmental effects of the ribosome and codon-dependent conformational changes of the anticodon loop leading to altered stacking interactions of the dye within the tRNA molecule. However, the observation that the presence of the codon influences both the mobility and the fluorescence spectrum of a probe next to the anticodon strongly suggests that there is anticodon-codon interaction taking place in the P-site. Preliminary spectroscopic measurements on A-site-bound fluorescent Phe-tRNA have shown that the tRNA's in the two sites may be distinguished by their different emission spectra, an important requirement for translocation experiments.

The fluorescence changes have been utilized to monitor the kinetics of tRNA binding to ribosomes in stopped-flow experiments (Wintermeyer et al. 1979). In the ab-

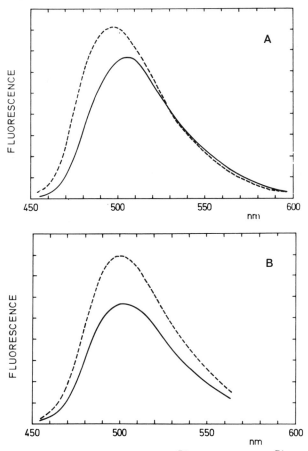

Fig. 5A, B. Emission spectra of tRNA$_{Prf}^{Phe}$ 37 (**A**) and tRNA$_{Prf}^{Phe}$ 16/17 (**B**) and their complexes with poly(U)-programmed *E. coli* ribosomes. Spectra were measured in 50 mM Tris-HCl, pH 7.4, 90 mM NH$_4$Cl, 50 mM KCl, 20 mM MgCl$_2$, 1 mM dithioerythritol at 7°C before (*solid line*) and after (*broken line*) the addition of a twofold molar excess of 70S tight couple ribosomes from *E. coli* MRE 600 which had been incubated with poly(U) before. Fluorescence was measured upon excitation at 436 nm after passing a cut-off filter (Schott KV 450) and is given in arbitrary units. Blank spectra of buffer or ribosomes, respectively, are subtracted

sence of poly(U), binding occurred in a very rapid process which could not be resolved. Upon complex formation with poly(U)-programmed ribosomes, four relaxation times ranging from 100 ms to 50 s could be resolved in addition to an unresolved rapid step. The resolved steps were dependent on the presence of the cognate codon; no slow step was observed when poly(A) was used as codon, nor when a fluorescent derivative of tRNA^{Ser} was bound to poly(U)-programmed ribosomes.

The rates of the slow steps need some further comment. Obviously these rates are much too slow in view of the in vivo rate of protein synthesis. It has to be kept in mind, however, that the experiments reported here have been performed under artificial conditions in a highly simplified system. In order to approach the in vivo situation one certainly has to work at conditions which are closer to the physiological ones, e.g., to study the factor-dependent binding of aminoacyl tRNA.

Although at present there are not enough data to develop a consistent model which correlates the kinetic steps with biochemically defined events, the kinetic data suggest that the binding of the tRNA to a functional ribosomal site occurs in several steps. In the unresolved, probably diffusion-controlled association step, a tRNA-ribosome complex is formed in a codon-independent manner which is highly reversible. In the following steps, which depend on the presence of the proper codon, the initial complex undergoes transitions until the final state is reached. This picture corresponds to what one may expect on the basis of proof-reading models (Ninio 1973; Kurland et al. 1975), in which the codon-anticodon interaction acts as an allosteric signal, triggering conformational transitions of the tRNA-ribosome complex.

D. Summary

The use of fluorescent derivatives of tRNA^{Phe} (yeast) in studies on tRNA conformation and on tRNA-ribosome recognition is described. Evidence is presented which indicates that under physiological conditions with respect to ionic strength and Mg^{2+} concentration, tRNA^{Phe} exists in at least two conformations. The functional significance of this behavior is discussed on the basis of aminoacylation experiments. The investigation of the ribosome complexes of tRNA^{Phe} labeled in the anticodon and D-loops has provided evidence suggesting that the presence of the codon, although not appreciably altering the apparent association constant, leads to qualitatively different complexes in which the tRNA appears to be rigidly bound to the codon even in the P-site of the ribosome. From kinetic experiments, it is concluded that binding of the tRNA to the ribosome occurs in several steps, which take place only in the presence of the proper codon. One or more of these steps may represent codon-induced conformational changes of the tRNA molecule, which constitute the molecular basis of the highly specific binding of the tRNA to the ribosome.

References

Altman S (ed) (1978) Transfer RNA. MIT Press, Cambridge/Mass London
Beardsley L, Cantor CR, Tao T (1970) Studies on the conformation of the anticodon loop of phenylalanine transfer ribonucleic acid. Effect of environment on the fluorescence of the Y base. Biochemistry 9: 3524–3532
Ehrenberg M, Rigler R, Wintermeyer W (1979) On the structure and conformational dynamics of yeast phenylalanine-accepting transfer ribonucleic acid in solution. Biochemistry 18: 4588–4599
Fairclough RH, Cantor CR, Wintermeyer W, Zachau HG (1979) Fluorescence studies of the binding of a yeast tRNA^{Phe} derivative to *Escherichia coli* ribosomes. J Mol Biol 132: 557–573
Johnston PD, Redfield AG (1977) An NMR study of the exchange rates for protons involved in the secondary and tertiary structure of yeast tRNA^{Phe}. Nucleic Acids Res 4: 3599–3615

Kurland CG, Rigler R, Ehrenberg M, Blomberg C (1975) Allosteric mechanism for codon-dependent tRNA selection on ribosomes. Proc Natl Acad Sci USA 72: 4248–4251

Ladner JE, Jack A, Robertus JD, Brown RS, Rhodes D, Clark BFC, Klug A (1975) Structure of yeast phenylalanine transfer RNA at 2.5 A resolution. Proc Natl Acad Sci USA 72: 4414–4418

Ninio J (1973) Recognition in nucleic acids and the anticodon families. Prog Nucleic Acid Res Mol Biol 13: 301–337

Odom OW, Hardesty B, Wintermeyer W, Zachau HG (1975) Efficient polyphenylalanine synthesis with proflavine and ethidium labeled tRNA^{Phe} from yeast in the reticulocyte ribosomal system. Biochim Biophys Acta 378: 159–163

Olson T, Fournier MJ, Langley KH, Ford NC (1976) Detection of a major conformational change in transfer ribonucleic acid by laser light scattering. J Mol Biol 102: 193–203

Pachmann U, Cronvall E, Rigler R, Hirsch R, Wintermeyer W, Zachau HG (1973) On the specifity of interactions between transfer ribonucleic acids and aminoacyl-tRNA synthetases. Eur J Biochem 39: 265–273

Quigley GJ, Wang AHJ, Seeman NC, Suddath FL, Rich A, Sussman JL, Kim SH (1975) Hydrogen bonding in yeast phenylalanine transfer RNA. Proc Natl Acad Sci USA 72: 4866–4870

Rich A, RajBhandary UL (1976) Transfer RNA: molecular structure, sequence and properties. Annu Rev Biochem 45: 805–860

Rigler R, Ehrenberg M, Wintermeyer W (1977) Structural dynamics of tRNA — a fluorescence relaxation study of tRNA$_{yeast}^{Phe}$. In: Molecular biology biochemistry biophysics, vol 24, pp 219–244. Springer, Berlin Heidelberg New York

Robertson JM, Kahan M, Wintermeyer W, Zachau HG (1977) Interactions of yeast tRNA^{Phe} with ribosomes from yeast and *E. coli*. Eur J Biochem 72: 117–125

Schimmel P, Söll D, Abelson J (eds) Transfer RNA. Cold Spring Harbor Laboratory

Schleich HG, Wintermeyer W, Zachau HG (1978) Replacement of wybutine by hydrazines and its effect on the active conformation of yeast tRNA^{Phe}. Nucl Acids Res 5: 1701–1713

Schwarz U, Gassen HG (1977) Codon-dependent rearrangement of the tertiary structure of tRNA^{Phe} from yeast. FEBS Lett 78: 267–270

Wintermeyer W, Zachau HG (1971) Replacement of Y base, dihydrouracil and 7-methylguanine in tRNA by artificial odd bases. FEBS Lett 18: 214–218

Wintermeyer W, Zachau HG (1975a) Tertiary structure interactions of 7-methylguanosine in yeast tRNA^{Phe} as studied by borohydride reduction. FEBS Lett 58: 306–309

Wintermeyer W, Zachau HG (1975b) Characterization of fluorescent derivatives of tRNA^{Phe} by experiments in the ribosomal system. Mol Biol (Russian) 9: 63–69; (English) 9: 49–53

Wintermeyer W, Zachau HG (1979) Fluorescent derivatives of yeast tRNA^{Phe}. Eur J Biochem 98: 465–475

Wintermeyer W, Robertson JM, Weidner H, Zachau HG (1979) Studies on tRNA conformation and ribosome interaction with fluorescent tRNA derivatives. In: Schimmel P, Söll D, Abelson J (eds) Transfer RNA, Part 1, pp 445–457. Cold Spring Harbor Laboratory, Cold Spring Harbor

Wong KL, Kearns DR, Wintermeyer W, Zachau HG (1975) NMR investigation of the effect of selective modifications in the anticodon loop on the conformation of yeast transfer RNA

4. Structure and Evolution of Ribosomes

H.G.WITTMANN

A. Summary

Ribosomes are multicomponent particles on which biosynthesis of proteins occurs in all organisms. The best-known ribosome, namely that of *E. coli*, consists of three RNA's and 53 different proteins. All proteins have been isolated and characterized by chemical, physical, and immunological methods. The primary sequences of 49 *E. coli* ribosomal proteins have so far been determined. Studies of the shape, as well as of the secondary and tertiary structure, of the proteins are in progress.

Various techniques, e.g., immune electron microscopy and cross-linking of neighboring components in situ, give information about the architecture of the ribosomal particle. The first technique resulted in illustrative and detailed knowledge not only on the shape of the ribosomal subunits but also about the location of many proteins on the surface of the particles. The analysis of cross-links between ribosomal proteins and/or RNA's has in several cases been pursued to the level of elucidating which amino acids and/or nucleotides are cross-linked together in situ.

Reconstitution of a fully active *E. coli* 50S ribosomal subunit from its isolated RNA and protein components can be accomplished by means of a two-step incubation procedure. From the analysis of the intermediates occurring during the reconstitution process it has been concluded that the in vitro reconstitution process resembles the in vivo assembly of 50S subunits in many respects.

E. coli mutants with alterations in almost all ribosomal proteins have been isolated. Their biochemical and genetic analyses are very useful tools for obtaining information about the structure, function, and biosynthesis of ribosomes, as well as about the location of the genes for these proteins on the chromosome.

From comparative electrophoretic, immunological, protein-chemical, and reconstitution studies on ribosomes from various species it has become clear that there is little homology between ribosomal proteins from prokaryotes and those from eukaryotes. This finding is surprising since there is no essential difference in the way in which pro- and eukaryotic ribosomes function in protein biosynthesis.

Max-Planck-Institut für Molekulare Genetik, 1000 Berlin-Dahlem, FRG

B. Introduction

Biosynthesis of proteins takes place on ribosomes which are multicomponent particles occurring in all organisms. There are two classes of ribosomes in nature: the 80S ribosomes present in the cytoplasm of eukaryotic cells, and the 70S ribosomes in prokaryotes. Ribosomes in chloroplasts and mitochondria are related to the prokaryotic ribosomes, although their sedimentation coefficients differ in many cases from 70S.

C. Components

The best-known ribosome is that of the bacterium *E. coli*. It consists of two subunits of unequal size (30S and 50S). Besides three RNA molecules (5S, 16S, and 23S) there are many ribosomal proteins. Their number was first determined by two-dimensional polyacrylamide electrophoresis. 21 protein spots, named S1–S21, were found in the small, and 34 spots (L1–L34) in the large subunit (Kaltschmidt and Wittmann 1970). When the ribosomal proteins were isolated and characterized by chemical, physical, and immunological methods (Wittmann 1974; Stöffler 1974), it turned out that each of the 21 spots on the two-dimensional electropherogram corresponds to an individual protein, whereas in the 50S subunit one of the spots, L8, reflects a quite stable aggregate of three proteins: L7, L10, and L12 (Pettersson et al. 1976).

Immunological and protein-chemical studies on the isolated protein showed that, with two exceptions, there are no cross-reactions and no homologous structures among the *E. coli* ribosomal proteins. The two exceptions are: proteins L7 and L12 from the 50S subunit, which differ only at the N-terminus, and S20 from the 30S and L26 from the 50S subunit, which have identical primary structures (Stöffler 1974; Wittmann and Wittmann-Liebold 1974). Up to now the complete amino acid sequences of the following 49 ribosomal proteins have been determined in our laboratory or in collaboration with other groups: S3, S4, S5, S6, S7, S8, S9, S10, S11, S12, S13, S14, S15, S16, S17, S18, S19, S20, S21, L1, L2, L4, L5, L6, L7, L10, L11, L12, L13, L14, L15, L16, L18, L19, L20, L21, L22, L23, L24, L25, L26, L27, L28, L29, L30, L31, L32, L33, and L34 (reviewed in Wittmann et al. 1979). The elucidation of the sequences of the other *E. coli* ribosomal proteins is now in progress and can be expected to be completed in the near future.

Four computer programs, developed for the prediction of the secondary structure of a protein based on its amino acid sequence, have been applied to those ribosomal proteins whose primary structures are known (Wittmann-Liebold et al. 1977a, b; Dzionara et al. 1977a, b). An example is shown in Fig. 1. It remains to be seen how far these predictions can be confirmed experimentally, e.g., by CD-studies currently in progress. For these and other studies on the structural properties of the ribosomal proteins, it is advantageous to isolate them by methods other than those (e.g., by the use of urea, 67% acetic acid, lyophilization) that can lead to denaturation. When several physical techniques (proton magnetic resonance, Raman spectroscopy, circular dichroism, low-angle X-ray scattering, and hydrodynamic methods) were used to compare proteins isolated under nondenaturing conditions (Dijk and Littlechild 1978) with

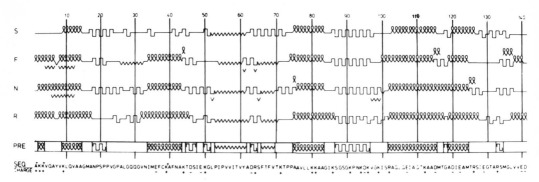

Fig. 1. Secondary structure of protein L11 as predicted according to four different methods (S, F, N, R). The symbols represent residues in helical (𝕏𝕏𝕏𝕏), turn or bend (ⅉⅉⅉⅉ), extended (⋀⋀⋀) or coil (——) conformational states, respectively. The line *PRE* summarizes the secondary structure obtained when at least three of the four predictions are in agreement. The amino acid sequence of protein L11 is shown in the *bottom line* in the one-letter code

those isolated in the presence of urea and acetic acid, a higher degree of secondary and tertiary structure was found with the former than with the latter proteins (e.g., Morrison et al. 1977). The presence of structural domains in ribosomal proteins is further indicated by experiments in which the proteins were subjected to limited proteolysis. Stable fragments were obtained for a number of proteins, e.g., S1, S4, S5, S8, S20, L6, L11, L16, L17, and L24 (see Wittmann et al. 1979).

The shape of the ribosomal proteins has been determined by low-angle X-ray scattering and by hydrodynamic methods, i.e., sedimentation, diffusion, and viscosity. It was found that many proteins have a slightly, or even highly, elongated shape whereas several ribosomal proteins are globular (reviewed by Wittmann et al. 1979). Information about the shape of the ribosomal proteins in situ derives from immune electron microscopical studies (see below).

D. Architecture of the Ribosome

Knowledge of the ribosomal architecture, i.e., the shape and contours of the ribosome and its subunits, as well as the spatial arrangement of the many components (proteins and RNA's) within the particles, is an essential requirement for an understanding of the role of the ribosomes in the various steps of protein biosynthesis at the molecular level.

During the last several years several approaches were used to obtain information on the architecture of the *E. coli* ribosome and its subunits. The most efficient of them are the following: (1) cross-linking by bifunctional reagents between neighboring proteins within the particle (Traut et al. 1974); (2) cross-linking between a given protein and a RNA region in its neighborhood (Brimacombe 1978); (3) cross-linking between RNA regions which are in close proximity in the ribosome (Wagner and Garrett 1978; Zwieb et al. 1978); (4) examination of protein binding sites on the three

RNA's (Zimmermann 1974; Erdmann 1976); (5) assembly map (Nomura and Held 1974); (6) subunit fragmentation (Brimacombe et al. 1976); (7) neutron scattering (Moore et al. 1974); (8) singlet energy transfer between fluorescently labeled proteins (Cantor et al. 1974); and last, but not least (9) immune electron microscopy (Stöffler and Wittmann 1977). The advancements in these fields during the last few years have been summarized in recent reviews (Brimacombe 1978; Brimacombe et al. 1976; Stöffler and Wittmann 1977; Kurland 1977; Brimacombe et al. 1978).

I. Immune Electron Microscopy

This technique combines the application of electron microscopy with the use of antibodies against individual ribosomal proteins, and it has already given the most illustrative and detailed information not only on the shape of the ribosomal subunits but also about the location of most proteins on the surface of the particles.

The morphology of the *E. coli* ribosome and its subunits has been examined by many authors (reviewed by Stöffler and Wittmann 1977; van Holde and Hill 1974). The electron micrographs of the small ribosomal subunit published by various investigators show a great deal of similarity on the gross morphology. A rod-like particle with an axial ratio of 2:1 (long axis of approx. 220 Å) is divided by a hollow or cleft which is arrranged perpendicular to the long axis of the 30S subunit and separates the "head" from the larger "body" (see Fig. 1). Two lobes (Tischendorf et al. 1974, 1975) or a single narrow platform (Lake 1976) protrude from the body. These protrusions break the apparent symmetry of the 30S subunit, independently of the different interpretations. The principal structural elements are also contained in each of the three-dimensional 30S subunit models which have been proposed so far (Tischendorf et al. 1975; Lake 1976; Vasiliev 1974; Boublik et al. 1977).

Electron micrographs of 50S ribosomal subunits show particles in several forms, resembling crowns, maple leaves, kidneys with an asymmetric notch, and rounded structures. The fundamental features of electron micrographs of 50S subunits, as published by several investigators, are similar (see Stöffler and Wittmann 1977; van Holde and Hill 1974). Each of the proposed models shows a plane with a spherical base and three protuberances. The differences in the models are concerned with the origin of these protuberances from the plane and with their sizes and shapes (see Brimacombe et al. 1978).

70S ribosomes are relatively globular or polygonal particles. The small subunit is arranged horizontally transverse with respect to the 50S subunit. The hollow of the 30S subunit is oriented towards the plane of the 50S particle; thus a channel is formed between the two subunits (Tischendorf et al. 1975; Lake 1976; Boublik et al. 1977).

The location of proteins on the subunit surface is being determined as follows: Specific, non cross-reacting antibodies directed against each of the proteins of *Escherichia coli* are bound to intact ribosomal subunits. Complexes of two subunits connected by one antibody are purified, stained with uranyl acetate and examined by electron microscopy. The position of the bound antibody, specific for a particular protein, is determined. It indicates the location of the corresponding protein on the ribosomal surface. The binding sites for antibodies to each of the proteins from the 30S subunit,

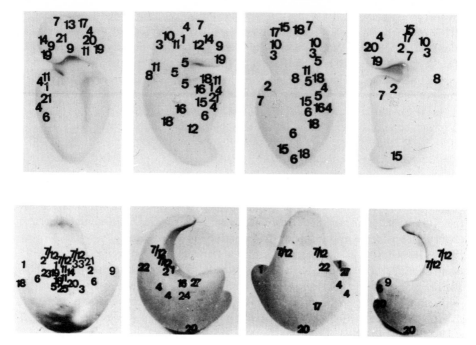

Fig. 2. Three-dimensional model of *E. coli* ribosomal subunits with the locations of antibody binding sites for individual ribosomal proteins. *Above:* Four views of the 30S subunit model, obtained by successive rotation of the model by an angle of 90°. The antibody binding sites for the 21 ribosomal proteins are indica9ed by *numbers. Below:* Three-dimensional model of the 50S subunit with the centers of the antibody binding sites

and to 30 proteins of the 50S subunit, have been localized on the surface of the corresponding subunits in this way (Fig. 2; Stöffler and Wittmann 1977; Tischendorf et al. 1975; Wittmann et al. 1978). Each of approximately 20 proteins showed antibody binding sites at remote positions on the subunit surface. Some of the complexes are simultaneously connected by two or three antibody molecules, which are specific for a single protein indicating highly extended conformations of the protein in situ.

Immune electron microscopical studies have also been performed by other investigators. Six 30S proteins were localized by Lake et al. (1974) as well as by Lake and Kahan (1975). Proteins L7 and L12 were mapped by Boublik et al. (1976). A specific region of the 16S RNA, the 6-N,N dimethyl-adenine residue near the 3'-terminus of 16S RNA, was also located by immune electron microscopy (Politz and Glitz 1977; Dieckhoff et al., manuscript in preparation).

Despite certain important differences between the results of several groups, a reasonably good agreement exists with regard to the distribution of the mapped antigenic sites. The main discrepancies are caused by the different interpretations of the three-dimensional subunit structures by the various authors. These discrepancies will probably be solved by the development of new methods which provide a better structure preservation during specimen preparation and by an improvement of the electron optical resolution. Both these conditions are satisfied by a new technique for tungsten shadowing of freeze-dried particles with a low-specimen damage electron impact evaporator (Tesche 1975, 1978). This procedure has been applied to the study of the morphology of *Escherichia coli* ribosomes (Tesche et al. 1978).

Another goal of these studies is to localize the various functionally important domains on the ribosome surface. An attempt has recently been made (Stöffler and Wittmann 1977) by combining data on the physical neighborhood of ribosomal components (as derived from immune electron microscopy, neutron scattering, fluorescence labeling, cross-linking, etc.) with data on the functional contribution of individual ribosomal proteins. This attempt is complicated considerably by the finding that many ribosomal proteins have elongated structures. For example, proteins S1, S4, S18, and S21 have been shown by various techniques to be in the neighborhood of the ribosomal decoding site (see Brimacombe et al. 1978). Each of these four proteins is elongated and occurs on two widely separated regions on the 30S subunit: one of them is located on the "head" and the other on the "body" of the subunit (see Fig. 1). Hence, the location of the decoding site cannot be unambigously determined.

To overcome this difficulty, experiments were made which allow a more direct localization of functional domains. The position of the chloramphenicol binding site on the ribosome was thus mapped by employing specific antisera to this drug, and it was found to be in a domain that comprises ribosomal proteins L16, L2, and L27. This is in agreement with the results from reconstitution and affinity-labeling techniques. Other ribosomal drug binding sites are being studied in a similar manner. Since it is known which of the various steps during protein biosynthesis are inhibited by these antibiotics, a correlation can be made between the location of ribosomal components on the subunit and the ribosomal function in which these components are involved. In a similar way, the direct visualization of the binding sites of various protein synthesis factors, e.g., IF-3, EF-Tu, and EF-G, and of aminoacyl-tRNA should give direct information about the location of functional domains (Lührmann et al. 1979).

II. Cross-Linking of Ribosomal Components

Protein-protein cross-linking techniques have been used for some years in studies on the topography of the *E. coli* ribosome (for reviews see Traut et al. 1974; Brimacombe et al. 1976; Kurland 1977). This principle has now been extended to include RNA-protein and RNA-RNA cross-linking within the ribosomal subunits. However, it has at the same time become clear that an identification of the ribosomal components involved in the cross-link is no longer sufficient to give useful structural information. The complexity of the ribosomal particles, and in particular the elongated shapes of many of the proteins, demand that the cross-linking analysis be pursued to the level of defining which amino acids or nucleotides are cross-linked together.

In the case of protein-protein cross-linking this has been accomplished for proteins S5 and S8 cross-linked by dimethyl (C^{14})-suberimidate in the 30S subunit. After its isolation the protein pair was digested with trypsin and the two cross-linked peptides were purified. Sequence analysis established that the lysine residue in position 166 of protein S5 was cross-linked to lysine-93 in protein S8 (Allen et al. 1979). Therefore, the distance between these two amino acids within the 30S subunit cannot exceed 8 Å, i.e., the length of the bifunctional reagent dimethyl suberimidate (Fig. 3).

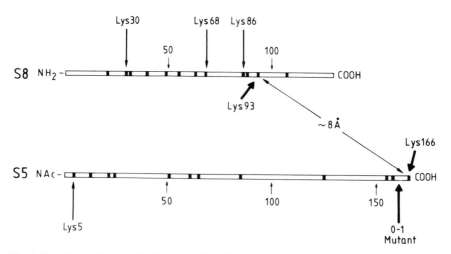

Fig. 3. Protein-protein cross-linking in the *E. coli* ribosome: Amino acid residue Lys-93 of protein S8 has been identified to be cross-linked by dimethylsuberimidate (Length: 8 Å) to amino acid residue Lys-166 of protein S5

1. Ultraviolet Irradiation

2. 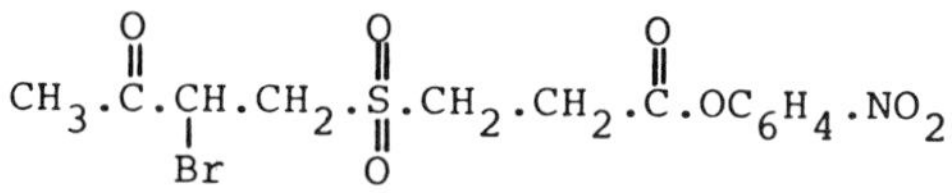 $CH_3.\overset{O}{\overset{\|}{C}}.\underset{Br}{CH}.CH_2.\overset{O}{\underset{\|}{\underset{O}{S}}}.CH_2.CH_2.\overset{O}{\overset{\|}{C}}.OC_6H_4.NO_2$

3.

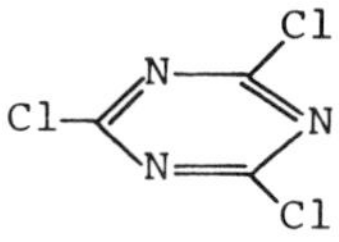

4. 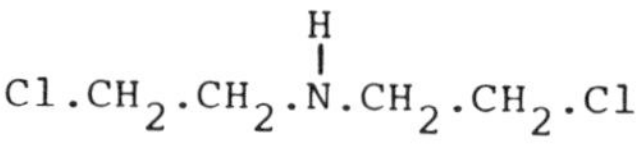 $\overset{H}{Cl.CH_2.CH_2.\overset{|}{N}.CH_2.CH_2.Cl}$

5. 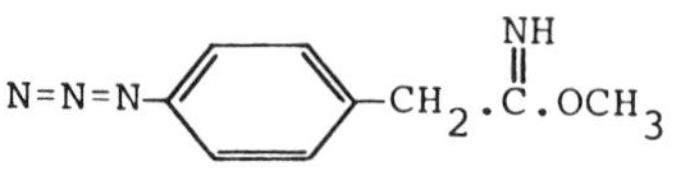 $N=N=N-\langle\rangle-CH_2.\overset{NH}{\overset{\|}{C}}.OCH_3$

Fig. 4. RNA-protein cross-linking agents

The studies, which are primarily concerned with RNA-protein cross-linking, have been divided into two parts, namely a search for suitable bifunctional reagents (Fig. 4), and secondly the development of methodology for the detailed analysis of the cross-linking points. This latter problem has been approached using a simple cross-linking system in which protein S7 (and no other protein) is covalently linked to 16S RNA in the 30S subunit by low doses of UV irradiation (Möller and Brimacombe 1975). To analyze the cross-linked peptide, a ^{32}P-labeled S7-oligonucleotide complex is isolated, and, after digestion with trypsin, a peptide containing ^{32}P-label can be positively iden-

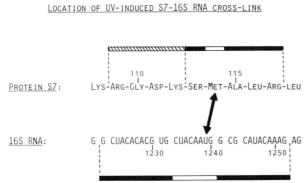

LOCATION OF UV-INDUCED S7-16S RNA CROSS-LINK

PROTEIN S7:

16S RNA:

Fig. 5. RNA-protein cross-linking in the *E. coli* ribosome. Amino acid residue Met-114 of protein S7 has been identified to be cross-linked by UV irradiation to nucleotide U-1239 in the 16S RNA

tified, and its amino acid content and sequence determined (Möller et al. 1978). For the cross-linked oligonucleotide, a ribonucleoprotein fragment is first isolated, containing the cross-linked protein. This is applied to a special polyacrylamide gel system, in the presence of a nonionic detergent, which deproteinizes the fragment and at the same time separates RNA from the cross-linked RNA-protein complex. The latter can be isolated, subjected to fingerprint analysis, and its position in the 16S RNA sequence determined. The results showed an RNA region from which a single characteristic oligonucleotide was absent; this oligonucleotide was confirmed as containing the site of cross-linking by direct analysis of the short oligonucleotide remaining attached to the protein (Zwieb and Brimacombe 1978). Further studies established that the uracil residue in position 1239 of the 16S RNA is cross-linked to methionine-114 in protein S7 (Zwieb and Brimacombe 1979). Therefore, these two points must be in close contact within the 30S subunit (Fig. 5).

The methods just mentioned have been developed with the intention that they should be applicable in the case of cross-linking by bifunctional chemical reagents, where it would be expected that several proteins are simultaneously cross-linked to the RNA within the 30S particle. It is also important to note that the methodology requires that the cross-link be kept intact throughout the analysis, and that therefore reversible cross-linking reagents are of no advantage in this type of experiment. A suitable bifunctional reagent has been found to be the short, symmetrical compound bis-(2-chloroethyl)-amine. This leads to stable cross-linking of a number of proteins to RNA in reasonable yield at low concentrations of reagent under physiological conditions, in both 30S and 50S subunits (Ulmer et al. 1978). Some of these proteins have been identified on two-dimensional gels, and the stability of the cross-link can be demonstrated by the use of ^{32}P-labeled subunits, in which case the isolated cross-linked proteins contain ^{32}P-label. The methodology described above is currently being applied to identify the cross-linking sites on several of these proteins (Rinke et al. 1979).

The bis-(2-chloroethyl)-amine is a symmetrical reagent and should therefore also cause some RNA-RNA cross-linking. This has been found to be the case (Zwieb et al. 1978), and in particular the reagent can be used to generate 16S-23S RNA cross-links within the 70S ribosome. RNA-RNA cross-linking within the 16S and 23S molecules can be shown to be induced by low doses of UV irradiation. Other reagents for both RNA-protein and RNA-RNA cross-linking are currently being developed, and one of them, 1,4-phenyl-diglyoxal, has already led to the identification of an RNA-RNA cross-link within the 5S RNA (Wagner and Garrett 1978): an intramolecular cross-link between guanosine-2 and guanosine-112 in the stem region of the 5S RNA molecule was identified.

III. 5S RNA and 5S RNA-Protein Complexes

In order to obtain more detailed information about the secondary and tertiary structure of the 5S RNA, studies on oligonucleotides binding (Wrede et al. 1978) and with infra-red spectroscopy (Appel et al. 1978) were carried out. Neither investigation supported any of the many structural models proposed for 5S RNA (for a recent review see Erdmann 1976). It was found that *E. coli* 5S RNA contains single-stranded regions around positions 30, 60, and 70. Similar results were obtained with 5S RNA from other organisms, e.g., *Bacillus stearothermophilus* and yeast, although the nucleotide sequences of the regions mentioned above differ among the 5S RNA's from these organisms.

Because the sequences of the 5S RNA, and of the three proteins L5, L18, and L25 which bind to the 5S RNA, are known, the RNA-protein complex is an excellent system for studying the interaction between RNA and protein. The binding sites of the three proteins on the 5S RNA from *E. coli* and *Bacillus stearothermophilus* have been localized (Zimmermann and Erdmann 1978). Neutron scattering experiments with the complex between the 5S RNA and proteins L18 and L25 showed that the RNA and the proteins are almost evenly distributed and that the mass centers of both proteins are separated by 115 Å (May et al. 1978). After binding of the proteins to the 5S RNA, its conformation is drastically altered, as revealed by oligonucleotide binding studies (Erdmann et al. 1973).

Reconstitution experiments, in which it was attempted to incorporate the 5S RNA from various organisms into the 50S subunit of *Bacillus stearothermophilus,* led to biologically active 50S particles only if the 5S RNA's originated from prokaryotes. No eukaryotic 5S RNA's were incorporated (Wrede and Erdmann 1973). Similarly, heterologous 5S RNA-protein complexes could only be obtained with RNA's and proteins from prokaryotic cells, and not with 5S RNA's from eukaryotes and proteins from prokaryotes (Horne and Erdmann 1972). The interesting finding that the *E. coli* proteins L18 and L25 bind specifically not only to the 5S RNA from prokaryotes, but also to the 5.8S RNA from eukaryotes, implies that the eukaryotic 5.8S RNA (and not the 5S RNA) is homologous to the 5S RNA from prokaryotes (Wrede and Erdmann 1977).

E. Assembly

After the total reconstitution of 30S subunits from *E. coli* (Traub and Nomura 1968) and of 50S subunits from *Bacillus stearothermophilus* (Nomura and Erdmann 1970), it took several years before 50S subunits from *E. coli* could be reconstituted to fully

Assembly of the 50S subunit of *E. coli* in vitro

23 S + 5 S RNA + 18/19 proteins	L1 L2 L3 L4 L5 L8/9 L10 L11 L13 L17 L19 L20 L21 L22 L23 L24 L29 L33	$\xrightarrow{0^{\circ}C,\ 4\ mM\text{-}Mg^{2+}}$ $RI_{50}(1)$	(33 S)
	$RI_{50}(1)$	$\xrightarrow{44^{\circ}C,\ 4\ mM\text{-}Mg^{2+}}$ $RI_{50}^{*}(1)$	(41 S)
$RI_{50}^{*}(1)$ + 8 proteins	L6 L15 L16 L25 L27 L28 L30 L32	$\xrightarrow{44^{\circ}C,\ 4\ mM\text{-}Mg^{2+}}$ $RI_{50}(2)$ or $50^{\circ}C,\ 20\ mM\text{-}Mg^{2+}$	(48 S)
	$RI_{50}(2)$	$\xrightarrow{50^{\circ}C,\ 20\ mM\text{-}Mg^{2+}}$ rec. 50S	
Not assigned:	L7/L12, L14, L18, L26, L31 and L34		

Fig. 6. Flow-diagram of the 50S reconstitution procedure. Proteins important for RI_{50}^{*} (1) formation are enclosed by a *square*

active particles from dissociated RNA and protein fractions (Nierhaus and Dohme 1974). The procedure consists in a two-step incubation: In the first incubation 23S and 5S RNA and the total proteins (TP50) are incubated at $44^{\circ}C$ under defined ionic conditions (4 mM Mg^{2+}, etc.), after which the Mg^{2+} concentration is raised to 20 mM Mg^{2+} and the incubation is continued at $50^{\circ}C$ (see Fig. 6). During this reconstitution procedure 30 S subunits and polyamines are not required (Nierhaus and Dohme 1974; Dohme and Nierhaus 1976a).

Firstly, it was determined whether the in vitro process leads to a continuous S-value shift from 23S (RNA) to 50S (the active particle), or whether discrete reconstitution intermediates are formed in the course of assembly. It was shown that the assembly in vitro occurs in discrete steps. Three reconstituted intermediates were found which were designated $RI_{50}(1)$, $RI_{50}^{*}(1)$, and $RI_{50}(2)$, respectively. Further analysis revealed that $RI_{50}(1)$ and $RI_{50}^{*}(1)$ particles contain the same complement of ribosomal components, namely 23S and 5S RNA, and 18–19 proteins.

However, these particles differ drastically in their S-values (33S and 41S, respectively). The $RI_{50}^{*}(1)$ particle binds the remaining proteins, forming the $RI_{50}(2)$ particle (48S). This particle is totally inactive in poly(Phe) synthesis, and changes its conformation to an active 50S particle exclusively during the second step (Dohme and Nierhaus 1976a). Thus, the two conformational changes $RI_{50}(1) \rightarrow RI_{50}^{*}(1)$, and $RI_{50}(2) \rightarrow$ 50S, occur under different conditions of temperature and ionic strength in the course of the in vitro assembly. Therefore, a two-step incubation procedure is necessary for the reconstitution of active 50S subunits from *E. coli*.

Kinetic analysis performed at various temperatures during the first or second incubation revealed that the rate-limiting step in either incubation is a first-order reaction.

Dilution experiments confirmed further that the rate-limiting step is a unimolecular reaction in the second incubation and, most probably, also in the first. Hence, the Arrhenius activation energy can be calculated for both rate-limiting steps, and this was determined as 290 and 225 kJ/mol for the first and the second incubation, respectively. Interestingly, when the preparation technique for the RNA isolation was changed (acetic acid extraction instead of standard phenol treatment), the activation energy for the first incubation decreased to 220 kJ/mol, suggesting that the acetic acid RNA has a conformation more closely resembling the RNA conformation present in the 50S subunit than that of the phenol RNA (Sieber and Nierhaus 1978).

Although 5S RNA is present in the $RI_{50}^*(1)$ particle, this species plays no role in generating the $RI_{50}^*(1)$ conformation (Dohme and Nierhaus 1976b). Therefore, we tested whether all the proteins present in the $RI_{50}^*(1)$ particle are important for the formation of the $RI_{50}^*(1)$ conformation. A detailed analysis demonstrated that 23S RNA and five proteins (L4, L13, L20, L22, and L24) are required for $RI_{50}^*(1)$ formation; L3 has a stimulatory effect. With these seven components a particle with the $RI_{50}^*(1)$ conformation could be formed (Spillmann et al. 1977).

When native 50S subunits are incubated with 4 M LiCl (Homann and Nierhaus 1971), the resulting 4.0c core contains about 10 out of the 34 proteins. This core can be reconstituted to an active particle by the second incubation of the two-step procedure alone (Nierhaus and Dohme 1974), i.e., the conformation of this core represents that of the $RI_{50}^*(1)$ or $RI_{50}(2)$ particle. Surprisingly, the 4.0c core does not contain protein L24, which is one of the essential proteins for generating the $RI_{50}^*(1)$ conformation. It follows that L24 is essential for the formation, but not for the maintainance of the $RI_{50}^*(1)$ conformation. A protein fraction complementary to the 4.0c core and containing a full complement of L24 was fractionated in order to remove this latter protein. With this protein fraction and the 4.0c core, a particle lacking L24 could be reconstituted, which was fully active in translating artificial and natural mRNA. Thus L24 is essential for the early assembly but dispensable for the subsequent assembly steps and the function of the mature 50S subunit (Spillmann and Nierhaus 1978). Recently, a similar analysis revealed that L20 is also an early assembly protein which plays no role in the late assembly and in the functioning 50S subunit (Nowotny and Nierhaus, manuscript in preparation).

An "assembly map" of the proteins within the 50S subunit (Fig. 7) has been determined under the conditions of the first-step incubation. Two categories of binding dependence were established and these are indicated in Fig. 7, which contains the results so far obtained (Roth and Nierhaus 1979; Röhl and Nierhaus, unpublished). If the presence of protein Lx in the reconstitution mixture is sufficient for the binding of Ly, this is indicated by a *thick arrow* leading from Lx to Ly. However, if Lx only stimulates the binding of Ly, then these proteins are connected by a *thin arrow*. The proteins which bind under conditions of the first-step incubation to 23S RNA are located in Fig. 7 according to their approximate binding sites on the large fragments (13S, 8S, and 12S), into which 23S RNA can be subdivided. This map contains all the 50S proteins except L28, L30, L31, L32, L33, and L34.

When two proteins are interrelated in the assembly map, the simplest interpretation is that these proteins are neighbors, and this topographical correlation has been found for the 30S assembly map. However, the reverse conclusion need not always be

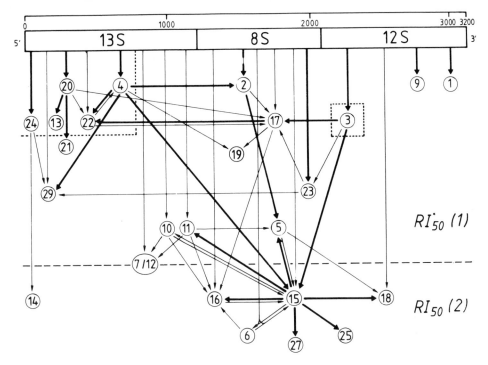

Fig. 7. Assembly map of the 50S subunit determined under the conditions of the first step incubation. For details see review by Nierhaus (1979)

valid: topographical neighborhood between two proteins may not necessarily imply an assembly dependence.

The importance of the assembly map is more than just the topographical aspect. Some previous observations can be substantiated by assembly relationships. A comparison of the in vitro assembly with the sequence in which the proteins can be washed off the 50S subunit by increasing LiCl concentrations suggested that the one sequence was roughly the reverse of the other (Dohme and Nierhaus 1976a). This interrelationship is strikingly reflected by the assembly map. The proteins are indeed split off the 50S subunit in just the opposite sequence to that in which they are assembled.

Inspection of the assembly map reveals assembly domains in the 50S subunit. One domain is located around protein L20 and consists of L4, L13, L20, L21, L22, L24, and possibly L29. Three of these proteins bind strongly to the RNA, and their binding sites are located near to the 5'-end. All the proteins essential for $RI_{50}^{*}(1)$ formation (L4, L13, L20, L22, and L24) are present in this group, and, as outlined above, at least two of these proteins (L20 and L24) are exclusively involved in the early assembly, and play a role neither in the late assembly nor in the functions of the mature 50S subunit. Thus, the "L20 domain" forms a core or nucleus in the early assembly. A second domain surrounds L15 and consists of L5, L6, L7/L12, L10, L11, L15, L16, L18, L25, and L27. None of these proteins binds tightly to the 23S RNA, indicating that the "L15 domain" is a late assembly domain. In fact, most of the late assembly group

(L6, L15, L16, L25, and L27, but without L28, L30, and L32, which are not yet integrated in the assembly map) is present in this domain. The "L15 domain" contains functionally important proteins, e.g., L7/L12, which are involved in the binding of elongation factors, and L15, L16, and L18, which are essential for peptidyltransferase activity. Thus, the "L15 domain" represents a late assembly domain and a functional cluster.

As mentioned above, the 5S RNA is not involved in the early assembly stages. However, this RNA species is important for the last assembly step, i.e., the conformational change $RI_{50}(2) \rightarrow 50S$ (Dohme and Nierhaus 1976b). A two-step reconstituted particle lacking 5S RNA was fully active in EF-G dependent GTPase activity. This finding strikingly demonstrates that 5S RNA is not involved in this function. Residual but significant activity (10 to 25%) was found in the peptidyltransferase activity and IF-dependent binding of tRNA to the P-site, indicating that 5S RNA is not essential for these functions either. However, the 5S RNA lacking particle did not enzymatically bind tRNA to the A-site. Probably, the 5S RNA participates directly in the binding of tRNA to the A-site, in good agreement with evidence that the G-T-Ψ-C region of the tRNA interacts with the G-A-A-C sequence of the 5S RNA in the A-site (for review, see Erdmann 1976). Surprisingly, the artificial and natural mRNA-dependent protein synthesis (during both activities aminoacyl-tRNA is bound transiently to the A-site) proceeded at a low but significant level in the absence of 5S RNA. This suggests that the cooperativity of the elongation cycle as a whole can overcome the poor tRNA binding to the A-site, which only becomes evident when this function is tested separately (Dohme and Nierhaus 1976b).

The assembly in vitro resembles that in vivo in many respects. Three RI particles were found in vitro, and three precursor particles exist in vivo with similar S-values: 33S, 41S, 48S, and 32S, 43S, ~50S respectively (for review, see Schlessinger 1974). Furthermore, the proteins of the early in vivo assembly (Pichon et al. 1975) bind to the 5'-half of the 23S RNA (Chen-Schmeisser and Garrett 1976). Thus, these proteins possibly determine conformational changes during early stages of assembly, before the transcription of the 23S RNA is finished. In vivo, the assembly follows a "gradient" along the 23S RNA towards the 3'-end. Therefore, such a high entropy never exists in vivo as we generate in vitro when we mix together 23S and 5S RNA and all ribosomal proteins. This feature contributes to the high activation energy of the assembly process in vitro, in contrast to that in vivo. Other features, which may in addition be important for the low activation energy in vivo, are processing of RNA, methylation of the proteins, and "assembly proteins" not present in the mature 50S subunit (see Sieber and Nierhaus 1978). The specific role of such features in the course of assembly remains to be elucidated..

F. Mutants

The isolation of mutants with altered ribosomal proteins is a very useful tool for obtaining more information about the structure, function, and biosynthesis of ribosomes as well as about the location of the genes for the proteins on the chromosome. *E. coli* mutants with alterations in ribosomal components isolated up to 1974 have been described elsewhere (Wittmann and Wittmann-Liebold 1974). In the last few years the number of such mutants, in which one or more of the ribosomal proteins have been found to be altered, drastically increased. At present, mutants with alterations in almost all of the ribosomal proteins are available. This has been achieved by the following two methods of isolation: (1) When streptomycin-independent revertants from the *E. coli* mutant VT, which shows a novel type of streptomycin dependence and has an alteration in protein S8, were isolated, it was found that they contain a great variety of altered ribosomal proteins. Alterations in 20 proteins from the small, and in 30 proteins from the large subunit have so far been detected by two-dimensional gel electrophoresis (Dabbs and Wittmann 1976; Dabbs 1978). (2) A similar set of mutants with alterations in many ribosomal proteins was obtained when more than 2000 temperature-sensitive mutants were screened by two-dimensional gel electrophoresis for altered ribosomal proteins. About 300 mutants were found to have alterations in one or more of these proteins (Isono et al. 1976, 1977, 1978).

The genetic analysis of mutants with alterations in those ribosomal proteins whose structural genes have not been mapped by other techniques has led to the discovery of several new sites for ribosomal proteins on the *E. coli* chromosome. In addition, the genes coding for the modifying enzymes which acetylate proteins S5, S18, and L12 have been localized (Isono and Kitakawa 1977; Kitakawa and Isono 1977; Isono 1978; Isono and Kitakawa 1978). By combining these results with those obtained from genetic hybrids (Sypherd and Osawa 1974) and from physical mapping (Nomura et al. 1977), the locations of the genes for almost all of the ribosomal proteins have been mapped on the *E. coli* chromosome. There are at least 12 loci, and among them several gene clusters, where the genes for ribosomal proteins and their modifying enzymes are located (Fig. 8).

Protein-chemical analyses have been carried out on many mutants with altered ribosomal proteins. Most of these studies were performed with the mutationally altered proteins S5 and S12. Figure 9 summarizes the amino acid replacements revealed so far in protein S5. The amino acid exchanges in various spectinomycin-resistant mutants are clusterered in positions 19–21 of this protein. Clustering of amino acid replacements has also been found in neamine-resistant mutants altered in protein S5 and in erythromycin-resistant mutants altered in protein L4.

Figure 10 illustrates the amino acid exchanges in protein S12. The following two points can be concluded from our protein-chemical studies: (1) All replacements are clustered in only two short regions of protein S12, namely in position 42 and within the region from positions 85–91. No amino acid exchanges in any of the many mutants analyzed so far have been detected outside of these two short regions. (2) The replacement of a single amino acid in protein S12 suffices to make the ribosome streptomycin-resistant. Since it has been shown by reconstitution tests (Ozaki et al. 1969) that protein S12 alone confers streptomycin resistance, it follows from the protein-chemical results that the replacement of only one out of the 8000 amino acids present in the *E. coli* ribosome induces the altered ribosomal reaction towards the antibiotic.

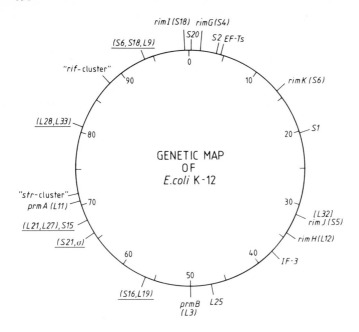

Fig. 8. Genetic map of the *E. coli* chromosome with the location of the genes for ribosomal proteins and for enzymes modifying ribosomal proteins. For details see Isono (1979)

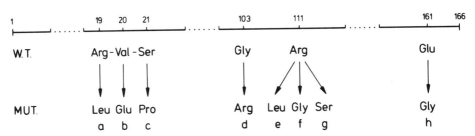

Fig. 9. Amino acid replacements in protein S5 of *E. coli* mutants. For details see Wittmann et al. (1979)

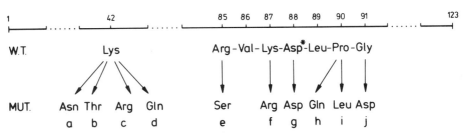

Fig. 10. Amino acid replacements in protein S12 of *E. coli* mutants. For details see Wittmann et al. (1979)

G. Evolution of Ribosomes

Extensive comparisons have been made between the structure of ribosomes from *E. coli* and those of other organisms. Two-dimensional gel electrophoresis showed very similar protein patterns among the *Enterobacterioaceae*, a bacterial family to which *E. coli* belongs. As expected, the differences increase as more distantly related species of bacteria are compared (Geisser et al. 1973a, b); finally, there are no detectable similarities between *E. coli* ribosomes and those of eukaryotes, e.g., yeast, mammals, or plants. The same general conclusion can be drawn from immunological studies using specific antibodies against individual *E. coli* ribosomal proteins. However, there are a few proteins, especially L7 and L12, that seem to be conserved during evolution. In addition to immunological cross-reaction between these two proteins from *E. coli* and their analogs from eukaryotes, e.g., yeast and rat, it was found that protein L7/L12 from *E. coli* can replace their eukaryotic equivalents in functional studies (reviewed by Stöffler 1974; and Wool and Stöffler 1974).

More quantitative information about the degree of homology can be obtained from comparative studies of the primary structure of ribosomal components isolated from different species. Using the data on the N-terminal regions of ribosomal proteins from various bacteria, relatively strong homology has been found between *E. coli* and various *Bacillus* species (Yaguchi et al. 1973; Isono et al. 1973; Higo and Loertscher 1974; Yaguchi et al. 1974). This finding is consistent with the possibility of reconstituting hybrid 30S ribosomal subunits from *E. coli* RNA and *Bacillus* proteins and vice versa (Nomura et al. 1968). It is also possible to reconstitute functionally active 30S ribosomes in which an individual *Bacillus* protein is replaced by its *E. coli* equivalent (Higo et al. 1973).

Recently, the complete primary sequences of a ribosomal protein from *Bacillus subtilis* (Itoh and Wittmann-Liebold 1978a), from yeast (Itoh and Wittmann-Liebold 1978b), and from the brine shrimp *Artemia salina* (Amons et al. 1979) have been determined. These are the first cases in which the complete amino acid sequences or ribosomal proteins from other species than *E. coli* have been elucidated. Comparison of

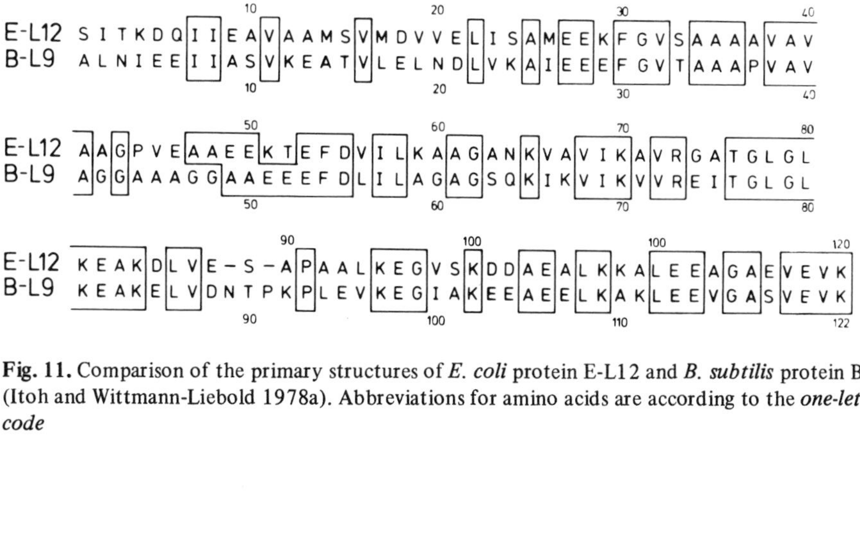

Fig. 11. Comparison of the primary structures of *E. coli* protein E-L12 and *B. subtilis* protein B-L9 (Itoh and Wittmann-Liebold 1978a). Abbreviations for amino acids are according to the *one-letter code*

the primary structures of these proteins with those of *E. coli* ribosomal proteins shows a strong homology between *E. coli* and *B. subtilis* (Fig. 11), but no significant homology between *E. coli* and yeast or *Artemia salina*. On the other hand, a significant degree of homology exists between ribosomal protein from yeast and rat liver (Wittmann-Liebold et al., manuscript submitted), as well as between L7/L12-type proteins from yeast and the brine shrimp *Artemia salina* (Amons et al. 1977, 1979).

From the immunological, protein-chemical, and reconstitution experiments it can be concluded that the fundamental organization of ribosomes from many bacteria is similar and has been conserved among these organisms. This is also true for the thermophilic *Bacillus stearothermophilus,* whereas the ribosomes from the halophilic *Halobacterium cutirubrum,* which are only stable in near-saturated salt conditions, differ from those of all other bacteria that have been studied in that they contain acidic instead of very basic proteins (Visentin et al. 1972). It is interesting that ribosomes from *Halobacterium* contain a protein, HL20, whose sequence is homologous not only to that of the *E. coli* proteins L7/L12 (Oda et al. 1974; Matheson et al. 1978; Visentin et al. 1979) but also to that of the *Artemia salina* protein eL12 (Amons et al. 1979; Matheson et al. 1978; Visentin et al. 1979), whereas there is little homology between the *E. coli* proteins L7/L12 and the *Artemia* protein eL12 (Amons et al. 1977, 1979). It therefore appears that the halophilic bacterium *Halobacter* has an interesting phylogenetic position with properties of both pro- and eukaryotes (Matheson et al. 1978; Visentin et al. 1979).

Similar comparative studies, as summarized for the ribosomal proteins, have also been carried out with ribosomal 5S RNA's. They have been reviewed in detail (Erdmann 1976). Recently, it was concluded from reconstitution experiments that 5S RNA from *E. coli* is homologous to 5.8S RNA (and not to 5S RNA) from eukaryotic ribosomes (Wrede and Erdmann 1977).

From the structural studies on both the ribosomal proteins and RNA's of organisms belonging to different classes it is clear that the primary structures of ribosomal components have changed during evolution to such an extent that there is only little homology between ribosomes from pro- and eukaryotes. On the other hand, no essential difference has been found in the way in which pro- and eukaryotic ribosomes function in protein biosynthesis. This apparent discrepancy raises the question as to how the components in such a complex structure as the ribosome, in which numerous proteins and several RNA's interact with each other, can change so drastically without impairing the vital function. To answer this question, more information is necessary, especially on ribosomes other than those of *E. coli*. No doubt, ribosomes are an ideal system to study this and many other problems of great scientific interest.

References

Allen G, Capasso R, Gualerzi C (1979) Identification of the amino acid residues of proteins S5 and S8 adjacent to each other in the 30S ribosomal subunit of *Escherichia coli*. J Biol Chem 254: 9800–9806

Amons R, van Agthoven A, Pluijms W, Möller W, Higo K, Itoh T, Osawa S (1977) A comparison of the amino-terminal sequence of the L7/L12-type proteins of *Artemia salina* and *Saccharomyces cerevisiae*. FEBS Lett 81: 308–310

Amons R, Pluijms W, Möller W (1979) The primary structure of ribosomal protein eL12/eL12-P from *Artemia salina* 30S ribosomes. FEBS Lett 104: 85–89

Appel B, Erdmann VA, Stulz J, Ackermann T (1978) Determination of A-U and G-C base pairing in prokaryotic and eukaryotic 5S RNAs by infrared spectroscopy. 12. FEBS Meet, Dresden, Abstract 1627

Boublik M, Hellmann W, Roth HE (1976) Localization of ribosomal proteins L7 and L12 in the 50S subunit of *Escherichia coli* ribosomes by electron microscopy. J Mol Biol 107: 479–490

Boublik M, Hellmann W, Kleinschmidt AK (1977) Size and structure of *E. coli* ribosomes by electron microscopy. Cytobiology 14: 293–300

Brimacombe R (1978) The structure of the bacterial ribosome. Symp Soc Gen Microbiol 28: 1–26

Brimacombe R, Nierhaus KH, Garrett RA, Wittmann HG (1976) The ribosome of *Escherichia coli*. Prog Nucl Acid Res Mol Biol 18: 1–44

Brimacombe R, Stöffler G, Wittmann HG (1978) Ribosome structure. Annu Rev Biochem 47: 271–303

Cantor CR, Huang KH, Fairclough F (1974) Fluorescence spectroscopy: approaches to the study of three-dimensional structure of ribosomes. In: Nomura M, Tissières A, Lengyel P (eds) Ribosomes, pp 587–599. Cold Spring Harbor Press, Long Island, N.Y.

Chen-Schmeisser U, Garrett RA (1976) Distribution of protein assembly sites along the 23S ribosomal RNA of *E. coli*. Eur J Biochem 69: 401–410

Dabbs ER (1978) Mutational alterations in 50 proteins of the *Escherichia coli* ribosome. Mol Gen Genet 165: 73–78

Dabbs ER, Wittmann HG (1976) A strain of *E. coli* which gives rise to mutations in a large number of ribosomal proteins. Mol Gen Genet 149: 303–309

Dieckhoff J, Bald R, Tischendorf GW, Stöffler G (manuscript in preparation)

Dijk J, Littlechild J (1978) Purification of ribosomal proteins from *Escherichia coli* under nondenaturing conditions. Methods Enzymol 59: 481–502

Dohme F, Nierhaus KH (1976a) Total reconstitution and assembly of 50S subunits from *E. coli* ribosomes in vitro. J Mol Biol 107: 585–599

Dohme F, Nierhaus KH (1976b) 5S RNA in assembly and function of the 50S subunit from *E. coli*. Proc Natl Acad Sci USA 73: 2221–2225

Dzionara M, Robinson SML, Wittmann-Liebold B (1977a) Secondary structures of proteins from the 30S subunits of the *Escherichia coli* ribosome. Hoppe-Seyler's Z Physiol Chem 358: 1003–1019

Dzionara M, Robinson SML, Wittmann-Liebold B (1977b) Prediction of secondary structures of ten proteins from the 50S subunit of the *Escherichia coli* ribosome. J Supramol Struct 7: 191–204

Erdmann VA (1976) Structure and function of 5S and 5.8S RNA. Progr Nucl Acid Res Mol Biol 18: 45–90

Erdmann VA, Sprinzl M, Pongs O (1973) The involvement of 5S RNA in the binding of tRNA to ribosomes. Biochem Biophys Res Commun 54: 942–948

Geisser M, Tischendorf GW, Stöffler G, Wittmann HG (1973a) Immunological and electrophoretical comparison of ribosomal proteins from eight species belonging to *Enterobacteriaceae*. Mol Gen Genet 127: 111–128

Geisser M, Tischendorf GW, Stöffler G (1973b) Comparative immunological and electrophoretic studies on ribosomal proteins of *Bacillaceae*. Mol Gen Genet 127: 129–145

Higo KI, Loertscher J (1974) Amino-terminal sequence of some *E. coli* 30S ribosomal proteins and functionally corresponding *Bac. stearothermophilus* ribosomal proteins. J Bacteriol 118: 180–186

Higo KI, Held W, Kahan L, Nomura M (1973) Functional correspondence between 30S ribosomal proteins of *E. coli* and *Bac. stearothermophilus*. Proc Natl Acad Sci USA 70: 944–948

Holde KE van, Hill WE (1974) General properties of ribosomes. In: Nomura M, Tissières A, Lengyel P (eds) Ribosomes, pp 53–91. Cold Spring Harbor Press, Long Island, N.Y.

Homann HE, Nierhaus KH (1971) Protein compositions of biosynthetic precursors and artificial subparticles from ribosomal subunits in *E. coli* K12. Eur J Biochem 20: 249–257

Horne JR, Erdmann VA (1972) Isolation and characterization of 5S RNA-protein complexes from *Bac. stearothermophilus* and *E. coli* ribosomes. Mol Gen Genet 119: 337–344

Isono K (1978) Genes encoding ribosomal proteins S16 and L19 from a gene cluster at 56.4 min in *Escherichia coli*. Mol Gen Genet 165: 265–268

Isono K (1979) Genetics of ribosomal proteins and their modifying and processing enzymes in *E. coli*. In: Chambliss G et al. (eds) Ribosomes, pp 641–669. University Park Press, Baltimore

Isono K, Kitakawa M (1977) A new ribosomal protein locus in *Escherichia coli*: The gene for protein genes in *Escherichia coli* containing genes for proteins S6, S18 and L9. Natl Acad Sci USA 75: 6163–6167

Isono K, Isono S, Stöffler G, Visentin LP, Yaguchi M, Matheson AT (1973) Correlation between 30S ribosomal proteins of *Bacillus stearothermophilus* and *Escherichia coli*. Mol Gen Genet 127: 191–195

Isono K, Krauss J, Hirota Y (1976) Isolation and characterization of temperature-sensitive mutants of *E. coli* with altered ribosomal proteins. Mol Gen Genet 149: 297–302

Isono K, Cumberlidge AG, Isono S, Hirota Y (1977) Further temperature-sensitive mutants of *Escherichia coli* with altered ribosomal proteins. Mol Gen Genet 152: 239–243

Isono S, Isono K, Hirota Y (1978) Mutations affecting the structural genes and the genes coding for modifying enzymes for ribosomal proteins in *Escherichia coli*. Molec Gen Genet 165: 15–20

Itoh T, Wittmann-Liebold B (1978a) The primary structure of *Bacillus subtilis* acidic ribosomal protein B-L9 and its comparison with *Escherichia coli* proteins L7/L12. FEBS Lett 96: 392–394

Itoh T, Wittmann-Liebold B (1978b) The primary structure of protein 44 from the large subunit of yeast ribosomes. FEBS Lett 96: 399–402

Kaltschmidt E, Wittmann HG (1970) Number of proteins in small and large ribosomal subunits of *E. coli* as determined by two-dimensional gel electrophoresis. Proc Natl Acad Sci USA 67: 1276–1282

Kitakawa M, Isono K (1977) Localization of the structural gene for ribosomal protein L19 (rplS) in *Escherichia coli*. Mol Gen Genet 158: 149–155

Kurland CG (1977) Structure and function of the bacterial ribosome. Annu Rev Biochem 46: 173–200

Lake JA (1976) Ribosome structure determined by electron microscopy of *E. coli* small subunits, large subunits and monomeric ribosomes. J Mol Biol 105: 131–159

Lake JA, Kahan L (1975) Ribosomal proteins S5, S11, S13 and S19 localized by electron microscopy of antibody-labeled subunits. J Mol Biol 99: 631–644

Lake JA, Pendergast M, Kahan L, Nomura M (1974) Localization of *Escherichia coli* ribosomal proteins S4 and S14 by electron microscopy and antibody-labeled subunits. Proc Natl Acad Sci USA 71: 4688–4692

Lührmann R, Bald R, Tesche B, Tischendorf GW, Stöffler G (1979) The localization of functional domains on the *E. coli* ribosome as determined by immuno electron microscopy. Hoppe-Seyler's Z Physiol Chem 360: 320

Matheson AT, Yaguchi M, Nazar RN, Visentin LP, Willick GE (1978) The structure of ribosome from moderate and extreme halophilic bacteria. In: Caplan SR, Ginzburg M (eds) Energetics and structure of halophilic microorganisms, pp 481–501. Elsevier/North-Holland, Amsterdam

May R, Stöckel P, Strell I, Hoppe W, Lorenz S, Erdmann VA, Wittmann HG, Crespi HL, Katz JJ (1978) Determination of quaternary structures of multicomponent macromolecules by the label triangulation method: Ribosomes and ribosomal partial complexes. 12. FEBS Meet, Abstract 1643, Dresden

Möller K, Brimacombe R (1975) Specific cross-linking of proteins S7 and L4 to ribosomal RNA by UV-irradiation of *E. coli* ribosomal subunits. Mol Gen Genet 141: 343–355

Möller K, Zwieb C, Brimacombe R (1978) Identification of the oligonucleotide and oligopeptide involved in an RNA-protein cross-link induced by ultraviolet irradiation of *E. coli* 30S ribosomal subunits. J Mol Biol 126: 489–506

Moore PB, Engelman DM, Schoenborn BP (1974) Neutron scattering studies of the *E. coli* ribosome. In: Nomura M, Tissières A, Lengyel P (eds) Ribosomes, pp 601–613. Cold Spring Harbor Press, Long Island, N.Y.

Morrison CA, Bradbury EM, Littlechild J, Dijk J (1977) Proton magnetic studies to compare *Escherichia coli* ribosomal proteins prepared by two different methods. FEBS Lett 83: 348–352

Nierhaus KH (1980) Analysis of the assembly and function of the 50S subunit from *Escherichia coli* ribosome by reconstitution. In: Chambliss G et al. (eds) Ribosomes, pp 267–294. University Park Press, Baltimore

Nierhaus KH, Dohme F (1974) Total reconstitution of functionally active 50S ribosomal subunits from *E. coli*. Proc Natl Acad Sci USA 71: 4713–4717

Nomura M, Erdmann VA (1970) Reconstitution of 50S ribosomal subunits from dissociated molecular components. Nature 288: 744–748

Nomura M, Held WA (1974) Reconstitution of ribosomes. In: Nomura M, Tissieres A, Lengyel P (eds) Ribosomes, pp 193–223. Cold Spring Harbor Press, Long Island, N.Y.

Nomura M, Traub P, Bachman H (1968) Hybrid 30S ribosomal particles reconstituted from components of different bacterial origins. Nature 219: 793–799

Nomura M, Morgan EA, Jaskunas SR (1977) Genetics of bacterial ribosomes. Annu Rev Genet 11: 297–347

Nowotny V, Nierhaus KH (manuscript in preparation)

Oda G, Strøm AR, Visentin LP, Yaguchi M (1974) An acidic, alanine-rich 50S ribosomal protein from *Halobacterium cutirubrum:* Amino acid sequence homology with *Escherichia coli* proteins L7 and L12. FEBS Lett 43: 127–130

Ozaki M, Mizushima S, Nomura M (1969) Identification and functional characterization of the protein controlled by the streptomycin-resistant locus in *E. coli*. Nature 222: 333–339

Pettersson I, Hardy SJS, Liljas A (1976) The ribosomal protein L8 is a complex of L7/L12 and L12. FEBS Lett 64: 135–138

Pichon J, Marvaldi J, Marchis-Mouren G (1975) The in vivo order of protein addition in the course of *E. coli* 30S and 50S subunit biogenesis. J Mol Biol 96: 125–137

Politz SM, Glitz DG (1977) Ribosome structure: Localization of N,N-dimethyladenosine by electron microscopy of a ribosome-antibody complex. Proc Natl Acad Sci USA 74: 1468–1472

Rinke J, Zwieb C, Meinke M, Ulmer E, Maly P (1979) RNA-Protein Quervernetzung innerhalb des *E. coli* Ribosoms. Hoppe-Seyler's Z Physiol Chem 360: 353

Roth HE, Nierhaus KH (1979) Assembly map of the *Escherichia coli* 50S ribosomal subunit covering the proteins present on the first reconstitution intermediate. Manuscript submitted

Schlessinger D (1974) Ribosome formation in *E. coli*. In: Nomura M, Tissières A, Lengyel P (eds) Ribosomes, pp 393–416. Cold Spring Harbor Press, Long Island, N.Y.

Sieber G, Nierhaus KH (1978) Kinetic and thermodynamic parameters of the assembly in vitro of the large subunit from *E. coli* ribosomes. Biochemistry 17: 3505–3511

Spillmann S, Nierhaus KH (1978) The ribosomal protein L24 of *E. coli* is an assembly protein. J Biol Chem 253: 7047–7050

Spillmann S, Dohme F, Nierhaus KH (1977) Assembly of the 50S subunit from *E. coli* ribosomes. Proteins essential for the first heat-dependent conformational change. J Mol Biol 115: 513–523

Stöffler G (1974) Structure and function of the *Escherichia coli* ribosome: immunochemical analysis. In: Nomura M, Tissières A, Lengyel P (eds) Ribosomes, pp 615–667. Cold Spring Harbor Press, Long Island, N.Y.

Stöffler G, Wittmann HG (1977) Primary structure and three-dimensional arrangement of proteins within the *E. coli* ribosome. In: Weissbach H, Pestka S (eds) Molecular mechanisms of protein biosynthesis, pp 117–202. Academic Press, New York

Sypherd P, Osawa S (1974) Ribosome genetics revealed by hybrid bacteria. In: Nomura M, Tissières A, Lengyel P (eds) Ribosomes, pp 669–678. Cold Spring Harbor Press, Long Island, N.Y.

Tesche B (1975) Ein Elektronenstoß-Verdampfer mit geringem Leistungsbedarf für die Herstellung dünner Schichten aus hochschmelzenden Materialien. Vak Tech 24: 104–110

Tesche B (1978) Ein präparatschonender Elektronenstoß-Verdampfer für die Elektronenmikroskopie. Mikroskopie (Wien) 34: 29–34

Tesche B, Tischendorf GW, Stöffler G (1978) Morphology of *E. coli* ribosomes as obtained by high resolution shadow cast. In: Sturgess JM (ed) Electron microscopy, vol II, pp 240–241. Internat Congr Electron Microscopy, Toronto

Tischendorf GW, Zeichhardt H, Stöffler G (1974) Location of proteins S5, S13 and S14 on the surface of the 30S ribosomal subunit from *E. coli* as determined by immune electron microscopy. Mol Gen Genet 134: 209–223

Tischendorf GW, Zeichhardt H, Stöffler G (1975) Architecture of the *E. coli* ribosome as determined by immune electron microscopy. Proc Natl Acad Sci USA 72: 4820–4824

Traub P, Nomura M (1968) Structure and function of *E. coli* ribosomes. V. Reconstitution of functionally active 30S ribosomal particles from RNA and proteins. Proc Natl Acad Sci USA 777–784

Traut RR, Heimark RL, Sun TT, Hershey JWB, Bollen A (1974) Protein topography of ribosomal subunit from *Escherichia coli*. In: Nomura A, Tissieres A, Lengyel P (eds) Ribosomes, pp 271–308. Cold Spring Harbor Press, Long Island, N.Y.

Ulmer E, Meinke M, Ross A, Fink G, Brimacombe R (1978) Chemical cross-linking of protein to RNA within the intact ribosomal subunits from *E. coli*. Mol Gen Genet 160: 183–193

Vasiliev VD (1974) Morphology of the ribosomal 30S subparticle according to electron microscopic data. Acta Biol Med Germ 33: 779–793

Visentin LP, Chow C, Matheson AT, Yaguchi M, Rollin F (1972) *Halobacterium cutirubrum* ribosomes. Properties of the ribosomal proteins and ribonucleic acid. Biochem J 130: 103–110

Visentin LP, Yaguchi M, Matheson AT (1979) Structural homologies in alanine-rich acidic ribosomal proteins from pro- and eucaryotes. Can J Biochem 57: 719–726

Wagner R, Garrett RA (1978) A new RNA-RNA cross-linking reagent and its application to ribosomal 5S RNA. Nucleic Acids Res 5: 4065–4076

Wittmann HG (1974) Purification and identification of *E. coli* ribosomal proteins. In: Nomura M, Tissieres A, Lengyel P (eds) Ribosomes, pp 93–114. Cold Spring Harbor Press, Long Island, N.Y.

Wittmann HG, Wittmann-Liebold B (1974) Chemical structure of bacterial ribosomal proteins. In: Nomura M, Tissieres A, Lengyel P (eds) Ribosomes, pp 115–140. Cold Spring Harbor Press, Long Island, N.Y.

Wittmann HG, Nierhaus KH, Sieber G, Stöffler G, Tesche B, Tischendorf GW, Wittmann-Liebold B (1978) Structure and assembly of the *Escherichia coli* ribosome. In: Sturgess JM (ed) Electron microscopy, vol III, pp 459–469. Internat Congr Electron Microscopy, Toronto

Wittmann HG, Littlechild J, Wittmann-Liebold B (1979) Structure of ribosomal proteins. In: Chambliss G et al. (eds) Ribosomes, pp 51–88. University Park Press, Baltimore

Wittmann-Liebold B, Robinson SML, Dzionara M (1977a) Prediction of secondary structure in proteins from the *Escherichia coli* 30S ribosomal subunit. FEBS Lett 77: 301–307

Wittmann-Liebold B, Robinson SML, Dzionara M (1977b) Predictions for secondary structures of six proteins from the 50S subunit of *Escherichia coli* ribosome. FEBS Lett 81: 204–213

Wittmann-Liebold B, Geissler W, Collatz E, Lin A, Tsurugi K, Wool IG (1980) The amino-terminal sequence of rat liver ribosomal proteins S4, S6, S8, L6, L7', L18, L27, L30, L37, L37' and L39. J Supramol Struct 12: 425–433

Wool I, Stöffler G (1974) Structure and function of eukaryotic ribosomes. In: Nomura M, Tissieres A, Lengyel P (eds) Ribosomes, pp 417–460. Cold Spring Harbor Press, Long Island, N.Y.

Wrede P, Erdmann VA (1973) Activities of *Bac. stearothermophilus* 50S ribosomes reconstituted with prokaryotic and eukaryotic 5S RNA. FEBS Lett 33: 315–319

Wrede P, Erdmann VA (1977) *Escherichia coli* 5S RNA binding proteins L18 and L25 interact with 5.8S RNA but not with 5S RNA from yeast ribosomes. Proc Natl Acad Sci USA 74: 2706–2709

Wrede P, Pongs O, Erdmann VA (1978) Binding oligonucleotides to *Escherichia coli* and *Bacillus stearothermophilus* 5S RNA. J Mol Biol 120: 83–96

Yaguchi M, Roy C, Matheson AT, Visentin LP (1973) The amino acid sequence of the N-terminal region of some 30S ribosomal proteins. Can J Biochem 51: 1215–1217

Yaguchi M, Matheson AT, Visentin LP (1974) Procaryotic ribosomal proteins: N-terminal sequence homologies and structural correspondence of 30S ribosomal proteins from *E. coli* and *Bac. stearothermophilus*. FEBS Lett 46: 296–300

Zimmermann J, Erdmann VA (1978) Identification of *Escherichia coli* and *Bacillus stearothermophilus* ribosomal protein binding sites on *Escherichia coli* 5S RNA. Mol Gen Genet 160: 247–257

Zimmermann RA (1974) RNA-protein interactions in the ribosome. In: Nomura M, Tissieres A, Lengyel P (eds) Ribosomes, pp 225–269. Cold Spring Harbor Press, Long Island, N.Y.

Zwieb C, Brimacombe R (1978) RNA-protein cross-linking in *E. coli* 30S ribosomal subunits: a method for the direct analysis of the RNA regions involved in the cross-links. Nucleic Acids Res 5: 1189–1206

Zwieb C, Brimacombe R (1979) RNA-protein cross-linking in *E. coli* 30S ribosomal subunits. Nucleic Acids Res 6: 1775–1790

Zwieb C, Ross A, Rinke J, Meinke M, Brimacombe R (1978) Evidence for RNA-RNA cross-link formation in *E. coli* ribosomes. Nucleic Acids Res 5: 2705–2720

E. Philosophical Reflexions

1. Molecular Biology, Culture, and Society

R. MONRO

A. Introduction

In 1962 Lipmann wrote a short paper entitled *Disproportions Created by the Exponential Growth of Knowledge*. The core of his thesis was as follows:

"... changes have increased with unusual rapidity in degree and impact through our success in the natural sciences. ... When we plot progress against time, we get a soaring curve, the upper one in our figure (Fig. 1). ... this describes only part of the overall picture of our man-dominated earth. The missing part is ... the social relationship between men ... On the average, ... this change has been, at best, not exponential but rather proportional with time, and slow at that, as indicated by the lower curve of the figure. *These two curves,* the one running away from the other, I believe dramatically *present the problem of our time.* ... They reflect the schizophrenic condition of man as a rational animal overtaking ... man as a social animal. The only remedy is for the lower curve to adjust; ... can we remedy the situation before irreparable damage has been done?"[1] (Emphasis added)

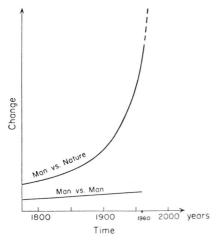

Fig. 1. (Reproduced from Lipmann, 1962)

Department of Biophysics, University of London, King's College, 26–29 Drury Lane, London WC2B 5RL, Great Britain

1 Quoted from Lipmann (1962)

When Lipmann wrote this, eighteen years ago, he was particularly concerned about nuclear armaments. Since then, some relatively new "disproportions" have emerged. Amongst these, computer engineering and bio-engineering are particularly striking, adding new dimensions to our capacity to change nature (including ourselves). "The problem of our times" becomes all the more poignant since Lipmann, himself, has in a sense contributed to the second of these developments, through the very success of his work.

Our problem is far too complex to be resolved by any single, cut-and-dried approach. What I try to do here is to draw attention to certain aspects which are relatively little appreciated.

B. Interrelatedness of Science and Society

The central thesis I wish to explore is that the curves in Fig. 1 are interrelated: that science influences the surrounding culture *and* is itself influenced by that culture; that science and society evolve together in closely interrelated ways; and that we consequently need to consider the nature of science, as well as of society, in approaching "our problem".

Most scientists accept that the growth of scientific knowledge (and related technologies) affects society. They also accept that the *rate* of growth of science is influenced by social factors. But they balk at the idea that the actual *character* of scientific knowledge is influenced by culture. This idea challenges the deeply-rooted feeling that science is objectively and unambiguously describing reality.

C. Uncertainty of Scientfic Knowledge

However, it is now almost universally accepted by philosophers of science that *all* knowledge is uncertain. Karl Popper has been particularly influential in promoting this view (Popper 1968, 1969). Popper was deeply affected in his youth, when Newtonian physics, which had been the unquestioned foundation of science for nearly three centuries, was "falsified" by Einstein and replaced by relativity. Popper came to the conclusion that all scientific knowledge involves generalisations which may include false presuppositions. At the same time, he accepted that science does, over the course of its

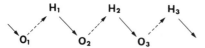

Fig. 2. Observation, O_1, together with other considerations (including imaginative leaps) leads (*broken arrow*) to the formulation of an hypotheses, H_1. H_1 entails (*solid arrow*) the occurrence of observations, O_2. Empirical investigation corrects, refines and extends O_2, leading to formulation of an improved hypotheses, H_2. The cycle is repeated again and again, resulting in progressive improvements in the 'fit' between theories and empirical data. The actual process of testing and reformulation of hypotheses is more complex than the diagram suggests

development, increase in comprehensiveness and explanatory power. He believed that science approaches reality, but recognised that, owing to the overpresent possibility of revolutionary changes in its basic presuppositions (Fig. 2), its progress may be uneven, and its closeness to reality unascertainable. He emphasised the importance of falsifiability (or, more generally, testability) as a criterion of good scientific theories.

D. Paradigms

Another highly influential philosopher of science, Thomas Kuhn (1970), considers scientific progress to be more ambigous. His model for the evolution of science is shown in simplified, schematic form in Fig. 3, and consists in the following main points.

Initially there is no clearly-defined field. Instead, there are many, competing, ill-defined views, and there is no community of practitioners or concerted approach. Fundamental issues are continually mixed and confused with technical and superficial ones. Discoveries and theories are repeatedly made and forgotten. The process is non-cumulative. This is what Kuhn terms *pre-paradigm science.*

In certain conditions a dominant, unifying *paradigm* emerges from the morass of pre-paradigm science. The term, paradigm, can be taken to denote the "matrix of ideas" underlying a discipline[2]. Major paradigms correspond roughly to world views, and underly large episodes of science and of culture generally, while minor paradigms underly particular disciplines or branches thereof.

Once a paradigm has emerged, a community of practitioners can form, held together by its common acceptance. Members of such a community tacitly agree over basic issues, concerning both the nature of the universe and what the ultimate laws and explanations must be like. They are thus freed from time-consuming metaphysical speculation. They can get down to "puzzle-solving" within a relatively fixed framework, and are motivated to tackle the laborious "nitty-gritty" problems necessary for most successful exploratory work. A new field is thus opened up. If the paradigm is appropriate, discoveries follow in rapid succession, and the paradigm is systimatically applied and *articulated* in a variety of areas. This is what Kuhn terms *normal science.* In contrast to pre-paradigm science, normal science provides favourable conditions for concerted programmes of research and the cumulative growth of knowledge.

As the research in a field of normal science proceeds, anomalies begin to appear. Most of these are shown to be due to methodological errors, or eliminable through minor refinements of the paradigm. However, a hard core of unresolved anomalies often

---

2 Kuhn uses the term, paradigm, in several different ways in his *Structure of Scientific Revolutions*, but later, in response to criticism, he classified these into two broad types: (i) denoting the "matrix of ideas" underlying a discipline, and (ii) denoting specific "exemplars" or "shared rules" which members of a discipline agree upon (Kuhn 1974). The first of these is very broad, and includes the second as one of its subsets, along with "symbolic generalizations" and "models". Paradigm is used in both these senses here, even though Kuhn, himself, now prefers term, disciplinary matrix to denote the former

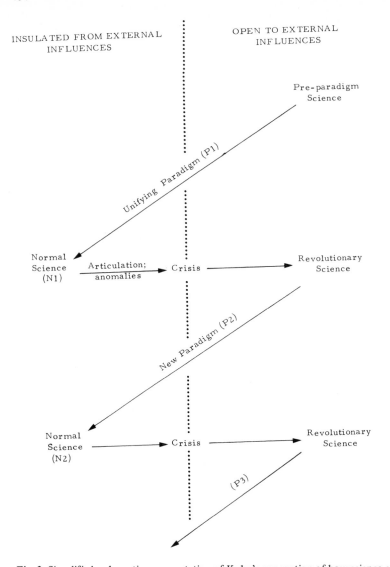

Fig. 3. Simplified, schematic representation of Kuhn's conception of how science evolves

persists and grows, eventually leading to a state of *crisis*. The paradigm is felt to be failing in some peculiar way, and a few of the more far-sighted scientists begin to look at its basic presuppositions and to explore other paradigms. Out of this process a new paradigm emerges, resolving (some of) the anomalies. This is what Kuhn terms a *revolution*. Major revolutions are not just superficial affairs: they involve changes of paradigm which are so radical that the very nature of the universe and the problems connected with it seem to change (e.g., the Copernican revolution). The paradigms before and after such a revolution are *incommensurable* with one another.

Once a new paradigm has been established, a fresh period of normal science ensues,, and the process of paradigm-articulation, accumulation of anomalies, crisis, and revolution recurs. Kuhn considers this whole repeating cycle to be the characteristic form of modern (*mature*) science. Disciplines centre around the major paradigms, and change their form (or new ones are born) when revolutionary paradigm-changes occur. In contrast, pre-paradigm science has a constant admixture of many vague, partially overlapping paradigms, and has no real revolutions and no well-defined disciplines.

Perhaps the most significant feature of Kuhn's philosophy is his emphasis on the prevalence of normal science at most times, and on associated tendencies towards dogmatism. The transition from pre-paradigm science to normal science involves a narrowing down of interests. A paradigm necessarily represents only a partial view: it is just *because* it does so that it galvanises scientists into successful action. Francis Bacon wrote, "Truth emerges more readily from error than from confusion". Similar views have been expressed by Medawar (1967), who describes science as "The Art of the soluble" (the "soluble being what can be assimilated into the prevailing structure of science"), and Jacob (1974), who writes:

". . . There are perhaps other possible coherences in descriptions. But *science is enclosed in its own explanatory system,* and cannot escape from it. . . Today the world is messages, codes, and information. Tomorrow what analysis will break down our objects and reconstitute them in a new space?"[3] (Emphasis added)

Jacob's use of the term space can be equated with what Kuhn would consider to be a field of normal science opened up by a given paradigm.

A dominating paradigm not only fosters ordered research programmes, but also, through its very rigidity, provides a sensitive indicator for anomalies which do not fit with it. Just as in a room that is kept very tidy anything out of order is soon noticed, so in normal science anything which does not fit is readily detectable. Thus, while normal science is inherently conservative, it also has, built into it, the seeds of self-criticism.

Nevertheless, the transition from normal to revolutionary science is often slow and painful. One a paradigm has become established, a variety of factors — metaphysical, psychological, sociological — tend to perpetuate it. A price must be paid for the successful blossoming of research programmes in normal science. The research worker often becomes so entrenched in the prevailing paradigm that he forgets the continuing persistence of deeper, unresolved problems. The typical "normal scientist" considers philosophical reflection on such problems to be a waste of time, irrelevant to the "genuine scientist" who wants to get down to the "hard facts" and the "real problems". What the hard facts and the real problems are is taken as self-evident.

When anomalies do begin to accumulate in a given discipline, it is usually only a few of the members who become seriously concerned. The majority stick dogmatically to their paradigm, unable — or unwilling — to conceive the possibility of alternatives to it. Thus, revolutions are often pioneered not by the established members of a discipline but by individuals who are new to it (either through youth or through a change of career). Even after a revolution it is not uncommon for many members of a scientific community to persist doggedly with the outmoded paradigm for the rest of their

---

3 Quoted from Jacob (1974), p. 324 (Copyright 1974 by Allen Lane, London)

lives. It thus may take a generation or more for a new paradigm to become thoroughly established.

It is not surprising that a community practicing normal science often overlooks results and ideas which are potentially important to it but do not fit readily with its paradigm. The history of science is littered with such cases, both in the form of "premature discovery" and as blindness to knowledge in other fields.

D. Lipmann' Paradigm

Lipmann's work illustrates several points in Popper's and Kuhn's philosophies. Most scientists who have worked in Lipmann's laboratory will testify to the fascination of his approach. My own two years there have influenced the rest of my life's work (despite a radical change of direction). Lipmann likewise testifies to the revolutionary influence of his own two teachers, Warburg and Meyerhof. The tradition he represents has included some of the greatest names in organic chemistry and biochemistry (Fig. 4). What is it that characterises this tradition? In Lipmann's own words:

"Meyerhof's was a spiritual presence in his unit. At first I rather shunned the sometimes unwieldy muscle energetics he was so fond of. In later life, however, this developed into the profoundest influence. And then Meyerhof was really a pupil of Warburg. He had begun with a strong tendency for philosophical generalisation, but a short period of collaboration with Warburg was a turning point for him. It seemed to have convinced him that *the best hope of coming near to explaining the life phenomenon was in physico-chemical exploration.* Warburg's puritanical, pragmatic approach which rejected compromises and speculation filtered through to us from the fourth floor. He became our hero; his stern insistence on *letting experiments speak and keeping interpretation to a minimum* has dominated our generation. It has borne fruit for it developed a foundation that permitted biology to grow into an exact discipline and now allows us to move slowly into a more theoretical biology."[4] (Emphasis added)

The tradition thus combines daring goals with high critical standards and an emphasis on the physico-chemical approach.

More particularly, Lipmann's approach has been to investigate the functions of tissues and cells by taking them apart *and* reconstituting particular activities by recombining the purified components. The emphasis of this approach contrasts with those of both the Hopkins' school of biochemistry, which tended to concentrate on the structure and function of whole organisms and of highly purified components therefrom[5],

---

4 Quoted from Lipmann (1971), p. 7 Copyright 1971 by John Wiley & Sons, Inc., N.Y.)

5 This is an oversimplification, but I believe it is essentially correct. The investigation of chemical activities of cell-free extracts grew out of the German organic chemistry tradition (Fig. 4), Buchner being the first (after decades of attempts) to obtain an active cell-free extract for alcoholic fermentation (in 1897). Warburg pioneered the application of quantitative methods to such systems. Harden and Young, in England, also made important contributions, but were not in the Hopkins "School". For the history of studies on alcoholic fermentation, up to 1930, see Harden (1932); for histories of the Hopkins' and molecular genetics' "schools", see Needham and Baldwin (1949), and Cairns et al. (1966), respectively

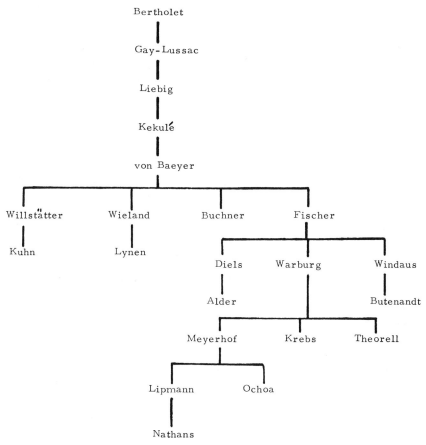

Fig. 4. "Family tree" of organic chemists and biochemists deriving from the German 'school' (adapted from Krebs 1967). The names from von Baeyer onwards have been chosen on the basis of being Nobel Laureates, thus giving a *relatively* objective (but also very incomplete) selection. A fuller family tree, dating back to Liebig, contains some sixty outstanding scientists (von Dechend 1965)

and the molecular genetics school, which used the "black box" approach of making mutations and seeing what changes followed in whole organisms and their purified components. All three of these approaches have much the same goals, even though they sprang from different sources; and it is their fusion in recent years that has provided the great power and success of modern molecular and cell biology.

In the above quotation, Lipmann speaks of Warburg's "insistence on letting experiments speak and keeping interpretations to a minimum". Newton, in a similar vein, insisted that *"hypotheses non fingo"* (I do not touch hypotheses). However, it is easy to show that the experimental investigations of both these scientists were guided by theories. What their insistence on letting experiments speak really amounted to was being willing to discard a theory, and explore others, if observations did not adequately tally with it.

Even the most empirical research is guided by theories — and often a theoretical approach is more effective than an empirical one in "picking up" key observations. For instance, prior to 1961 clusters of ribosomes had been recorded in electron micrographs, and an especially active, rapidly-sedimenting fraction of ribosomes was observed in the ultracentrifuge in early 1961, but it was only after the messenger RNA theory emerged, and shortly thereafter the concept of *polysomes*, that serious attention was paid to them.[6]

Indeed, one of the most attractive features of Lipmann's approach is his imaginative use of theories. I shall try to show how his concept of "energy rich bonds" has the characteristics of a Kuhnian paradigm.

Lipmann introduced the concept of "energy rich bonds" in a major review in 1941 (Lipmann 1941). Having been stimulated by Lohman's work on different types of phosphate linkages[7], he developed the idea that many processes in organisms, from biosynthesis to absorption and muscular contraction, are driven by reactions in which a chemical group is transferred from a position of high to low "potential". He envisioned phosphate groups as playing a central role in this process, the energy rich terminal phosphate of ATP being generated from low potential inorganic phosphate (and ADP) by catabolism, and then used for many different purposes (Fig. 5). The idea of coupling between energy-yielding and energy-requiring reactions (Fig. 6) played a central role in his scheme, energy rich groups other than phosphate also driving biosynthetic reactions.

Lipmann's paradigm was criticised (and largely ignored) by chemists (Lipmann 1971, p. 37) as being thermodynamically ill-defined. Lipmann had, amongst other things, inadvertantly used a terminology which clashed with the standard one and, even when this was clarified, thermodynamic problems remained; so that acceptance by chemists took many years and even then was only partial. Nevertheless, this paradigm was revolutionary for biochemistry, helping to integrate diverse observations, and

---

6 In 1963 four different groups all reported evidence for polysomes, interpreting it firmly in terms of the "messenger" hypothesis: W. Gilbert, J. Mol. Biol. **6**, 374 (1963); A. Gierer, ibid,, p. 148; J.R. Warner, P. Knopf, A Rich, Proc. Natl. Acad. Sci USA **49**, 22 (1963); F.O. Wettstein, T. Staehelin, H. Noll, Nature **197**, 430 (1963). G. von Ehrenstein had already observed a particularly active, rapidly-sedimenting fraction of ribosomes from reticulocytes in 1961 in Lipmann's Laboratory (personal recollection), but none of us was quick enough to interpret this in terms of the messenger hypothesis, which was only just emerging at that time and was still widely doubted. Clustering of ribosomes was thought at that time to be artefactual, or related in some way to membranes. Ribosome clusters had been observed by Pallade and other workers in the electron microscope for several years

7 Lipmann's ideas on energy rich bonds appear to have matured in Copenhagen in the late 1930's, where at the time there was a remarkable concentration of talent (including Linderstrøm-Lang, Kalckar, Hotchkiss, Zamecnik, Fruton and Delbrück). Discussions between Lipmann and Kalckar (to which Hotchkiss listened) were particularly significant. From Meyerhof's work they knew that the energy for muscular contraction derived from the conversion of glucose to lactic acid, a definite amount of energy being liberated in this conversion. But, *how* was this energy utilised? Hotchkiss believes that emergence of the idea of energy rich bonds was closely connected to the feeling that the rather "homogeneous" glucose molecule would have to be converted into intermediates having uneven distributions of "energy" (giving rise to unstable, or energy rich bonds) in order for the energy of degradation to be utilised. (Hotchkiss, personal communication)

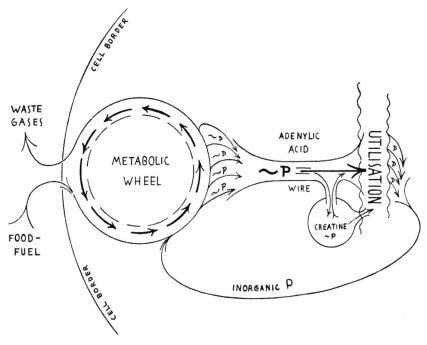

Fig. 5. Imaginative scheme from Lipmann's 1941 review. In his own words: "The metabolic dynamo generates ∼P-current. This is brushed off by adenylic acid, which likewise functions as a wiring system, distributing the current. Creatine ∼P, when present, serves as P-accumulator" (Lipmann 1941)

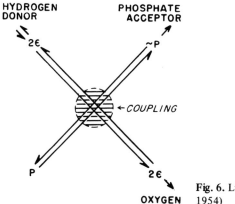

Fig. 6. Lipmann's coupling scheme (Hoch and Lipmann 1954)

stimulating very fruitful lines of research. It led Lipmann to the discovery of acetyl CoA, carbamyl phosphate, "active" sulphate, and, via Zamecnik's group and Berg, to the mechanism of amino acid activation in protein biosynthesis (Lipmann 1971). And in biochemistry, generally, it promoted studies on photophosphorylation, active transport, muscle action, and biosynthesis. Even today it is still finding new applications, as for instance in the idea that extra "energy" is needed to ensure accuracy in protein biosynthesis and nucleic acid biosynthesis, and to promote the assembly of microtubules (Kurland 1978; Kaziro, this volume; see also CIBA 1975).

   This example illustrates how a field of science may be opened up and guided by an idea which is incommensurable with the theoretical frameworks in bordering fields. Also, as a corrollary, it illustrates how the theoretical frameworks of well-established fields may, through their limitations, impede the emergence of a new field.

E. Cultural Influences on Science

If some of the paradigms, or guiding principles, underlying science are relatively ill-defined, and unintegrated with other parts of science, then the possibility arises that at any given time there may be various alternative routes open for the development of science; and, furthermore, that the actual routes taken may be governed partly by cultural factors.

   Such effects are difficult to perceive in the short-term growth of knowledge within particular fields, but become more apparent from wider perspectives. For instance, the growth of organic chemistry, and more recently of biochemistry and molecular biology, has been greatly promoted by the link-up with industry. One consequence has been

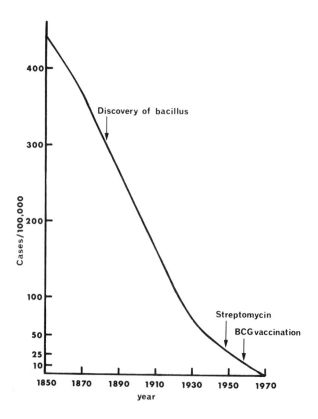

Fig. 7. Respiratory tuberculosis in the U.K. (from Robinson et al. 1974). For the earlier years there are doubts about the accuracy of diagnosis and recording. However, it is unlikely that the general picture of the rate of decline, and its relationship to the introduction of treatments and preventative measures, is seriously distorted. For further discussion and criticism see Richards (1980) and Clarke (1980)

an emphasis in medicine on the use of drugs, and a relative neglect of alternative approaches.

That there are other possible approaches to health and disease, is illustrated by Fig. 7, which shows that drugs played a relatively small — albeit very significant — part in the eradication of tuberculosis: improvements in sanitation, nutrition, and general living standards were of greater overall significance. Nor can such alternative approaches be replaced by drugs, as is apparent from the continued persistence, today, of tuberculosis among poverty-stricken populations in various parts of the world. Many other diseases also have major environmental components, which cannot adequately be dealt with by the drug approach alone. Overemphasis on the drug approach in medicine has quite frequently led to neglect of socio-economic dimensions of disease, and also (of no less importance) of psychosomatic aspects. Such neglect must be held to be at least partly responsible for the persistence of "premature" mortality among middle aged males in most industrialised countries.

The link-up between chemistry, biology, and industry has thus led scientific medicine to emphasise one approach and neglect others. It has concomitantly stimulated certain lines of development in chemistry and biology. Other lines of development, with different social implications, are also possible, and in recent times have fortunately been gaining increasing recognition and support. Chemistry and biology can back these up just as effectively as they backed up the drug approach, provided certain false presuppositions, and commercial and other vested interests, are recognised. For instance, studies on intermediary metabolism and on immunology have contributed much to improvements in nutrition (and continue to do so); chemists and biologists are helping to track down and eliminate carcinogens in our environments and food; and, in quite a different field, physiological studies on acupuncture are substantiating the effectiveness of this ancient form of therapy[8], which represents a radical alternative to certain uses of drugs but has been largely shunned by "scientific" medicine (Fig. 8).

F. Frames of Mind

The way we see things is strongly influenced by our "mental sets". One mental set brings out certain features, while another brings out others: it is difficult, perhaps impossible, to see all the features simultaneously (Fig. 9).

Our intellectual "perceptions" also display fixations, being especially influenced by our cultural background (Fig. 10). It is very difficult for any one person to operate in more than a few different mental modes; and the more established one's thinking becomes in a particular mode, the more difficult it becomes to see things in other ways.

This is the psychological basis for conservatism in science. It partly explains why the members of a well-established discipline often fail to recognise the importance of

---

8 For a review of recent physiological studies on acupuncture, see Gwei-Djen and Needham (1979); see also Scientific American (1979)

THORAX OPERATIONS

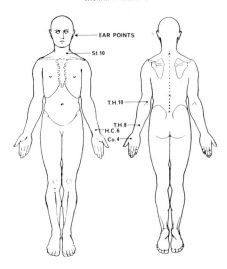

OPERATIONS ON UPPER ABDOMEN.

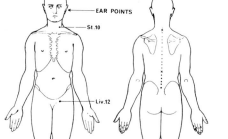

OPERATIONS ON LOWER ABDOMEN

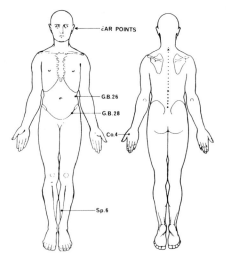

Fig. 8. Acupuncture points used to produce analgesia for major surgery of the thorax, and of the upper and lower abdomen, in Herget's "combination method". Electro-acupuncture is used in combination with much lower than normal concentrations of drugs, and patients are conscious during operations. This method has been used for thousands of operations in W. Germany. Diagrams constructed by Richard Clark. For further details, see Monro (1981; Fig. 4.6.6.)

new, emerging disciplines, based on different paradigms (e.g., the chemist's response to Lipmann's energy rich bonds). It also shows that changing science through recognising false presuppositions, as proposed above, may be less easy than at first it sounds.

Fig. 9. This shape can be seen *either* as a rabbit *or* as a duck. It transforms itself in some subtle way when the duck's beak becomes the rabbit's ears and brings the otherwise neglected spot into prominence as the rabbit's mouth. It is impossible to see the shape apart from its interpretation, or to see both readings simultaneously (reproduced from Gombrich 1968)

Rabbit or duck?

TΛE CΛT

Fig. 10. We read the middle letters of the two words as *H* and *A* respectively, because of their intellectual context. A person who did not know English would not distinguish between them

There are several major discontinuities in the theoretical framework of contemporary science. Some of the most socially relevant of these occur in biomedical science: e.g. between physico-chemical and psychological approaches, and between modern medicine and traditional therapies (e.g. acupuncture). If science is to function in harmony with society, it must come to terms with such discontinuities. The tendency of most people who try to interrelate incommensurable approaches is to interpret them all in terms of their own paradigm. This is unsatisfactory because they consequently miss important features of the other paradigms. Moreover, if everyone does this, there is no chance for dialogue between people following different approaches (since no one is prepared to modify his views). Such an impasse makes for a disharmonious and fragmented society.

To interrelate two, different, incommensurable approaches adequately, it is necessary to explore each in its own terms, and then by switching back and forth between them to work towards a synthesis which does justice to both. While this is difficult, I believe it is possible, and that it is an essential prerequisite for the harmonious functioning together of science and society. It can also be rewarding from scientific viewpoints.

G. Conclusions

The outstripping of our social capacities by our scientific knowledge presents one of the greatest contemporary human problems. To approach this problem adequately, we need, *inter alia* to recognise the close interrelatedness of scientific knowledge and its cultural context. Various paths are open for the evolution of science, each having different social implications. We need to seek syntheses between science and culture which do justice to all aspects of human experience. To do this effectively it is necessary to have an open, enquiring mind, free from dogmatic fixations, preconceptions, and corruption by vested interests. The topics discussed in this paper are more fully considered and documented in another publication (Monro 1981).

References

CIBA (1975) Energy transformation in biological systems. Ciba Foundation Symposium 31 (New series). Elsevier, Excerpta Medica, North-Holland, New York

Cairns J, Stent GS, Watson JD (eds) (1966) Phage and the origins of molecular biology. Cold Spring Harbor Laboratory for Quantitative Biology, New York

Clarke CA (1981) Chapter 4.2.7 in: ref. Monro

Dechend H von (1965) In: Justus von Liebig. Verlag Chemie, Weinheim

Gombrich EH (1968) Art and illusion, 3rd ed. Phaidon Press, London

Gwei-Djen L, Needham J (1979) The celestial lancets: History and rationale of acupuncture and moxa. Cambridge University Press, Cambridge

Harden A (1932) Alcoholic fermentation, 4th ed. Longmans, Green, London

Hoch FL, Lipmann F (1954) The uncoubling of respiration and phosphorylation by thyroid hormones. Proc Natl Acad Sci USA 40: 909–921

Jacob F (1974) The logic of living systems: a history of heredity. Allen Lane, London

Krebs HA (1967) The making of a scientist. Nature 215: 1441

Kuhn TS (1970) The structure of scientific revolutions, 2nd ed. Chicago Univ. Press, Chicago

Kuhn TS (1974) Second thoughts on paradigms. In: Suppe F (ed) The structure of scientific theories, pp 459–482. Univ. Illinois Press, Urbana (Reprinted in: Kuhn TS (1977) The essential tension. Univ Chicago Press, Chicago)

Kurland CG (1978) The role of guanine nucleotides in protein biosynthesis. Biophys J 22: 373

Lipmann F (1941) Metabolic generation and utilization of phosphate bond energy. Adv Enzymol 1: 99

Lipmann F (1962) Disproportions created by the exponential growth of knowledge. Perspectives in biology and medicine, vol V, no 3. University of Chicago, Chicago (Reprinted in: see ref. Lipmann, 1971, p 227)

Lipmann F (1971) Wanderings of a biochemist. Wiley-Interscience, New York London

Medawar PB (1967) The art of the soluble. Methuen, London

Monro RE (1981) In preparation

Needham J, Baldwin E (eds) (1949) Hopkins and biochemistry. Cambridge University Press, Cambridge

Popper KR (1968) The logic of scientific discovery, rev. edn. Hutchinson, London

Popper KR (1969) Conjectures and refutations: The growth of scientific knowledge, 3rd ed. Routledge, London

Richards MPM (1981) Chapter 4.2. in: see ref. Monro

Robinson DA, Radford AJ, Fendall NRE (1974) Fallacies in comparing international disease trends. J Biosoc Sci 6: 279–292

Scientific american (1969) The chemistry of acupuncture, p 69. Scientific American

2. Personal Recollections on Fritz Lipmann During the Early Years of Coenzyme a Research

G.NOVELLI

A. Introduction

My relationship with Prof. Fritz Lipmann has been unique in many respects. During this brief presentation I would like to review with you my personal recollection of seven years, 1946 to 1953, of close association with Fritz Lipmann that will illustrate his unusual creative ability and intellectual capacity which spans a great era of biochemistry and medicine. Lipmann has been called the "scientist's scientist." As with other great men, we find in Lipmann that happy combination of mind and soul. Lipmann is a modest, soft-spoken man of great humility who is not only admired and respected for his scientific abilities, but who is loved by his associates for the greatness of his human character.

Lipmann (1971) has written in a charming book, *Wanderings of a Biochemist,* a personal account of his trials and tribulations as he learned to develop into an accomplished biochemist.

I. Events Leading to my Joining Lipmann's Laboratory

For my part, I would like to write this as a similar tale of Lipmann as seen by his first Ph.D. student who literally learned biochemistry at the feet of the master.

My arrival in Lipmann's laboratory came about quite by accident, and what a happy and fortunate accident it turned out to be.

I obtained a Master's degree in microbiology at Rutgers University in 1942 and then proceeded to Scripps Institute of Oceanography in La Jolla, California, where I worked toward a Ph.D. in marine microbiology. Before I could complete these studies, I was drafted into the Army of the United States in 1944. During my military service, I became disenchanted with marine microbiology and, largely influenced by Lipmann's (1941) classical paper on *Metabolic generation and utilization of phosphate bond energy,* I decided to forego the study of marine microbiology and to seek the possibility of obtaining a doctorate in biochemistry. I have already published the circumstances in the military that prompted my seeking advice from Lipmann as to where to go to complete my studies (Novelli 1966), so I will not repeat these details here. I should state,

Biology Division, Oak Ridge National Laboratory, Oak Ridge, Tn. 37830, USA

however, that in my letter to Lipmann describing my previous background and training and asking his advice, I did by chance add a postscript asking if it would be possible to do a Ph.D. thesis with him. In his reply he wrote:

"Dear Novelli,

Yes, it is possible to do a Ph.D. thesis with me for which to apply you should better hurry." (Lipmann, private communication).

II. Arrival in Lipmann's Laboratory

I joined Lipmann's laboratory in the Department of Surgery at the Massachusetts General Hospital in August 1946, just one month after my discharge from the Army.

There I found working with Lipmann, L. Constance Tuttle, who was both secretary and technician, and Nathan Kaplan, a postdoctoral fellow who had joined the laboratory about eight months earlier.

The two years of military service had left we with a two-year blank period regarding the present status of the science of biochemistry. I found that I had to learn much and rapidly in order to catch up with the "state of the art." This was accomplished by personal teaching from Lipmann in the laboratory and through daily discussions with Lipmann and Kaplan, sitting at the circular lunch table in the doctors' cafeteria at "Mass General." Each day Lipmann and Kaplan would discuss a relevant article from a current issue of one of the biochemical journals. Usually I would be quite uninformed regarding the issue under discussion, and I would hasten to the library to read the article they had been discussing, only to find that in order to understand the significance of the article I would have to go back in history to read a half dozen background articles; and so it would proceed day after day. I soon reached the point where I was unafraid to ask questions and I know that most of these were naive. It was at this time that I learned how kind, gentle, considerate, and knowledgeable a man Fritz Lipmann is. Those of you who know him and who have read his book *Wanderings of a Biochemist* know that he studied biochemistry with some of the giants of his time during a period of rapid accumulation of the basic elements of intermediary metabolism. I realized how fortunate I was to be taught personally by this great man who not only taught me the facts, but could tell me of the where and why and the thinking that had been involved in the discoveries that had occurred during the heroic period through which he had lived.

B. The Trail of an Idea

I. From the Pasteur Effect to the Discovery of Coenzyme A

I would now like to share with you my personal reconstruction of the intellectual processes that took place in Lipmann's mind during almost 20 years that ultimately culminated in the logical discovery of coenzyme A. This reconstruction is based primarily on numerous discussions with Lipmann regarding the background thinking of the various leaders in the field of the rapidly emerging facts of glycolysis and intermediary metabolism.

In Otto Meyerhof's laboratory both at Berlin-Dahlem and then later in Heidelberg, Lipmann worked on various aspects of glycolysis; but perhaps the most instructive work he did, which was to play a highly significant role in his own future work, was a collaboration with Karl Lohman in which he learned much about phosphate ester chemistry.

After leaving Meyerhof's laboratory in 1930, Lipmann obtained his first job as an assistant in Albert Fischer's tissue culture laboratory in Berlin-Dahlem, in a biology laboratory close to where Meyerhof's had been and not far from the laboratories of Otto Warburg. Thus, Lipmann entered the newly emerging field of tissue culture. His first assignment was to attempt to use the measurement of some metabolic function to replace the generally used method of measuring tissue expansion planimetrically as a function of cell growth. This he succeeded in doing by measuring the evolution of CO_2 in bicarbonate buffer because of aerobic glycolysis. This came about at the time when reports from Warburg's laboratory were beginning to show that a high rate of aerobic glycolysis of cultured cells showed a high correlation to their degree of malignancy, a subject that Lipmann never actively pursued.

During the period of 1931 and 1932, while a new laboratory was being built in Copenhagen for Albert Fischer, Lipmann spent the time at the Rockefeller institute. Here he worked in P.A. Levine's laboratory on phosphoproteins, and he succeeded in isolating serine phosphate as the amino acid to which the phosphate in phosphoproteins is bound. He also learned a good deal about the chemistry of the serine hydroxyl-phosphate bond that was quite different from the bound phosphate with which he had worked with Lohman in Meyerhof's laboratory. Some 30 years later he would again return to the study of phosphoporteins and make some highly significant contributions, some of which will be discussed at this Symposium.

When Lipmann returned to Copenhagen to rejoin Albert Fischer in his new laboratory, his fascination with his previous metabolic studies with fibroblasts, with their unusually strong anaerobic glycolysis paired with a high respiration, caused Lipmann to turn his interest to the Pasteur effect. This effect is essentially the inhibition of alcoholic fermentation by yeast when oxygen is introduced into the culture, and also happens with other tissues that have a high rate of anaerobic glycolysis, such as fibroblasts, muscle extracts, etc. This was one of the first regulatory reactions that created much attention and controversy among biochemists of the early 1930's. This inhibition of glycolysis by respiration yields about ten times as much metabolically useful energy per mol of glucose utilized.

In looking for an explanation of the Pasteur effect, Lipmann first looked for a direct inhibitory effect of oxygen on various glycolytic enzymes unsuccessfully. After much frustration, Lipmann in 1935 reasoned that the point of divergence between anaerobic and aerobic metabolism more than likely occurred at the pyruvate step. Thus, a better understanding of the aerobic oxidation of pyruvate would perhaps throw light on the Pasteur effect.

He first attempted to study respiration and glycolysis in brain slices from normal and thiamine-deficient pigeons, since it had been shown in Sir Rudolf Peters' laboratory that pyruvate piled up in the brain of such vitamin-deficient birds. He confirmed the fact that pyruvate oxidation could be restored to normal in slices from such brains by the addition of thiamine. All attempts to obtain a soluble system that would oxidize pyruvate, and thereby permit its dissection into its component parts as had been done with the glycolytic system, failed. At that time, mitochondria were not yet known. Calling upon a lession he had learned in Meyerhof's laboratory that if a mammalian system is unsatisfactory one should try microorganisms as an alternate source, he found a paper by Davis (1933a, b) in which an acetone powder prepared from *Lactobacillus delbrueckii acidificans longissimus* was shown to contain a powerful pyruvate oxidation system. Lipmann was able to quickly make a soluble extract from such acetone powders and by appropriate techniques was able to dissect it into its component parts. He showed that the enzyme system required the newly discovered thiamine pyrophosphate and a flavin-adenine compound. The surprising, and what eventually proved to be the most important, finding was that the oxidation of pyruvic acid was absolutely dependent upon inorganic phosphate. However, he was unable to demonstrate the accumulation of a phosphorylated intermediate that he suspected to be acetylphosphate. Calling again on a trick he had learned while working with Lohman, he added adenylic acid to the oxidation system and was able to demonstrate the disappearance of inorganic phosphate into an acid labile form, most likely ATP. Subsequently, the phosphorylated intermediate was identified as acetyl phosphate and his earlier failure to show the accumulation of acetyl phosphate as the phosphorylated intermediate was because acetyl phosphate is unstable under the ordinary Fisk-Subbarow procedure for measuring inorganic phosphate.

Prior to this time, there was a strong belief amongst biochemists that phosphorylation was quite specifically coupled with the glycolytic reaction, but this new system of pyruvate oxidation leading to uptake of phosphate clearly showed that phosphorylation could also be coupled with a respiratory pathway. This, together with Kalckar's (1939) observation that oxidation reactions by kidney cortex "brei" were strongly coupled to phosphorylation, started Lipmann thinking of "the rather sweeping biochemical significance of the transformation of electron potential, respiratory or fermentative, to phosphate bond energy and therefore to a wide range of biosynthetic reactions" (Lipmann 1941). These thoughts rolled around in Lipmann's mind for several years and resulted in the publication in 1941 of his now famous paper *Metabolic generation and utilization of phosphate bond energy.*

This started a new generation of biochemists to begin to take a more serious look at biosynthetic reactions in general.

II. Search for the "Active Two-Carbon Fragment"

During the early 1940's evidence was beginning to accumulate from a number of laboratories, notably that of Shoenheimer at Columbia (see review by Rittenberg and Shemin 1946) that complex molecules like fatty acids, cholesterol, and other sterols were biosynthesized by the stepwise condensation of a two-carbon fragment, and, since biosynthetic reactions require the input of metabolic energy, the fragment was variously called "active two-carbon fragment" or "active acetate." Lipmann, having discovered acetyl phosphate, found in studying the chemistry of the molecule that it not only carried an energy-rich phosphoryl radical but was even more impressive because of its energy-rich acetyl moiety. Lipmann believed that he had indeed discovered the elusive "active acetate." However, he ran into two unexpected disappointments. In the first place, he found that most animal tissues contain a very potent and heat-stable acetyl phosphatase; even in tissues with low levels of the enzyme acetyl phosphate was inert and was not synthesized during pyruvate oxidation in mammalian systems. However, Lipmann was so convinced of his concept of group potential as a pivotal one for biosynthetic reactions that he continued to search for a mammalian system in which the activation of acetate could be studied. This is where his combined background in medicine and biochemistry served him well.

In the mid-1930's, pneumonia was the number one cause of death in the United States, as well as in most other countreis. Then sulfanilamide was introduced into medical practice about 1936 or 1937. There was an immediate and precipitous drop in deaths from pneumonia. The drug was being hailed as the long sought "magic bullet."

In those days there were not the rigorous regulations regarding the use of drugs that prevail today in the United States, so the drug was tried on almost every type of infectious disease. Soon trouble developed in pediatric practice. Many babies were dying of uremic poisoning. Autopsies showed that the cause of death was blockage of kidney tubules because of crystals of acetylsulfanilamide. It turned out that the acetylated form of the drug was less soluble than the parent compound. At this point, a pediatrician, Dr. Jerome Harris (who was later to become my Commanding Officer during my military service), did perfusion experiments with dogs (Klein and Harris 1938) and discovered that the drug was acetylated by the liver. Further experiments showed that rabbit liver slices would also acetylate the drug in Warburg vessel experiments, but the acetylation only took place in air or oxygen; it would not take place when the slices were incubated under N_2. They gave no explanation for this but simply recorded the observation. If one looks at the chemistry of this acetylation, there is no obvious requirement for oxygen to participate in the reaction, but it is known that the drug can be chemically acetylated with acetic anhydride, which is an activated form of acetate. Lipmann, having shown that activated acetate (acetyl phosphate) can be produced by the oxidation of pyruvate and having become inbued with the idea that phosphorylation is coupled to respiration, reasoned that the requirement for oxygen in the liver slice experiment indicated that acetate must be activated before it can acetylate the sulfanilamide. He repeated and confirmed the rabbit liver slice experiment. He tried to make soluble preparations from rabbit liver, even going through the acetone powder technique that he had been successful with in bacterial preparations, but once again he was disappointed that he could not find a soluble system that could be separated to study the mechanisms of the acetylation reaction.

Again, Lipmann called on his earlier experience where he had studied pyruvate oxidation in slices from the pigeon. A good acetylation of a sulfanilamide was found to occur in cell-free pigeon liver preparations. Although acetyl phosphate could not serve as the acetyl donor, acetate and ATP resulted in a vigorous acetylation of the aromatic amine. (It would be five or six years before the explanation for the failure of acetylphosphate to act as an acetyl donor in mammalian systems would be found.) In attempting to dissect the cell-free pigeon liver acetylation system, Lipmann discovered that either dialysis overnight or simply allowing the homogenate to stand for 4 or 5 h at room temperature resulted in the complete loss of the acetylation reaction (Lipmann 1945b). However, the addition of a small amount of boiled, fresh liver extract to the inactive extract completely restored the acetylation reaction, and thus the requirement of a heat-stable cofactor became evident (Lipmann 1945a). Using this "aged" extract of acetone dried pigeon liver, a quantitative assay for the purification of coenzyme A was established and used in measuring the degree of purification achieved by a series of procedures (Kaplan and Lipmann 1948). A series of fractional precipitations with heavy metals resulted in a preparation containing about 700 times the activity of the starting material. Since it was shown to be required for the acetylation of choline in the synthesis of acetylcholine (Lipmann and Kaplan 1946), it was named coenzyme A (for acetylation). The best preparation (Lipmann 1946a–c) contained adenine, ribose, and phosphate in the proportion of 1 to 1 to 3. That the third phosphate was not in pyrophosphate linkage was shown by acid hydrolysis. The adenylic acid content of almost 50% suggested the compound to be a dinucleotide with the second phosphate linked to an unidentified component. The cofactor could not be replaced by any of the then known coenzymes, such as cocarboxylase, NAD, NADP, pyridoxal phosphate, or folic acid conjugate. Lipmann obtained his biochemical education during a period when a number of coenzymes were discovered and all of them were shown to have one of the B vitamins as a component part of their structure.

Since none of the known B vitamin-containing coenzymes could replace the coenzyme for acetylation, Lipmann had hopes that he had a new coenzyme that might have a function for one of the metabolically still unidentified B vitamins. Accordingly, samples were sent out to various commercial laboratories and two research laboratories of pharmaceutical companies for vitamin analysis. The results came back universally negative. In spite of these disappointments, Lipmann had a hunch that pantothenic acid might be present in the coenzyme. He decided to send a sample to the laboratory of Roger Williams, the discoverer of pantothenic acid. Dr. Beverly Guiard carried out the analysis. By direct microbiological assay for pantothenic acid there was almost no response. Nevertheless, knowing that pantothenic acid often occurred in bound form and that it contained β-alanine, Dr. Guiard hydrolyzed the coenzyme with acid, used a microbiological assay for β-alanine, and found that it contained about 10% β-alanine equivalence of pantothenic acid. It was at this point that I arrived in Lipmann's laboratory and was given the task of finding out what compound the β-alanine was linked to. I spent the next two months in a series of frustrating and disappointing experiments trying various conditions of acid and alkaline hydrolysis, followed by microbiological assays independently for pantothenic acid and β-alanine. I could generally find β-alanine but only obtained tantalizing hints of the possible presence of pantothenic acid. From independent measurments we knew that the coenzyme A activity could be destroyed by incubation with intestinal phosphatase and, of course, by preincubation with fresh pigeon liver extract because this was the procedure routinely used to prepare the apoenzyme for the enzymatic acetylation reaction. Finally, Kaplan and I decided to try to liberate pantothenic acid from the coenzyme by enzymatic hydrolysis. Preincubation of the coenzyme with alkaline phosphatase gave us the encouraging result of the liberation of 1.4% pantothenic acid by microbiological assay, but we knew that by acid hydrolysis we could obtain β-alanine equivalent to 10% pantothenic acid. Independent preincubation with fresh pigeon liver enzyme increased the pantothenic acid release to 4.8%, still far short of what we expected from the β-alanine content. Furthermore, the release of even this small amount of pantothenic acid by fresh liver enzyme was completely blocked when the incubation was carried out in the presence of NaF, suggesting that in addition to the liver enzyme the action of a phosphatase was also required.

At this point, in November 1946, Lipmann returned to England, his first trip to Europe following the end of World War II. While he was traveling to England by ship, we digested the coenzyme with *both* fresh pigeon liver enzyme *and* alkaline phosphatase. We obtained the liberation of 9.0% pantothenic acid which was quite close to the expected value of 10% as determined by the β-alanine content. We next tested in the same preincubation system nine different samples of CoA that varied in activity from 9 to 132 CoA units/mg. This study gave us a constant of 0.7 μg pantothenic acid per unit of CoA activity and pretty much convinced us that a bound form of the vitamin was an integral part of CoA (Lipmann et al. 1947a, b). We immediately sent Lipmann a cablegram giving him the results of the foregoing experiments. An interesting and little known aspect of this story is that when Lipmann arrived in England he visited Marjory Stevenson's laboratory where she and her people were studying the metabolism of *Lactobacillus plantarum,* an organism used in the preparation of sauerkraut, that curiously produced large amounts of acetylcholine. Since most *Lactobacilli* required an exogenous source of pantothenic acid, Elizabeth Rowatt was immediately able to demonstrate the loss of synthesis of acetylcholine with pantothenate-deficient *L. plantarum,* which was promptly restored upon the addition of pantothenic acid to the deficient culture. Thus, our finding pantothenic acid in CoA was independently confirmed before we had published our findings (Stevenson and Rowatt, unpublished; F. Lipmann 1971).

The finding of pantothenic acid in CoA immediately changed the complexion of Lipmann's laboratory at MGH. Prior to that finding, we were frequently chided by our biochemical colleagues who would state: "Why are you grown men fooling around with a coenzyme required by a pigeon to acetylate a synthetic drug that it would never encounter in its lifetime?" The finding that most, if not all, bound pantothenic acid in tissues existed as CoA (Kaplan and Lipmann 1948), together with the terrifying list of pathological symptoms attributed to pantothenic acid deficiency (Williams 1943), moved CoA from the status of an idle curiosity to the center of the stage of intermediary metabolism at that time. This resulted in a steadily increasing interest in the role of coenzyme A in metabolism and a corresponding increase in the number of postdoctoral students and visiting investigators to Lipmann's laboratory.

C. Further Research Activities in the Lipmann Laboratory

Three major projects were now simultaneously investigated in Lipmann's laboratory. These were: (1) studies on the chemical structure of CoA, (2) the metabolic functions of the coenzyme, and (3) the biological production of high energy phosphate bonds.

I. The Chemical Structure of CoA

Through the use of controlled enzymatic degradation of coenzyme, the structure was elucidated (Novelli and Lipmann 1950; Novelli 1953). The positioning of the third phosphate on the 3' hydroxyl of the ribose moiety was demonstrated by Kaplan and his co-workers (Shuster and Kaplan 1953), and finally the structure was confirmed by chemical synthesis by Baddiley et al. (1953).

II. Metabolic Functions of CoA

The finding that almost all of the cellular pantothenic acid could be accounted for as CoA (Kaplan and Lipmann 1948) made it clear that CoA was the only functional form of the vitamin in all living tissues. Since the previously discovered vitamin-containing coenzymes indicated that the vitamin played an important role in the metabolic function of the coenzyme, we focused our attention on the pantothenic acid portion of the molecule and tended to overlook the sulfur-containing moiety β-mercaptoethylamine. This is an excellent example of how a preconceived idea can be misleading.

The first attempts to obtain further evidence for a metabolic function for CoA (other than in the acetylation of aromatic amines and of choline, which were already known) was through the use of pantothenic acid-deficient microorganisms. We followed up the observation of Dorfman et al. (1942) and Hills (1943) that pantothenate stimulated the oxidation of pyruvate by panthothenate-deficient cells of *Proteus morganii*, because we considered this to be suggestive of the participation of CoA in the synthesis of citric acid. We were able to demonstrate a parallelism between CoA levels and pyruvate oxidation in *P. morganii* (Novelli and Lipmann 1947). This observation was then extended to the demonstration of a CoA dependence for acetate oxidation by pantothenate-deficient yeast (Novelli and Lipmann 1948).

Around this time, Morris Soodak, Hubert Chantrenne, and John Gregory joined the laboratory.

The next advance in demonstrating the generality of the metabolic function of CoA was the demonstration of the requirement for CoA in the synthesis of acetoacetate (Soodak and Lipmann 1948).

In 1949, Earl Stadtman came to Lipmann's laboratory to spend a year as a postdoctoral fellow, bringing with him that marvelous microorganism *Clostridum kluyverii* which he had used in his Ph.D. studies with H.A. Barker. It is not within the scope of this article to discuss the overall fermentation of ethanol and acetic acid by this organism that leads to the production of a variety of fatty acids. This has all been thoroughly reviewed (Barker 1956; Stadtman 1954). Suffice it to say that during his graduate work in Berkeley he had discovered that extracts of *C. kluyverii* carried out a vigorous arsenolytic decomposition of acetyl phosphate. It was postulated that the arsenolysis of acetyl phosphate might be brought about by the reversible transfer of the (acetyl moiety to an acceptor "X," and then by substitution of arsenate for phosphate in the reverse reaction would yield acetyl arsenate which would spontaneously hydrolyze in aqueous media (Stadtman and Barker 1950). At first they considered that "X" might be the transacetylase itself. Stadtman and Barker discussed the possibility that

"X" could be CoA, but they were unable to obtain evidence for the participation of CoA in this reaction. However, shortly after Stadtman's arrival at MGH, the reason for the failure to demonstrate a CoA dependence for the arsenolysis reaction, as well as for fatty acid synthesis, became obvious. By direct assay it was shown that *C. kluyverii* had the highest content of CoA of any tissue we had ever measured, and it appeared virtually impossible to resolve enzymes from *C. kluyverii* free of CoA. Fortunately, at that time I had resolved an enzyme preparation from *Escherichia coli* that carried out the CoA-dependent condensation of acetate, ATP, and oxaloacetate to form citric acid (Novelli 1949; Novelli and Lipmann 1950). Furthermore, in contrast to animal tissues, acetyl phosphate was more than twice as active as an acetyl donor than acetate plus ATP. This surprising finding made Lipmann so happy that finally his beloved acetyl phosphate had gained respectability that he danced a little jig in the laboratory.

This result then prompted Stadtman and me to test this *E. coli* preparation for his arsenolysis reaction, and indeed the preparation was able to catalyze the arsenolytic decomposition of acetyl phosphate, albeit much less vigorously than Stadtman's *C. kluyverii*. However, the important point was established that the arsenolysis of acetyl-phosphate was absolutely CoA dependent and established the previously postulated "X" to be indeed CoA. This finding then gave us the needed impetus to make a heroic effort to resolve the enzyme, now named phosphotransacetylase, from Stadtman's *C. kluyverii* extracts. This proved to be successful and led to the rapid development of the metabolic roles of CoA (Stadtman et al. 1951a, b). At this point I would like to quote from a publication by Earl Stadtman (1976) who expressed the situation more eloquently than I can:

"The year in Lipmann's laboratory was an unusually rich and satisfying experience for me, both from the personal and scientific view. . . .here there was an intangible benefit to be gained by close association with a man of unusual imagination and perception and one who possessed a gentle warmth of character and concern for others that endeared him to all his associates. It was here that I discovered that productive laboratories are not merely the reflection of good scientific discipline and expert direction but depend almost as much on the establishment of a congenial atmosphere in which science can flourish as a consequence of free thought, unguarded exchange of ideas, critical discussion and a respectful interaction among all of its personnel. Such was Lipmann's laboratory.

Unfortunately, I cannot discuss here the important contributions of other works in Lipmann's laboratory which were important in establishing the acetyl carrier functions of CoA. Few observations were more decisive, however, than the demonstration by G.D. Novelli and myself that CoA is absolutely required for the phosphotransacety-lase catalyzed arsenolysis of acetyl-P, as well as for the exchange of orthophosphate into acetyl-P."

Soon after this finding, Stern and Ochoa (1949) showed a CoA-dependent citrate synthesis with a pigeon liver fraction similar to the one previously used by Soodak for acetoacetate synthesis. Later on during Stadtman's postdoctoral year in Lipmann's laboratory, he visited Ochoa's laboratory in New York bringing with him the purified phosphotransacetylase. After a minor problem with the transacetylase, they did succeed in showing that acetyl CoA generated in the transacetylase system could be successfully coupled with the crystalline citrate condensing enzyme (Stern et al. 1951).

Meanwhile, T.C. Chou was fractionating the pigeon liver sulfanilamide acetylating system. By acetone fractionation he was able to obtain two fractions, one that separated out with 40% acetone and another that came out at 60% acetone, neither of which could carry out the acetylation reaction alone but could do so when combined (Chou and Lipmann 1952). Now a further possibility for demonstrating the acetyl carrier function of CoA became possible when we could substitute Chou's A-40 with Stadtman's purified phosphotransacetylase for the generation of acetyl CoA and couple it with the A-60 fraction from pigeon liver and acetylate sulfanilamide (Chou et al. 1950).

During Stadtman's year in Lipmann's laboratory Michael Douderoff spent a few weeks with us, and he and Stadtman made two observations that were a key to the understanding of fatty acid metabolism. They were able to demonstrate that acetoacetate synthesized in a coupled reaction containing phosphotransacetylase and acetoacetate synthetase from pigeon liver involved the condensation of two molecules of acetyl CoA. The other important observation they made was that CoA is required for the synthesis of butyryl phosphate from acetyl-P and butyrate (Stadtman et al. 1951). During this same period another contribution concerning the metabolic role of CoA was made by Hubert Chantrenne. Chantrenne and Lipmann (1950) demonstrated a requirement for CoA in the reversible exchange between formate and pyruvate in extracts of *E. coli*.

The foregoing studies on the metabolic functions of CoA attracted much interest in the problem, and other laboratories started to work on CoA. Other postdoctoral fellows and visiting investigators were attracted to Lipmann's laboratory and the work continued mainly on the biosynthesis of CoA and on the mechanism of the formation of acetyl CoA.

III. Biosynthesis of CoA

In this effort we were joined by John Gregory, Leon Leventow, Ruth Flynn, Frank Schmetz, Mahlon Hoagland, and Werner Mass.

We were able to establish that the sulfhydryl-containing compound in CoA was 2-mercaptoethylamine (Gregory et al. 1952). At about this time Lipmann's laboratory was astounded to learn that acetyl CoA had been isolated from yeast juice in Lynen's laboratory and was shown to be a relatively stable thioester (Lynen et al. 1951). We had been misled by focusing attention on the pantothenate portion of CoA as the possible acetyl carrier and also believed that the compound would be too labile to permit its isolation.

After Snell et al. (1950) had shown by chemical synthesis that the *Lactobacillus bulgaricus* growth factor (LBF) consisted of pantothenic acid joined through a peptidic link to 2-mercaptoethylamine that was named pantetheine, we (Leventow and Novelli 1954) succeeded in purifying pantetheine kinase from hog liver. From phospho-pantetheine, we were able to bring about the complete synthesis of CoA (Novelli et al. 1954; Hoagland and Novelli 1954). Somewhat earlier Werner Maas and I (1953) had been studying the synthesis of pantothenate from pantoic acid and β-alanine in the presence of ATP and made the surprising finding that the synthesis was carried out by splitting inorganic pyrophosphate from ATP, rather than the usual split of orthophosphate. In a similar way Hoagland and I (1954) showed that the condensation between phosphopantetheine and ATP to make dephospho-CoA also led to the split of pyrophosphate from ATP. The other well-known reaction at that time that led to the splitting out of inorganic pyrophosphate during an enzymatic synthetic reaction was the synthesis of DPN shown by Kornberg (1950).

IV. The Mechanism of the Formation of Acetyl CoA

In the early experiments of Chou and Lipmann (1952) with the pigeon liver aromatic amine acetylation system in which hydroxylamine was used as the acetyl acceptor, there was the puzzling observation that hydroxamic acid formation yielded less than the expected inorganic phosphate. After Mary Ellen Jones and Simon Black joined the laboratory and became involved in this study, they shifted to the much more active yeast fraction; again very little inorganic phosphate was released in the ATP-CoA-acetate-hydroxylamine reaction when the reaction was carried out in the presence of fluoride. When the fluoride was omitted, the formation of hydroxamic acid was accompanied by about twice as much inorganic phosphate. Since yeast pyrophosphatase is strongly inhibited by fluoride, the explanation became clear. Here again, there was a synthetic reaction that involved the split of inorganic pyrophosphate from ATP (Lipmann et al. 1952).

During this period Fedor Lynen was a guest in Lipmann's laboratory, and we had many lively discussions regarding the enzymatic mechanism whereby acetyl-CoA was produced from acetate, ATP, and CoA. Intensive studies were undertaken using a variety of exchange reactions with ^{32}P-labeled pyrophosphate. I think that this period and the explanation of the results obtained were best expressed by Fritz Lipmann himself in 1971, and I would like to quote what he said in his book *Wanderings of a Biochemist:*

"After this, for various reasons, we again missed the true intermediate of this activation — acetyl adenylate — and Paul Berg (1955) found it. Our inability to see the right mechanism after discovering the unusual products adenylic acid and PP was due to our unawareness of an unusually strong binding of acetate to proteins. In our assay, after 'exhaustive' dialysis, an addition of acetate to the enzyme was not needed to obtain reaction of ATP with CoA. It was this mistake that made us propose CoA-pyrophosphate as a possible intermediary.

All this sums up to teaching one how difficult it is to see the new because it is new and how badly one may be handicapped by preconceived notions. We thougth ourselves lucky to have found a coenzyme that contained a new vitamin and may be excused for having fixed our minds on the appearance of this vitamin in the coenzyme that we neglected the SH-function in CoA in our search for an energy-rich attachment to the vitamin part, even after we had clearly realized that coenzyme A was the carrier of active acetate. We were rather satisfied at finding adenylic acid in the pyrophosphate link to the vitamin part of the coenzyme, which conformed to the general experience with DPN and flavin adenine dinucleotide; in some manner we expected a direct functional participation of the vitamin."

D. Generation of Energy-Rich Phosphate Bonds

During the coenzyme A period from 1945 to 1953, although the major emphasis in
the Lipmann laboratory was directed to the unraveling of the CoA story, Lipmann had
maintained an active interest in the metabolic generation of high-energy phosphate
bonds. In 1948 W.F. Loomis joined the Lipmann laboratory after having spent a period of time in D.E. Green's laboratory studying the respiration of mitochondria, which
in those days was called the "Cyclophorase System" (Green et al. 1949). Although this
system had adequate oxidative power, it was not known whether this oxidation was
coupled to phosphorylation since ATP did not accumulate during the oxidation. Lipmann suggested to Loomis that he try to trap any ATP formed during the oxidation
by adding yeast hexokinase and glucose to the system. This proved to work and a P:O
ratio of 2.2 was demonstrated. It was known from the work of Clifton (1946) that di-
nitrophenol (DNP) in low concentrations completely blocked growth and synthetic reactions while leaving respiration intact. When DNP was tested in the kidney homogenate, oxygen uptake remained the same, but the P:O ratio decreased from 2.2 to 0.2
(Loomis and Lipmann 1948). This uncoupling of oxidation from the generation of
high-energy phosphate bonds went a long way toward establishing ATP as a universal
carrier of biological energy. Later on other investigators in the laboratory continued
studies on the uncoupling of oxidative phosphorylation. These were Bob Crane, Her-
man Niemeyer, Gene Kennedy, Fred Hoch, S.H. Mudd, and Jane Park. Collectively
and at different times, these people showed a stimulation of respiration upon the addition of the phosphate acceptor, glucose plus hexokinase (Niemeyer et al. 1951). The
uncoupling of respiration and phosphorylation by thyroid hormones (Hoch and Lipmann 1954; Mudd et al. 1955) demonstrated a magnesium antagonism of the uncoupling of oxidative phosphorylation by iodo-thyronines.

   The incisiveness of Lipmann's intellectual power is exemplified in the synthesis of
carbamyl phosphate. We had attended a symposium on Amino Acid Metabolism at the
McCollum-Pratt Institute of Johns Hopkins University. Lipmann was particularly fas-
cinated by independent reports from three laboratories that were in essential agreement that the decomposition of citrulline to ornithine by extracts from different bacteria led to the production of 1 mol of ATP (Oginsky 1955; Korzenovsky 1955; Slade
1955). The requirements for the reaction with cell-free extracts were shown to be inorganic phosphate, adenylic acid or ADP, and divalent ions. On the ride back to
Boston, Lipmann kept ruminating about this curious reaction and believed that there
should be a high energy phosphorylated intermediate before its conversion to ATP. He
compared, in his mind, this reaction to the well-known phosphoroclastic split of pyruvate that yields acetyl phosphate as the high energy intermediate, and he thought that
intermediate might be a phosphorylated compound of CO_2 and PO_4. On his return to
the laboratory at MGH, he convinced Lenny Spector and Mary Ellen Jones to join
forces to solve the problem. Spector was able to work out a relatively simple synthesis
of carbamyl phosphate, and Mary Ellen Jones prepared an extract from *Streptococcus
faecalis* that contained an active citrulline phosphorolysis. The extract was shown to
be able to synthesize citrulline when incubated with synthetic carbamyl phosphate,
ornithine, and Mg^{++}, and was also able to transfer the phosphate group from carbamyl
phosphate to ADP to form ATP (Jones et al. 1955).

E. Conclusion

I. Leaving Lipmann, Leaving Coenzyme A, the Start of a New Career

It was with great reluctance, in September 1953, that I left Lipmann's laboratory where I had spent seven glorious years learning biochemistry from one of the greatest biochemists of this century. In those seven years we had grown very close to each other; I considered him not only as my scientific father but also as a true father because of his compassion and great consideration for the feelings and nonscientific problems of other people in general (my real father had died when I was still a youth). The position offered by Lester Krampitz at Western Reserve University School of Medicine appeared too good to turn down and Fritz reluctantly suggested that I accept it. A strong motivating force was that Ruth and I had our first child, and the financial support at MGH and Harvard was tenuous at best and Federal financing of basic research was only beginning to build up to what it was to become. Therefore, I accepted the position at Western Reserve and my very close association with Lipmann came to an end. However, from that time until today, we have maintained a close personal relationship through correspondence, phone calls, and visiting his laboratory, now in New York. Whenever I had a particular problem, I always felt that I could call on him for advice.

II. Leaving CoA Research

When I left Lipmann's laboratory in September 1953, we felt that the work on CoA was drawing to a cleaning-up stage and decided to shift to other fields of research. Of course Lipmann and all of his associates were especially delighted to learn that the Swedish scientific community decided to confer on him the Nobel Prize in October 1953, not only for his discovery of coenzyme A but also for establishing that the generation of high-energy phosphate bonds in the formation of ATP was the "currency" with which the performance of biosynthetic reactions and biological work was purchased.

III. The Start of a New Career

In his 1941 article on the *Metabolic generation and utilization of phosphate bond energy,* he proposed that the phosphorylation of the carboxyl group of amino acids might be a logical step in their activation for peptide bond formation. During the CoA period, he often compared the acetylation of the amino group of sulfanilamide with that of the formation of a peptide bond. In 1949 he wrote an article on the further possibility of carboxyl activation of amino acids in the mechanism of peptide bond formation, in which further examples of compounds of a peptidic nature were considered. Then Chantrenne, working at the University of Wisconsin, showed that CoA was involved in the synthesis of hippuric acid (Chantrenne 1951). Lipmann considered this another example of the formation of a peptidic link.

These discoveries in many biosynthetic reactions, already alluded to, suggested a preliminary step in which ATP underwent a pyrophosphorolytic cleavage that became the driving force for the subsequent synthetic step. At a symposium on "The Mechanism of Enzyme Action" at Johns Hopkins University, Lipmann proposed a theoretical model cycle for protein synthesis. The essence of the model was that if the early stages of the reactions were reversible then one should see an amino acid-dependent exchange of inorganic pyrophosphate with ATP (Lipmann 1954a).

Shortly thereafter I moved to Western Reserve University and Hoagland rejoined Paul Zamecnik's laboratory on the floor below Lipmann's laboratory, in which an active investigation of protein synthesis in mammalian liver was in progress. Hoagland (1955) confirmed the amino acid-dependent exchange between ATP and PP using an extract from rat liver. At about the same time De Moss and Novelli (1955) confirmed a similar reaction with a variety of microbial extracts. Thus, the saga of protein synthesis was launched. The rest is now recorded history, some of which we will hear at this Symposium. Lipmann's laboratory now entered the field of protein synthesis and in particular the biosynthesis of polypeptide antibiotics.

Although Lipmann's path and ours in the studies of protein synthesis took different courses, we continued to keep in fairly close contact with one another. In later years after my move to ORNL, where I established a fairly respectable Pilot Plant, it became my turn to provide him from time to time with rather large quantities of bacteria and various parts thereof.

In conclusion, I can state that the most glorious years of my scientific career were the seven years with Fritz at MGH.

Acknowledgment. Research sponsored by the Office of Health and Environmental Research, U.S. Department of Energy, under contract W-7405-eng-26 with the Union Carbide Corporation.

References

Baddiley J, Thain EM, Novelli GD, Lipmann F (1953) Structure of coenzyme A. Nature 10: 76
Barker HA (1956) Bacterial fermentations. Wiley, New York London
Berg P (1955) Participation of adenyl-acetate in the acetate activating system. J Am Chem Soc 77: 3163–3164
Chantrenne H (1951) The requirement for coenzyme A in the enzymatic synthesis of hippuric acid. J Biol Chem 189: 227–234
Chantrenne H, Lipmann F (1950) Coenzyme A dependence and actyl donor function of the pyruvate-formate exchange system. J Biol Chem 187: 757–767
Chou TC, Lipmann F (1952) Separation of acetyl transfer enzymes in pigeon liver extract. J Biol Chem 196: 89–103
Chou TC, Novelli GD, Stadtman ER, Lipmann F (1950) Fractionation of coenzyme A-dependent acetyl transfer reactions. Fed Proc 9: 160
Clifton CE (1946) Microbial assimilations. Adv Enzymol 6: 269–308
Davis JG (1933a) Über Atmung und Gärung von Milchsäurebakterien. Biochem Z 265: 90–104
Davis JG (1933b) Über Atmung und Gärung von Milchsäurebakterien II. Biochem Z 267: 357–359
De Moss JA, Novelli GD (1955) An amino acid dependent exchange between inorganic pyrophosphate and ATP in microbial extracts. Biochim Biophys Acta 18: 592–594
Dorfman A, Berkman S, Koser SA (1942) Pantothenic acid in the metabolism of *Proteus morganii.* J Biol Chem 144: 393–400

Green DE, Loomis WF, Auerbach VH (1948) Studies on the cyclphorase system I. The complete oxidation of pyruvic acid to carbon dioxide and water. J Biol Chem 172: 389–403

Gregory JD, Novelli GD, Lipmann F (1952) The composition of coenzyme A. J Am Chem Soc 74: 854

Hills GM (1943) Experiments on the function of pantothenate in bacterial metabolism. J Biochem 37: 418–425

Hoagland M (1955) An enzymic mechanism for amino acid activation in animal tissues. Biochim Biophys Acta 16: 288–289

Hoagland MB, Novelli GD (1954) Biosynthesis of coenzyme A from phosphopantetheine and of pantetheine from pantothenate. J Biol Chem 207: 767–773

Hoch FL, Lipmann F (1954) The uncoupling of respiration and phosphorylation by thyroid hormones. Proc Natl Acad Sci USA 40: 909–921

Jones ME, Spector L, Lipmann F (1955) Carbamyl phosphate, the carbamyl donor in enzymatic citrulline synthesis. J Am Chem Soc 77: 819–820

Kalckar HM (1939) Phosphorylations in kidney extracts. Enzymology 5: 365–371

Kaplan NO, Lipmann F (1948) The assay and distribution of coenzyme A. J Biol Chem 174: 37–44

Klein JR, Harris JS (1938) The acetylation of sulfanilamide in vitro. J Biol Chem 124: 613–626

Kornberg A (1950) Reversible enzymatic synthesis of diphosphopyridine nucleotide and inorganic pyrophosphate. J Biol Chem 182: 779–793

Korzenovsky M (1955) Metabolism of arginine and citrulline by bacteria. In: McElroy WD, Glass H (eds) A symposium on amino acid metabolism, pp 309–320. Hopkins, Baltimore

Leventow L, Novelli GD (1954) The synthesis of coenzyme A from pantetheine: preparation and properties of pantetheine kinase. J Biol Chem 207: 761–765

Lipmann F (1935) Über die Hemmung der Mazerationssaftgärung durch Sauerstoff in Gegenwart positiver Oxydoreduktionssysteme. Biochem Z 268: 205–213

Lipmann F (1939) An analysis of the pyruvic acid oxidation system. Cold Spring Harbor Symp Quant Biol 7: 248–259

Lipmann F (1941) Metabolic generation and utilization of phosphate bond energy. Adv Enzymol 1: 99–162

Lipmann F (1945a) Acetylation of sulfanilamide by liver homogenates and extracts. J Biol Chem 160: 173–190

Lipmann F (1945b) The mechanism of sulfanilamide acetylation. Fed Proc 4: 97

Lipmann F (1946a) Acetyl phosphate. Adv Enzymol 6: 231–267

Lipmann F (1946b) Metabolic progress patterns. In: Green DE (ed) Currents of biochemical research, pp 137:148. Interscience, New York

Lipmann F (1946c) Enzymatic acetylation and the coenzyme of acetylation. Biol Bull 91: 239

Lipmann F (1949) Mechanism of peptide bond formation. Fed Proc 8: 597–602

Lipmann F (1954a) On the mechanism of some ATP-linked reactions and certain aspects of protein synthesis. In: McElroy WD, Glass B (eds) The mechanism of enzyme action, pp 599–607. Hopkins, Baltimore

Lipmann F.(1954b) Development of the acetylation problem, a personal account. Science 120: 855–865

Lipmann F (1971) Wanderings of a biochemist. Wiley-Interscience, New York London

Lipmann F, Kaplan NO (1946) A common factor in the enzymatic acetylation of sulfanilamide and choline. J Biol Chem 162: 743–744

Lipmann F, Kaplan NO, Novelli GD (1947a) Chemistry and distribution of the coenzyme for acetylation (coenzyme A). Fed Proc 6: 272

Lipmann F, Kaplan NO, Novelli GD, Tuttle LC, Guirard BM (1947b) Coenzyme for acetylation, a pantothenic acid derivative. J Biol Chem 167: 869–870

Lipmann F, Jones ME, Black S, Flynn RM (1952) Enzymatic pyrophosphorylation of coenzyme A by adenosine triphosphate. J Am Chem Soc 74: 2384–2385

Loomis WF, Lipmann F (1948) Reversable inhibition of the coupling between phosphorylation and oxidation. J Biol Chem 173: 807–808

Lynen F, Reichert E, Ruff L (1951) Zum biologischen Abbau der Essigsäure. IV. Activierte Essigsäure, ihre Isolierung aus Hefe und ihre chemische Natur. Liebigs Ann Chem 574: 1–32

Maas WK, Novelli GD (1953) Synthesis of pantothenic acid by depyrophosphorylation of adeno-
    sine triphosphate. Arch Biochem Biophys 43: 236–238
Mudd SH, Park JH, Lipmann F (1955) Magnesium antagonism of the uncoupling of oxidative
    phosphorylation by iodothyronines. Proc Natl Acad Sci USA 41: 571–576
Niemeyer H, Crane RK, Kennedy EP, Lipmann F (1951) Observations on respiration and phos-
    phorylation with liver mitochondria of normal, hypo-, and hyperthyroid rats. Fed Proc 10:
    229
Novelli GD (1949) Studies concerning the chemistry and metabolic function of coenzyme A (a
    pantothenic acid derivative). Doctoral Dissertation, Harvard University
Novelli GD (1953) Enzymatic synthesis and structure of CoA. Fed Proc 12: 675–681
Novelli GD (1966) From ~P to CoA to protein biosynthesis. In: Kaplan NO, Kennedy EP (eds)
    Current aspects of biochemical energetics, pp 183–197. Academic Press, New York
Novelli GD, Lipmann F (1947) Bacterial conversion of pantothenic acid into coenzyme A (acety-
    lation) and its relation to pyruvic oxidation. Arch Biochem Biophys 14: 23–27
Novelli GD, Lipmann F (1948) Respiratory metabolism in yeast and coenzyme A levels effect of
    phenyl panthetone on coenzyme A synthesis. Fed Prod 7: 341
Novelli GD, Lipmann F (1950) The catalytic function of coenzyme A in citric acid synthesis. J
    Biol Chem 182: 213–228
Novelli GD, Schmetz FJ Jr, Kaplan NO (1954) Enzymatic degradation and resynthesis of coen-
    zyme A. J Biol Chem 206: 533–545
Oginsky EL (1955) Mechanism of arginine and citrulline breakdown in microorganisms. In: Mc-
    Elroy ED, Glass HB (eds) A symposium on amino acid metabolism, pp 300–308. Hopkins,
    Baltimore
Rittenberg D, Shemin D (1946) Isotope technique in the study of intermediary metabolism. In:
    Green DE (ed) Currents in biochemical research, pp 261–276. Interscience, New York
Shuster L, Kaplan NO (1953) A specific b nucleosidase. J Biol Chem 201: 535–545
Slade HD (1955) The metabolism of citrulline by bacteria. In: McElroy WD, Glass HB (eds) A
    symposium on amino acid metabolism, pp 321–334. Hopkins, Baltimore
Snell EE, Brown GM, Peters VJ, Craig JA, Wittle EL, Moore JA, McGlohon JA, Bird OD (1950)
    Chemical nature and synthesis of the *Lactobacillus bulgaricus* factor. J Am Chem Soc 72:
    5349–5350
Soodak M, Lipmann F (1948) Enzymatic condensation of acetate to aceto-acetate in liver extracts.
    J Biol Chem 175: 999–1000
Stadtman ER (1954) The biochemical mechanism of fatty acid oxidation and synthesis. Rec Chem
    Progr 15: 1–17
Stadtman ER (1976) The *Clostridium kluyveri* – Acetyl CoA epoch. In: Kornberg A, Horecker
    BL, Cornudella L, Oro J (eds) Reflections of biochemistry, pp 161–172. Pergamon, New York
Stadtman ER, Barker HA (1950) Fatty acid synthesis by enzyme preparations of *Clostridium
    fluyveri*. VI. Reactions of acyl phosphates. J Biol Chem 184: 769–793
Stadtman ER, Novelli GD, Lipmann F (1951a) Coenzyme A function in acetyl transfer by the
    phosphotransacetylase system. J Biol Chem 191: 365–376
Stadtman ER, Douderoff M, Lipmann F (1951b) The mechanism of acetoacetate synthesis. J Biol
    Chem 191: 377–382
Stern JR, Ochoa S (1949) Enzymatic synthesis of citric acid by condensation of acetate and oxal-
    acetate. J Biol Chem 179: 491–492
Stern JR, Shapiro B, Stadtman ER, Ochoa S (1951) Enzymatic synthesis of citric acid. III Revers-
    ability and mechanism. J Biol Chem 193: 703–720
Williams RJ (1943) The chemistry and biochemistry of pantothenic acid. Adv Enzymol 3: 253–
    287

Volume 27

Effects of Ionizing Radiation on DNA

Physical, Chemical und Biological Aspects

Editors: A. J. Bertinchamps (Coordinating Editor), J. Hüttermann, W. Köhnlein, R. Téoule
1978. 74 figures, 48 tables. XXII, 383 pages
ISBN 3-540-08542-4

For the first time, the three essential approaches to research on the effects of ionizing radiation on DNA and its constituents have been described together in one book, providing an overall view of the fundamental problems involved. A result of the European study group on "Primary Effects of Radiation on Nucleic Acids", this book contains the current state of knowledge in this field, and has been written in close collaboration by 27 authors.

Volume 28: A. Levitzki

Quantitative Aspects of Allosteric Mechanisms

1978. 13 figures, 2 tables. VIII, 106 pages
ISBN 3-540-08696-X

This book provides a concise but comprehensive treatment of the basic regulatory phenomena of allostery and cooperativity. It critically evaluates the differences between the allosteric models and their applicability to real situations. For the first time the full analysis of the different allosteric models is given, and compared with the pure thermodynamic approach. The treatment of the subject of allostery in this book is of great value to enzymologists, receptorologists, pharmacologists and endocrinologists, as it provides the basic rules for the study of ligand-protein and ligand-receptor interactions.

Volume 29: E. Heinz

Mechanics and Energetics of Biological Transport

1978. 35 figures, 3 tables. XV, 159 pages
ISBN 3-540-08905-5

This book presents the interrellations of mechanistic models on the one hand and the kinetic and energetic behavior of transport and permeatin processes on the other, using the principles of irreversible thermodynamics. The advantages of each method are compared. The special aim is to show how to appropriate formulas can be transformed into each other, in order to recognize in what way the kinetic parameters correspond to those of irreversible thermodynamics.

Volume 30: D. Vázquez

Inhibitors of Protein Biosynthesis

1979. 61 figures, 13 tables. X, 312 pages
ISBN 3-540-09188-2

This is the first comprehensive treatment of how antibiotics and other compounds inhibit protein biosynthesis. Various antibiotics and compounds are analyzed to illustrate the mode of action and selectivity of a number of drugs used medically as antibacterial or antitumor agents. Antibiotics are also studied to shed light on ribosomal structure and the process of translation.
This research offers valuable information for general microbiologists, pharmacologists, biochemists, molecular biologists and all specialists working on the problems of protein biosynthesis and ribosomal structure.

Volume 31

Membrane Spectroscopy

Editor: E. Grell
1980. 146 figures, approx. 46 tables.
Approx. 512 pages
ISBN 3-540-10332-5

The aim of the book is to introduce the reader to the application of spectroscopic methods to the study of membranes. Each chapter summarizes the experimental and theoretical principles of a particular technique and the special applications of that technique to the investigation of membranes. The contributions critically review the current exploitation of the technique by considering the results obtained on membrane constituents, simple model membranes and on biological membrane systems of a highly complex nature. A common aspect in all the chapters is the intensive search for a detailed understanding of the structures and functions of biological membranes at a molecular level.

Springer-Verlag
Berlin
Heidelberg
New York

FEI